PAINT AND SURFACE COATINGS:
Theory and Practice

ELLIS HORWOOD SERIES IN
APPLIED SCIENCE AND INDUSTRIAL TECHNOLOGY

Series Editor: Dr. D. H. SHARP, OBE, former General Secretary, Society of Chemical Industry; formerly General Secretary, Insitution of Chemical Engineers; and former Technical Director, Confederation of British Industry

This collection of books is designed to meet the needs of technologists already working in fields to be covered, and for those new to the industries concerned. The series comprises valuable works of reference for scientists and engineers in many fields, with special usefulness to technologists and entrepreneurs in developing countries.

Students of chemical engineering, industrial and applied chemistry, and related fields, will also find these books of great use, with their emphasis on the practical technology as well as theory. The authors are highly qualified chemical engineers and industrial chemists with extensive experience, who write with the authority gained from their years in industry.

Published and in active publication

PRACTICAL USES OF DIAMONDS
A. BAKON, Research Centre of Geological Technique, Warsaw, and A. SZYMANSKI, Institute of Electronic Materials Technology, Warsaw
POTTERY SCIENCE: Materials, Processes and Products
A. DINSDALE, lately Director of Research, British Ceramic Research Association
MATCHMAKING: Science, Technology and Manufacture
C. A. FINCH, Managing Director, Pentafin Associates, Chemical, Technical and Media Consultants, Stoke Mandeville, and S. RAMACHANDRAN, Senior Consultant, United Nations Industrial Development Organisation for the Match Industry
OFFSHORE PETROLEUM TECHNOLOGY AND DRILLING EQUIPMENT
R. HOSIE, formerly of Robert Gordon's Institute of Technology, Aberdeen
MEASURING COLOUR
R. W. G. HUNT, Visiting Professor, The City University, London
PAINTS AND SURFACE COATINGS: Theory and Practice
Editor: R. LAMBOURNE, Technical Manager, INDCOLLAG (Industrial Colloid Advisory Group), Department of Physical Chemistry, University of Bristol
CROP PROTECTION CHEMICALS
B. G. LEVER, Development Manager, ICI plc Plant Protection Division
HANDBOOK OF MATERIALS HANDLING
Translated by R. G. T. LINDKVIST, MTG, Translation Editor: R. ROBINSON, Editor, Materials Handling News. Technical Editor: G. LUNDESJO, Rolatruc Limited
FERTILIZER TECHNOLOGY
G. C. LOWRISON, lately of Fison's Fertilisers Ltd.
NON-WOVEN BONDED FABRICS
Editor: J. LUNENSCHLOSS, Institute of Textile Technology of the Rhenish-Westphalian Technical University, Aachen and W. ALBRECHT, Wuppertal
MICROCOMPUTERS IN THE PROCESS INDUSTRY
E. R. ROBINSON, Head of Chemical Engineering, North East London Polytechnic
QUALITY ASSURANCE: The Route to Efficiency and Competitiveness
L. STEBBING, Technical Director, Bywater Technology Limited
INDUSTRIAL CHEMISTRY: Volume 1
E. STOCCHI, Milan, with additions by K. A. K. LOTT and E. L. SHORT, Brunel
REFRACTORIES TECHNOLOGY
C. STOREY, Consultant, Durham, former General Manager, Refractories, British Steel Corporation
PERFUMERY TECHNOLOGY 2nd Edition
F. V. WELLS, Consultant Perfumer and former Editor of Soap, Perfumery and Cosmetics, and M. BILLOT, former Chief Perfumer to Houbigant-Cheramy, Paris, Présidenté d'Honneur de la Societe Technique des Parfumeurs de la France
THE MANUFACTURE OF SOAPS, OTHER DETERGENTS AND GLYCERINE
E. WOOLLATT, Consultant, formerly Unilever plc

PAINT AND SURFACE COATINGS:
Theory and Practice

Editor:
R. LAMBOURNE
Technical Manger
INDCOLLAG (Industrial Colloid Advisory Group)
Department of Physical Chemistry
University of Bristol

ELLIS HORWOOD LIMITED
Publishers · Chichester

Halsted Press: a division of
JOHN WILEY & SONS
New York · Chichester · Brisbane · Toronto

First published in 1987 by
ELLIS HORWOOD LIMITED
Market Cross House, Cooper Street,
Chichester, West Sussex, PO19 1EB, England
The publisher's colophon is reproduced from James Gillison's drawing of the ancient Market Cross, Chichester.

Distributors:

Australia and New Zealand:
JACARANDA WILEY LIMITED
GPO Box 859, Brisbane, Queensland 4001, Australia

Canada:
JOHN WILEY & SONS CANADA LIMITED
22 Worcester Road, Rexdale, Ontario, Canada

Europe and Africa:
JOHN WILEY & SONS LIMITED
Baffins Lane, Chichester, West Sussex, England

North and South America and the rest of the world:
Halsted Press: a division of
JOHN WILEY & SONS
605 Third Avenue, New York, NY 10158, USA

© **1987 R. Lambourne/Ellis Horwood Limited**

British Library Cataloguing in Publication Data
Paint and surface coatings: theory and practice. —
(Ellis Horwood series in applied science and industrial technology)
1. Coatings
I. Lambourne, R.
667'.9   TP935

Library of Congress Card No. 86–27595

ISBN 0–85312–692–5 (Ellis Horwood Limited)
ISBN 0–470–20809–0 (Halsted Press)

Phototypeset in Times by Ellis Horwood Limited
Printed in Great Britain by The Camelot Press, Southampton

Table of contents

3   Pigments for paints

Miss J. F. Rolinson

4   Solvents, thinners, and diluents
R. Lambourne

5   Additives for paint
R. A. Jeffs and W. Jones

6   The physical chemistry of dispersions
A. Doroszkowski

7   Particle size and size measurement
A. Doroszkowski

8   The industrial paint-making process
F. K. Farkas

10 Automotive paints

D. A. Ansdell

11 Automotive refinish paints

A. H. Mawby

12 General industrial paints

G. P. A. Turner

16 Mechanical properties of paints and coatings
T. A. Strivens

17 Appearance qualities of paint — basic concepts
T. R. Bullett

18   Specification and control of appearance
T. R. Bullett

19   Durability testing
R. Lambourne

List of Contributors

D. A. Ansdell	Technical Manager, Automotive Products, ICI Paints Division, Slough, SL2 5DS.
J. Bentley	Senior Scientist, Research Department, ICI Paints Division, Slough, SL2 5DS.
T. R. Bullett	Retired. Formerly Research Director, Paint Research Association, Teddington, Middx.
A. Doroszkowski	Research Department, ICI Paints Division, Slough, SL2 5DS.
F. K. Farkas	Consultant, FKF & Associates,
J. A. Graystone	Technical Manager, Research Department, ICI Paints Division, Slough, SL2 5DS.
R. A. Jeffs	Retired. Formerly with Research Department, ICI Paints Division, Slough, SL2 5DS.
W. Jones	Research Department, ICI Paints Division, Slough, SL2 5DS.
R. Lambourne	Technical Manager, Industrial Colloid Advisory Group, University of Bristol, Bristol, BS8 1TU.
A. H. Mawby	Director, Refinish Marketing, PPG Industries (UK) Ltd, Ladywood, Birmingham, B16 0AD.
Miss J. F. Rolinson	European Technical Manager for Products and Materials, ICI Paints Division, Slough, SL2 5DS.
T. A. Strivens	Senior Scientist, Research Department, ICI Paints Division, Slough, SL2 5DS.
G. P. A. Turner	Technical Manager, Industrial Coatings Research, ICI Paints Division, Slough, SL2 5DS.

Editor's preface

For many years I have felt that there has been a need for a book on the science and technology of paints and surface coatings that would provide science graduates entering the paint industry with a bridge between academia and the applied science and technology of paints. Whilst there have been many excellent books dealing with the technology there have not to my knowledge been any that have sought to provide a basic understanding of the chemistry and physics of coatings. Many of the one-time standard technological texts are now out of date (and out of print), so it seemed appropriate to attempt to produce a book that will, I hope, fill a gap. Nevertheless, it was with some trepidation that I undertook the task of editing a book covering such a diverse technology. The diversity of the technology is such that rarely will an acknowledged expert in one aspect of the technology feel confident to claim expertise in another. It therefore seemed to me that a work produced by a single author would not meet the objectives I had in mind, and I sought the help of friends and colleagues in the industry to contribute individual chapters on subjects where I knew them to have the requisite expertise. Fortunately, I was able to persuade sufficient contributions from individuals for whom I have the highest regard in respect of their knowledge and years of experience within the paint industry to satisfy myself of the ultimate authenticity of the book as a whole.

However, because of limitations of space it is impossible for a book of this kind to be completely comprehensive. Thus I have had to make decisions regarding content, and have adopted a framework which gives more space, for example, to the physics of paint and the physical chemistry of dispersions than most books of this kind. In doing so I have had to reduce the breadth (and in some cases the depth) of treatment of specific technologies. Thus, whilst the chapters on automotive painting and architectural paints are fairly detailed, the treatment of general industrial finishing is less an 'in depth' account of specific technologies, but is intended to illustrate the very wide range of requirements of manufacturing industry and the problems the paint technologist may encounter as a result of this.

In chapters dealing with the fundamental principles underlying the technology authors have been invited to provide critical accounts of the science and technology as it stands today. This is reflected in the extensive lists of references to original work mostly published within the last decade. It is hoped that readers wishing to delve further to increase their understanding will find these references a valuable source of information.

It is important to record that apart from the authors, a number of individuals have contributed to the production of this book. I would like to record my thanks to Dr Gordon Fettis, Research Manager of ICI Paints Division, for his support and encouragement from its inception, and for the use of many of the facilities of ICI in the production of the manuscript. Thanks are also due to Mrs Millie Cohen (of ICI) and Mrs Kate Slattery (of Bristol University) who between them typed the major part of the manuscript.

R. Lambourne
Chalfont St Giles, April 1986

1

Paint composition and applications — a general introduction

R. Lambourne

1.1 A SHORT HISTORY OF PAINT

Primitive men are credited with making the first paints about 25 000 years ago. They were hunters and cave dwellers and were probably inspired by the rock formations of their cave walls to outline and colour the shapes of the animals they hunted. It is possible that by creating these images they thought their power over their prey would be increased.

Chemical analysis of cave paintings discovered at Altamira (Spain) and Lascaux (France) show that the main pigments used by Palaeolithic artists were based upon iron and manganese oxides. These provide the three fundamental colours found in most cave paintings, namely black, red, and yellow, together with intermediate tints. Carbon from burnt wood, yellow iron carbonate, and chalk may also have been used. Surprisingly, there is no trace of a white pigment (the commonest pigment in use today) at Lascaux, where the natural colour of the rock was used as a pale background. However, white pigments do occur in some prehistoric paintings in Africa.

These earth pigments were ground to a fine powder in a pestle and mortar. Naturally hollowed stones are thought to have been used as mortars and bones as pestles, following the finds of such articles stained with pigments. The powdered pigments were probably mixed with water, bone marrow, animal fats, egg white, or vegetable sugars to form paints. They were applied by finger 'dabbing', or crudely made pads or brushes from hair, animal fur, or moss. Cave paintings have survived because of their sheltered positions deep in caves which were subsequently sealed off. These paints have very poor durability, the binders serving merely to make the pigments stick to the cave walls.

The Egyptians developed the art of paint-making considerably during the period circa 3000–600 BC. They developed a wider colour range of pigments which included the blues, lapis lazuli (a sodium silicate — sodium sulphide mixed crystal), and azurite (chemically similar to malachite). Red and yellow ochres (iron oxide), yellow orpiment (arsenic trisulphide), malachite green (basic copper carbonate), lamp-black, and white pigment gypsum (calcium sulphate) all came into use during this period. The first synthetic pigment, known today as Egyptian Blue, was produced almost 5000 years ago. It was obtained by calcining lime, sodium carbonate malachite, and silica at a temperature above 830°C. The Egyptians also developed the first lake pigments. These were prepared by precipitating soluble organic dyes onto an inorganic (mineral) base and 'fixing' them chemically to form an insoluble compound. A red dye obtained from the roots of the madder plant was used in the first instance. This is no longer used other than in artists' colours ('rose madder') because it fades rapidly on exposure to sunlight, and it has been replaced by alizarin. Lake pigments still, however, represent an important group of pigments today. Red lead was used in preservative paints for timber at this time, but was more extensively used by the Romans. The resins used were almost all naturally occurring gums; waxes which were applied molten as suitable solvents were unknown. Linseed and other drying oils were known, but there is no evidence that they were used in paints.

The Greeks and Romans in the period 600 BC–AD 400 almost certainly appreciated that paint could preserve as well as decorate objects. Varnishes incorporating drying oils were introduced during this period. However, it was not until the thirteenth century that the protective value of drying oils began to be recognized in Europe. During the Middle Ages much painting, especially on wood, was protected by varnishing. The varnish was made by dissolving suitable resins in hot linseed, hempseed, or walnut oil, all of which tend to darken with time.

By the late eighteenth century, demands for paints of all types had increased to such an extent that it became worthwhile for people to go into business to make paint and varnishes for others to use. In 1833, J. W. Neil advised varnish makers always to have an assistant present during the varnish making process, for safety. 'Never do anything in a hurry or a flutter... a nervous or timorous person is unfit either for a maker or assitant, and the greatest number of accidents occur either through hurry, fear or drunkeness.' This admonition is indicative of the increase in scale of manufacture and the dangers of use of open-pan varnish kettles.

The industrial revolution had a major effect on the development of the paint industry. The increasing use of iron and steel for construction and engineering purposes resulted in the need for anti-corrosive primers which would delay or prevent rusting and corrosion. Lead- and zinc-based paints were developed to fulfil these needs. It is interesting to note that one of the simplest paints based upon red lead dispersed in linseed oil is still probably one of the best anti-corrosive primers for structural steel. Lead-based paints are being superseded not because better products have been produced, but because of the recognition of their toxicity and the hazards attendant upon their use.

An acceleration of the rate of scientific discovery had a growing impact on the development of paints from the eighteenth century to the present day. Prussian blue, the first artificial pigment with a known chemistry, was discovered in 1704. The use of turpentine as a paint solvent was first described in 1740. Metal driers, for speeding up the drying of vegetable oils, came into use about 1840.

The basis of formaldehyde resin chemistry was laid down between 1850 and 1890 although it was not used in paints until the twentieth century. Likewise, it was discovered in 1877 that nitrocellulose could be made safe to use as a plastic or film, by plasticizing it with camphor, but it was not until after the First World War that it was used in any significant amount in paints. The necessary impetus for this to happen came with the mass production of the motor car. Vast quantities of nitrocellulose were manufactured for explosives during the war. At the end of the war, with the decline in the need for explosives, alternative outlets for nitrocellulose needed to be found, and the mass production of motor cars provided the necessary market. The war had accelerated the exploitation of the discoveries of chemistry and the growth of the chemical industry. New coloured pigments and dyestuffs, manufactured synthetically, became available, and in 1918 a new white pigment, titanium dioxide, which was to replace white lead completely, was introduced. Titanium dioxide improved the whiteness and 'hiding' or obliterating power of paint, but when originally introduced it contributed to more rapid breakdown of paints in which it was used because of its photoactivity. Subsequent research has overcome this problem and ensured that the modern pigmentary forms of titanium dioxide can be used in any type of composition without suffering any disadvantage of this kind.

Subsequent chapters of this book will be largely concerned with developments that have taken place in the twentieth century, and most of which have occurred within the last fifty years.

1.2 PAINT OR SURFACE COATING?

The terms 'paint' and 'surface coating' are often used interchangeably. Surface coating is the more general description of any material that may be applied as a thin continuous layer to a surface. Paint was traditionally used to describe pigmented materials as distinct from clear films which are more properly called lacquers or varnishes. We shall be most concerned with paint in the context of this book; but, as we shall see, modern painting processes may include composite systems in which a total paint system comprises several thin films some, but not all, of which may be pigmented. We shall use both terms as appropriate to the context in which specific paint compositions are being discussed.

The purpose of paints and surface coatings is two-fold. They may be required to provide the solution to aesthetic or protective problems, or both. For example, in painting the motor car the paint will be expected to enhance the appearance of the car body in terms of colour and gloss, and if the body is fabricated out of mild steel it will be required to give protection against corrosion. If the body is formed from glass fibre reinforced plastic the paint will only be required for aesthetic purposes. There are obviously very sound economic reasons why it is attractive to colour only the outer surface of articles that might otherwise be self-coloured by using materials of fabrication e.g. plastics that are pigmented, particularly if a wide choice of coloured effects is required. This topic will be developed in the chapters on paints for specific markets (Chapters 9–13).

In considering the nature of paints it will become abundantly clear that the relationship between the coating and the substrate is extremely important. The requirements for a paint that is to be applied to wood are different from those of a paint to be applied to a metal substrate. Moreover, the method by which the paint is applied and cured (or dried) is likely to be very different. In formulating a paint for a particular purpose it will be essential for the formulator to know the use to which the painted article is to be put, and physical or mechanical requirements are likely to be called for. He will also have to know how it is to be applied and cured. Thus, a paint for an item made from cast iron may call for good resistance to damage by impact (e.g. chipping), whilst a coating on a beer can will call or a high degree of flexibility. These different requirements will be described in Chapters 9–13 which will deal with specific areas of paint usage.

It has long been recognized that it is difficult, if not impossible, to meet the requirements of many painting processes by the use of a single coat of paint. If one lists the requirements of a typical paint system it is easy to see why. Many, if not all, of the following are likely to be required: opacity (obliteration); colour; sheen (gloss); smoothness (or texture); adhesion to substrate; specific mechanical or physical properties; chemical resistance; corrosion protection; and the all-embracing term 'durability'. Durability is an important area that we shall return to in many contexts. The number of different layers that comprise the paint system will depend on the type of substrate and in what context the coated object is used. A typical architectural (gloss) paint system might consist of a 'primer', an 'undercoat', and a 'topcoat'. All three are likely to be pigmented compositions, and it is probable that there will be more than one coat (or layer) of each of these paints. An architect may well specify one coat of primer, two coats of undercoat, and two coats of topcoat. The purpose of these individual layers and hence their composition is likely to be very different. The primer is designed largely to seal the substrate and provide a means of achieving good adhesion between substrate and undercoat. It may contribute to opacity, but this will not be its main purpose. The undercoat will be used for two purposes: to contribute significantly to the obliteration of the substrate and to provide a smooth surface upon which to apply the topcoat. The smooth surface is obtained by abrading the dried undercoat (after each coat has dried) with fine tungsten carbide paper. The topcoat is then applied to complete the process of obliteration and to provide the appropriate aesthetic effect (i.e. colour and sheen). The system as a whole would be required to give protection to the wood or metal substrate to which it is applied.

The interrelationship between these multilayers is worth considering. The mechanical and physical properties of the individual coatings will often be very different. The function of the primer in promoting adhesion has already been mentioned. It may also be required to relieve stresses that are built up within the coating system as a result of hardening and ultimately embrittlement of the topcoat (and undercoat) as a result of ageing, or to accommodate stresses imposed by the differential movement of the substrate. The softwoods used in the construction of window frames are known to expand and contract between the dry (summer) and wet (winter) conditions by a least 10% across the grain, but much smaller changes are observed in the direction of the grain. The undercoat will be formulated in a colour close to that of the topcoat, but it may serve this purpose to several closely related topcoat colours. It will normally be highly pigmented, in contrast to the topcoat which will not. The reason for the latter is the need, for example, to maximize gloss and extensibility. The use of the type and concentration of pigmentary material in the undercoat would not be conducive to maximizing these properties in the topcoat.

The primer will frequently be required to contribute to corrosion protection.

Those formulated for use on steel are likely therefore to incorporate a chemically active anti-corrosive pigment. Corrosion protection may be achieved by yet another means, the chemical treatment of the substrate. Thus many industrial coating processes involve a chemical pretreatment of metal, mainly aluminium or ferrous substrates. The latter is most frequently treated with a phosphate solution that produces a crystalline phosphate layer. Subsequent paint application, i.e. priming, is therefore to a crystalline inorganic layer and not directly to an (uncoated) pure metal surface.

Surfaces are seldom what they seem. With the exception of the noble metals almost all surfaces that will be commonly regarded as 'metal' surfaces will present to the paint a surface that is not a metal, but an oxide layer. Even so the purity or cleanliness of the surface may well be an unknown quantity. Since this surface will have an important effect on such properties as the adhesive performance of the paint system it is important to appreciate this point. Just as most surfaces will be 'dirty' and thus be ill-defined, it is necessary to produce paint systems that can accommodate the contamination and general variability of surfaces. These types of system must be 'tolerant' to all but excessive contamination and are often described as 'robust' if the required degree of tolerance can be achieved. This is not to say that industrial coating processes do not require certain pretreatments such as degreasing, and may involve the chemical pretreatments indicated above.

The deterioration of paints which occurs in many situations is largely due to changes in the chemical nature of the film former with consequent changes in its mechanical properties, and research continues unabated to formulate polymers and resins to improve the performance of paints in use. The development of new improved pigments may contribute to improvements in durability, but in most cases the weakest link in the system is the film former. One consequence of this is the development of systems for specific end uses. Such approaches are adopted when it is practicable to avoid the compromises that are otherwise likely to be required for a general-purpose product. For economic and marketing reasons the best product may not be available for a specific end use, and a compromise of cost and performance may be required. Indeed the cost-effectiveness of a particular coating composition will usually dominate other considerations particularly in the industrial paint markets.

1.3 THE COMPONENTS OF PAINT

The composition of a paint is indicated in Table 1.1 which also indicates the function of the main components.

Not all paints have every ingredient, as will be indicated in subsequent specialist chapters. For example gloss paints will not contain extenders which are coarse particle inorganic materials. These are used in matt paints such as the surfacers or primer surfacers used in the motor industry.

Major differences occur between the polymers or resins that are used in paints formulated for different purposes. This is because of differences between the methods of application and cure, the nature of the substrate, and the conditions of use. Thus architectural ('decorative' or 'household') paints will be required to be applied *in situ* at ambient temperatures (which may be between 7°–30°C depending

Table 1 — The compostion of paints

Components			Typical function
Paint	Vehicle (Continuous phase)	Polymer or Resin (Binder)	Provides the basis of continuous film, sealing or otherwise protecting the surface to which the paint is applied. Varies in chemical composition according to the end use.
		Solvent or Diluent	The means by which the paint may be applied. Avoided in a small number of compositions such as powder coatings and 100% polymerizable systems.
	Pigment (Discontinuous phase)	Additives	Minor components, wide in variety and effect e.g. catalysts, driers, flow agents.
		Primary pigment (fine particle organic or inorganic)	Provides opacity, colour, and other optical or visual effects. Is thus most frequently used for aesthetic reasons. In primers the pigment may be included for anti-corrosive properties.
		Extender (coarse particle inorganic matter)	Used for a wide range of purposes including opacity/ obliteration (as an adjunct to primary pigment); to facilitate sanding e.g. in primer surfacers.

on climate and geographical location). They will 'dry' or 'cure' by one of two mechanisms: (i) atmospheric oxidation or (ii) the evaporation of diluent (water) accompanied by the coalescence of latex particles comprising the binder. Many industrial finishing processes will require the use of heat or other forms of radiation (UV, IR, electron beam) to induce chemical reactions, such as free radical or condensation polyermization, to convert liquid polymers to highly crosslinked solids. The most common of these processing methods uses 'thermosetting' polymers which will frequently be admixtures of two quite different chemical types of material, e.g. alkyd combined with amino resin. There is a similarity between both the oxidative drying and industrial thermosetting processes in so far that in both cases the polymers used are initially of low molecular weight and the curing process leads to crosslinking of these polymers to yield highly complex extremely high molecular weight products. In contrast, it is possible to produce coatings without the need for crosslinking, for use in both of these distinctive markets. In the case of decorative or architectural paints this is exemplified by the emulsion paint, in which the binder is in the form of high molecular weight polymer particles suspended in an aqueous medium. Lacquers used in the motor industry may be based upon high molecular weight polymer in solution. Both systems avoid the need for crosslinking to achieve satisfactory film properties.

1.3.1 Polymer or resin film formers

The organic chemistry of film formers is described in detail in Chapter 2. It will be useful, however, to indicate here some of the range of polymers and resins that have come to be used as film formers, and to indicate their general areas of application. Film formers or binders may be classified according to their molecular weight. Thus low molecular weight polymers that will not form solid films normally without further chemical reaction form one class. High molecular weight polymers that will form useful films without further chemical reaction form the second class. Examples of polymers and resins classified by this means are shown in Table 1.2.

Table 1.2

Low molecular weight	High molecular weight	
Oleoresinous binders	Nitrocellulose	
Alkyds	Solution vinyls	
Polyurethanes	Solution acrylics	
Urethane oils	Non-aqueous dispersion polymers (NADs)	
Amino resins	PVA	⎫
Phenolic resins	Acrylic	⎬ latexes
Epoxide resins	Styrene/butadiene	⎭
Unsaturated polyesters		
Chlorinated rubber		

1.3.1.1 Low molecular weight film formers
Oleoresinous binders

These are prepared by heating vegetable oils with naturally occurring resins such as wood rosin, fossil resins such as Congo Copal and Kauri gum. They would also include oil-modified phenolics. To a large extent these types of resinous binder have been displaced by alkyd resins and the like, but many are capable of providing products that in performance are the equal of their successors, particularly in specific applications such as in architectural undercoat paint. They are less reproducible than condensation polymers like alkyds and are less attractive in terms of modern manufacturing processes.

Alkyds
Alkyds are polyesters derived as the reaction products of vegetable oil triglycerides, polyols (e.g. glycerol), and dibasic acids or their anhydrides (e.g. phthalic anhydride). They are generally formulated with very different end uses in mind, and classified according to vegetable oil content (described by the term 'oil length') in three broad categories: short oil, medium oil, and long oil, roughly corresponding to <45%, 45–60%, >60% respectively. The variation in oil length is usually coincident with changes in the nature of the vegetable oil used and consequently the end use. Thus, for architectural gloss paint of maximum exterior durability, the alkyd binder will be a long oil alkyd based upon a drying oil such as linseed or soya bean oils (i.e. an unsaturated triglyceride oil). The drying oil provides the means by which the film former dries. In this case the conversion from a low molecular weight liquid polymer to a highly crosslinked solid arises as a result of oxidation. One characteristic of the long oil alkyd is that it is soluble in aliphatic hydrocarbons. In contrast the short oil alkyd is likely to be based on a saturated triglyceride (such as coconut oil). It will not be soluble in aliphatic hydrocarbons and will normally be dissolved in a high boiling aromatic. Although the short oil alkyd may be capable of forming a lacquer-like film it would have a low softening point, and it is necessary to crosslink it in order to achieve a satisfactory film. In this case it is usually combined with an amino resin and crosslinked by condensation in a heat-curing process. It is debatable in this type of system whether the amino resin is crosslinking the alkyd or the alkyd is plasticizing the highly crosslinked amino resin. The former explanation is usually preferred because the proportion of the alkyd is invariably greater than that of the amino resin. The alkyd/amino resin ratio usually falls between 2:1 and 4:1 by weight. These types of system are used in industrial finishing. Alkyd/melamine formaldehyde compositions have found use in the automotive market for many years. Alkyd/urea formaldehyde compositions have found use in the domestic appliance market, although in both cases there have been new products developed that have displaced these materials to some extent, particularly when more stringent performance requirements have to be met.

Polyurethanes, urethane alkyds, and urethane oils
Structurally, these materials resemble alkyds in which polyester linkages are replaced partially or totally by urethane $-NH-\overset{\underset{\|}{O}}{C}-O-$ linkages.

Polyurethanes also include two-pack compositions for the industrial and refinish markets in which the curing is achieved by reaction between free isocyanate groups in one component and hydroxyl groups in the second component. The advantages of

paint primer, or used simply in an unpigmented form as an oleoresinous varnish. Phenolic-based compositions find use in chemically resistant systems such as are required for pipes and tank linings.

Epoxide resins

The use of the epoxide or oxirane group $-CHCH_2$ (with $\underset{O}{\diagdown\diagup}$) as a means of synthesizing resins and as a means of crosslinking binders is now well established. A large group of epoxide resins is based upon the reaction products of epichlorhydrin and bisphenol A (diphenylolpropane). These resins may be esterified with unsaturated fatty acids to give epoxy esters. These are film formers in their own right and resemble air-drying alkyds. They exhibit better chemical resistance than alkyds but are less durable than long oil alkyds in some circumstances. The way in which they break down may, however, be turned to good use, for example when used as 'self-cleaning' coatings. In this case the films may be pigmented with uncoated titania in the anatase form so that degradation in ultraviolet radiation causes erosion of the surface layers of the film, otherwise known as chalking. Such films gradually weather away, always presenting a white surface to the elements.

Epoxide resins may also be used in conjunction with MF or phenolics, or they may be formulated into epoxy–alkyds, i.e. where they are effectively being used as polyols in admixture with less functional polyols such as glycerol. The epoxide group offers great versatility in curing, and a wide range of two-pack compositions are possible. One of the most popular methods of crosslinking uses the reaction with polyamides. This is the same method of cure as used in epoxy adhesive compositions. The crosslinking occurs as an addition of terminal amino groups of the polyamide to the epoxy group. This reaction occurs slowly at room temperature. Crosslinking through the epoxide group can also arise from the use of polyamines or by means of the acid-catalysed polymerization to give ether crosslinks. It will be apparent that most of these products will be used in industrial applications.

Unsaturated polyesters

Unlike the previously described binders, unsaturated polyesters offer the benefit of totally polymerizable systems because the solvent in which they are dissolved is a polymerizable monomer. The simplest and most common polyesters are prepared from maleic anhydride/phthalic anhydride mixtures esterified with glycols such as propyleneglycol. The resins so produced are dissolved in styrene or vinyl toluene. The free radical copolymerization of the vinyl monomer and the maleic unsaturation in the polyester is usually initiated by a transition metal/organic hydroperoxide system at ambient temperature or by the use of the thermal scission of a diacyl peroxide at higher temperatures. Unsaturated polyesters have found extensive use both pigmented and as clears in the wood finish market. They are capable of widely divergent uses depending on their composition. Chemically resistant finishes for tank linings, for example, can be formulated on polyesters derived from isophthalic and terephthalic acids. Another class of chemically resistant finish is based upon chlorinated polyesters. In this case the polyester incorporates chlorendic or HET anhydride in place of the more common phthalic anhydride.

Chlorinated rubber
Chlorinated rubber is a film-forming resin that is available in a wide range of molecular weights, ranging from 3500 to about 20 000. It is prepared by chlorinating rubber in solution, the commercial product containing about 65% of chlorine. It is used as the main binder in air-drying paints which are to be used in situations requiring a chemically resistant product of great durability. Because the polymer is a brittle solid, in paint applications chlorinated rubber requires plasticization. Chlorinated rubbers are also used in conjuction with other resins with which they are compatible, such as alkyds. Paints based on chlorinated rubber have been used for building, masonry, swimming pools, road marking, and marine purposes.

1.3.1.2 High molecular weight film formers
Almost all high molecular weight polymers are produced by the free radical initiated polymerization of mixtures of vinyl, acrylate, or methacrylate monomers. They may be polymerized in solution, in suspension or dispersion. Dispersion polymerization may be in hydrocarbon diluents (non-aqueous dispersion, NAD) or in aqueous media ('emulsion polymers'). The reaction processes differ considerably between these systems as will become apparent when the subject is developed in Chapter 2.

One major exception to the foregoing is nitrocellulose. This material is formed by the direct nitration of cellulose in the presence of sulphuric acid. It is available in grades determined by the degree of nitration which, in turn, determine its solubility in various solvents. The grades used in refinish paints, wood-finish lacquers, etc. require the molecular weight of the original cellulose to be reduced somewhat, to meet viscosity requirements in the solvents commonly used.

In most cases the high molecular weight polymers do not need to be crosslinked in order to develop adequate film properties. A small number of solution polymers of moderately high molecular weight are, however, crosslinked through reactive groups incorporated into the polymer chain. The physical properties of films produced from high polymers may be only marginally affected by the manner in which they were prepared or their physical form at the stage of film application. Thus automotive finishes derived from solution acrylics and NADs are virtually indistinguishable, albeit that the methods of application, processing conditions, etc. may be significantly different. In most cases the product that is selected will be dependent on the economics of the process overall rather than the product cost alone. The need to meet end use specification is the main reason for the similarity in film properties and performance of alternative product formulations.

Aqueous latexes (emulsion polymers) as a group have been one of the fastest-growing sectors of the paint market. Starting with the use of homopolymer polyvinyl acetate as a binder in matt and low-sheen decorative wall paints we have seen the development of more robust systems using internal plastization (i.e. the incorporation of a plasticizing comonomer) and, more recently, the use of acrylic and methacrylic copolymer latexes. Improved performance has enabled the introduction of these paints into exterior masonry applications. One of the reasons for the growth of emulsion paints is the desire by communities to reduce or eliminate atmospheric pollution, and much legislation has been enacted to protect the environment.

1.3.2 Pigments

1.3.2.1 Primary pigments

Primary pigments comprise solid particulate material which is dispersed in the binder or film former described previously. We shall distinguish between them and supplementary pigments, extenders, fillers, etc. in that they contribute one or more of the principal functions, namely colour, opacification, and anti-corrosive properties. The supplementary pigments, extenders, although important, do not in general contribute to these properties to a major extent. Their function is related to reductions in cost, but they can contribute properties to a paint that may be less obvious than colour or opacity. Thus they may enhance opacity, control surface sheen, and facilitate ease of sanding.

The principal pigment in use is titanium dioxide. This is primarily because of reasons of fashion, e.g. in the decorative paint market there has been a tendency for white and pastel shades to gain greater acceptance over strong colours in recent years. The same is true to a lesser extent in other fields, such as the motor car industry. In this case early motor cars were painted black, and it was difficult to obtain cars in other colours. Nowadays it is the black car that is a comparative rarity. Titanium dioxide in its pigmentary form is a highly-developed pigment. The high refractive index rutile form is most commonly used, and the pigment is manufactured within very close limits with respect to particle size and particle size distribution. Although normally regarded as being chemically inert the surface of the titanium dioxide crystal is photoactive, and the pigmentary form is surface coated to minimize photochemical activity at the binder/pigment interface. Typical coatings contain silica and alumina in various proportions, and in addition to reducing photochemical activity they can improve the dispersibility of the pigment in the binder. Considerable research has been carried out into titanium dioxide pigments, and manufacturers are now able to offer a range of grades that are suitable for use in different types of paint media. The uses of titanium dioxide and other white pigments are detailed in Chapter 3, and its optical properties are discussed in Chapters 17 and 18.

Coloured pigments fall into two main groups, inorganic pigments and organic pigments. As a result of legislation governing the handling and use of toxic materials, many of the inorganic pigments traditionally used by the paint industry have been replaced by other less toxic materials. Thus except in a few industrial compositions lead pigments of all types (e.g. chromate, oxide, carbonate) have to a large extent been replaced, and it is to be expected that lead in all of its forms and chromates will ultimately be eliminated from all paints. Their removal from the scene gives rise to problems if a coloured pigment is used for the dual purposes of providing colour and contributing positively to the anti-corrosive properties of the paint film.

Some commonly-used pigments are shown in Table 1.3. The selection and use of pigments is covered in detail in Chapter 3. The physics of colour are treated in Chapter 18 which includes a treatment of the principles and practice of colour matching. At one time colour matching was an art carried out visually by very skilful individuals. Whilst the skilled colour matcher has not been entirely replaced, there has been a gradual change to instrumental colour matching, a trend that has been accelerated by the development of sophisticated colorimeters and the use of the computer.

In automotive paints there has been a growth in the use of aluminium flake as a

Table 1.3 — Some typical primary pigments

Colour	Inorganic	Organic
Black	carbon black copper carbonate manganese dioxide	aniline black
Yellow	lead, zinc, and barium chromates cadmium sulphide iron oxides	nickel azo yellow
Blue/violet	ultramarine Prussian blue cobalt blue	phthalocyanin blue indanthrone blue carbazol violet
Green	chromium oxide	phthalocyanin green
Red	red iron oxide cadmium selenide red lead chrome red	toluidine red quinacridones
White	titanium dioxide zinc oxide antimony oxide lead carbonate (basic)	—

pigment to provide attractive metallic finishes. The dimensions of the flakes and their orientation within the paint film give rise to optical effects that cannot be achieved in other ways. Inevitably such paints are subject to problems in quality control, application, and colour matching that do not exist in solid colours. Nevertheless these problems are sufficiently well understood to be kept under control, and their continued use is amply justified aesthetically.

1.3.2.2 Extenders, fillers, and supplementary pigments
All three names have been applied to a wide range of materials that have been incorporated into paints for a variety of purposes. They tend to be relatively cheap materials, and for this reason may be used in conjunction with primary pigments to achieve a specific type of paint. For example it would be technically difficult and prohibitively expensive to produce satisfactory matt white emulsion paint using titanium dioxide as the only pigment. Titanium dioxide is not cost effective as a matting agent and indeed is not designed for this purpose. It is much more economic to use a coarse particle extender such as calcium carbonate in conjunction with TiO_2 to achieve whiteness and opacity in a matt or semi-matt product (e.g. a matt latex

decorative paint, an undercoat or primer). Extenders do not normally contribute to colour, and in most cases it is essential that they be colourless. The particle sizes of extenders range from submicrometre to a few tens of micrometres; their refractive index is generally close to that of the organic binders in which they are used, and so they contribute little opacity from light scattering. Platelet type extenders such as wet-ground mica can influence the water permeability of films, and many therefore contribute to improved corrosion resistance. Talcs are often used (e.g. in automotive primer surfacers) to improve the sanding of the paint prior to the application of a topcoat. Many of the extenders in common use are naturally-occurring materials that are refined to varying extents according to the use to which they are put. Whilst every attempt is made to ensure that they are reproducible they tend to be more variable and with a greater diversity of particle shape, size, and size distribution than primary pigments. A list of typical inorganic extenders is given in Table 1.4.

Table 1.4 — Some typical extenders

Chemical nature	Type
Barium sulphate	Barytes Blanc fixe
Calcium carbonate	Chalk Calcite Precipitated chalk
Calcium sulphate	Gypsum Anhydrite Pecipitated calcium sulphate
Silicate	Silica Diatomaceous silica Clay Talc Mica

In recent years there have been several attempts to make synthetic (polymeric) extenders to meet special needs. In particular to replace some of the titanium dioxide in the paint film. One such material, 'Spindrift', originating from Australia, is in the form of polymer beads (spherical particles up to 30μm in diameter which incorporate submicrometre air bubbles and a small proportion of pigmentary titanium dioxide). The introduction of air bubbles into the polymer that forms the beads influences the light-scattering power of the titanium dioxide in one of two ways. It can reduce the average effective refractive index of the polymer matrix and so enhance the light-scattering of the titanium dioxide, if the bubbles are very small (<0.1 μm). If the bubbles are large (~0.8 μm) they are able to scatter light in their own right. Other methods of introducing voids into paints in a systematic way to take advantage of the scattering of light by bubbles have included the emulsification of small droplets of a volatile fluid into aqueous latex paints which leave cavities within the film by evaporation after coalescence ('Pittment'); yet another is the use of non-coalescing latex particles in combination with coalescing latexes to give rise to cusp-like voids within the film (Glidden). All of these methods seek to achieve a cost-effective contribution to opacity without sacrifice of film integrity and other properties. These methods have received limited acceptance, mainly owing to economic reasons. Nevertheless, they do increase the opportunities open to the paint formulator to formulate a paint for a given purpose.

1.3.3 Solvents
Solvents are used in paint compositions for two main purposes. They enable the paint to be made, and they enable it to be applied to surfaces. This may seem to be stating the obvious, but it is important to appreciate that so far as the paint film performance is concerned the solvent plays no long-term role in this. This is not to say that in the early life of the film solvent retention does not affect hardness, flexibility, and other film properties.

The term solvent is used frequently to include liquids that do not dissolve the polymeric binder, and in these cases it is more properly called a diluent. The function of the diluent is the same as a solvent, as stated above. In water-based systems the water may act as a true solvent for some components, but be a non-solvent for the main film former. This is the case in decorative emulsion paints. More often in these cases it is common to refer to the 'aqueous phase' of the composition, acknowledging that the water present, although not a solvent for the film former, is present as the major component of the liquid-dispersing phase.

A wide range of organic liquids are used as paint solvents, the type of solvent depending on the nature of the film former. A detailed account of solvency, solvent selection, etc. is given in Chapter 4.

Considerable research effort has gone on into the thermodynamics of solutions within the last two decades. This has provided the paint formulator with much more precise methods of solvent selection. The improved methods are based upon a better knowledge and understanding of molecular attractions in liquids and a recognition of the additivity of molecular attractions in mixed solvent systems. It is very rare for a single solvent to be acceptable in most situations, and the newer methods based upon solubility parameter concepts enable the more rational selection of solvent mixtures to meet a particular need.

Solvency alone is not the only criterion upon which solvent choice is made. Other important factors include evaporation rate, odour, toxicity, flammability, and cost. These factors assume different degrees of importance depending on how the paint is used. If the paint is applied under industrial manufacturing conditions it is likely that problems associated with odour, toxicity, and flammability may be under control, but this is by no means certain. The need to install expensive extraction equipment or after-burners may preclude the use of some solvents and therefore some types of paint composition. Closely related to considerations of toxicity are those of pollution. Many countries have enacted legislation to protect the individual and the environment. This legislation has had a profound effect on the development of the paint industry and has influenced both the raw material supplier on the one hand and the paint user on the other. In North America there has been an enormous growth in water-borne systems at the expense of solvent-borne paints. This is a trend that is likely to continue.

Other alternatives to solvent-based paints are 100% polymerizable systems and powder coatings. In the former case a polymerizable monomer such as styrene fulfils the role of solvent for the composition, being converted into polymer in the curing reaction. With powder coatings, solvents may be used in the early stages of the paint-making process, but they are removed and recycled, so do not provide a hazard or problem for the user.

1.3.4   Paint additives

The simplest paint composition comprising a pigment dispersed in a binder, carried in a solvent (or non-solvent liquid phase) is rarely satisfactory in practice. Defects are readily observed in a number of characteristics of the liquid paint and in the dry film. These defects arise through a number of limitations both in chemical and physical terms, and they must be eliminated or at least mitigated in some way before the paint can be considered a satisfactory article of commerce.

Some of the main defects worth mentioning are settlement of pigment and skinning in the can; aeration and bubble retention on application; cissing, sagging, and shrivelling of the paint film; pigment flotation; and flooding. These defects represent only a small number of defects that can be observed in various paints. It is perhaps worthwhile to describe cissing, shrivelling, sagging, flotation, and flooding here. They will be dealt with in more detail in Chapter 5, which describes some of the more common types of paint additives.

'Cissing' is the appearance small, saucer-like depressions in the surface of the film:

'Shrivelling' is the development of a wrinkled surface in films that dry by oxidation:

'Sagging' is the development of an uneven coating as the result of excessive flow of a paint on a vertical surface:

'Floating' is the term used for the colour differences that can occur in a paint film because of the spontaneous separation of component pigments after application:

'Flooding' (also known as 'brush disturbance') is the permanent colour change of a paint subject to shear after application.

To overcome these defects their cause requires to be understood and a remedy found. In some cases the defect may be overcome by minor reformulation. Shrivelling, for example, is commonly due to an imbalance between the surface oxidative crosslinking of a film and the rate of crosslinking within the film. This can usually be overcome by changing the drier combination, which consists of an active transition metal drier such as cobalt which promotes oxidation and a 'through' drier such as lead or zirconium which influences crosslinking, but does not *per se* catalyse the oxidation process. In other cases simple reformulation will not provide a remedy, and specific additives have been developed to help in these cases. Thus anti-settling agents, anti-skinning agents, flow agents, etc. are available from specialist manufacturers for most defects and for most paint systems.

The problems of 'floating' and 'flooding' are associated with colloidal stability of the pigment dispersion and may arise from a number of different causes. The differential separation of pigment illustrated by floating occurs as a result of the differences in particle size of the component pigment and may be overcome by coflocculation of the pigments in the system. Another method of curing the condition may be to introduce a small proportion of a very fine particle extender such as alumina, of opposite surface charge to the fine particle pigment, to coflocculate with the latter.

The flooding (or brush disturbance) problem is indicative of flocculation occurring as a film dries. Under shear, as the brush disturbs the paint, the pigment is redispersed and the paint becomes paler in shade. This is because an increase in the back-scattering of incident light occurs, owing to the white pigment becoming deflocculated.

Cissing and sagging are illustrative of other aspects of physical properties associated with surface chemistry and rheology. In the former case, the effect is caused by a localized change in the surface tension of the film. In extreme cases this can give rise to incomplete wetting of the substrate, often distinguished by the term 'crawling'. Sagging, on the other hand, is a bulk property of the film that may be influenced by the colloidal stability of the composition. Ideal, colloidally stable dispersions tend to exhibit Newtonian behaviour, i.e. their viscosity is independent of shear rate. This means that on a vertical surface a Newtonian liquid that is of a suitable viscosity to be spread by a brush, i.e. with a viscosity of about 5 poise, will flow excessively unless the viscosity rises rapidly as a result of solvent loss. Alternatively, the paint formulator may aim to induce non-Newtonian behaviour such that the low shear viscosity of the product is very high. Thus, sagging may be avoided by either or a combination of these effects.

1.4   METHODS OF APPLICATION

There are four main methods of applying paint:

(i)   by spreading, e.g. by brush, roller, paint pad, or doctor blade;
(ii)   by spraying, e.g. air-fed spray, airless spray, hot spray, and electrostatic spray;
(iii)   by flow coating, e.g. dipping, curtain coating, roller coating, and reverse roller coating;
(iv)   by electrodeposition.

The methods adopted depend on the market in which the paint is used, each type of paint being formulated to meet the needs of the application method. Spreading by brush or hand-held roller is the main method for applying decorative/architectural paints and the maintenance of structural steelwork and buildings generally. It is also important in marine maintenance, although other methods (e.g. airless spray) may be used during the construction of a ship.

Application by spraying is the most widely applicable method. It is used for painting motor cars in the factory and by refinishers following accident damage; it is used in the wood-finishing industries (e.g. furniture) and in general industrial paints (e.g. domestic appliances). The various forms of spray painting make it a particularly versatile method of application. The flow coating methods are limited essentially to flat stock (e.g. chipboard) and coil coating (aluminium or steel coil) where they are much valued because of the high rates of finishing that can be achieved.

Electrodeposition as a method of painting has become a growth area during the last two decades. It has become established as the main method of priming the steel body shells of motor cars. The total process which involves degreasing, phosphate treatment, electrodeposition of primer, and then spray application of surfacer and finishing coats has raised the standards of corrosion resistance and general appearance considerably during this period. Electrodeposition may take place with the car body acting as either the anode or the cathode. In recent years it has been claimed that cathodic forms of electropaint give the better corrosion protection.

The methods of application will be described in the appropriate chapters, discussing each of the main market areas.

2

Organic film formers

J. Bentley

2.1 INTRODUCTION

The first chapter has indicated the major types of surface-coating resins used, and this chapter will describe their chemistry in more detail, including their preparation.

In this introductory section an outline is presented of the theory of polymer formation and curing; the mechanism specific to each type of resin is then covered more fully in the subsequent sections of this chapter. The classes of resins available and the factors that decide the choice of resin for a particular use are also indicated. The properties and uses of each type of resin, again, are detailed more fully in the later relevant sections.

A number of terms are used interchangeably to describe the film-forming component of paint, as will already be apparent. 'Film former', 'vehicle', or 'binder' relate to the evident fact that this component carries and then binds the particulate components together, and that this provides the continuous film-forming portion of the coating. Resin or varnish are older terms relating to the previous more prevalent use of natural resins in solution or 'dissolved' in oils as the film former, and date from when the chemistry and composition of these components was far less well understood. Nowadays, with our better knowledge of the materials used, along with the wide application of the sophisticated polymers used also in the plastics and adhesives industries but tailored to our own use, it is strictly more correct to refer to this component as the polymeric film-forming component. The interchangeable use of old and new nomenclature is also found in the manufacture of film formers where 'kettle' refers to the polymerization reactor, and 'churn' to the thinning tank normally part of the manufacturing plant.

Film-forming polymers may or may not be made in the presence of solvent; however, since the polymers in solvent-free form generally range from highly viscous liquids to hard brittle solids, they are practically always handled in storage and in the paint-making process in solution (or in dispersion) with significant quantities of solvent or diluent included. The only exceptions are the liquid oligomeric materials for high solids finishes, and solid resins for the very specialized application of powder coatings.

Most polymers used will be in true solution, with solvent being the other component. However, in certain cases, either for reasons connected with the polymer preparation or with its final use, the polymer will exist in the form of a fine-particle dispersion in non-solvent; this is true for aqueous emulsions, non-aqueous dispersions, and for the emulsified materials used in electrodeposition and certain other waterborne applications. In some cases the systems may be mixed solution/dispersion, for example a solution containing micellar polymer dispersion, micro-emulsion, or microgel. A particularly striking consequence of whether the polymer is dissolved or dispersed is the viscosity; dispersions are invariably more fluid at comparable solids contents, than solutions. Most significantly while solution viscosity increases as molecular weight rises, the viscosity of emulsions or dispersions is independent of molecular weight. For any given polymer in solution or dispersion, viscosity will increase as the solids increases; the viscosity will decrease if the temperature increases.

As with any utility product, the paint user is concerned mostly with the ability of the material to provide final protective and decorative effects and has little regard for composition, except in so far as it guides him to the ability of the material to satisfy those needs. Table 1.1 in Chapter 1 has listed the function of paint components and Table 2.1 shows the contribution that the three major components — resin, pigment and solvent — make to the most important properties of a typical gloss paint.

Table 2.1 — Contribution of major paint components to final paint properties

Property	Film former	Pigment	Solvent
Application	Major	Minor	Major
Cure rate	Major	None	Significant
Cost	Major	Major	Minor
Mechanical properties	Major	Minor	—
Durability	Major	Major	—
Colour	Minor	Major	—

Informed readers will see limitations in the above, but it is primarily intended to highlight the broader binder/solvent contributions.

Generally all polymer types can provide a spectrum of compositions covering a span of properties at varying cost, and so given user criteria may be satisfied by selection from a number of resin types. The principal final choice for the user, whose application and cure conditions will probably have been determined by scale, ultimately would appear to concern balancing performance and cost, true cost including the total of paint cost, labour and equipment cost, and energy for cure.

The industry is constantly striving for higher performance and novel products, and factors influencing system design may well include current trends to guarantee performance in such diverse applications as decorative maintenance paints and in automobile and coated coil products. External forces which apply are legislation to control pollution, availability and cost of energy supplies, and periodic abundance or shortage of natural and oil-based raw materials. New requirements continue to emerge, such as the change of motor car bodies from all metal to metal/plastic composite, requiring high-performance coatings for plastic substrates.

The choice of solvent or diluent used will depend on the nature of the polymer and the method of application; the quantity of solvent (solids) and final viscosity will then depend on the latter. The method of application generally imposes constraints regarding solvent boiling point and evaporation rate, for example, to ensure good spray or brush application. If the polymer can be prepared in the presence of little or no solvent, the solvent necessitated by the method of paint application has little practical significance to the resin chemist, i.e. an alkyd or polyester may easily be thinned at end point with high *or* low boiling solvent. However, for an acrylic resin it is usually necessary to use a solvent or solvent blend of low chain transfer properties, and with a boiling point such that the reaction mixture can be refluxed to remove heat of polymerization and such that an initiator system is available at that temperature capable of efficient conversion of monomer to polymer.

The mechanical properties of any given polymer 'improve' as molecular weight increases up to a value at which no change is seen with further increase. In contrast, the viscosity of the solution of a polymer continues to increase with molecular weight without a break. This imposes the constraint in designing a surface-coating system, that if the polymer is to be made for optimum application *and* final properties, then molecular weight needs careful specification and control. Furthermore, since for many polymer systems the molecular weight necessary for good mechanical properties and durability will be high, considerable amounts of solvent will be needed to obtain good application properties, if the polymer is to be applied in solution at that molecular weight. The kind of system this describes is a lacquer, drying by solvent evaporation alone to leave a film of polymer with useful properties, with no subsequent change of molecular weight or further reaction occurring. Practically this system, initially used with varnishes such as Shellac varnish and French polish, now continues with plasticized nitrocellulose and with thermoplastic acrylic lacquer systems used both in automotive manufacture and refinishing systems.

The simplest way of avoiding the molecular weight/viscosity conflict is the use of dispersion rather than solution systems as with decorative aqueous emulsions now made in high volume, and also with the dispersed polymer used in high-solids systems, non-aqueous dispersion (NAD) systems, and organosols. The use of dispersions can also allow the use of cheaper, less polluting diluents, and is particularly valuable where legislation exists controlling the VOC (volatile organic content) of factory exhaust emissions. However, these polymers are complex to formulate and more restricted in composition than solution polymers.

Many surface-coating systems avoid the molecular weight constraint by utilizing 'crosslinking' or 'curing' reactions; the system is assembled with one or more reactive polymer components of relatively low molecular weight, capable of further reaction after application to high or infinite molecular weight. Crosslinking implies a multifunctionality such that each initial component molecule links to a number of other molecules, so that an infinite network is formed in the final coating.

Methods used for film formation with typical polymer systems are indicated in Table 2.2. In every case except the lacquer system, the polymer system assembled

Table 2.2 — Methods of film formation for typical polymer systems

Method	External agent	Typical polymer system
Solvent evaporation	None or heat	Lacquer systems
Environmental cure	Oxygen	Oil-modified alkyd
	Moisture	Moisture-curing urethane
Vapour phase curing	Amine	Hydroxy acrylic/isocyanate blend
2-pack	None or heat	2-pack epoxy/amine
Radiation	IR/UV/EBC	Photocuring unsaturated polyester
Thermosetting	Stoving oven	Alkyd/nitrogen resin blend
		Thermosetting acrylic

includes in its architecture free reactive groups appropriate to the method of curing chosen. The curing reactions are discussed later in this chapter under the appropriate resin types. Apparent overlap of techniques may be found; for example, stoving may be used to accelerate solvent evaporation so blurring the distinction between thermoplastic and thermosetting acrylics; equally, stoving may accelerate curing of an alkyd capable of slower but ultimately satisfactory air-drying unaided.

Electrodeposited films normally require a subsequent heat cure; electrodeposition insolubilizies the polymer on deposition, but the film is soft and not fully resistant to other agents until crosslinking occurs.

Polymers are conveniently categorized by their method of polymerization from monomer, which may be by stepwise or functional group polymerization, or by chain addition polymerization. These terms are more satisfactory than the other common terms, condensation and addition polymerization, being both more precise and more generally applicable.

Stepwise polymerisation is identified where the monomer units initially present are nearly all incorporated into larger molecules at an early stage in the reaction; these larger molecules remain reactive and continue to join together so that the average molecular weight increases with time, but once the reaction is under way the yield of polymer species is not a function of reaction time. The reaction usually involves two separate monomer species with different but co-reactive groups. Most usually, though not invariably, some small molecule such as water will be eliminated as each step occurs. Alkyds and polyesters are typical stepwise polymers. The concept of functionality is particularly relevant to this type of polymerization in that a minimum requirement of each 'monomer' molecule is that it shall possess two reactive functional groups, if the product of each reaction step is to be able to participate in further reactions.

Chain addition polymerization has the characteristic that high molecular weight polymer is formed from the start by a chain reaction, and monomer concentration decreases steadily throughout the polymerization period; thus the yield of polymer increases with time, unlike stepwise polymerization. Polymerization is initiated by some active species capable of rupturing one of the bonds in the monomer, and may be radical, electrophilic, or nucleophilic in character. Only free radical chain addition polymerization is generally practised in production by surface-coating manufacturers though they may purchase polymers made by other techniques. Acrylic polymers are the best known chain addition polymers.

Table 2.3 shows the broad types of polymers used in coatings, partly in terms of

Table 2.3 — Classification of polymer types used in coatings

Type	Sub group
Natural resins	Resins, gums, rosin
Modified natural resins	Cellulose, starch, nitrocellulose
Stepwise (condensation) polymers	Polyester, alkyd resins
	Formaldehyde resins (UF, MF, PF)
	Epoxy resins
Chain addition polymers	Acrylic polymers and copolymers
	Vinyl polymers
Ether polymers	Polyethylene oxides and glycols

the above classification. Ether polymers are separated into a group of their own, in part to acknowledge that materials such as polyethylene glycol can be prepared by both stepwise and addition polymerization, and also to avoid further confusion. Epoxy resins are formed by stepwise polymerization, but it should be noted that the epoxide group may take part in both stepwise and chain addition polymerization reactions.

2.2 NATURAL POLYMERS

The vehicles used as surface coatings were originally based on natural oils, gums, and resins, giving, when combined, a range of both lacquer and autoxidatively drying products. The naturally occurring triglyceride oils still find considerable use in oil modified alkyds and to a lesser extent in other fatty acid modified products such as epoxy esters, and these are described under separate headings. The use of other natural products has now diminished to very minor amounts, in part owing to limited availability; some mention of those natural resins still used is made in the section on oleoresinous vehicles.

Under the description of modified natural products mention must be made of cellulose derivatives, particularly 'nitrocellulose', which is the major binder in cellulose lacquers which still find considerable use. Nitrocellulose, more accurately named cellulose nitrate, is obtained by the nitration of cellulose under carefully specified conditions, which control the amount of chain degradation and the extent of nitration of the hydroxyl groups of the cellulose.

The nitration reaction in essence is

$$HC-OH \xrightarrow{HNO_3} HC-ONO_2$$

Organic esters of cellulose can also be produced, and the mixed ester cellulose acetate butyrate is used as a modifying resin additive, particularly for acrylic lacquers. The mixed ester is used in preference to the softer higher esters, or the acetate which has poorer dimensional stability. (See 2.7).

Cellulose derivatives such as hydroxy ethyl cellulose and the salts of carboxy methyl cellulose also find use as protective colloids in the emulsion polymerization of vinyl monomers. (See 2.8).

2.3 OILS AND FATTY ACIDS

Vegetable oils and their derived fatty acids play nearly as important a role in surface coatings today as they did in the past because of their availability as a renewable resource, their variety and their versatility.

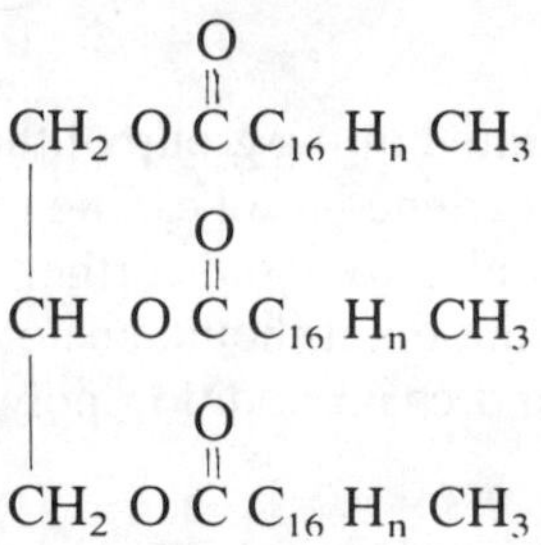

$$
\begin{array}{l}
\quad\quad\ \ \overset{\displaystyle O}{\overset{\|}{}} \\
CH_2\ O\ C\ C_{16}\ H_n\ CH_3 \\
|\quad\quad\ \ \overset{\displaystyle O}{\overset{\|}{}} \\
CH\ \ O\ C\ C_{16}\ H_n\ CH_3 \\
|\quad\quad\ \ \overset{\displaystyle O}{\overset{\|}{}} \\
CH_2\ O\ C\ C_{16}\ H_n\ CH_3
\end{array}
$$

n may be 32, 30, 28 or 26

Fig. 2.1 — Representative structure of naturally occuring oil.

Oils are mixed glycerol esters of the long chain (generally C_{18}) monocarboxylic acids known as fatty acids (Fig. 2.1); unrefined oils also contain free fatty acids, lecithins, and other constituents, the first two finding their own application in coatings.

Oils useful in coatings (Table 2.4) include linseed oil, soya bean oil, coconut oil, and 'tall oil'. Many countries will find some application of their own indigenous oils as well as those detailed here.

When chemically combined into resins, oils contribute flexibility and, with many oils, oxidative crosslinking potential, one of the properties most exploited in paint.

Table 2.4 — Composition of major oils used in surface coatings

	Saturated acid	Oleic acid	9,12 Linoleic acid	9,12,15 Linolenic acid	Conjugated acid
Tung	6	7	4	3	80***
Linseed	10	20–24	14–19	48–54	0
Soya bean	14	22–28	52–55	5–9	0
Castor oil	2–4	90–92*	3–6	0	0
Dehydrated castor oil	2–4	6–8	48–50	0	40–42**
Tall	3	30–35	35–40	2–5	10–15***
Coconut	89–94	6–8	0–2	0	0

* Principally ricinoleic acid, not oleic.
** Conjugated 9,11 linoleic acid.
*** Conjugated linoleic acid, isomerized linolenic acid, and 9,11,13 eleostearic acid, proportions dependent on source and degree of refinement.

Oils are classified as drying, semi-drying, or non-drying, and this is related to the behaviour of the unmodified oil, depending on whether it is, on its own, able to oxidize and crosslink to a dry film. This behaviour is directly related to the concentrations of the various fatty acids (Table 2.5) contained in the structure. These

Table 2.5 — Structures of the commonly occuring fatty acids

	Lauric	$CH_3 (CH_2)_{10} COOH$
	Stearic acid	$CH_3 (CH_2)_{16} COOH$
9,12	Linoleic acid	$CH_3 (CH_2)_4 CH=CH CH_2 CH=CH (CH_2)_7 COOH$
9,12,15	Linolenic acid	$CH_3 CH_2 CH=CH CH_2 CH=CH CH_2 CH=CH (CH_2)_7 COOH$
9,11,13	Eleostearic acid	$CH_3 (CH_2)_3 CH=CH CH=CH CH=CH (CH_2)_7 COOH$
	Ricinoleic acid	$CH_3 (CH_2)_5 \underset{\underset{OH}{\mid}}{CH} CH_2 CH=CH (CH_2)_7 COOH$

fatty acids contain, with a few exceptions, 18 carbon atoms including the terminal carboxyl. Fatty acids may be saturated (no double bonds), unsaturated (one double bond), or polyunsaturated (two or more double bonds); and to be considered 'drying' an oil must contain at least 50% of polyunsaturated fatty acids.

Thus, linseed oil, a good drying oil, contains over 60% of the polyunsaturated linoleic and linolenic acids, while soya bean oil, usually classified as semi-drying, contains just over 50% of linoleic acid. By contrast, coconut oil, a non-drying oil, contains 90% saturated lauric acid and less than 10% unsaturated fatty acid.

Polyunsaturated fatty acids may be conjugated or non-conjugated, and tung oil, now in short supply, contains nearly 80% of the conjugated eleostearic acid. By contrast 9,12,15 linolenic acid has three non-conjugated double bonds, while 9,12 lineoleic acid has two non-conjugated double bonds. High conjugation is reflected in higher drying ability and greater tendency to heat body over that expected from the double bond content; thus, tung oil can in fact readily be gelled by heating, and for this reason — if incorporated into an alkyd resin — would not be processed by the monoglyceride process, this being considered too risky.

The drying oils, tung and linseed, may find application either unmodified or only heat bodied in coating compositions in simple blends with other resins. In contrast, most other oils find little application alone but only after chemical modification; soya bean oil, for example, finds its major application in chemically combined form in long oil alkyds. Oils such as coconut oil do not dry or heat body and have no application alone, but only find use when reacted into alkyds. In this case the oil component provides plasticization. Coconut alkyds have good colour and are used for automotive stoving finishes and appliance enamels.

The two most important oils to the industry nowadays are probably soya bean oil and 'tall oil', the latter misnamed, as it is a fatty acid mixture containing over 50% polyunsaturated fatty acid, principally 9,11 linoleic acid. Tall oil is a by-product of the paper industry, and to be useful in coatings must be refined to reduce its rosin content to 4% or less from the higher concentration in the crude product. However, it is the fractionally distilled form [1] which can now find interchangeability with soya fatty acids (and in alkyds where soya bean oil is replaced with tall oil and polyol) depending on seasonal fluctuations of availability and price. Tall oil of Scandinavian origin may be more favoured than that from America owing to its higher unsaturated fatty acid content. Tall oil is probably more variable in quality, dependent on source and refinement, than any other oil or fatty acid, and hence needs careful evaluation and specification for any critical use.

An oil with a more distinctive fatty acid is castor oil which contains over 90% of ricinoleic acid, or 12-hydroxy oleic acid. It has some use as a polyol for polyurethane preparation, but a major use by exploiting the hydroxyl content, is in the preparation of higher hydroxyl containing alkyds for crosslinking with nitrogen resins or isocyanates, and for use as plasticizing alkyds for nitrocellulose where high polarity is necessary for compatibility.

Oils are generally refined by acid or alkali refining for the coatings industry in order to remove materials such as free fatty acids and lecithins, so as to give a neutral clear product and to improve colour. When further reaction is required, oils may be separated into constituent fatty acids and glycerol, and though more expensive than the oil, the free fatty acids of all the common oils are currently available to the coatings industry.

2.3.1 Modified oils
Drying oils may be heat-bodied at temperatures in the region of 290°C to produce 'stand oils', where the viscosity increase is due principally to dimerization reactions through double bonds of unsaturated fatty acid moieties. This reaction may be followed through decrease in the iodine value. 'Stand oils' find application in oleoresinous vehicles and in alkyd resins. The separated dimer fatty acids also find specialized uses, for example in polyamide resin manufacture.

So-called boiled oils were once prepared by heating metal oxides in oil until fatty acids released during bodying solubilized the oxides as soaps. Nowadays a 'boiled oil' may be prepared more conveniently by blending commercial driers with a 'stand oil'.

Dehydration of castor oil is important, since the removal of the hydroxy group and an adjacent hydrogen atom creates an additional double bond, so increasing the residual unsaturation. Dehydrated castor oil (DCO) has distinctive characteristics since the second double bond generated is generally positioned relative to the original one to give conjugated rather than non-conjugated polyunsaturation. Some bodying may occur on dehydration, although this can be minimized by use of a good vacuum and suitable catalysts, thus reducing the time necessary for the reaction. By virtue of its unsaturated fatty acid content DCO is classed as a drying oil. DCO finds major application in stoving alkyds and in alkyds required for further modification by vinylation; it is little used in air-drying finishes because of the incidence of surface 'nip' (tackiness) in dried films. DCO fatty acids are also available to resin formulators.

Castor oil may also be hydrogenated, and hydrogenated castor oil (HCO) has some application in alkyd resins for stoving application with MF resins; HCO fatty acid, more usually named 12-hydroxy stearic acid, finds application particularly for graft copolymers [2] for use in dispersants and surfactants (see 2.9).

Isomerization of oils and fatty acids is used to improve the drying properties particularly of medium oil length alkyds [3]. Isomerization means a transformation of the double bonds of non-conjugated fatty acids (principally linoleic) to conjugated form, and in certain cases to the *trans* isomer; a number of processes have been claimed for carrying this out [4, 5]. Improvements from using these oils/fatty acids include better colour, initial dry, and resistance to water, acid, and alkali. Weathering and gloss retention are also said to be improved.

Oils and fatty acids containing unsaturation can be reacted with maleic anhydride. Where conjugated double bonds are present a Diels Alder reaction is possible, and this proceeds exothermically. Where the bonds are non-conjugated (linolenic or linoleic) the first addition is by the Ene reaction; this has the consequence of moving the double bonds from a non-conjugated to a conjugated configuration with the possible side effect observed that a sluggish first Ene reaction (Fig. 2.2) may be

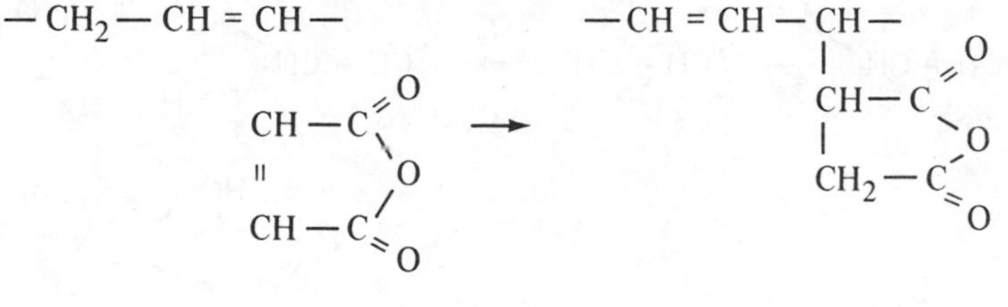

Fig. 2.2 — Ene reaction.

followed by a faster exothermic Diels Alder second reaction (Fig. 2.3).

Maleinization is carried out to increase acid functionality of fatty acids, for example for subsequent water solubilization (see 2.5 and 2.16). It also occurs *in situ* in alkyd preparations where maleic anhydride is included in the formulation, and its urethane oils and urethane alkyds are derived from the resistance of the urethane link to hydrolysis. In decorative (architectural) paints it is common practice to use binders which are a mixture of a long oil alkyd and urethane alkyd for maximum durability.

Amino resins
The most common types of amino resin are reaction products of urea or melamine (1:3:5 triamino triazine) and formaldehyde. The resins are prepared in alcoholic media, which enables the molecular weight and degree of branching to be controlled within practically determined limits related to the end use of the resin. The effect of this modification is shown in the solubility and reactivity of the resins so produced. The polymers produced are generally regarded as being derived from the hydroxy-methyl derivatives of melamine and urea respectively; subsequent addition conden-sation and etherification reactions lead to complex, highly-branched polymeric species. Curing or crosslinking to solid films (usually in combination with an alkyd or other polymer) can be achieved thermally (oven-curing) or at room temperature. In both cases the presence of an acid catalyst is essential if adequate and rapid cure is to be obtained.

The crosslinking capability of amino resins may also be utilized as a means of curing acrylic resins. In this type of film former a minor proportion (usually a few percent) of a monomer such as *N*-butoxymethyl acrylamide is incorporated into the polymer. This provides reactive sites which enable the acrylic copolymer to be crosslinked. Suitable choice of monomers allows the acrylic resin to be plasticized internally, so that the use of added plasticizers is avoided. These resins are of particular interest where high levels of performance and particularly good adherence and flexibility are required.

Phenolic resins
The reaction of formaldehyde with a phenol gives rise to a range of resins that, in combination with other resins or drying oils, find use in industrial coatings. Broadly two main types of phenolic are produced, novalacs and resoles. Novalacs are low molecular weight linear condensation products of formaldehyde and phenols that are alkyl substituted in the *para* position. If the substituent alkyl group contains four or more carbon atoms (i.e. butyl or above   in the homologous series) the resin is likely to be oil-soluble. Resoles are products of the reaction of unsubstituted phenols with formaldehyde. Since the *para* position on the phenolic ring is available for reaction as well as the *ortho* position these resins are highly-branched and can with continued reaction be converted into hard intractable glassy solids. Phenolics tend to confer chemical resistance to the compositions in which they are used. They are always used in combination with other resinous film formers. In some cases they may be pre-reacted with the other resin components, or they may be simply blended together. Thus a phenolic resin (e.g. a novalac) may be reacted with rosin or ester gum and then blended with a bodied (heat-treated) drying oil to form the binder in an architectural

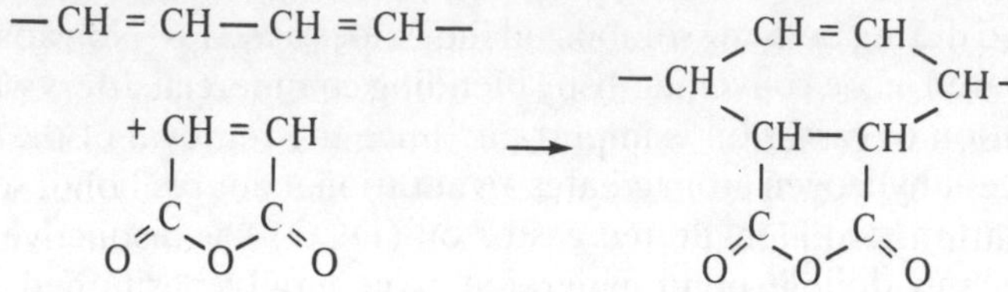

Fig. 2.3 — Diels–Alder reaction.

effect is to increase viscosity by increasing acid functionality. Care is needed in formulation since a small maleic addition can for this reason have an unexpectedly large effect.

True 'urethane oils' are obtained where a monoglyceride produced by reaction of an oil with glycerol (see 2.5) is reacted with a diisocyanate; most 'polyurethanes' of this type are, however, urethane alkyds where some condensation of monoglyceride with dibasic acid is first carried out prior to diisocyanate reaction.

Oils may be vinylated, but this alone has little application, though it may be a first step in the preparation of a vinylated alkyd (see 2.5.7).

2.3.2   Drying oil polymerization

As previously stated, drying oil fatty acids may dimerize under heat treatment by Diels Alder or Ene reaction of unsaturated groups. However, the most exploited reaction of these unsaturated groups is autoxidative crosslinking, accelerated by the presence of heavy metal driers and also with the aid of heat in stoving compositions.

The theories of autoxidative dry have been well reviewed [6], and only an outline is given here. It is now accepted that the first steps in autoxidative drying involve hydroperoxide formation, and it is believed that in the process double bonds may shift from the *cis* to the *trans* structure; for linoleates hydroperoxidation can lead to a movement in the double bond adjacent to the position of attachment of the hydroperoxide leading to a conjugated structure (Fig. 2.4) which stabilizes the product [7].

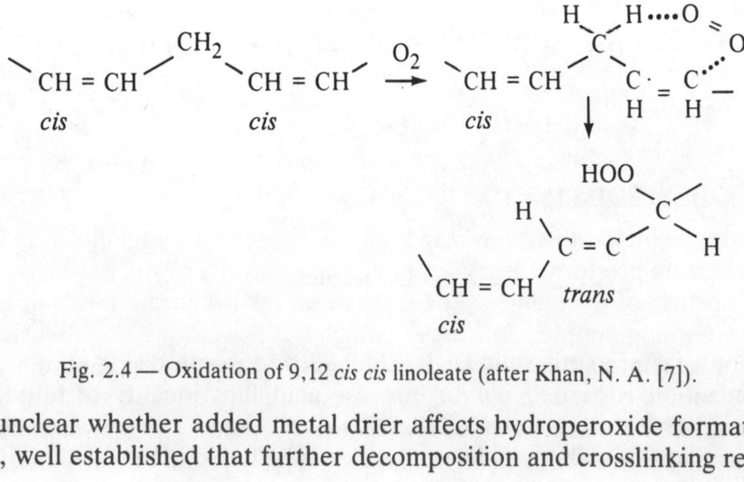

Fig. 2.4 — Oxidation of 9,12 *cis cis* linoleate (after Khan, N. A. [7]).

It is unclear whether added metal drier affects hydroperoxide formation. It is, however, well established that further decomposition and crosslinking reactions of hydroperoxides are accelerated by the presence of metallic driers (particularly cobalt). The use of a hydroperoxide with cobalt drier to initiate the cure of unsaturated polyesters is another example of this reaction. Just as this latter reaction is identifiable as free radical in nature, so the autoxidative crosslinking reaction is also at least partly free radical in mechanism, and this may be exploited by the copolymerization of high boiling or involatile vinyl monomers into an autoxidatively drying system to modify final properties [8, 9].

$$2 \text{ ROOH} \xrightarrow{\text{Co}^{2+} \text{ Soap}} \text{RO} \cdot + \text{RO}_2 \cdot + \text{H}_2\text{O}$$

The actual crosslinking reactions following radical formation are complex, and as well as the desired high molecular compounds, scission products giving characteristic drying smells are also produced. Polymerization reactions certainly involve both radical dimerizations and reactions further involving residual unsaturation.

Metallic catalysts or accelerators are principally derivatives of heavy metals, particularly cobalt, with added lead, calcium and other metals to enhance through drying properties in paint compositions; metals are now normally added as soaps of long chain acids so that they are fully soluble in the media [10]. Naphthenates have now been largely displaced by synthetic branched acids. New combinations continue to be proposed such as the use of zirconium and aluminium to enable lead reduction or elimination in paint systems.

2.3.3 Characterizing oils

The iodine value of an oil is the number of grammes of iodine absorbed by 100 g of oil; since iodine adds across double bonds this is a measure of the unsaturation present. The Wijs method is the most commonly used, though this can be inaccurate with conjugated systems.

The saponification value gives an indication of the molecular weight of the component fatty acid chains, assuming the original oil is intact and entirely triglyceride.

It must be noted that oils, being derived from annual crops of typically leguminous plants, can vary from season to season in quality (that is in fatty acid distribution); crop yields also vary, causing supply and price fluctuations. Geographical variations also occur as cited earlier in the difference between Scandinavian and American tall oils.

2.4 OLEORESINOUS MEDIA

Oleoresinous vehicles are those manufactured by heating together oils and either natural or certain pre-formed synthetic resins, so that the resin dissolves or disperses in the oil portion of the vehicle. The equipment for preparing this type of product may be essentially simple, and these vehicles were among the first used in the coatings industry after the exploitation of simple gum solutions and natural oils in either unmodified or bodied state. The manufacturing process involves heating oil and resin together until such time as the product becomes clear (this may be tested by cooling a small sample 'pill' of material on a cold plate) and of the required viscosity.

The temperatures used are frequently those at which oil bodying occurs, that is around 240 degrees C or higher, and the best performance is often obtained when the resin is taken close to gelation point; for this reason great skill is required of the maker of this type of vehicle. Reaction near to end point may be rapid, and thus quick though subjective tests such as bodying to 'a short string' may have to be employed. Following attainment of end point the product may be cooled and thinned, often in one operation.

Oleoresinous vehicles continue in use for a number of applications, though they have been displaced by alkyds and other synthetic resins for many other uses. For example, clear varnishes, primers and undercoats, aluminium paints, and marine coatings still use oleoresinous vehicles to advantage, and a high residual use persists in printing inks, though this is not a topic considered further in this chapter.

Oils used are exclusively drying and semi-drying oils, with the more highly unsaturated oils, i.e. tung, linseed, and oiticica preferred. The resins used have included a number of natural 'fossil' resins such as rosins, copals, shellac, etc. of which only 'tall' rosin is now in good supply. (Rosin consists of a mixture of acids including abietic acid.) Bitumen is a natural resin, and black japan which contains bitumen is, in fact, an oleoresinous vehicle which is prepared as described above. Synthetic resins now used include rosin derivatives, and phenolic and epoxy resins. The latter two retain considerable importance for marine and insulating varnishes.

An important concept first used in oleoresinous vehicles and now used in defining alkyd formulations is that of oil length. In this context it refers to the oil content of the final resin; different nomenclatures may be used, either US gallons of oil per 100 lb of resin or in a more precise definition, the weight of oil % in the total non-volatile varnish produced. Short oil vehicles contain less than 67% oil and are fast drying, giving hard films but lacking in flexibility; they are used for floor varnishes and gold size. Long oil vehicles (more than 67% oil) are slower drying but more flexible and are used for exterior varnishes and undercoats.

2.5 ALKYD RESINS

The alkyd resin was one of the first applications of synthetic polymer synthesis in surface coatings technology; it was successful in chemically combining oil or oil-derived fatty acids into a polyester polymer structure, thus enhancing the mechanical properties, drying speed, and durability of these vehicles over and above those of the oils themselves and the oleoresinous vehicles then available. Though now surpassed by more sophisticated polymers for the more exacting applications, the alkyd, because of its partial reliance on renewable natural resource, and especially because of the enormous variety of compositions possible, still accounts for a very high volume of total surface coating resin produced. The long oil alkyd, optionally used blended with other modified alkyds to increase toughness or alter rheology, remains the major resin used in brush applied solvent based decorative coatings. Short oil alkyds still have considerable use in automotive and general industrial stoving compositions where they are combined with melamine/formaldehyde hardening resins.

Oils are not directly reactive with polyester components and therefore cannot be incorporated into an alkyd structure without prior modification. We can, however, start with fatty acids saponified from the oil, though for many applications this will be an expensive route, especially if glycerol is intended to be included in the formulation. The method evolved to achieve oil modification is to carry out, as a first stage, a so-called monoglyceride preparation. In this process glycerol (or other polyol) and oil in the molar ratio of $2:1$ are reacted at around 240°C in the presence of a basic catalyst (sodium hydroxide, litharge, etc.) forming 'monoglyceride', the mono fatty acid ester of glycerol (Fig. 2.5).

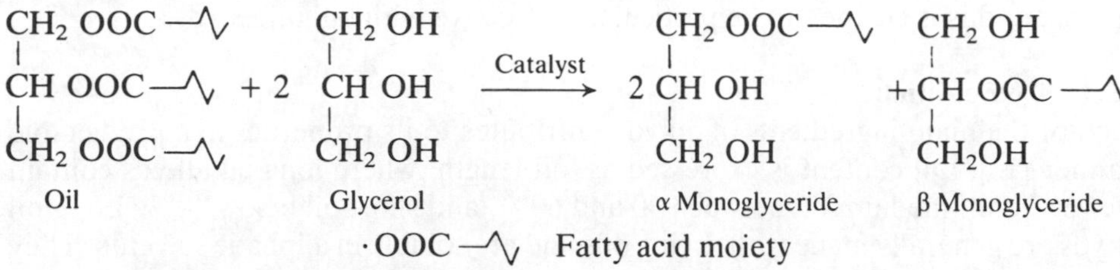

$$
\begin{array}{c}
CH_2\ OOC - \wedge \\
| \\
CH\ OOC - \wedge \\
| \\
CH_2\ OOC - \wedge \\
\text{Oil}
\end{array}
\; + 2 \;
\begin{array}{c}
CH_2\ OH \\
| \\
CH\ OH \\
| \\
CH_2\ OH \\
\text{Glycerol}
\end{array}
\quad \xrightarrow{\text{Catalyst}} \quad
2
\begin{array}{c}
CH_2\ OOC - \wedge \\
| \\
CH\ OH \\
| \\
CH_2\ OH \\
\alpha\ \text{Monoglyceride}
\end{array}
\; + \;
\begin{array}{c}
CH_2\ OH \\
| \\
CH\ OOC - \wedge \\
| \\
CH_2OH \\
\beta\ \text{Monoglyceride}
\end{array}
$$

$$\cdot OOC - \wedge \quad \text{Fatty acid moiety}$$

Fig. 2.5 — Schematic representation of monoglyceride formation.

Practically, the reaction does not go to completion, and an equilibrium distribution of species is present including oil, polyol and mono and di-glycerides. If polyols other than glycerol are to be used, such as pentaerithritol (PE) in long oil alkyds, a 'monoglyceride' stage may still be carried out using the appropriate amount of polyol to give a notional dihydroxy functional product for this stage. The satisfactory attainment of sufficient randomization of structure may be tested for by an alcohol tolerance test, when no separation of free oil should be observed. Sometimes with polyols such as PE the alcohol test is not satisfactory, in which case reaction is carried out on a small scale in a wide-necked tube with the correct proportions of 'monoglyceride' sample and polybasic acid to test for the ability to derive a clear final product without gelation resulting. With some oils it is important not to prolong the time at temperature for this stage in order to minimize oil bodying; for the same reason it is good practice to standardize on heat-up rate and time for the normal reaction, to ensure reproducible and predictable behaviour in the following stage.

Preparation of the final resin is carried out by the addition of further polyol and dibasic acid in the so-called bodying stage; in this stage, multiple esterification reaction occurs (Fig. 2.6) and water formed as a by-product of reaction must be removed. Esterification results in the steady building-up of polymer structure and hence viscosity increases; the stage is taken to a degree of reaction at which the desired viscosity at a specified concentration (in solvent) and temperature has been attained.

While the monoglyceride method as described is used for long oil alkyds, the simpler so-called 'fatty acid process' is practised, with direct charging of all ingredients including fatty acid (not oil), and is frequently used for shorter oil alkyds where polyols other than glycerol are required. Other techniques are possible, for example an acidolysis route where oil is first equilibrated with polybasic acid prior to reaction with polyol in an apparent reversal of the monoglyceride process; this method is not widely favoured. Where it is necessary to incorporate tung oil into an alkyd, when to carry out a monoglyceride state would risk excessive bodying, a variant is the fatty acid/oil process. Here the tung oil is taken along with a good proportion of fatty acid, polyol, and dibasic acid, and a one-stage reaction carried out. Providing the oil concentration is not high, transesterification of sufficient magnitude takes place during processing to react the oil into the resin and ensure a clear final product. Another technique proposed for alkyd manufacture is the 'high polymer' technique [11, 12] where by stepwise addition of ingredients, more favourable molecular weight and reactive group distributions are said to be obtained and certain properties enhanced; this is claimed to be particularly effective with tall oil alkyds.

2.5.1 Composition
Each of the main ingredients of alkyd contributes to its properties in a predictable manner [13]. Oil content is expressed as 'oil length' where long oil alkyds contain over 60% oil, medium oil between 40 and 60%, and short oil below 40%. Long oil alkyds are generally made with drying oils and are soluble in aliphatic solvents. They are low viscosity resins that air-dry slowly to give soft flexible films with poorer gloss retention and durability.

Shorter oil length generally results in the need for aromatic solvent, giving higher viscosity resins at lower solids. If formulated with drying or semi-drying oils, and of medium or long oil length, the alkyd can air dry oxidatively at room temperature, otherwise it will not do so, and to form hard films must be stoved with a crosslinking resin. Medium and short oil alkyds give hard films with good gloss retention and chemical resistance; they are less flexible than long oil alkyds and are normally dried by stoving.

Aromatic acids such a phthalic anhydride or isophthalic acid, and maleic anhydride contribute hardness, chemical resistance, and durability to alkyds; in contrast long chain dibasic acids such as azelaic acid, are sometimes used to plasticize alkyds and provide flexibility, not hardness. Phthalic anhydride dominates in alkyd formulations because of its low cost and ready availability. It should be noted that it has added economy in use, in that by virtue of its anhydride structure, water evolution is halved compared to a dibasic acid; the first stage of its reaction is exothermic, with so-called 'half ester' formation.

Polyols used are generally at least trifunctional to permit branching or crosslinking and can provide the alkyd with hydroxyl groups for further reaction. In general they contribute good colour and colour retention, but vary in their chemical and weathering resistance (see also 2.6). Most general purpose alkyds are formulated on phthalic anhydride with properties modified by polyol variation. The effects of polyol change may be readily related to structure so that in a series where glycerol is replaced by trimethylol ethane and then by trimethylolpropane, aliphatic solubility increases, viscosity falls, and flexibility increases, contributed to by increased side chain length; otherwise the effect of varying the polyol component may have a significant effect, not least because with varying polyol MW the phthalic content of the final resin is affected, and hence the hardness of the final alkyd.

Rosin was originally included in alkyds as a modifier by reason of its low cost. However, it may also be added as may other monobasic acids such as benzoic acid, to enable oil length to be shortened without causing increased viscosity. Rosin itself

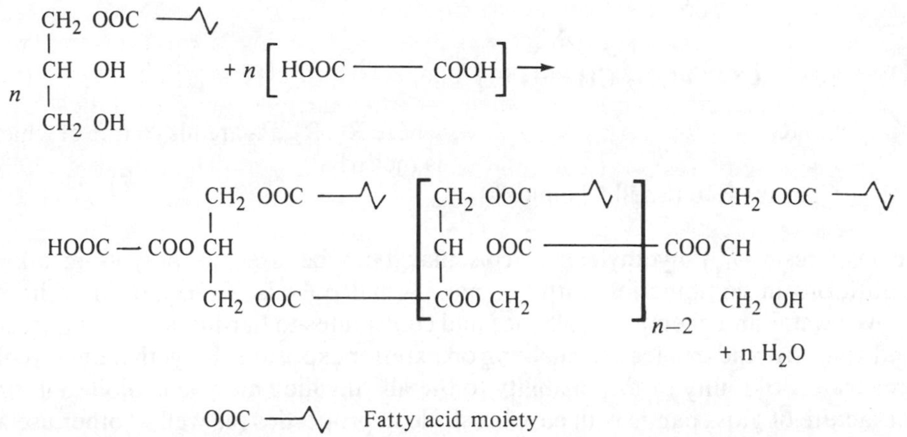

$$n \begin{array}{l} CH_2\ OOC \longrightarrow\!\!\!\!\vee \\ |\\ CH\ \ OH \\ |\\ CH_2\ OH \end{array} \quad + n \left[HOOC \longrightarrow COOH \right] \longrightarrow$$

$$HOOC - COO\ \begin{array}{l} CH_2\ OOC \longrightarrow\!\!\!\!\vee \\ |\\ CH \\ |\\ CH_2\ OOC \longrightarrow \end{array} \left[\begin{array}{l} CH_2\ OOC \longrightarrow\!\!\!\!\vee \\ |\\ CH\ \ OOC \longrightarrow COO \\ |\\ COO\ CH_2 \end{array}\right]_{n-2} \begin{array}{l} CH_2\ OOC \longrightarrow\!\!\!\!\vee \\ |\\ COO\ CH \\ |\\ CH_2\ OH \end{array}$$

$$+\ n\ H_2O$$

OOC $\longrightarrow\!\!\!\!\vee$ Fatty acid moiety

Fig. 2.6 — Schematic representation of alkyd polymer formation.

contributes to faster dry and film hardness, but degrades weathering performance; its use is now restricted to alkyds for primer applications.

Where short oil length and aliphatic solubility present conflicting requirements, trimethylol propane can be used as polyol, though with some penalty of increasing softness, and also *para* tertiary butyl benzoic acid used as modifying monobasic acid, both being effective because of the contribution of their aliphatic side groups.

Isophthalic acid gives higher molecular weight resins than orthophthalic, with better drying characteristics and harder, more durable films. However, it is more difficult to incorporate, remaining solid in the reaction mixture until at least one of its acid groups has reacted. Like orthophthalic acid it can sublime, but unlike orthophthalic it can lead to overhead blockages in manufacturing plant because of its lower solubility in xylene which is used as the entraining solvent for the removal of water of reaction. Terephthalic acid has little use in alkyds because of its extremely low solubility in reaction mixtures and its very slow reactivity.

Special mention must be made of Cardura E10 (Shell Chemicals) which is the glycidyl ester of the branched Versatic Acid (Shell Chemicals), the latter being a 'tertiary' or trialkyl acetic acid with a total of ten carbon atoms. By virtue of the epoxide group, it reacts rapidly and completely with carboxyl groups at temperatures above 150°C; when incorporated into an alkyd, the effect is that which would be contributed by a non-drying oil or fatty acid component. Cardura E10 contributes special properties of good colour, gloss retention, and very good resistance to yellowing and staining [14, 15]. Derived alkyds can be used in stoving enamels or in nitrocellulose lacquers. Care should be taken in the manufacture of alkyds which include both Cardura E10 and phthalic anhydride since the reaction is both rapid and exothermic (around 110 kJ per epoxy equivalent). Special formulating techniques are available for this type of alkyd which also allow control of polymer architecture [16, 17]. Cardura E10 has special use for reducing the acid value in resins as a final treatment, for example where zero acid value is required for resins for use with photocatalyst, in metallic paints or with certain drier systems.

R$_1$
                                        O
                                      /   \
R$_2$ —— CO O CH$_2$ CH —— CH$_2$

R$_3$ where R$_1$, R$_2$, R$_3$ are alkyl, one of which
                                                      is methyl
              Cardura E10 (Shell Chemicals)

Epoxy resin or polyethylene glycols may both be used as polyols in alkyd compositions in combination with the more usual polyols. Epoxy resin addition improves water and chemical resistance and contributes to hardness and toughness, but adds to cost and can lead to chalking on exterior exposure. Polyethylene glycols confer water solubility or dispersibility to the alkyd which may be exploited in the manufacture of gloss paints with easy brush clean properties (as well as other use as surfactant) [18].

Another route to water solubility or to enable self-emulsification is to formulate the alkyd to high acid value using trimellitic anhydride (TMA) or dimethylol propionic acid (DMPA), followed by partial neutralization with alkali, ammonia, or amine. The most common method in processing water soluble alkyds containing TMA is first to react all ingredients except the anhydride to form a polymer with an acid number below 10, and then add the TMA at a reduced temperature of around 175°C. Under these conditions the TMA is attached by a single ester link to hydroxyl groups on the preformed polyester, and the two remaining carboxyls are available for salt formation and subsequent water solubilization. This is termed the 'ring-opening' technique from the single stage reaction of the anhydride group of the TMA.

Careful formulation of water soluble or emulsified resins is necessary to avoid both water sensitivity in the final film and poor storage stability of the composition. DMPA is an alternative ingredient available to formulate high acid value alkyds, though it is less used; it has been claimed that derived resins are more stable against hydrolysis. Drying oil fatty acids are generally included in this type of resin, the compositions finding application in air drying paints or with water soluble melamine/formaldehyde resins in stoving compositions.

Since alkyds are soluble in either aromatic or aliphatic solvent dependent on their oil length, they are tolerant of additional stronger alcohol, ester, or ketone solvents for special applications. Alkyds may also be formulated specially for use in industrial dip tanks, thinned in trichlorethylene.

2.5.2 Alkyd reactions and structure
The major polymer-forming reaction in alkyd preparation is the esterification reaction between acid and alcohol, and hence the mechanism is through a type of stepwise polymerization (Fig. 2.6). In alkyds this reaction is generally not catalysed. Side reactions which can occur include etherification of polyol especially in pro-longed monoglyceride stages.

Pentaerithritol is particularly prone to etherification, and grades should be chosen with low dipentaerithritol content.

A reason for lower molecular weight with orthophthalic alkyds compared with those prepared with iso- or terephthalic acids is the tendency to form intramolecular ring structures in *o*-phthalate esters. Because of this, in calculations the effective functionality of phthalic anhydride is always taken as less than two.

It has been claimed that alkyds owe some of their special properties, including application and dry, to the presence of 'microgel' in the alkyd [19]. While this seems to be proven, it would appear practically difficult to exploit in formulating for a microgel content unless a 'microgel' prepared separately were to be deliberately blended with the composition [20, 21]. It has been suggested that gelation of an alkyd occurs through separation of microgel particles, until particles become crowded enough to coalesce into a 'macrogel'.

2.5.3 Alkyd reaction control

Alkyd preparations must be controlled to a viscosity end point by sampling, with acid value also monitored. Viscosity measurement may be simply attained by using Gardner bubble tubes with thinned samples; however, improved speed of attainment of results and higher accuracy may be achieved by the use of a heated cone and plate viscometer [22]. The sampling interval will often need to be reduced as the end point is approached, owing to an increased rate of rise in viscosity, especially with short oil alkyds.

Where behaviour has been characterized from previous batches, a 'flare path' guide may be provided to follow the viscosity and acid value change; if reference to this shows deviation from normal progress, adjustment may be made by a small addition of acid or polyol, or where solvent process and oil bodying is occurring, by changing the reflux rate, providing this is carried out at an early enough stage in the process. It is also possible to make temperature changes during processing to modify the bodying rate.

On attainment of end point, cooling and thinning are necessary to halt the reaction; thinning is also essential since the alkyd resin, in the absence of thinning solvent, will either be extremely viscous or solid when cold.

For new formulations, process temperature will be established by consideration of the need for higher temperatures for oil bodying and faster reaction in long oil alkyds, or for lower temperature for easier control of end point in shorter oil alkyds. Acid value at gelation (AV Gel) prediction may be possible as the preparation proceeds (see 2.5.5). The result of experiment will allow modification in both composition and processing conditions.

2.5.4 Describing alkyds

In specifications for alkyds, oil type, oil length, and solvent will be stated, and sometimes the dibasic acid and its % in the composition.

Oil length refers to the % weight of glyceride oil calculated on the theoretical yield of solid alkyd; fatty acid content for non-glycerol-containing compositions may alternatively be quoted. In both cases this is the major indication of the flexibility, solubility, and — where containing drying oil — drying potential to be expected for the resin.

Solids, viscosity, and acid value (AV) will be determined as for other resins, and hydroxyl value (OHV) may also be measured. Both the latter are quoted in units of weight of potassium hydroxide in milligrams, equivalent to 1 gram of solid resin.

Excess hydroxyl content, in contrast to OHV, which can be found experimentally, is calculated from the molar formula and expressed as the excess hydroxyl % over that required to react completely with the polybasic acid moieties present. Excess hydroxyl % may be used as a formulating parameter to derive formulations with a required AV gel or average functionality.

Acid value at gelation (AV Gel) may be calculated theoretically, and may also be determined practically by extrapolation of a graph of acid value against 1/viscosity as an alkyd preparation proceeds to its end point.

2.5.5 Alkyd formulating

Practical alkyd formulating involves a process of calculation, application of previous practical experience, and trial preparation to arrive at the desired product. Principally oil length and oil/fatty acid type will be specified for the application, and polyol and dibasic acid dictated by a combination of other considerations including solubility, availability, and cost.

The average functionality concept is one method of formulating

$$F_{average} = 2 \times \frac{\text{Total acid equivalents}}{\text{Total mols}},$$

where for gelation to occur F should be two or more. When applying this equation to oil-containing formulations, the mols of oil should be split into constituent fatty acids and glycerol before applying the equation.

An alternative method useful in arriving at a trial formulation involves calculating the extent of reaction at gelation, most usefully expressed as AV Gel, which should be around five units below that desired in the final resin. It will be found necessary to vary the hydroxyl content, expressed as excess OH in order to alter the AV Gel. Various equations have been produced as gelation theory has been developed, by for example Carothers, Flory, and Stockmeyer, and formulating guides are available which review the whole field of alkyd calculation [23, 24]. Frequently, one of the forms of the Stockmeyer equation will be used, such as:

$$P^2 = \frac{(\Sigma \text{ g B})^2}{(\Sigma \text{ f}^2 \text{ A} - \Sigma \text{ f A}) (\Sigma \text{ g }^2\text{B} - \Sigma \text{ g B})}$$

where

      p = degree of reaction of acid groups at gelation
      f, g = functionality of carboxy and hydroxy moieties respectively
      A, B = mols of carboxy and hydroxy moieties present

The effectiveness of the various equations has been established experimentally [25], and it should be noted that it is necessary to consider phthalic anhydride as having a functionality of less than two because of its less than ideal behaviour. The equations offered by Weiderhorn [26] may be found a simpler and equally practical alternative, being less complex to calculate in the circumstances to which they apply; the reader will find worked examples of all the possible equations in the references cited.

In formulating short oil alkyds, monobasic acid may need to be included, dictated by solubility and hydroxyl content requirements. The use of bodied oil in long oil alkyds will be determined by the need to increase molecular weight and obtain higher viscosity, when this is not attainable by other means; the effect will be to raise the practical AV Gel above that calculated.

With the availability of microcomputers, calculations of greater complexity can be handled, limited only by one's ability to specify the equations used and to write programs in the language of the machine. Thus it is perfectly feasible for a microcomputer program to be written with a database containing the parameters of all the commonly used raw materials, in order to formulate alkyds on oil length and hydroxyl content, and to calculate AV Gel, average functionality, and a number of other items using the equations mentioned above. Of enormous value is the ability with a computer, especially if the equations cannot otherwise be defined, to work iteratively i.e. to re-cycle calculations quickly and to vary the factors and examine effects before commitment to practical experiment. Programs of high complexity and speed have been written [27].

2.5.6 Alkyd preparation equipment
Alkyds may be prepared on both laboratory and production scales by both the so-called 'fusion' and 'solvent' process techniques. In the former, ingredients are heated in vented stirred pots; problems encountered are mainly due to variable losses of reactant, particularly phthalic anhydride which is volatilized along with the water of reaction. In the latter process, more sophisticated equipment is needed, namely reactors fitted with condensers, water separation equipment, and solvent return systems. Organic solvent, typically xylene, is added to the resin in the bodying stage (2–5%) and re-cycled via the overheads, water being removed by azeotropic distillation. It can be of advantage to fit a solvent process plant with a vacuum facility which, in addition to providing assistance with loading, can be used, for example, to reduce the residual solvent content at the end of processing a batch.

Problems of phthalic anhydride sublimation in solvent process may be minimized by careful plant design or avoided by the use of bubble cap scrubbers or fractionating equipment [28]. Inert gas is normally supplied in all alkyd processing to improve colour and increase safety. Process temperatures may be between 180°C and 260°C. Where consistent practice such as heating rate, gas flow, and ventilation can be achieved, it must be stressed that the fusion process can be a satisfactory process for almost all classes of alkyd, provided that environmental problems caused by fume can be controlled.

2.5.7 Modified alkyds
Urethane alkyds may be prepared by the replacement of part of the dibasic acid in the alkyd formulation with a diisocyanate such as toluene diisocyanate (TDI); the formulating calculations applicable are those used for alkyds. Long oil urethane alkyds are used in decorative paint formulations to impart greater toughness and quicker drying characteristics. While total replacement of dibasic acid is possible (thus producing a urethane oil), it is more normal to formulate an alkyd with a reduced dibasic acid content, processing this to low acid value, but a high residual

hydroxyl content and then to treat the thinned resin with diisocyanate. The isocyanate groups react with hydroxyl groups to form urethane links, but, unlike acids in ester formation, do so exothermically at lower temperatures and without by-product evolution. It is essential to complete the reaction of all isocyanate groups present because of the high toxicity associated with free isocyanates. In-process testing may be by titration with amine, with final verification being by infrared spectroscopy. Small remaining residues of isocyanate may be removed by deliberate addition of low molecular weight alcohol. For the preparation of urethane alkyds, a monoglyceride catalyst should be chosen for the alkyd stage which does not promote unwanted isocyanate reactions, e.g. allophanate formation, and such catalysts as calcium oxide or a soluble calcium soap should be used.

Another modification of long oil alkyd is that with polyamide resin, used to impart thixotropy in the final resin solution. Polyamide resins used are typically those derived from dimer fatty acid and a diamine. The physical properties of these alkyds are very delicately balanced, and such characteristics as gel strength, gel recovery rate, and non-drip properties of derived paints are intimately connected with both the composition and the processing of the alkyd, along with the proportion and type of polyamide and degree of reaction [29]. If this type of alkyd is being prepared, or when formulating paint compositions containing them, care should be taken to avoid the use of, or contamination with, polar, e.g. alcoholic diluents; these destroy the rheological structure to which hydrogen bonding almost certainly provides a considerable contribution. Typical levels of modification are around 5%, and care in processing is necessary to stop the reaction in its final stage at such a point as to achieve clarity and the optimum and desired gel properties.

Silicone modification of long oil alkyds has been claimed to impart outstanding gloss retention for their use in maintenance paints. The same hydroxy or methoxy functional silicone resins are applicable as used for modification of polyesters. In general, the degree of enhancement in properties is in proportion to the level of modification with these expensive materials, and formulators need to weigh carefully the balance of cost versus durability enhancement achieved.

Phenol formaldehyde resins of both reactive and non-reactive oil soluble types can be used as alkyd modifiers. They increase film hardness and improve resistance to water and chemicals, but cause yellowing. Such modified alkyds are useful in primer systems as cheaper but less chemically resistant alternatives to epoxy resins.

Vinyl modification of alkyds is used to impart faster dry, increase hardness, and give better colour, water, and alkali resistance. However, decreased gloss retention and poorer resistance to solvents result, the latter making formulation for early recoatability difficult to achieve. Vinylated alkyds are typically used for metal finishes. Styrene and vinyl toluene are the usual modifiers, methyl methacrylate being less common. The modification stage involves free radical chain addition polymerization of the monomer in the presence of the alkyd under conditions where grafting to the unsaturated oil component of the alkyd is encouraged. This is typically by the use of at least a portion of dehydrated castor oil in the alkyd composition since the conjugated fatty acid moieties then present encourage the necessary co-reaction; the typical use of peroxide catalysts such as di-tert butyl peroxide also encourage grafting by hydrogen abstraction.

A number of techniques for preparing vinylated alkyds are possible; most common is post-vinylation of the thinned alkyd by feeding monomer and initiator into the resin held at 150–160°C. Less common alternatives are to vinylate the oil prior to alkyd manufacture or to vinylate the monoglyceride prior to esterification with dibasic acid. These latter techniques can allow final dilution in low boiling solvent or chlorinated solvents in which the vinylation reaction would have been impossible. Where difficulties in achieving adequate grafting are experienced, resulting in cloudy resin preparations, possibilities exist of including maleic anhydride or methacrylic acid in the alkyd polyester structure to provide additional grafting sites.

To the list of modifications must be included the possibility of precondensation with melamine/formaldehyde (MF) resins. Normally for a coating composition which is multicomponent, viscosity will be minimized by keeping the components unreacted until the curing stages; however, for some systems, advantages of increased stability, compatibility, or improvement in cure properties are seen by partial pre-reaction, even as in this case where the two polymers are compatible. For low bake curing alkyd/MF systems, precondensation of alkyd and MF can enhance final cure properties, while reducing the tendency of the MF component to self-condense when highly acidic catalyst systems are used.

2.6 POLYESTER RESINS

In the chemical sense, the term polyester embraces saturated polyesters, unsaturated polyesters, and alkyds. However, the term alkyd is normally reserved for the oil-modified alkyd discussed above. Similarly, unsaturated polyesters containing vinyl unsaturation, which most typically are maleic-containing resins partially or totally thinned with vinyl or acrylic monomer, are referred to as such; hence the term polyester is generally reserved for oil-free, acid, or hydroxy functional polyester resins. Polyester resins are typically composed mainly of co-reacted di- or polyhydric alcohols and di- or tri-basic acid or anhydride, and will be thinned with normal solvents. Much of the following applies nonetheless to both saturated and unsaturated polymers, the special features of unsaturated polyesters being referred to at the end of this section. The use of monobasic acids is not excluded from our definition of polyester. For the purpose of this section it is convenient to call a polyester containing lauric or stearic acid an alkyd, but to consider as polyesters those containing only shorter chain aliphatic acids, including the lower molecular weight branched synthetic fatty acids, and monobasic aromatic acids, such as benzoic acid.

Polyesters can be formulated both at low molecular weight for use in high solids compositions, and at higher molecular weight, and can be both hydroxy and acid functional. Since they exclude the cheaper oil or fatty acid components of the alkyd and are hence normally intrinsically more expensive, much effort has gone into understanding their formulation, in order to exploit the higher performance of which they are capable; also into the development of new raw materials. The latter are in the main new polyols to suit high-durability applications such as automotive topcoats and the painting of steel coil strip. The concept of the 'anchimeric' effect has been exposed [30], which requires that for hydrolysis resistance and stability of the ester, there shall be no hydrogen atoms on the carbon atom in the position beta to the hydroxy group and subsequent ester group. This requirement is met by such polyols as cyclohexane dimethanol (CHDM) and 2,2,4 trimethyl 1,3 pentane diol (TMPD) where the beta hydrogen content is reduced or hindered, and in materials such as trimethylol propane (TMP), neopentyl glycol (NPG), and Ester Diol 204 (Union Carbide) (Fig. 2.7).

$$HO\ CH_2\langle\bigcirc\rangle CH_2\ OH$$

CHDM

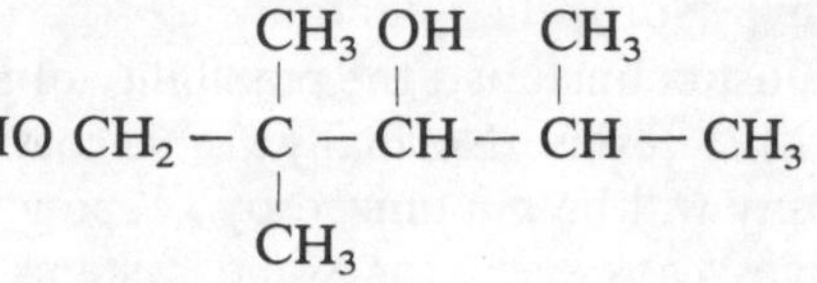

$$HO\ CH_2 - \overset{\overset{\displaystyle CH_3}{|}}{\underset{\underset{\displaystyle CH_3}{|}}{C}} - \overset{\overset{\displaystyle OH}{|}}{CH} - \overset{\overset{\displaystyle CH_3}{|}}{CH} - CH_3$$

TMPD

$$HO\ CH_2 - \overset{\overset{\displaystyle CH_3}{|}}{\underset{\underset{\displaystyle CH_3}{|}}{C}} - CH_2\ O\ CO - \overset{\overset{\displaystyle CH_3}{|}}{\underset{\underset{\displaystyle CH_3}{|}}{C}} - CH_2\ OH$$

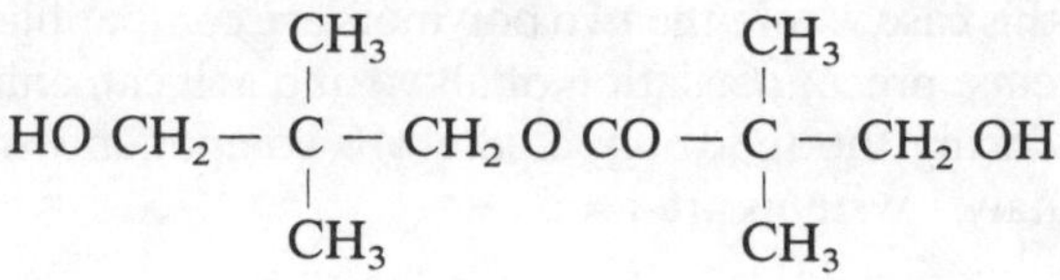

Ester Diol 204 (Eastman Chemicals)

Fig. 2.7 — Typical polyol structures with hindered β positions relative to OH group.

2.6.1 Formulation

Polyesters are formulated in similar fashion to alkyds by making calculations of average functionality, supplemented by calculation of acid value of gelation by the Stockmeyer method [23]. In addition, for a theoretical hydroxy functional polyester, molecular weight can be calculated from the formula

$$MW = \frac{\text{Formula weight}}{\Sigma M - \Sigma E_A + \left(\dfrac{\text{Formula weight} \times AV}{56100}\right)}$$

where ΣM is the total mols. present and ΣE_A the total initial number of equivalents of acid present.

The hardness/flexibility of polyesters may be adjusted by either blending 'softer' aliphatic dibasic acids with the 'harder' aromatic acids or by the inclusion in the polyol blend of the more rigid CHDM in place of 'softer' aliphatic polyols. With the general use of isophthalic acid (IPA) rather than orthophthalic anhydride, crystallization problems can occur with the final polyester solutions, and one technique of controlling this is to include a small proportion of terephthalic acid (TPA) to disrupt chain symmetry.

Solvent-borne polyesters for such applications as coil coating, automotive basecoat or clearcoat, and wood finish, may be formulated from a blend of the 'hard' and 'soft' acids, IPA and adipic acid, with diols and triols such as NPG and TMP. Formulation variants can include phthalic anhydride for cheapness, and longer chain glycols added for extra flexibility. Monobasic acids such as pelargonic or benzoic acid may be included. These polyesters will generally be crosslinked by MF resin in stoving applications or in non-bake applications as a 2-pack formulation with polyfunctional isocyanate adduct. The triol included can contribute to branching and to in-chain hydroxy/functionality depending on the formulation; for some applications such as coil coating where high molecular weight may be an advantage, triol content will be minimized.

With polyesters for high solids coatings, the total polyol in the composition is increased, though triol concentration may be reduced, and the resin processed to lower molecular weight and higher hydroxyl content. Water-soluble polyesters are formulated to a high acid value, often by a 2-stage technique, as described for the preparation of water-borne alkyds, where ring opening with trimellitic anhydride is carried out as a stage subsequent to polymerization. Dimethylol propionic acid has also been used to prepare high acid value water soluble polyesters.

Polyesters for powder coatings, since they must have a softening point typically >40°C, will generally not contain long chain plasticizing dibasic acids, but one or more aromatic acids with possibly a simple diol. They may be of high acid value for epoxy resin cure or low acid value and high hydroxyl content for melamine/formaldehyde resin cure.

2.6.2 Polyester preparation

For polyesters the general techniques used in alkyd manufacture are used, but with refinements appropriate to the materials used. All-solid initial charges will arise with many formulations, requiring careful initial melting. Difficulties in achieving clarity, even with complete reaction, can be encountered when isophthalic acid is present, and this may be exacerbated when the even less soluble and less reactive terephthalic acid is present. Since a number of the glycols used in polyester manufacture are volatile (e.g. ethylene glycol, neopentyl glycol) and hence are easily lost with the water of reaction, any serious attempts to prepare polyesters containing these components reproducibly demand a reactor fitted with a fractionating column [31, 32]. Pressure processing has been advocated as a means of raising reaction temperatures and hence reaction rates, without serious loss of glycol, particularly where ethylene and propylene glycols are included in the formulations. Where high molecular weight polyesters are desired it can also be useful to have vacuum available for use later in the reaction to strip final residual water of reaction and achieve increased molecular weight (the technique is that used in the manufacture of polyesters of fibre forming molecular weight [33]). For high molecular weight polyesters, particular care needs to be taken with regard to glycol loss, since small losses can cause sufficient imbalance in the formulation to restrict the molecular weight.

Catalysts are frequently used to increase reaction rate in polyester preparation, choice of catalyst needing care since colour may be adversely affected. Tin catalysts are found particularly useful [34].

2.6.3 Modification

Silicone modification is used to enhance the properties of polyesters, particularly with respect to durability. Modification of both solvent-borne [35] and water-borne [36] polyesters are possible, and with the former has become firmly established practice for coil application. In solvent-borne compositions, siliconization is carried out as a second stage following polyester preparation, using a silicone level of typically 20% to 50%; in the case of high acid value water-borne resins, the siliconization stage may be carried out after esterification and before the TMA addition stage. Silicone resins in common use are methoxy functional, and methanol evolution occurs in reaction, which under some conditions may be used to monitor the reaction. The degree of reaction in terms of methoxy functionality may be of the order of 70% to 80%.

2.6.4 Unsaturated polyesters

The established unsaturated polyester resins are polyesters derived from polyols and dibasic acids, which also include some constituent containing a double bond, and thinned in polymerizable monomer. By far the majority of these polyesters are linear and contain co-condensed maleic anhydride as the source of unsaturation. In surface coatings these resins have been used mainly in wood finish and refinish 2-pack finishes, including a significant use in refinish putties. In thin-film coatings, difficulties were found in controlling air inhibition of the curing reaction at the surface of the film, which limited their use. Recent additions to the monomer range available, and increased use of radiation curing, have enabled this problem to be overcome to a large extent, and their use has risen. For radiation cure, unsaturated polyesters may be of the post-acrylated type acrylated type rather than the maleic type described below. For this application they are, however, now in competition with acrylic and urethane oligomers (see 2.19).

Generally, unsaturated polyesters are formulated with ortho and isophthalic acids along with maleic anhydride, with appropriate aliphatic glycols, such as diethylene glycol, chosen as co-reactants to control flexibility. The actual amount of maleic in proportion to other acids present may be from 25% to 75% molar. During preparation, a degree of isomerization of cis maleate residues to trans fumarate occurs (Fig. 2.8), this being important to final cure as copolymerization with styrene

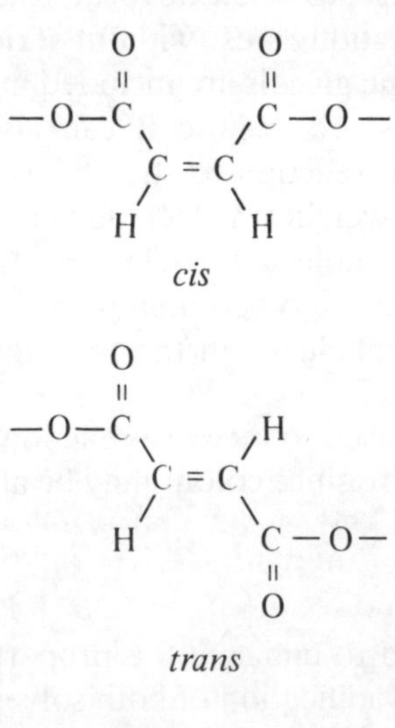

Fig. 2.8 — Cis and trans configurations of maleate and fumarate.

in particular is favoured with the trans configured fumarate. This isomerization can be virtually 100%, but varies with reaction mixture composition and reaction conditions.

The post-acrylation method referred to above involves the preparation of a saturated polyester with excess hydroxyl groups which are esterified with acrylic acid as a final stage in the preparative process.

Preparation of unsaturated polyesters is straightforward and is carried out in like manner to saturated polyesters [29]. However, where the slower reacting iso and terephthalic acids are used rather then o-phthalic anhydride it is claimed that multi-stage processing, reacting the aromatic acid in a first stage with subsequent reaction of maleic anhydride, gives better performing products [37, 38]. Since unsaturated polyesters are thinned in monomers capable of polymerizing thermally, care needs to be taken in this stage and as low a thinning temperature used as possible. For additional safety in thinning, and stability on storage, it is necessary to add an inhibitor such as para tert butyl catechol to the monomers in the thinning tank prior to resin dissolution. Alternatively, the resin may be allowed to solidify and the broken up solid subsequently dissolved in monomer.

The monomer traditionally used is styrene, but vinyl toluene, methyl methacrylate, and some allyl ethers and esters such as diallyl phthalate are also used. The ultimate polymerization mechanism is most commonly a redox initiated chain addition polymerization operating at ambient temperature, though thermal initiation is possible. The redox system is 2-pack, where an organic peroxide or hydroperoxide is used as the second component, with the reducing component (amine or metal soap) mixed with the resin. (See next section).

Oxygen is an inhibitor particularly for styrene polymerizations, and exposed styrene thinned unsaturated polyester films suffer air inhibition which shows up as surface tack. This can be minimized by either the mechanical exclusion of oxygen by wax incorporation, or by the introduction into the composition of chemical groups which react with oxygen. Allyl ether groups have been used in such a role either through the addition of allylated monomers [39] or by reacting allylated materials such as allyl glycidyl ether or trimethyl propane diallyl ether into the backbone of the polyester [40]. (The allyl oxy or allyl ether grouping is the only synthetic oxidatively crosslinking group to approach the reactivity of the natural drying oils.) Radiation-curing compositions probably suffer less from air inhibition because the higher radical flux allows the reaction to proceed faster than the rate at which oxygen dissolves in the film, and because with higher functionality in both monomers and oligomers, normal with this type of composition, the effect is less detrimental.

2.7 ACRYLIC POLYMERS

Acrylic polymers are widely used for their excellent properties of clarity, strength, and chemical and weather resistance. The term acrylic has come to represent those polymers containing acrylate and methacrylate esters in their structure along with certain other vinyl unsaturated compounds. Both thermoplastic and thermosetting systems are possible, the latter formulated to include monomers possessing additional functional groups that can further react to give crosslinks following the formation of the initial polymer structure. Vinyl/acrylic polymerization is particularly versatile, in that possibilities are far wider than in condensation polymerization of controlling polymer architecture and in introducing special features; for example, by modification stages following the initial formation. Various papers fully discuss the formulation and use of acrylic polymers in coatings [41, 42].

2.7.1 Radical polymerization

The kinetics and mechanism of radical-initiated chain addition polymerization [43, 44] are important in all those processes where unsaturated vinyl monomers are used, either in the resin preparation or in the final cure mechanism. In resin preparation, processes discussed in this chapter include preparation of solution acrylic thermoplastic and thermosetting resins, emulsion and dispersion polymers, and the vinylation of alkyds. These reactions are also involved in the cure of unsaturated polyesters and of radiation-curable polymers. Autoxidative cure of oils and alkyds also occurs by a free radical mechanisn, but not in the simple manner illustrated below. Radical polymerization is the most widespread form of chain addition polymerization.

The main polymer-forming reaction is a chain propagation step which follows an initial initiation step (Fig. 2.9). A variety of chain transfer reactions are possible

Initiator breakdown $\qquad$ $I{:}I \rightarrow I{\cdot} + I{\cdot}$

Initiation and propagation $\quad I{\cdot} + M_n \rightarrow I(M)_n{\cdot}$

Termination $\qquad$ $I(M)_n{\cdot} + {\cdot}(M)_m I \rightarrow I(M)_{m+n}I \qquad$ by combination

$\qquad\qquad\qquad I(M)_n{\cdot} + {\cdot}(M)_n I \rightarrow I(M)_{n-1}(M{-}H) + I(M)_{m-1}(M{+}H)$

$\qquad\qquad\qquad\qquad\qquad\qquad\qquad\qquad\qquad$ by disproportionation

Transfer $\qquad\qquad I(M)_n{\cdot} + \text{Polymer} \rightarrow I(M)_n H + \text{Polymer} {\cdot} \quad$ to Polymer

$\qquad\qquad\qquad I(M)_n{\cdot} + \quad\text{RSH} \quad \rightarrow I(M)_n H + \text{RS} {\cdot} \qquad$ to transfer agent

$\qquad\qquad\qquad I(M)_n{\cdot} + \quad\text{Solvent} \rightarrow I(M)_n H + \text{Solvent} {\cdot} \quad$ to solvent where

$\qquad\qquad\qquad\qquad\qquad\qquad\qquad\qquad\qquad\qquad\qquad\qquad\qquad\quad$ present

Fig. 2.9 — Representation of main reactions occuring in free radical chain addition polymerization.

before chain growth ceases by a termination step.

Radicals produced by transfer, if sufficiently active, can initiate new polymer chains where a monomer is present which is readily polymerized. Radicals produced by so-called chain transfer agents are designed to initiate new polymer chains; these agents are introduced to control molecular weight and are usually low molecular weight mercaptans e.g. primary octyl mercaptan.

Transfer to polymer is not normally a useful feature. At extremes such as with vinyl acetate polymerization, transfer at methyl groups on the acetate leads to grafted polymer side chains being present which are relatively easily hydrolyzed, leading to undesirable structural breakdown of the polymer in use.

Transfer to solvent generally does not occur to a very significant extent and does not lead to radicals initiating new polymer chains. Occasionally polymerization systems may be designed with, for example, alcohol solvent present where its chain-terminating effect can assist with molecular weight control. The nature of the solvents present can significantly affect the decompositon rates of most peroxy initiators.

Initiators (Table 2.6) used to give free radicals, are compounds typically breaking down under the influence of heat; two types predominate, those with an azo link (—N=N—) and those with a peroxy link (—O—O—).Initiator breakdown is characterized by 'half life' which varies with temperature.

Table 2.6 — Typical initiators used in polymer preparation

Name	Formula	$\frac{1}{2}$ life 30 mins
Azo di isobutyronitrile (AZDN)	$\begin{array}{ccc} CH_3 & & CH_3 \\ | & & | \\ CH_3{-}C{-}N{=}N{-}C{-}CH_3 \\ | & & | \\ CN & & CN \end{array}$	89°C
Di benzoyl peroxide	⟨◯⟩ CO OO OC ⟨◯⟩	98°C
T-butyl perbenzoate	*t* Bu OO CO ⟨◯⟩	130°C
Di *t*-butyl peroxide	*t* Bu OO *t* Bu	156°C

The 'half life' of an initiator is the time within which 50% of the material has decomposed at a specified temperature. For many processes an initiator/temperature combination will be chosen in order that the half life under reaction conditions is in the region of 15–30 minutes. This ensures the steady generation of radicals at such a rate that the heat of reaction can be safely contained and high conversion of monomer to polymer ensured.

Despite the handling problems caused by its solid nature and poor solubility, AZDN is often used for its non-grafting characteristics; di benzoyl peroxide, though of only slightly higher half life temperature conversely has a strong grafting tendency, a fact that may be deliberately exploited, for example to produce acrylic grafts onto other polymers such as epoxides. (Grafting is where the radical extracts a hydrogen atom from a site on the polymer chain, leaving a radical site onto which further polymer growth occurs.) This latter initiator is also a solid, normally supplied as a paste in plasticizer; for many uses it must first be dissolved and the solution assayed. It is therefore another less than ideal initiator. Both t-butyl perbenzoate and di t-butyl peroxide are, however, liquid initiators and are thus more readily handled. Many peroxide initiators produce acidic breakdown products which may have adverse affects in formulations.

It has been stated above that initiator decomposition may be affected by (solvent) environment, and reducing agents may in fact be deliberately added along with the initiator to produce Redox (reduction–oxidation) systems where initiator decomposition is induced at far lower temperatures than would otherwise be the case. These systems have a number of specialist applications in both polymerization and curing reactions. Fig. 2.10 shows the manner of decomposition of benzoyl peroxide by both

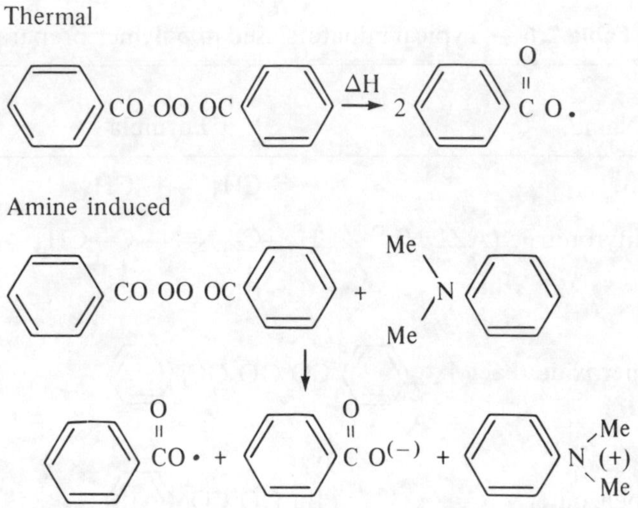

Thermal

Amine induced

Fig. 2.10 — Thermal and amine induces decomposition of benzoyl peroxide.

thermal and amine induced redox initiation routes to produce initiating benzoyl radicals.

2.7.2 Monomers and copolymerization

Monomers may be classified as 'hard', 'soft', or 'reactive', based on the properties they confer on the final polymer or copolymer. Hard monomers are, for example, methyl methacrylate, styrene, and vinyl acetate. The acrylates are 'softer' than methacrylates, and useful 'soft' monomers include ethyl acrylate and 2-ethyl hexyl acrylate, as well as the long chain methacrylates. Reactive monomers may have hydroxy groups, for example hydroxy ethyl acrylate. Acrylamide and glycidyl methacrylate possess exploitable reactivity, the latter being particularly versatile. Acidic monomers such as methacrylic acid are also reactive and are often included in small amounts in order that the acid groups may enhance pigment dispersion and provide catalysis for cure of the derived polymer.

Methyl methacrylate as a hard monomer imparts resistance to petrol, UV resistance, and gloss retention. It is therefore used in copolymers for topcoats, particularly for automotive application. Butyl methacrylate is a softer monomer imparting excellent low-bake humidity resistance, but its plasticizing effect is limited. It gives good intercoat adhesion and solvent resistance and excellent UV resistance and gloss retention. Ethyl acrylate has good plasticizing properties but as a monomer has a highly unpleasant and toxic vapour. Its copolymers are fairly resistant to UV and give good gloss retention.

Table 2.7 — Typical unsaturated monomers

Monomer	Structure	Polymer Tg °C
Butyl methacrylate	n Bu—O—C—C=CH$_2$ (with O double bond and CH$_3$)	22
Ethyl acrylate	CH$_3$—CH$_2$ OOC—CH=CH$_2$	−22
2-ethyl hexyl acrylate	C$_4$ H$_9$—CH—CH$_2$—OOC—CH=CH$_2$ (with C$_2$H$_5$)	−70
2-hydroxy propyl methacrylate	CH$_3$—CH—CH$_2$ OOC—C=CH$_2$ (with OH and CH$_3$)	76
Methacrylic acid	HO C C=CH$_2$ (with O double bond and CH$_3$)	210
Methyl methacrylate	CH$_3$—O—C—C=CH$_2$ (with O double bond and CH$_3$)	105
Styrene	CH$_2$=CH (phenyl)	100
Vinyl acetate	CH$_2$=CH O C CH$_3$ (with O double bond)	30

Practical coating polymer systems are rarely homopolymer but are copolymers of hard and soft monomers. Polymer hardness is characterized by glass transition temperature (Tg) and for any given copolymer the Tg of the copolymer may be estimated by the equation $1/TG = W_1/TG_1 + W_2/TG_2$ etc. where TG_1, TG_2 are the Tg's of homopolymers of the component monomers in °K, and W_1, W_2 their weight fraction present. (The calculated Tg will not be the final film Tg if the polymer is thermosetting since crosslinking will further raise the Tg, and this should be taken into account where appropriate.)

While monomers can combine during copolymerization in a variety of configu-rations (random, alternating, block, or graft), the vast majority of acrylic polymers actually used in coatings are random. This randomness also has the effect that the tacticity and crystallization phenomena so important in bulk polymer properties are generally not apparent or important in coating polymers, the most common structural effects experienced being phase separation and domain effects occuring either by accident or design.

The way different monomers react in copolymerization with other monomers depends on both their own structure and the nature of other monomers present [45]. Monomers of similar structure generally copolymerize readily and randomly. Thus all acrylates and methacrylates, including the parent acids, can be copolymerized together satisfactorily in almost any composition, though the longer chain monomers may be slower to polymerize and hinder attainment of complete conversion. Styrene can be included up to certain levels in acrylic polymers to reduce cost, and raise hardness and refractive index, but can be difficult to convert in higher concentrations. Maleic anhydride and maleate esters, though apparently able to copolymerize satisfactorily with other monomers when included in small amounts, do not homopolymerize but can form alternating structures with certain other monomers, notably styrene; they should be used with circumspection unless there is good understanding of this behaviour. Vinyl acetate and other vinyl esters are difficult to polymerize except in disperse systems.

When two monomers copolymerize, the rate may generally be slower than that of either polymerizing alone. The polymer radicals may react more readily with one monomer than another, and in fact four growth reactions can be envisaged, proceeding at different rates as shown by the following scheme:

$$R-M_1 . + M_1 \xrightarrow{k_{11}} R-M_1 .$$

$$R-M_1 . + M_2 \xrightarrow{k_{12}} R-M_2 .$$

$$R-M_2 . + M_1 \xrightarrow{k_{21}} R-M_1 .$$

$$R-M_2 . + M_2 \xrightarrow{k_{22}} R-M_2 .$$

where k_{11} and k_{22} are the reaction rate constants for chain ends with one monomer radical adding the same monomer, while k_{12} and k_{21} are for chain ends with one monomer radical adding the other monomer. The ratios k_{11}/k_{12} and k_{22}/k_{21} are referred to as the reactivity ratios, r_1 and r_2, and characterize the relative rates of reaction of the monomers with each other. Thus $r>1$ indicates a polymer radical reacting more rapidly with its own monomer, and $r<1$ indicates a polymer radical reacting selectively with the comonomer.

The product $r_1 r_2$ has interest in that if, for example, $r_1 r_2 = 1$ the monomers copolymerize truly randomly, while if $r_1 r_2$ approaches zero the monomers trend strongly towards alternating in an equimolar manner.

The kinetics of copolymerization have had considerable attention [46], and comprehensive tables of r_1 and r_2 values have been published [47].

A more general scheme, which can be used to predict the behaviour of all monomers for which values are available, and can be used for terpolymers and higher, is the Q–e scheme [48]. The Q parameter refers to the degree of resonance stabilization of the growing polymer chain, whereas the e-value represents its polarity. Similarity in values between comonomers predicts their copolymerization randomly; charge complex formation between monomers of opposite polarity (e) leads to alternating copolymerization. Monomers with large differences in Q-value and similar e-values are the most difficult to copolymerize.

Since monomers can polymerize thermally and by exposure to light, it is necessary to store monomers in opaque containers in cool conditions and also to ensure that an adequate inhibitor level is maintained to prevent adventitious polymerization. Inhibitors are typically phenolic, and for some monomers checks on levels may be necessary on prolongued storage. Monomers also have toxicity ratings ranging from the mild to very severe, and many are highly flammable.

2.7.3 Formulation and preparation

When formulating acrylic resins, monomer composition may be determined by durability and cost requirements, and by functionality [41, 42]. Hard and soft monomer ratios will be adjusted for Tg, which along with MW, may have fairly rigid definition for the intended application of the coating. The major task of the formulator may be to achieve these and to obtain conversion. The major factors affecting molecular weight in preparation are the choice of initiator and its concentration, temperature, and monomer concentration, and the concentration present of chain transfer species, if any. Normally polymerization will be carried out with up to 50% solvent present and at reflux, the temperature thus defined by the solvents used. Initiator contents vary between 0.1 and 4% and polymerization temperatures for thermal initiator from 90 to 150°C. Molecular weight (number average) is often found to vary inversely with the square root of initiator concentration at a given temperature.

Polymerization of acrylic monomers may be by bulk, suspension, emulsion, dispersion, or solution polymerization techniques; but for surface coating applications solution, emulsion, and dispersion techniques are most common. The first of these is discussed below, and the others elsewhere in this chapter.

Solution polymerization has a number of possible modes, the simplest being the one-shot technique where solvent, monomer, and catalyst are heated together until conversion is achieved. However, vinyl polymerization is highly exothermic (50–70 kJ/mol) and this heat is generally removed in a refluxing system by the condenser. The one-shot process may give rise to dangerously high exotherm peaks and thus processes where monomer and initiator, or initiator alone, are fed to the other refluxing reactants over 1–3 hours, moderate the reaction and the heat evolution rate, enabling the heat to be removed in a more controlled fashion. The process, where monomer and initiator are fed into refluxing solvent, also serves to give better control of molecular weight and molecular weight distribution, and is most favoured on technical as well as safety grounds. Mixed monomer and initiator can be fed together, or for ultimate safety, from separate vessels, provided good control of the two concurrent feeds can be achieved. While post-reaction thinning can be carried out, polymerization is always in the presence of at least 30–40% solvent in order that the viscosity of the polymerizing mixture can be low enough to allow good mixing and heat transfer. It is not essential to have a separator after the condenser in most processes, though useful to remove adventitious water. However, it is essential for resins where a condensation reaction proceeds concurrently with the addition polymerization, e.g. with some acrylamide-containing polymers.

The solvent present will depend on the exact nature of the acrylic polymer. Acrylic polymer solubility is affected by the nature of the side group, and shorter side chain polymers are relatively polar and require ketone, ester, or ether–alcohol solvents. As the side chain length increases aromatic solvent can be introduced. In the choice of solvent composition it may also be necessary to consider the crosslinking method and additional crosslinking agent which may be introduced, and any requirements dictated by the method of application.

Acrylic processes are usually followed by measuring the viscosity and solids of the reactor contents, principally to monitor conversion. In these processes no correction is possible once reaction has commenced, if molecular weight is found to be incorrect at any stage. For many formulations conversion does not continue unaided to completion after the end of the feed period, and one or more 'spikes' of initiator will be necessary to complete the process. Accurate measurement of molecular weight may be by Gel Permeation Chromatography (GPC) or may be deduced from Reduced Viscosity measurement in a U-tube viscometer using a dilute solution of the polymer in dichloroethane.

2.7.4 Thermoplastic acrylic resins

Thermoplastic acrylic resins find application particularly for automotive topcoats both for factory application, and for refinishing following accident damage repair. Molecular weight and molecular weight distribution need careful specification and control for lacquers, particularly with metallic formulations where rheological behaviour affects the orientation of the aluminium flakes in the paint film, so necessary for the flip tone effects required. Too high a molecular weight gives low spraying solids, and very high molecular weight 'tails' can result in cobwebbing on spraying; low molecular weight degrades film strength, mechanical properties, and durability. It has been found that the optimum molecular weight is in the region of 80 000 for polymethyl methacrylate lacquers [49].

Polymethyl methacrylate homo- or copolymers for these applications can be prepared using peroxide initiator in hydrocarbon/ketone blends, the reaction optionally being carried out at above atmospheric pressure by a one-shot process [50]. Polymethyl methacrylate films require plasticization to improve cold crack resistance, adhesion, and flexibility, and early compositions used such external plasticizers as butyl benzyl phthalate or a low molecular weight polyester. Phthalate plasticizers can be volatile, causing film embrittlement; and since problems of repair crazing and cracking are also found in externally plasticized films and post-polishing is necessary, further development became necessary. The use of cellulose acetate butyrate allowed the use of the bake-sand-bake process, but the most dramatic improvement was found when 2-phase polymer films were designed [51]. These can be achieved, for example, by the blending of a methylmethacrylate/butylacrylate (MMA/BA) copolymer with homopolymer polymethylmethacrylate solution when apparently clear but 2-phase films with enhanced resistance to crazing result.

In lacquer formulations containing MMA, plasticizing monomers may be introduced in order to tailor flexibility and hardness to application requirements. Also small amounts of acid monomer such as methacrylic acid may be included to introduce a degree of polarity to help pigment wetting and adhesion; for adhesion improvement amino monomers such as dimethyl amino methacrylate may alternatively be included, though these may introduce slight yellow coloration to otherwise colourless formulations.

Thermoplastic resins may be blended with nitrocellulose and plasticizing alkyd for use in low-bake automotive finishes, both for use by low volume motor manufacturers and for repair.

2.7.5 Thermosetting acrylics

Thermosetting acrylic resins were formulated to overcome the defects of thermoplastic compositions, particularly apparent in industrial applications. Advantages from thermosetting resins accrue in improved chemical and alkali resistance, higher application solids in cheaper solvents, and less softening at higher service temperatures. Thermosetting resins can be self-crosslinking or may require a co-reacting polymer or hardener to be blended with them; most fall into the latter category.

Thermosetting acrylics may have molecular weights of 20–30 000, and the hardeners and resins themselves will typically be polyfunctional (>2). The concentration of functional monomer used in the backbone will be in the range 5–25% by weight. The overriding principle in design of the complete curing system is that on crosslinking, an infinite network of crosslinks between polymer chains will be created. Thus polymer of low molecular weight and low viscosity at the time of application is converted to infinite molecular weight and total insolubility after curing.

Hydroxy functional thermosetting acrylic resins are not self-reactive but must be blended with, for example, nitrogen resins for stoving applications or with isocyanate adducts for use as 2-pack finishes for room temperature or low-bake refinish paint application.

Hydroxy functionality is introduced by the incorporation of hydroxy alkyl (meth)acrylates such as hydroxy ethyl acrylate at up to 25% weight concentration in the total monomer blend [52]. A small percentage of acidic monomer is usually introduced to give a final resin acid value of typically 5–10 in order to catalyse curing reactions with nitrogen resins. The constituents must be balanced to obtain the desired hardness, flexibility, and durability, and care must be taken to achieve the necessary compatibility and storage stability with the chosen amino resin. This type of resin with melamine formaldehyde resin (MF) is used for example in automotive topcoats, the ratio of acrylic to melamine/formaldehyde being typically between 80–20 and 70–30. The crosslinking reactions observed are those typical for amino resin with hydroxy-containing materials (see 2.10). Acrylic/MF compositions are superior to alkyd/MF for colour and exterior durability, but not necessarily for gloss. For repair of automotive topcoats in the factory, acid catalysts such as the half esters of dicarboxylic acid anhydrides or acid phosphates, are added to reduce the stoving temperature necessary, and acrylic/MF's are superior to alkyd/MF's under these conditions.

Hydroxy acrylic resins are generally prepared by the techniques previously described, typically in aromatic or ester solvent, feeding the monomer blend along with a peroxy initiator such as t-butyl perbenzoate into the refluxing solvent. Chain transfer agents may be used to achieve better MW control.

Hydroxy acrylics can also be prepared by *in situ* hydroxylation of acid-containing polymer with ethylene or propylene oxides. Added flexibility can be given to the crosslinks by extending the hydroxyl group away from the acrylic backbone by reaction of existing hydroxy residues from hydroxy monomer with caprolactone.

The most distinctive class of self-crosslinking thermosetting compositions are those from acrylamide interpolymers reacted with formaldehyde and then alkoxylated [53, 54]. The ultimate reactive group is the alkoxy methyl derivative of acrylamide, typically the butyl ether. This group, which reacts in near identical manner to the groups in nitrogen resins, also improves the compatibility of the polymer with alkyd and epoxy resins and nitrocellulose. These resins may hence also be blended with the acrylic resin, and co-reaction can then occur. Typical curing reactions are shown in Fig, 2.11.

Polymer—CO—NH—CH$_2$ O Bu+HO—Polymer→

$\qquad\qquad\qquad$ Polymer—CO NH CH$_2$ O—Polymer+BuOH

Polymer—CO—NH—CH$_2$ O Bu+BuO CH$_2$ NH CO—Polymer→

$\qquad\qquad\qquad$ Polymer—CO NH CH$_2$ NH CO—Polymer+BuO CH$_2$ O Bu

Fig. 2.11 — Typical curing reactions of *N*-butoxymethyl acrylamide copolymers.

Three methods of preparation are possible in the manufacture of these acrylic polymers. The monomers, including the acrylamide, may be copolymerized and the amide groups subsequently reacted with formaldehyde and then alcohol. Concurrent reaction may be carried out when solvents, including alcohol, monomers, and formaldehyde are charged with polymerization and methylolation taking place simultaneously. As a final possibility, *N*-alkoxy methyl acrylamide may be prepared as an intermediate prior to the polymerization stage. Formaldehyde in these compositions is most usually included in alcoholic rather than aqueous solution (e.g. 40% formaldehyde in butanol) or as paraformaldehyde. The amide/formaldehyde reaction may be acid or base catalysed. For the etherification reaction it is necessary to remove water of reaction by azeotropic distillation. Excess of the alcohol, necessarily present for acrylamide solvation as well as being a reactant, should remain at the end as solvent to give stability to the resin. Polymerization may be initiated by a peroxide initiator, and chain transfer agent will be present since the final polymer being crosslinking need only be of low molecular weight.

Thermosetting acrylamide polymers may contain styrene for alkali, detergent, and salt spray resistance, acrylate esters for flexibility, and acrylonitrile for improving toughness and solvent resistance. A wide range of compositions are hence possible, for example, for domestic appliance, general industrial, or post-formed coil coated strip application. Acid monomer is generally included to provide catalysis of cure. Epoxy addition has been used in appliance finishes to obtain extremely good detergent and stain resistance, and silicone modification can also be carried out to improve weathering performance. High acid value variants are possible, which when amine neutralized can be used in waterborne applications [55, 56].

Carboxy functional thermosetting acrylic resins have been prepared which may be crosslinked with diepoxides. These have been evaluated for automotive topcoats, and for a long time were used for industrial finishes for appliances, coil coatings, and drums [57]. Their use has now been largely superseded by the hydroxy and acrylamide thermosetting acrylics.

Post-modification of acrylic resins is possible; for example acrylic resin with a high acid content may be condensed with the epoxy intermediate Cardura E10 (Shell Chemicals) to link plasticizing side chains to the polymer [14]. (The pre-prepared Cardura E10/unsaturated acid adducts may alternatively be used as monomers in preparation.) Saturated or unsaturated fatty acid modification has also been carried out on hydroxy functional acrylic resins for either plasticization purposes or to introduce air-drying functionality.

2.8 EMULSION POLYMERS

Emulsion polymers are now probably the highest volume resins used by the coatings industry, principally because of the continued growth and high usage of aqueous emulsion paints for home decoration; for this application the polymer is in the non-functional, room temperature coalescing, thermoplastic form. Other uses do, however, also exist for aqueous emulsion polymers in waterborne crosslinking coating systems, where functional polymer is used. It may be noted that high usage exists for emulsions in the adhesive and textile industries, and emulsion polymerization is a route to polyvinyl chloride and synthetic rubber polymer preparation.

'Emulsions' are strictly two-phase systems of two immiscible liquids, where small droplets of one form the dispersed phase in the other, which is the continuous phase. In the terminology of the polymer industry, emulsion polymerization and emulsion polymer describe the process and end product of polymerizing addition monomers in water in the presence of surfactant, using water-soluble initiators to form fine-particle stable dispersions. The term latex is also interchangeably used with emulsion for the final polymer dispersion. The polymer particles are typically sub-micrometre (0.1–0.5 μm) such that one litre of emulsion may contain 10^{16} individual particles of surface area 2000 m^2. The monomers and fundamental chemistry of polymer formation are those of acrylic polymerization described earlier. Emulsion polymerization is a member of the polymerization class known as dispersion polymerization, of which the other major member used in surface coatings is Non-Aqueous Dispersion polymerization (NAD). A few other disperse polymers find minor application in the coatings industry [58, 59]. Among these may be mentioned non-ionically stabilized latices whose stabilizing groups may for example be poly ethylene glycol chains [60] providing steric stabilization in like manner to non-aqueous dispersions (see 2.9).

A typical emulsion polymerization formulation will contain [61], in addition to water, around 50% monomers blended for the required Tg, with surfactant, and often colloid, initiator, and usually pH buffer and fungicide. Hard monomers used in emulsion polymerization may be vinyl acetate, methyl methacrylate, styrene, and the gaseous vinyl chloride. Soft monomers used include butyl acrylate, 2-ethylhexyl acrylate, Vinyl Versatate (Shell Chemicals) [15], maleate esters, and the gaseous monomers ethylene and vinylidene chloride.

Most suitable monomers are those with low, but not very low, water solubility; monomers of very low solubility can be difficult to use satisfactorily. To use any of the gaseous monomers requires special plant; the techniques for handling these are briefly mentioned at the end of this section. Other monomers may be included in formulations, for example acids such as acrylic and methacrylic acids and adhesion-promoting monomers. It is important that films coalesce as the diluent evaporates, and the minimum film-forming temperature ($MFFT$) of the paint composition is a characteristic, closely related to Tg of the polymer, but also affected by materials such as surfactant present and by inhomogeneity of polymer composition at the surface [62]. Higher Tg polymers that would not otherwise coalesce at room temperature may be induced to do so by incorporating a transient plasticizer or 'coalescing agent' such as benzyl alcohol into the paint composition. $MFFT$ is normally determined in preference to Tg for emulsion compositions, since it is difficult to allow for these deviating effects; for the usual decorative applications it may typically be in the range 0–10°C. Lower $MFFT$'s are required with more highly pigmented and extended finishes.

Available surfactants are anionic, cationic, and non-ionic, and essential characteristics of these surface active materials are that their molecules possess two dissimilar structural groups, one a water-soluble or hydrophilic group and the other a water-insoluble hydrophobic moiety. The composition, solubility properties, location, and relative sizes of the dissimilar groups in relation to the overall structure, determine the surface activity of the compound.

The role of surfactants [63] is first of all to provide a locus for the monomer to polymerize, and secondly to stabilize the polymer particles as they form. When surfactants are dissolved in water at concentrations above a certain level the molecules associate to form 'micelles' instead of being present in true solution; these micelles can solubilize monomer. Mixed anionic and non-ionic surfactants are the combination frequently used in emulsion polymerization; cationic surfactants are rarely used (Fig. 2.12).

Colloids or 'protective colloids' are water-soluble polymers such as poly(meth)-acrylic acid or its copolymers, so-called polyvinyl alcohol (partly hydrolysed polyvinyl acetate), or substituted celluloses such as hydroxy ethyl cellulose. The properties of these colloids vary with molecular weight, degree of branching, and composition (proportion of water-soluble acid or hydroxy present). When used in emulsion polymerizations they can be grafted by growing chains of the polymer being formed, especially in the case of the celluloses, or may undergo chain scission. They assist in particle size control and in determining the rheology of the final paint, particularly in the derived gel structure and in the degree of thixotropy found.

A typical emulsion polymerization will involve two characteristic stages known as the seed stage and feed stage. In the seed stage, an aqueous charge of water, surfactant, and colloid will be raised to reaction temperature (85–90°) and 5–10% of the monomer mixture will be added along with a portion of the initiator (typically a water-soluble persulphate). In this stage the first polymer particles are formed.

The seed formulation contains monomer droplets stabilized by surfactant, initiator, and a small amount of monomer in solution, and surfactant, both in solution and micellar form. Radicals are formed in solution from initiator break-down, and these polymerize the traces of monomer dissolved in the water until the growing polymer chain enters a micelle, where the monomer present within the micelle also polymerizes. Termination can occur when a further growing polymer radical enters the micelle particle. In the seed stage, during which no further reactants are added, initial monomer and initiator will largely be converted to polymer with the particle number roughly corresponding to the number of micelles

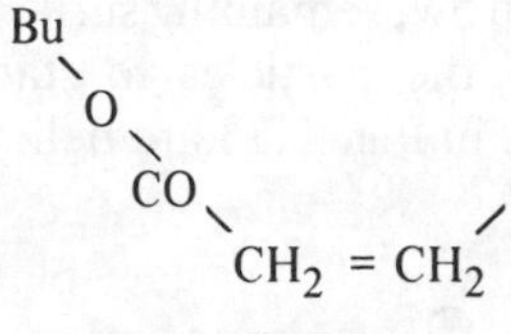

Dibutyl maleate

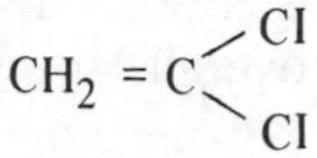

$CH_2 = C$ with Cl, Cl

Vinylidene chloride

R_1, R_2, R_3 with $CO\ O\ CH = CH_2$

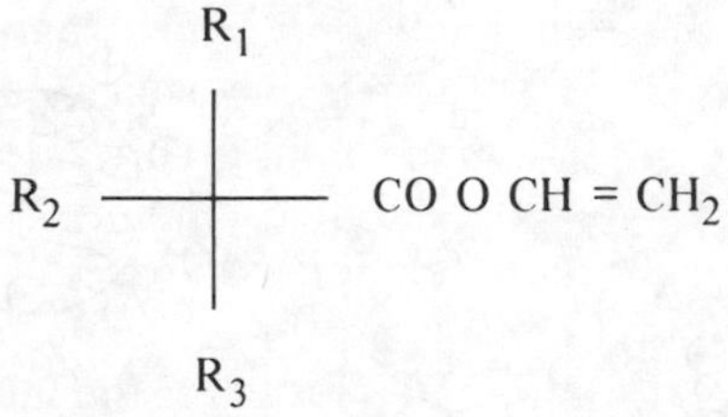

Where R_1, R_2, R_3 are alkyl, one of which is methyl.

Vinyl Versatate (Shell Chemicals)

Anionic surfactants
Stearate soaps $\qquad$ $CH_3\ C_{16}\ H_{32}\ COO^-\ Na^+$

Dodecyl benzene sulphonate $\qquad$ $CH_3\ C_{11}\ H_{22}$ — $SO_3^-\ Na^+$

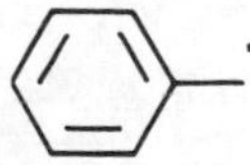

Sodium dioctyl sulpho succinate $\qquad$ $CH_3\ C_7\ H_{14}\ COO\ CH_2$
$CH_3\ C_7\ H_{14}\ COO\ CH\ SO_3^-\ Na^+$

Cationic surfactant
Cetyl trimethyl ammonium bromide $\qquad$ $(CH_3)_3\ N^+\ C_{15}\ H_{30}\ CH_3\ Br^-$

Nonionic surfactants
Polyethoxylated nonyl phenol $\qquad$ $CH_3\ C_8\ H_{16}$ — $O(C_2H_4O)_n\ H$

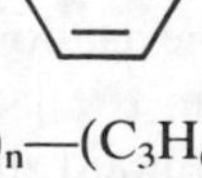

Polyethoxylated polypropylene glycol $\qquad$ $HO(C_2H_4O)_n—(C_3H_6O)_m—(C_2H_4O)_p\ H$

Fig. 2.12 — Typical surfactants used in emulsion polymerization.

of surfactant initially present; the concentration of surfactant remaining which is not associated with small polymer particles or any remaining monomer droplets will be small.

In the feed stage, remaining monomer and initiator solutions are fed together, the monomer reservoir in the reaction medium consisting of emulsified monomer droplets. Polymerization proceeds as monomer diffuses from the monomer droplets, through the water phase, into the already formed growing polymer particles. At the same time radicals enter the monomer-swollen particles causing both termination and re-initiation of polymerization. As the particles grow, remaining surfactant from the water phase is absorbed onto the surface of the particles to stabilize the dispersion. In the overall process the entities shown in Fig. 2.13 are believed to be

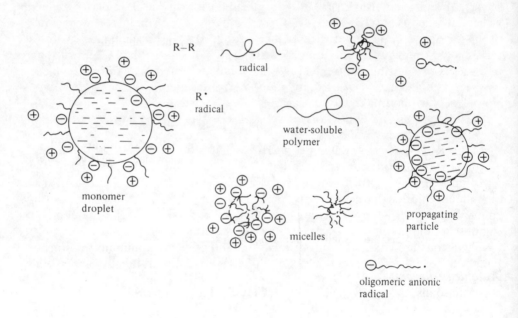

Fig. 2.13 — Entities involved in emulsion polymerization.

involved. Particles are stabilized from flocculation and coalescence by mutual repulsion of surface charges from the anionic surfactant.

The final stage of polymerization may include a further shot of initiator, either thermal or redox, to complete conversion, followed by cooling and addition of biocide if required.

Various factors make emulsion polymerization more difficult to control than normal acrylic polymerizations. Agitation is critical to get good emulsification of seed reactants, and in the later stages to get heat extraction through the reactor cooling surfaces and good incorporation of added monomer. At the same time formulations may be shear-sensitive, and excess agitation is to be avoided. Reflux generally can have an adverse effect, hence initial heating followed by progressive application of cooling is necessary to hold a steady temperature. Bit content and polymer build-up on reactor walls can cause problems.

The kinetics of emulsion polymerization are complex, and efforts to understand this and the overall mechanism have been extensive [64]. Techniques are now being established so that it is becoming possible to control the structure of particles to contain interpenetrating networks, to have separately distinguishable core and shell structures, or to provide microgel structures for either mechanical property modification or rheology control, as with solvent-borne microgels [58, 65, 66].

Though it is possible to prepare emulsions with hard monomer only, adding so-called external plasticizer to the paint formulation, all emulsions used nowadays are internally plasticized with copolymerized soft monomer. Styrene-containing emulsions find little use in coatings in the UK though they are used in continental Europe. General-purpose emulsions are often formulated with vinyl acetate plasticized with an acrylate such as butyl or 2-ethyl-hexyl acrylate, or a dialkyl maleate. Formulations of this type have good chalking resistance but poor alkali resistance and hydrolysis resistance; to improve these latter properties Vinyl Versatate (Shell Chemicals) may be used as plasticizing monomer [67]. All-acrylic formulations containing methyl methacrylate and acrylic plasticizing monomer give generally higher quality emulsions except for poorer chalking performance.

In order to use gaseous monomers, equipment is needed capable of working under pressure; these monomers can be used to give cheaper emulsions by copolymerizing vinyl chloride and ethylene with vinyl acetate. Ethylene in particular cheapens formulations but is so soft that typically vinyl chloride is incorporated as a hard monomer along with vinyl acetate. Polymers containing vinyl chloride and vinylidene chloride find particular application for anti-corrosive primers. Vinyl chloride monomer is carcinogenic, and final emulsions must be carefully treated by steam stripping to remove all traces of free monomer from the product.

Emulsions are characterized by solids and viscosity, by minimum film-forming temperature [68] and freeze–thaw stability. Film tests are additionally carried out to assess freedom from bits. pH may also need control particularly with vinyl acetate- and vinyl chloride-containing emulsions in order to prevent hydrolysis.

As well as the established use of autoclaves for the preparation of 'pressure' polymers, a technique capable of handling gaseous, as well as liquid, monomers is the loop reactor [69]. This is a form of continuous reactor whereby the reacting mixture is pumped around a heated/cooled loop, monomers and initiator being pumped in at one point in the loop, and at another point product 'overflows' from the loop at a similar rate to the incoming feeds. Claimed advantages include low installation, capital, and running costs, while disadvantages include inability to exploit, for example, the new concepts of core-and-shell morphology.

While minor monomer components may include acids to modify rheology, other monomers may alternatively be added to improve adhesion. Glycidyl methacrylate and amino monomers have been used, though where specially good performance is required, for example for primers to adhere well to wood and to old gloss paint, acetoacetic esters [70] and ureido [71] monomers have been proposed.

Polymers for thermosetting applications will be formulated and prepared similarly to those described above, but will contain hydroxy monomer and will generally be colloid free.

2.9 NON-AQUEOUS DISPERSION POLYMERIZATION

Bearing many superficial similarities to emulsion polymers which exist in aqueous media, the other major type of polymer dispersion used in the coatings industry is the non-aqueous dispersion (NAD). In this case the polymer, normally acrylic, is dispersed in a non-aqueous medium, typically aliphatic hydrocarbon [72]. Acrylic polymer NAD's currently find application chiefly in automotive systems, though a range of other applications have been described; however, it is possible to prepare both condensation and addition polymer dispersions with a variety of particle structures including layered, heterogeneous, and vesiculated [59].

NAD polymer dispersions may typically be submicron, of size 0.1–0.5 μm similar to aqueous polymers, and the chain addition polymerization mechanisms are again those earlier described under acrylic polymers. Unlike emulsion polymers, however, where particles are stabilized by charge-repulsion mechanisms derived from anionic surfactants, the only successful method available for stabilizing NAD's is by the process now generally known as steric stabilization involving the presence at the particle-medium interface of an adsorbed solvated polymer layer. Charge stabilization has an inadequate effect in the media used in NAD polymerization because of their dielectric constant, one or two orders of magnitude lower than that for water. In steric stabilization, the replusive forces which prevent flocculation when particles collide arise from the increase in local solvated polymer concentration arising at the point of contact; the system reacts to eliminate this local excess concentration by causing the particles to separate (Fig. 2.14). Steric stabilization can be effective in

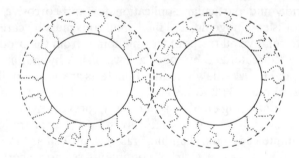

Fig. 2.14 — Representation of steric repulsion from soluble chains attached to dispersed particles.

media of high dielectric constant where charge stabilization is normally used, if nonionic stabilizers are used [60].

The most successful types of stabilizers used in dispersion polymerization have been based on block or graft copolymers [73], and special techniques have been developed to prepare amphipathic graft copolymers suitable for use in NAD preparation [74]. Preformed graft stabilizers based on poly 12-hydroxy stearic acid (PHS) are simple to prepare and effective in both addition and condensation NAD polymerizations. Commercial 12–hydroxy stearic acid contains 8–15% of palmitic and stearic acids which limit the molecular weight during polymerization so that on self-condensation a polymer of number average molecular weight 1500–2000 is produced, monofunctional with respect to carboxyl functionality. This oligomer may be converted to a 'macromonomer' by reacting the carboxy group with glycidyl methacrylate, and this 'macromonomer' is then copolymerized with an equal weight of methyl methacrylate or similar monomer to give a polymer of molecular weight 10–20000, which is then a 'comb' stabilizer with 5–10 soluble chains pendant from a polymeric anchor backbone (Fig. 2.15).

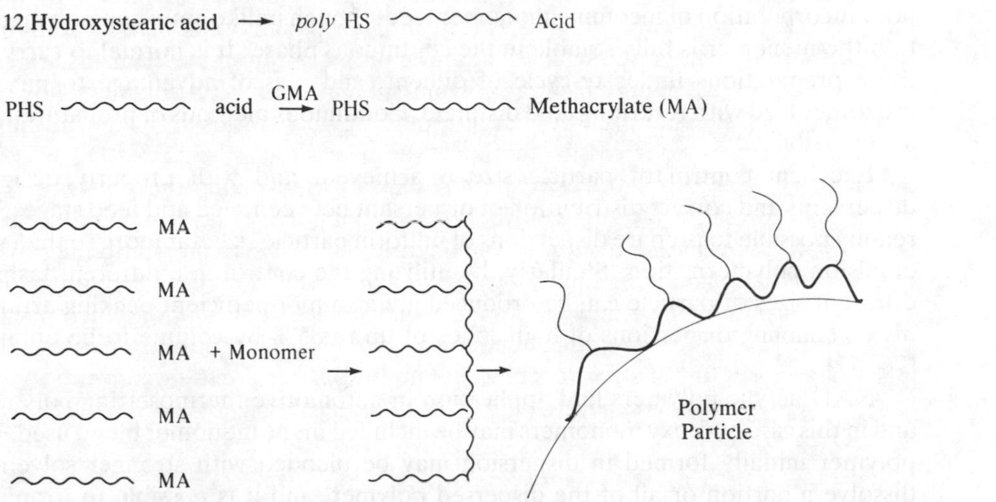

Fig. 2.15 — Schematic representation of p 12 hydroxystearic acid 'comb' stabilizer preparation.

The formation of the graft and its method of stabilization of dispersed particles by adsorption of the insoluble anchor component is shown diagrammatically above (Fig.2.15). Poly 12-hydroxy stearic acid derivatives also gain application more generally as dispersion agents [2]. The special techniques to prepare the 'macromonomers' necessary for preparing graft dispersants of defined structure have been reviewed [75].

Virtually the full range of addition polymers may be used in NAD polymerization, and it is of note that those found more difficult to polymerize in solution may be more readily polymerized in dispersion (as is also the case with aqueous emulsion polymerization). The major limitation for monomer composition is that the polymer produced shall be insoluble in the medium used, and this thus requires that if included, when in aliphatic media, the amount of longer chain plasticizing monomer shall be limited to satisfy this. As a consequence of nearly all polymerization taking place within particles, which consist of polymer swollen with monomer, the rate of polymerization is greatly accelerated compared with solution polymerization because of the reduction of the diffusion-controlled termination rate.

It is usual to carry out the preparation of NAD in two stages in like manner to an emulsion polymerization. In the seed stage, the diluent, together with a portion of the dispersant and monomer charge, is heated along with initiator to form an initial low-concentration fine dispersion. The remaining monomer, together with more dispersant and initiator, is then fed in over the course of several hours to complete the growth of the particles. A small amount of chain transfer agent is usually added to control molecular weight. Initiators may be either of the azo or peroxy type appropriate to the temperature of reaction.

Like emulsion polymerization, dispersion quality may be adversely affected by poor incorporation of incoming monomer feed, though unlike emulsion polymerization the monomer is fully soluble in the continuous phase. It is normal to carry out these preparations under re-cycle of diluent, and it is of advantage to mix the monomer feed with returning cold distillate. Continuous methods of preparation are also possible.

Excellent control of particle size is achieved, and with properly designed dispersants and correct distribution of dispersant between seed and feed stages, it is readily possible to prepare dispersions of uniform particle size, far more so than with emulsion polymerization. Similarly, by utilizing the control in a different fashion, different sizes of particle can be produced giving a more efficient packing arrangement, enabling dispersions of high solids of up to 85% by volume to be obtained [76, 77].

NAD acrylic polymers find application in automotive thermosetting polymers, and in this case hydroxy monomers may be included in the monomer blend used. The polymer initially formed in dispersion may be blended with stronger solvent to dissolve a portion or all of the dispersed polymer, and it is possible to formulate particle compositions to use in these circumstances which range in state from the polymer being in true solution, through solvent swollen gel, to stable unswollen polymer particles. The reason for the use of this type of NAD is that the presence of swollen insoluble polymer has a profound effect on the evaporation rate of the solvents used and on the rate of increase of viscosity during evaporation. These effects give wide control over the formation of thicker coatings by spray application, with little sagging and good control of metallic pigment orientation. It has, in fact, in recent years been found most useful to prepare organic 'microgels' for blending with solution prepared thermosetting hydroxy acrylic polymers to assist rheological control during application [66]. The monomers used here may include, for example, glycidyl methacrylate and methacrylic acid reacted *in situ* during the reaction, catalysed by added amine, such that the final particles become completely insoluble in organic solvents. These particles may be modified with auxiliary polymer subsequent to the microgel core preparation and finally diluted with a strong solvent blend.

2.10 AMINO RESINS

Amino or nitrogen resins are the condensation products of certain compounds with two or more amine groups, particularly urea and melamine, with formaldehyde. These condensation products are generally alkylated and may also be partly polymerized. The family of acrylamide-derived thermosetting acrylic resins bear considerable resemblance to amino resins in both their formation and crosslinking reactions.

Nitrogen resins are used as hardening or crosslinking resins with a range of hydroxy functional resins in both stoving and 2-pack room temperature curing systems; as such they are normally the minor component by weight. The special properties they confer are excellent colour and colour retention, hardness, and chemical resistance. Certain of the family of nitrogen resins are water-soluble. It may be noted that nitrogen resins find considerable use outside the coatings industry for bonding chipboard and as laminating adhesives.

The major amino resin types are those derived from melamine (MF) and urea (UF), though benzoguanamine and thiourea have been used and resins based on glycoluril [78] are commercially available (Fig. 2.16).

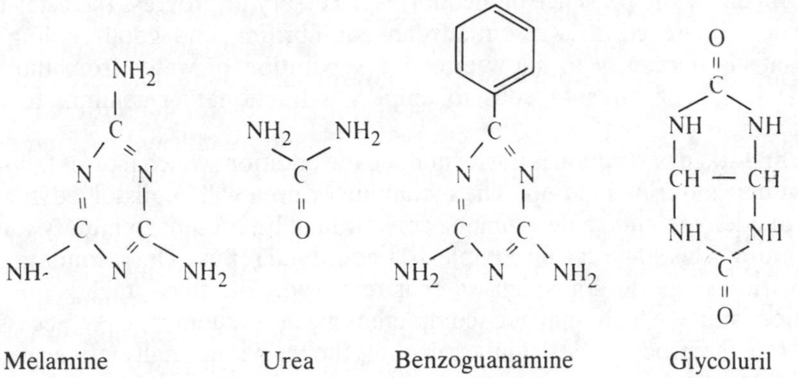

Melamine   Urea   Benzoguanamine   Glycoluril

Fig. 2.16 — Typical amino compounds used in formaldehyde condensate resins.

In all cases the first reaction in preparation is the condensation of formaldehyde with an amino group, as represented by:

$$\diagdown \!\!\!NH + HCHO \xrightarrow{H^+ \text{ or } OH^-} \diagdown \!\!\!N\,CH_2\,OH$$

In the case of melamine two molecules of formaldehyde can add on to each amino group; with urea only one molecule condenses under normal circumstances. Basic conditions are used for this stage for both urea and melamine resins, though for melamine resins acidic conditions are also effective. In all cases the reaction is exothermic. The preparation of both dimethylol urea and the fully reacted hexa methylol melamine are possible in a good state of purity.

The simple formaldehyde adducts which are water-soluble, though finding applications in other industries, have no use in surface coatings. For coatings these adducts will be at least partly alkylated with lower alcohols, and this confers compatibility with other resins and allows solubility in organic solvents. The alcohols mainly used are propanol, normal or isobutanols, and, only in the case of melamine, methanol; the etherification reaction may be represented by:

$$\underset{/}{\overset{\backslash}{N}} CH_2OH + H O \ Alkyl \rightarrow \underset{/}{\overset{\backslash}{N}} CH_2 \ O \ Alkyl + H_2O$$

The preparation of alkylated amino resins is normally 3-stage, and pH adjustment may be carried out for each of the reaction stages. It is essential to have solvent process plant available to re-cycle and distil off solvent and to remove water of reaction, with vacuum also necessary for some products. The detailed process will depend on the form of formaldehyde used which may be in either aqueous or alcoholic solution or as paraformaldehyde. Efficient removal of water (which is made more difficult by the presence of alcohol) is necessary to progress the esterification stage since the reaction is essentially an equilibrium, and good cooling of the condensate is necessary to allow effective separation of water from the solvent fraction. It can be advantageous to employ a fractionating column, to aid the separation.

The first stage of reaction is the formaldehyde addition, which if pH is favourable, will commence during heat-up. The melamine or urea will be dissolved in alcohol, and formaldehyde and some aromatic solvent may be present to modify the reflux temperature. This stage can be completed under total reflux. The second stage is the etherification stage during which water is removed. The third stage is the solvent distillation stage which may be completed using vacuum and is necessary to concentrate the product. With higher alcohols the resin is normally left in an alcohol-rich solvent blend in order to retain storage stability. Nevertheless storage life, especially in warmer conditions, may always be limited.

The simple reaction scheme above is complicated in practice by a possible range of side reactions leading to a degree of polymerization (Fig. 2.17).

$$-NH \ CH_2 \ OH + H_2 \ N- \ \rightarrow \ -NH-CH_2- + H_2O$$

$$-NH \ CH_2 \ OH + HO \ CH_2 \ NH- \ \rightarrow \ -NH \ CH_2 \ O \ CH_2 \ NH- + H_2O$$

Fig. 2.17 — Reactions leading to polymerization in the preparation of formaldehyde condensation polymers.

Because of these and the impossibility generally of any meaningful in-process sampling to control any particular stage, the key to successful nitrogen resin manufacture is strict control of time and temperature regimes, distillation rates, and distillate quantities.

The degree of polymerization when it occurs is not high, though gelation in preparation is possible if a resin is misformulated. It should be noted that in polymerization reactions the secondary NH groups remaining after first additions onto primary amine groups are themselves reactive; with urea, for example, a mean functionality of three may be observed as the overall reactivity.

The effect of variation of the rates and extent of different reactions may show in a number of ways. For example, when excessive alkylation occurs at the expense of polymerization more water is formed, the tolerance of the final resin for mineral spirits will increase, the solids content will be high, and the viscosity will be decreased.

2.10.1 Characterization
Mineral spirits or petroleum ether tolerance, which indicates the degree of alkylation and the compatibility with other resins, is expressed as the volume compatibility to a given weight of resin, determined as the threshold at which haze develops in the blend. Solids are normally measured at a longer time and lower temperature than for other resins in order to obtain more reproducible and meaningful results; more than with any other resin, though, the figure obtained is not a true figure since a degree of decomposition and crosslinking is unavoidable during solids determination. Water contents are normally measured as an assurance of correct processing, and pH is characterized since this may affect storage stability.

2.10.2 Formulating
The condensation reactions involved, particularly the main formaldehyde addition and etherification reactions, are reversible, and hence excess formaldehyde and alcohol concentrations are necessary to push equilibria towards the desired products. When formulating resins the ratio of formaldehyde to melamine or urea determines the degree of methylolation. With urea it may be necessary to charge 2.4 mols of formaldehyde to attain the normally required 2-mol addition; with melamine resins, where reactivities up to 6 are both possible and useful, a wide range of ratios may be used. The quantity of alcohol included, along with the processing conditions, then determines the degree of alkylation and polymerization, and the characteristics of the final resin. Again, a considerable excess of alcohol is normally required to be present, though its presence in the final resin assists storage stability. Reactivity of butylated melamine resins, for example, increases with decreasing formaldehyde to melamine ratio, and with decreasing degree of etherification.

Unlike other resins it is not possible to specify with accuracy the ratio of reactants actually incorporated into the resin; rather it may be more useful to specify the ratio of reactants charged initially.

2.10.3 Curing
Amino resins react on heating both with themselves, and with any other resin present containing functional groups, which will normally be hydroxy but equally possibly amino or carboxyl. The reactions are similar to those shown for formation, involving condensation and elimination reactions where water, alcohol, or formaldehyde may be the products evolved (Fig. 2.18); the reactions leading to polymerization in preparation also occur.

—NH CH$_2$ O Bu + HO − Polymer→—NH CH$_2$ O Polymer + BuOH

—NH CH$_2$ O H + HO − Polymer→—NH CH$_2$ O Polymer + H$_2$O

—NH CH$_2$ O Bu +

HOOC − Polymer→—NH CH$_2$ O CO Polymer + BuOH

Fig. 2.18 — Crosslinking reactions possible between formaldehyde condensate resins and other polymers.

The factors affecting reactivity are complex [79, 80], and as with formation, the curing rate and the relative importance of the above reactions are affected strongly by pH as altered by the presence of (typically) acid catalysts [81, 82, 83].

2.10.4 Uses

Melamine formaldehyde resins, because of their greater functionality, have significantly different properties to urea formaldehyde resins, having better chemical resistance, colour retention at elevated temperatures, better exterior durability, and shorter baking schedules when in combination with hydroxy resins. It should be noted that their reactivity may be less than that expected for steric reasons. They are also more expensive. UF resins are very satisfactory, however, for general-purpose industrial finishes and are extensively used in acid catalysed room temperature cured wood finishes.

While strong acid catalysis of blends containing UF resin proceeds satisfactorily, with melamine formaldehyde resins it may lead to self-reaction rather than co-reaction. For this reason for low-bake or room temperature cure with melamine formaldehyde resin blends, where high acid catalyst levels may be imperative, it is sometimes advantageous with, for example, alkyd resins, to prepare a precondensate with the MF resin.

Alkylated melamine resins will be chosen by their reactivity where high reactivity may mean high viscosity and rapid cure but low mineral spirit tolerance and hence poorer compatibility with alkyd resins. While the generally used resins are butylated, other alcohols are used. Secondary and tertiary alcohol modified resins are generally slower curing but may result in harder films. Mixed alcohol resins can provide additional formulating latitude.

Special mention must be made of hexamethoxymethyl melamine (HMMM) resins which have secured their own place in coating compositions. Though available as the pure material as a waxy solid, they are most widely used in slightly modified or condensed form when they are more easily handled viscous liquids. They are soluble in water and in all common organic solvents except aliphatic hydrocarbons, and are compatible with practically all resin media. They are usually available at 100% solids. Unlike other alkylated MFs they are considerably more stable. The superior adhesion and flexibility obtainable may be attributed to the lower tendency to self-condense and also to a more satisfactory film structure [80]. There is, however, a sharper optimum to the amount of HMMM required in any composition for the best properties. The resin finds particular use in high solids and water-borne compositions. As with other alkylated MFs their preparation requires a high alcohol excess during reaction to ensure complete reaction; the problems of preparation include the removal of methanol containing distillate.

2.11 PHENOL FORMALDEHYDE RESINS

In the early years of the coatings industry only naturally occurring resins were available to the varnish maker to enhance the properties of natural oils. The availability of the hard oil-soluble phenolic resin (PF) allowed a more scientific approach to varnish making. The development of the heat-reactive soluble phenolic resin in turn enabled the development of baking finishes with excellent solvent and corrosion resistance, and these resins still find application alone or in blends with alkyd or epoxy resins for can coatings, and tank and drum linings. However, they have in turn been supplanted as crosslinking resins by the amino resins because of their far superior colour for most applications.

Unmodified phenolics may be oil insoluble or soluble and may be heat-reactive or non-reactive. The initial reaction with formaldhyde and the subsequent reaction under acid conditions are shown in Fig. 2.19. The dimer shown can lose additional

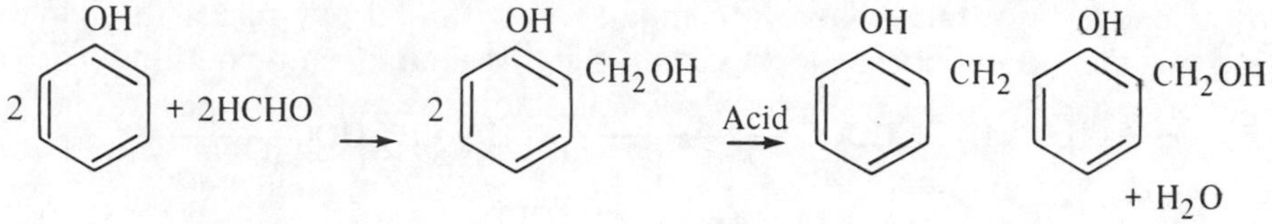

Fig. 2.19 — Novolac type phenolic resin formation.

water to form a resin. This kind of resin is a Novolac, and in simple form with unsubstituted phenol alone finds little use in coatings.

Alkaline-catalysed phenolic resins are most common for coatings since these condense by ether formation and hence provide softer resins with some residual methylol groups for further reaction (Fig. 2.20). This class of phenolic resin is known

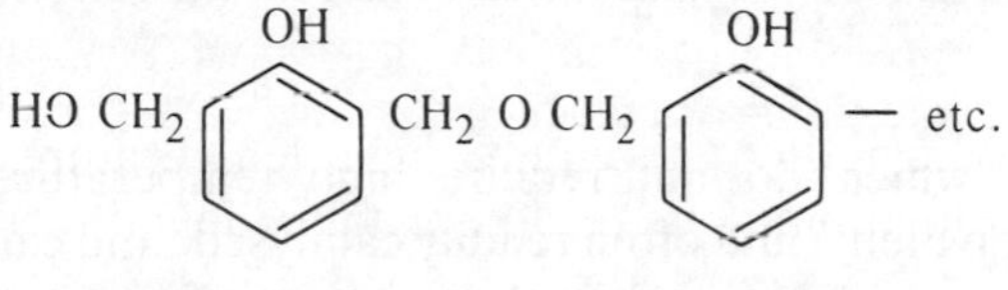

Fig. 2.20 — Resole type phenolic resin structure.

as a Resole, and made with phenol is only alcohol-soluble, but is thermo-hardening, unlike a Novolac, and capable of forming a crosslinked structure without any additional curing agent.

Other phenols may be used including cresol, 2,4-xylenol, and the para substituted phenols such as p-phenyl phenol, p,t-butyl phenol, and diphenylol propane.

The more useful PF resins of both classes are made with p-substituted phenols.

Oil-soluble and oil-reactive resins are obtained with p,t-butyl and p-phenyl phenols where acid-catalysed Novolacs are non-reactive and alkaline catalysed Resoles are oil reactive. It is also possible to etherify methylol groups with alcohol, so improving oil solubility.

Rosin-modified phenolics have found considerable application for oleoresinous vehicles, and are prepared by heating a Resole type PF with rosin when combination of unsaturated double bonds in the rosin and hydroxyl groups of the Resole occurs. The product is then esterified with a polyhydric alcohol such as glycerol or pentaer-ithritol to reduce the acid value. This type of resin can be either dissolved or cooked into oil for varnish preparation.

2.12   EPOXY RESINS

The epoxide or oxirane group has a number of reactions useful in resin chemistry, particularly those with carboxyl, hydroxyl, phenol, and amine (Fig 2.21).

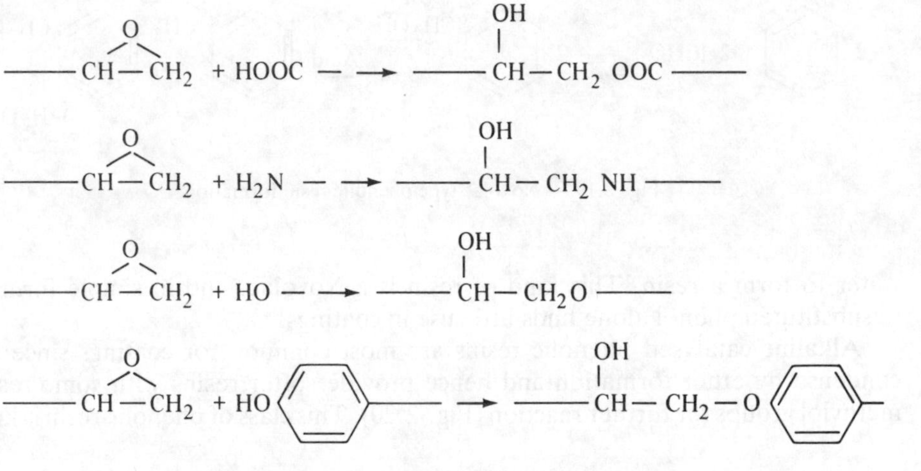

Fig. 2.21 — Typical reactions of the epoxide group.

These reactions which do not require high temperatures, are exothermic (70–80 kJ/epoxy equivalent) and often readily catalysed, and can be exploited in the assembly of polymers and in curing reactions; in certain circumstances novel multi-stage reaction routes can be devised. The amine–epoxide reaction is particularly used in the cure of epoxy resin (see 2.12.2). The reaction with dicarboxylic anhydrides, as well as occurring in the manufacture of alkyds containing Car-dura E10 (Shell Chemicals), is also used as a method of cure for powder coatings; in this reaction, requiring both initiation and catalysis, polyesters are produced both rapidly and exothermically without any water of reaction being evolved [84]. Since hydroxyl groups may always be present when epoxy resins are being cured, even when other reactive groups predominate, it should be noted that conditions of temperature and catalysis may enhance or suppress the possible etherification reaction with hydroxyls in relation to the other epoxy group reactions taking place [85]. For example, in the acid–epoxy reaction, base catalysis suppresses the etherification reactions that would otherwise occur.

The compound glycidyl methacrylate which has both vinyl unsaturation and an epoxy group is particularly useful as a bridge between condensation and addition stages in polymerizations. Possibilities occur to react either group initially, the second then reacting in a subsequent stage.

The most well-known epoxide containing materials are the range of preformed epoxy resins based on the reaction between diphenylol propane (Bisphenol A) and epichlorhydrin [86]; their general structure is shown in Fig. 2.22.

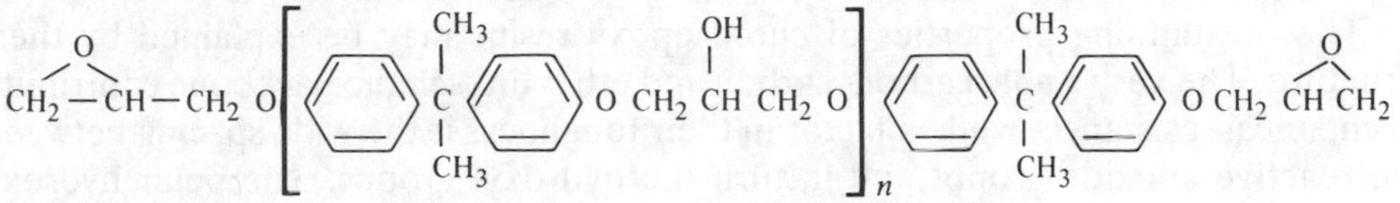

Fig. 2.22 — Generalized structure of Bisphenol epoxide resins

Where n aproaches 0, the resin approximates to the diglycidyl ether of diphenylol propane, and the product is liquid. As n increases from 2 up to about 13, solid forms with increasing melting points are encountered (Table 2.8). The lower-melting

Table 2.8 — Typical grades of bisphenol epoxide resins and properties

Number of repeat units (n)	Melting point °C	Epoxide equivalent
0.5	Viscous liquid	225–290
2	64–76	450–525
4	95–105	850–950
9	125–132	1650–2050
12	140–155	2400–4000

grades are often modified by pre-reaction before use in coatings, though the higher-melting grades may be used unmodified for can or drum lining applications.

Because of the problems in handling toxic epichlorhydrin and other practical difficulties it is normal for coatings manufacturers to purchase epoxy resins of this type; it is, however, possible to 'chain extend' liquid grades to higher molecular weight by reaction with diphenylol propane [87], and precatalysed liquid epoxy resin is available for this purpose. Adoption of this technique can reduce stocking of a range of grades and can provide access to intermediate solid grades; cost savings are also possible.

The formula above would indicate that all epoxy resins possess two terminal epoxide groups and all except that where $n=0$ possess in-chain hydroxy groups. Higher molecular weight epoxy resins may depart from this linear structure, however, owing to the incidence of side reactions, and residual chlorine from these may be an undesirable contaminant for some applications [88], and may, for example, deactivate amine catalyst in some reactions.

Considerable care is now taken in epoxy resin manufacture to completely eliminate free epichlorhydrin from the product because of its carcinogenic nature. Low molecular weight di-epoxides may also be carcinogenic, and for this reason the available range of low molecular weight epoxies is now limited, with certain low molecular weight aliphatic di-epoxides being no longer available.

The outstanding properties of cured epoxy resins may be explained by their structure. The very stable carbon–carbon and ether links in the backbone contribute to chemical resistance, while a factor in their toughness is the wide spacing between the reactive epoxide groups and in turn the hydroxyl groups. The polar hydroxy groups, some of which may always remain, also assist adhesion by hydrogen bonding. The aromatic ring structure enhances thermal stability and rigidity. Though these properties are attractive, aromatic epoxy resins yellow, and for this reason their major application is in primer and undercoat compositions where adhesion and corrosion resistance are particularly valuable, and also in can coatings for their good one-coat performance.

Epoxy resins may be formulated from other phenols than Bisphenol A, for example diphenylol methane (Bisphenol F). Epoxy type materials may also be glycidyl ethers of other resins such as PF novolacs [89].

Epoxide compounds are characterized by melting point and by their epoxide group content or epoxy equivalent which may be expressed in a number of ways.

2.12.1 Epoxy esters

Both the terminal epoxide groups, and the secondary hydroxy groups of solid epoxy resins, can be reacted with fatty acids to produce the so-called epoxy ester. In esterification reactions each epoxide group is equivalent to two hydroxy groups. Epoxy esters are prepared by heating the fatty acid and epoxy resin in an inert atmosphere, preferably under azeotropic conditions to remove water of reaction, with temperatures of between 240 and 260°C normally being used. The reaction may optionally be accelerated by the addition of, for example, calcium or zinc soaps at 0.1 or 0.2% weight on total charge.

Typically the epoxy resin where $n=4$ is used, and fatty acid contents chosen to esterify between 40 and 80% of the available groups including hydroxyl. Medium (50–70% modified) and long (over 70%) oil epoxy esters of drying oil fatty acids are used in air drying brushing finishes, while short (30–50%) oil drying or non-drying fatty acid esters are used in industrial stoving primers and finishes. Stoved epoxy resin ester films, especially when cured with melamine formaldehyde (MF) resin, are harder and of superior adhesion, flexibility, and chemical resistance than similar alkyd/MF formulations. Increased fatty acid content, as would be expected, imparts better aliphatic solubility, better exterior durability, and decreased hardness, gloss, and chemical resistance. Linseed, tall, and DCO esters are most usual, though all fatty acids and rosin can be used.

2.12.2   Other epoxy applications

Epoxy resins may be cooked into alkyd formulations, replacing part of the polyol. They also find a place unmodified as a third component in alkyd/MF compositions to upgrade the resistance properties of the films.

Epoxy resins react with phenol formaldehyde (PF) resins to form insoluble coatings, and well-formulated high MW epoxy/PF coatings meet the highest standards of chemical resistance. These products are suitable for the linings of food cans and collapsible tubes, coatings for steel and aluminium containers, and wire enamels. Curing probably involves the formation of polyether links between the hydroxyl groups of the epoxy and methylol groups present in PF resins of the resole type; the epoxy also reacts with phenolic hydroxyl groups on the PF. With some PF resins compatibility problems on cold blending may be solved by pre-condensation, involving refluxing the epoxy and PF resins together in solution, when some reactive groups combine, leaving the remainder free to react in the curing process.

Urea/formaldehyde resins or melamine/formaldehyde resins may be used to cure epoxy resins, giving stove films of paler colour but with a reduced level of chemical resistance compared with phenol/formaldehyde resins. Again, the higher molecular weight epoxy resins are preferred.

Two-pack epoxy/isocyanate finishes require separate solutions of high molecular weight epoxy resin and polyisocyanate adduct as the two components; the epoxy resin must be in alcohol-free solvent since the curing reaction is predominantly with in-chain hydroxyl groups on the epoxy resin. One-pack finishes can be formulated with blocked isocyanates.

Epoxy resins may also be used in two-pack compositions with polyamines or polyamides. The films obtained possess outstanding chemical resistance, hardness, abrasion resistance, flexibility, and adhesion. Low molecular weight solid epoxy resins are most used. Though primary or secondary amines such as triethylenetetramine may be used, in order to avoid the toxic hazards involved in handling amines, amine adducts with low molecular weight solid epoxy resin may also be used as hardeners. These adducts are prepared by the reaction of excess of an amine such as diethylene triamine with epoxy resin to produce fully amine-terminated adduct. Reactive polyamide resins formed from dimerized fatty acids with diamine are also used.

Reactive coal tar pitches may be incorporated into an epoxy resin base for curing with amine, amine adduct, or polyamide resin. The derived coatings have excellent chemical resistance and are not brittle; they hence find use as pipeline, tank and marine coatings.

Epoxy resin modification with silicone resin is possible to enhance water resistance; epoxy/silicone combinations are used in blends with other polymers with the cure mechanisms mentioned above.

2.13 ISOCYANATES

The isocyanate group (Table 2.9) is reactive at medium and room temperatures, and may be catalysed in its reactions. It reacts exothermically (40 kJ/mol) with many groups with active hydrogen atoms, in particular alcohols, amines, phenols, and water at room or moderately elevated temperatures [90]. Amide groups are also reactive.

Secondary alcohols and amines react similarly to primary but with less vigour. Both aromatic and aliphatic polyisocyanates are available, with the former generally more reactive than the latter.

Water reacts with the isocyanate group at similar rates to secondary alcohols; the initial product is an unstable carbamic acid (RNHCOOH) which decomposes to a primary amine and carbon dioxide, and this amine group is of course then available to react with additional isocyanate. This reaction is normally to be avoided in coating formulations, though it is exploited in the production of solid foams. For this reason, systems to be reacted with isocyanates should be dehydrated, and low moisture content urethane grade solvents used.

Isocyanates also react with themselves under certain conditions and in the presence of certain catalysts [90]; these reactions are generally undesirable. Tin compounds are strong catalysts for isocyanate reactions and are also generally 'clean' in not promoting side reactions. It will be evident that all reaction products of isocyanates contain active hydrogen groups attached to the nitrogen atom of the urethane or substituted urea reaction product; hence, steric factors permitting, reaction can continue with these products if additional isocyanate is present to give allophanates and biurets. All of the groups mentioned above may be present on simple molecules or attached to polymer.

Low molecular weight diisocyanates are used in coatings manufacture in preparing urethane alkyds and blocked isocyanate crosslinkers. Toluene diisocyanate (TDI) is the main diisocyanate used for urethane alkyds (normally the mixed 2,4 and 2,6 isomers), but a far broader range is used for blocked isocyanate preparation. Great care is necessary in the handling of all volatile isocyanates since their vapours are irritant and they are sensitizers at extremely low concentrations. Products derived from aromatic isocyanates such as TDI and diphenyl methane diisocyanate (MDI) tend to yellow, hence aliphatic isocyanates such as hexamethylene diisocyanate (HMDI) and isophorone diisocyanate (IPDI) find high usage (Fig. 2.23).

2.13.1 Blocked isocyanates and isocyanate adducts

The reaction of the isocyanate group with low molecular weight alcohols, phenols, and some other compounds such as lactams may be considered reversible, in that if an isocyanate that has been 'blocked' or masked in this way is heated with a polymer with reactive groups, the low molecular weight blocking material is released and the urethane or substituted urea bond is remade with the reactive group on the polymer. Whether free isocyanate groups are fully released in the reaction with polymer is a matter of conjecture. Blocked isocyanate curing agents are used in electrocoat compositions, various solvent applied finishes, and in powder coatings. Depending on the blocking agent, curing temperatures may be between 100 and 175°C [92, 93, 94]. For powder coatings blocked IPDI is frequently used.

Table 2.9 — Typical reactions and reaction conditions for isocyanates

Reaction		Temperature	Catalysts
R NCO + Alkyl OH	→ R NHCO O Alkyl	25–50°C	Varied
R NCO + Alkyl NH_2	→ R NHCO NH Alkyl	10–35°C	Not required
R NCO + Aryl OH	→ R NHCO NH Aryl	50–75°C	Tertiary amine

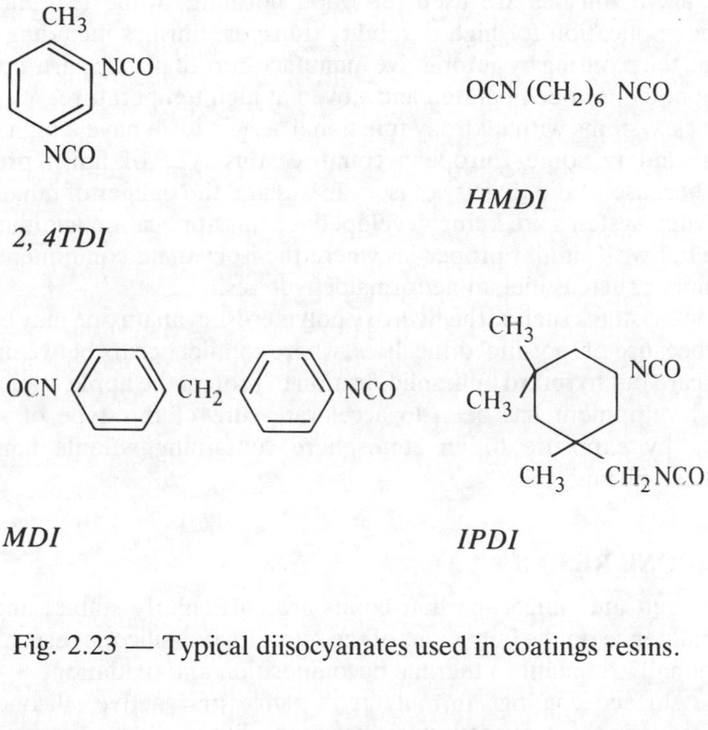
CH3
NCO
NCO
OCN (CH2)6 NCO
OCN
CH2
NCO
CH3
NCO
CH3
CH3
CH2NCO

2, 4TDI *HMDI* *MDI* *IPDI*

Fig. 2.23 — Typical diisocyanates used in coatings resins.

Owing to the toxicity of low molecular weight (volatile) isocyanates, polyfunctional (3+) isocyanate adducts of higher molecular weight are invariably used in two-pack applications of isocyanates. One frequently used isocyanate adduct is that from 3 mols of hexamethylene diisocyanate and 1 mol water, whose ideal biuret structure appears trifunctional:

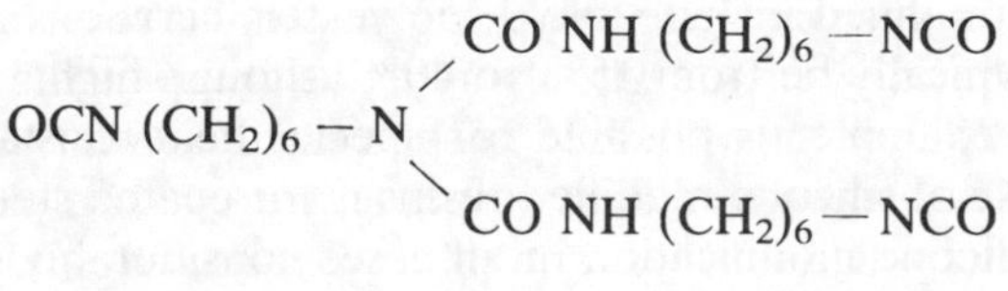
CO NH (CH2)6 — NCO
OCN (CH2)6 — N
CO NH (CH2)6 — NCO

Adducts derived from TDI and IPDI are also available for use in coatings formulations.

Moisture-curing products may also be used, based on stable isocyanate terminated polymers obtained from the reaction of excess low molecular weight di- or polyisocyanate with a polyfunctional hydroxyl terminated polyether. The pigmentation of moisture-curing finishes is difficult, though dehydrating agents are available to assist; the most frequent use of this type of product is for clear wood varnishes.

2.13.2 Polyurethane coatings

Two-pack finishes based on hydroxy functional resins and isocyanate adducts find application for tough high-solvent-resistant coatings curing at atmospheric temperature or under moderate stoving conditions. Resins which may be used are alkyd, polyester, polyether, epoxy, and acrylic. Alkyds used are frequently castor oil based. Two pack alkyd finishes are used for wood finishing, while two-pack polyester finishes find application for high-durability transport finishes including marine and aircraft, and the painting by automotive manufacturers of plastic parts fitted after the main metal body has been painted and stoved at high temperature.

Two-pack systems with hydroxy functional acrylic resin have a high usage in car refinishing, and in some European countries this type of finish predominates. However, because of the toxicity of isocyanates and the danger of inhaling droplets while spraying, systems are being developed for this application avoiding the use of isocyanate but with similar properties where the isocyanate component is replaced by a crosslinker such as melamine/formaldehyde resin.

Two-pack systems such as the hydroxy polymer/isocyanate one may be difficult to formulate because of 'pot-life' difficulties, where conflict occurs between the need to accelerate cure but to retard bulk solidification to protect the application equipment. A recent development has been to accelerate cure of this type of system after application, by exposure to an atmosphere containing volatile amine catalyst (vapour phase curing).

2.14 SILICONE RESINS

Silicone–oxygen and silicon–carbon bonds are particularly stable, and this has a beneficial influence on the behaviour of the semi-organic silicone resins, so that they are exceptionally resistant to thermal decomposition and oxidation.

For the surface coatings formulater, a range of reactive silicone resins are available for use in the preparation of silicone-modified polymers, and these may be either hydroxy or methoxy functional. Typical structures are shown in Fig. 2.24.

R may be phenyl or methyl, and both alkyl and aryl organo-silicones are currently available for resin modification. Reaction into a resin structure occurs through its available hydroxyl groups, when either water or methanol will be eliminated (Fig. 2.25); catalysis of these reactions is possible but not essential.

Polymers which may be modified include alkyd, polyester, acrylic, and epoxy. Silicone modification may typically be from 15% to 40%, though higher levels of modification of alkyds, for example, is possible for special heat-resistant applications. The single largest use of silicone-modified resins is for coated steel coil for building structure facing. Silicone modification in all cases considerably enhances durability (see 2.5, 2.6, and 2.16).

Pure silicone surface coatings are available for special application where very

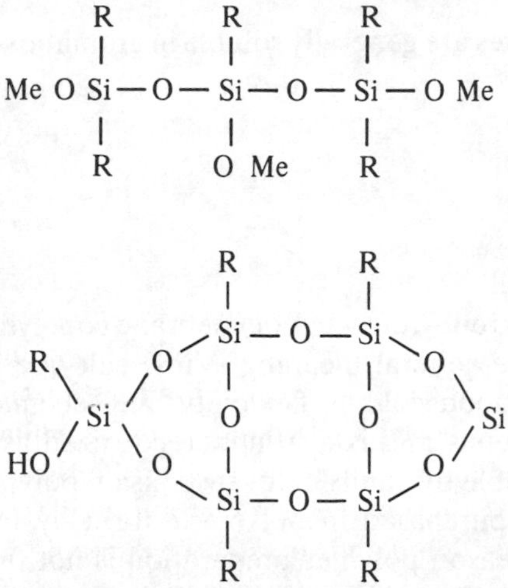

$$
\begin{array}{ccccc}
 & R & & R & & R \\
 & | & & | & & | \\
\text{Me O Si} & - & \text{O} - \text{Si} & - & \text{O} - \text{Si} & - \text{O Me} \\
 & | & & | & & | \\
 & R & & \text{O Me} & & R
\end{array}
$$

Fig. 2.24 — Typical silicone resins for polymer modification.

$$
\begin{array}{l}
| \qquad\qquad\qquad | \\
-\,\text{Si O H} + \text{HO Polymer} \rightarrow -\,\text{Si} - \text{O} - \text{Polymer} + \text{H}_2\text{O} \\
| \qquad\qquad\qquad\qquad\qquad\quad | \\[2mm]
| \qquad\qquad\qquad\qquad | \\
-\,\text{Si O Me} + \text{HO Polymer} \rightarrow -\,\text{Si} - \text{O} - \text{Polymer} + \text{Me OH} \\
| \qquad\qquad\qquad\qquad\qquad\qquad |
\end{array}
$$

Fig. 2.25 — Silicone resin reactions with hydroxy polymer.

high heat resistance properties are required. These resins cure by the same mechanisms as for modification, but in this case will be catalysed.

$$
\begin{array}{l}
| \qquad\qquad\quad | \qquad\qquad | \quad | \\
-\,\text{SiOH} + \text{HOSi} - \rightarrow -\,\text{SiOSi} - - + \text{H}_2\text{O} \\
| \qquad\qquad\quad | \qquad\qquad | \quad |
\end{array}
$$

Non-reactive linear silicones, e.g. poly dimethyl siloxanes of low viscosity, find frequent use in very low concentrations as flow control or marr aids in paint formulations.

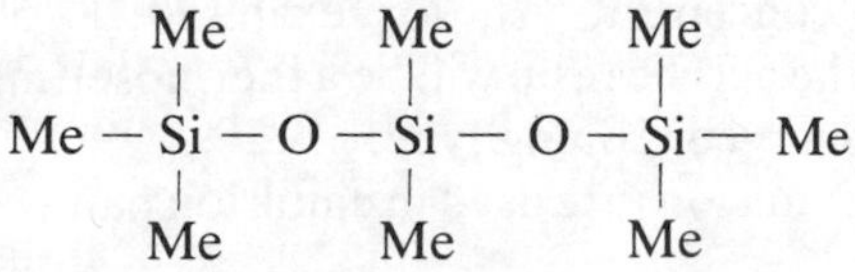

$$
\begin{array}{ccccc}
\text{Me} & & \text{Me} & & \text{Me} \\
| & & | & & | \\
\text{Me} - \text{Si} - \text{O} - \text{Si} & - & \text{O} - \text{Si} & - & \text{Me} \\
| & & | & & | \\
\text{Me} & & \text{Me} & & \text{Me}
\end{array}
$$

These particular silicones are generally soluble in aromatic solvents and in certain other 'strong' solvents.

2.15   VINYL RESINS

The term 'vinyl resin' commonly refers to polymers and copolymers of vinyl chloride, though the term has more general meaning. Vinyl chloride is a cheap monomer whose polymers possess good colour, flexibility, and chemical resistance. Vinyl chloride is, however, gaseous and is also now recognized as carcinogenic, hence though readily polymerizable by emulsion or suspension polymerization techniques in an autoclave, will be purchased in polymeric form by most surface coatings manufacturers. For this reason polymer preparation is not further discussed. The polymers are nevertheless quite widely used in coatings.

Both homopolymers, and copolymers usually with vinyl acetate, are used; small amounts of acidic monomer may also be present. A serious problem with vinyl chloride-containing polymers, if heated, is dehydrochlorination, and stabilization is often necessary with, for example, organo-tin compounds or group 2 metal carboxylates.

A major advantage of polyvinyl chloride (PVC) connected with its polar nature is its ability to accept plasticizers. Plastisols and organosols are dispersions of dried PVC particles in either plasticizer or mixed solvent/diluent. The basis of both of these types of coating compositon is that the particulate dispersion is stable until applied and heated, when the plasticizer or solvent swells the particles, softening them and allowing coalescence; this is resisted at room temperature because of the crystalline nature of the PVC particles. In the case of plastisols, plasticizer is retained in the coating after coalescence; with organosols the solvent and diluent are lost through evaporation.

The formulation of an organosol requires much care and a fine balance between diluent and solvent. If the mixture is too rich in diluent the particles flocculate and give a high-viscosity dispersion; if too rich in solvent the viscosity is too high, this time owing to solvent swelling of the particles.

Polyvinyl acetate/vinyl chloride copolymers are soluble in ketones or ketone/aromatic solvent blends, and this allows their use in lacquer formulations.

Vinyl resins are used in coatings for strip for venetian blinds and for bottle tops where extreme flexibility and extrusion properties are required. They are also used in heavy-duty and marine coatings where properties of toughness, elasticity, and water resistance are paramount, and have had application in coil coated strip for building facings. Polyvinylidene fluoride dispersion is now used in the highest durability coil coating finishes in blends with acrylic resin containing up to 80% vinyl polymer. Compositions containing lower concentrations, for example 30–50% of vinyl polymer, may also be used where the acrylic resin will be a thermosetting resin of, for example, the self-reacting acrylamide-containing type.

Polyvinyl chloride copolymer emulsions are used in emulsion paint compositions, and brief mention of formulations for this application has been made under emulsion polymerization.

2.16 WATER-BORNE SYSTEMS

Though water-borne systems have always found application, legislation in various countries has brought about increased use in order to reduce airborne pollution from solvents. Four main technologies compete to satisfy this constraint, the others being high solids, radiation-curing, and powder coatings. To reformulate a resin system to replace organic solvent by water requires an increase in the hydrophilicity of the polymer system either by the incorporation of water-soluble groups, or by the inclusion of surfactants, or both. The polymer may finally be either in solution or in dispersed form. Generally the use of water as the solvent or diluent increases the time of dry and can make humidity control of the booth necessary if spray-applied; because of the higher latent heat of water, if stoved, energy requirements may be greater. Unless very carefully formulated so that hydrophilic groups are destroyed or deactivated in cure, water sensitivity may be a problem in the final coating. Despite these problems the replacement of solvent by water can improve safety in application from the point of view of flammability and toxicity, and can reduce or eliminate environmental problems due to emissions on curing, and these benefits can outweigh the disadvantages. Electrodepositable coatings are a special class of water-borne systems.

Water-borne products prepared solely using surfactants are discussed in the section on emulsion polymerization; the following discussion is concerned with systems solubilized or dispersed with the aid of hydrophilic groups on the polymer backbone. In this case the polymer will normally be prepared in the absence of water, and only when at its terminal MW is it transferred into water. It will be recognized that polymers not containing soluble groups may be emulsified by added surfactant; this is not, however, common practice except in the case where resins are blended and one resin 'solubilizes' another by acting as emulsifier for all polymer species present; this may be done, for example, where the final cure requires the presence of an insoluble curing agent.

Polymers fully water-soluble without the addition of salt-forming additives exist; examples are polyethylene oxide or glycol, polyvinyl pyrrolidone, polyacrylamide, and copolymers containing a high proportion of these materials. Polyethylene glycol, by virtue of terminal hydroxyl groups, has a major use as a reactant in order to make alkyds water-soluble or dispersible; polyethylene oxide occurs as the water-soluble portion of many surfactants. Other water-soluble polymers, including polyvinyl alcohol, acrylic polymers with a high acid content, and modified celluloses, find application as colloid in emulsion polymerizations. Some formaldehyde condensate resins, including hexamethoxymethyl melamine and certain phenolics, are water-soluble, and hence are used in preference to other resins as crosslinking resins in water-borne systems [95].

The major route to water solubility/dispersibility, is by the preparation of polymer with acid or amine groups on their backbone, which may be solubilized by salt formation by the addition of a volatile amine or acid. When the resins dissolve in water, these salt groups ionize, the counter-ions being carried in the bulk solution. A particular advantage of the alkali or acid being volatile is that water sensitivity in the final film is reduced, owing to the loss of the salt-forming agent; this sensitivity can be further reduced or eliminated if the crosslinking reaction can then take place with these groups. As a general point it should be noted that many successful formulations are dispersions (colloidal or micellar), not solutions, simply because this can realize the highest solids and lowest viscosity. Resins may be prepared with a higher acid or amine content than necessary, the 'degree of neutralization' with salt-forming acid or amine then being about 40% to 70%, fine tuned to give the optimum viscosity, stability, freedom from settlement, and application properties.

The early water-borne vehicles were alkali neutralized maleic adducts, the simplest being maleinized oils. Maleinized fatty acids reacted with polymer backbones still play a considerable role in water-borne systems, and of epoxy resin esterified with maleinized linseed oil fatty acids alone or in combination with other fatty acids followed by neutralization, can provide vehicles for both normal spray or dip and for electrodeposited application. Polybutadiene may be maleinized and can then similarly form the basis of water-borne systems. Both oil and polybutadiene based systems may be cured by oxidation, usually stoved without the addition of further crosslinker. Water-borne alkyd and polyester systems have already been mentioned (2.5 and 2.6), specially formulated with a high acid value to give water solubility. Acidic acrylic systems are also preparable where, for example, maleic anhydride or acrylic acid are copolymerized with the other unsaturated monomers, and such thermosetting systems have already been described. Epoxy systems may be given acidic functionality by the ring-opening half ester reaction with dicarboxylic anhydrides, or trimellitic anhydride as described for alkyds and polyesters [96].

Water-borne resins may be self-crosslinking (alkyds, thermosetting acrylics) or may be crosslinked by added soluble or co-emulsified water-insoluble crosslinking resin. Fully soluble crosslinking resins may be blended into the system at any stage, and these include MF, UF, and phenolic resins, reacting as previously described for solvent-borne systems. A recent development also finding application is the β-hydroxyalkylamides, which are low-polluting crosslinkers specially applicable to carboxyl-containing water-borne systems. These are formed from the reaction of an alkyl ester (of a dicarboxylic acid) with an alkanolamine in the presence of basic catalyst (Fig. 2.26); a typical example would be prepared from the reaction of dimethyl adipate and di hydroxy propanolamine [97]. Their curing ability is comparable to amino resins, but with crosslinking being by ester formation with carboxyl groups (Fig. 2.26).

Epoxy resins can be formulated to be amine functional by reaction with secondary amines which will then be solubilized with, for example, acetic acid [96] (Fig. 2.27). Added Crosslinking resin must be added to these systems, for example MF, phenolic, or blocked isocyanate. Water-borne epoxy resins find particular application for metal primers and for can coatings [98], and for the latter usage water-reducible acrylic grafted epoxide resins find extensive application [99].

Specially formulated acrylic emulsions are also used in thermosetting water-borne systems, and these are prepared in the manner described in the section on emulsion polymerization. The monomers will include hydroxy functional monomer,

$$
\begin{array}{ccc}
& Me & & Me \\
& | & & | \\
R\ COO\ Me\ +\ NH\ CH_2\ CHOH & \rightarrow & R\ CO\ N\ CHOH\ +\ MeOH \\
| & & | \\
R'' & & R''
\end{array}
$$

Formation

$$
\begin{array}{ccc}
Me & & Me \\
| & & | \\
R\ CO\ N\ CHOH + HOOC-Polymer \rightarrow R\ CO\ N\ CH\ OOC-Polymer + H_2O \\
| & & | \\
R'' & & R''
\end{array}
$$

Cure

Fig. 2.26 — Formation and curing reactions of simple β hydroxyalkylamide.

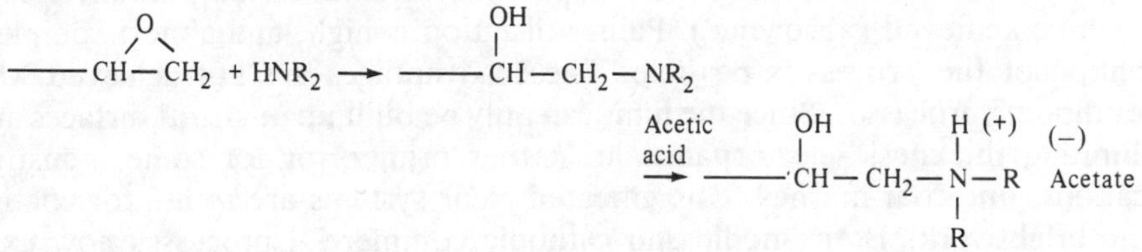

$$
\begin{array}{ccc}
& O & & OH \\
& / \backslash & & | \\
-\!\!-\!\!-CH - CH_2\ +\ HNR_2 & \longrightarrow & -\!\!-\!\!-CH - CH_2 - NR_2
\end{array}
$$

$$
\begin{array}{c}
\text{Acetic} \quad OH \qquad H\ (+) \\
\text{acid} \qquad | \qquad\quad | \qquad (-) \\
\longrightarrow \quad -\!\!-\!\!-CH - CH_2 - N - R \quad \text{Acetate} \\
| \\
R
\end{array}
$$

Fig. 2.27 — Epoxy resin/amine reaction and subsequent water solubilization.

and the preparation will generally be carried out without the inclusion of colloid, only surfactant being present. These are usually referred to as 'colloid free' latexes.

Water-soluble resins may be silicone modified, and the advantages of this have been demonstrated for alkyds and polyesters where siliconization is carried out by pre-reacting the silicone intermediate with polyol. For acrylic latex products siliconization of the latex may be carried out as a subsequent step to the emulsion polymerization [33].

Certain problems are found specific to water-borne products, and require additional care in formulation. Condensation products can be prone to hydrolysis. Curing may be adversely affected by the solubilizing agent either by pH effects or, in the case of air-drying alkyds, by neutralizing amine retarding oxidative crosslinking [100]. The presence of amines can also adversely affect colour retention.

Most water-soluble compositions contain a proportion of water-miscible co-solvent [101] which may have been a necessary component in the resin preparation state prior to water addition. However, its presence may also allow the paint formulator greater latitude to control properties such as stability, drying and rheology, and still meet any required VOC (volatile organic content) levels required by pollution legislation. Disperse compositions may contain organic solvent to aid coalescence. The presence of water-miscible solvents may also aid emulsification in disperse systems as is generally recognised with so-called 'microemulsions'.

2.17 RESINS FOR ELECTRODEPOSITION

Electrodepositable resins are a special class of water-borne resin [102]. The polymer is carried in an aqueous medium, and on application of current via suitable electrodes, polymer in the vicinity of one of the electrodes within the electrical boundary layer is destabilized and deposited on the electrode. The deposited polymer builds up to form an insulating layer which ultimately limits further deposition. The electrodeposition process has been called electrophoretic deposition, though it is now recognized that electrophororesis plays little part in the process.

In anodic electrodeposition negatively charged polymer is deposited on the anode, and in cathodic electrodeposition positively charged polymer is deposited on the cathode. Compared with conventional painting, very uniform coverage of external surfaces can be obtained, and deposition of paint inside partially closed areas can be achieved ('throwing'). Paint utilization is high, and almost complete automation of the process is possible. These advantages are not achieved with simpler dipping processes. Since the films can only be built up on metal surfaces and to a limiting thickness, electropaints are either primers or for some industrial applications, one-coat finishes. Unpigmented clear systems are in use for coating metallic brightwork. Both anodic and cathodic commercial processes now exist [103, 104, 105] with the more recent cathodic systems now surplanting anodic systems, particularly for automotive applications.

Almost all polymer types described earlier under water-borne coatings can be adapted to electrodeposition. In most cases, to be suitable for the electrodeposition process the resin will be held in stable particulate or micellar dispersion by the action of hydrophilic ionic groups which will provide colloidal stabilization. Non-ionically sterically stabilized dispersions, however, may also be electrodeposited [106].

The system should be designed such that the deposited film has high electrical resistance so that shielded areas can receive adequate coverage. The system may contain organic solvent to aid the dispersion process and dispersion stability, and act as a flow promoter during coating and curing. The role of the neutralizing acid or base is fundamental both to stability of the dispersion and to the electrocoating process. In practice, for anionic systems either alkali or amines may be used, and for cathodic systems lactic and acetic acids are usual choices for neutralizing agent.

The earliest anodic electrodeposition vehicles were based on maleinized oils and oil derivatives, and the chemistry of these developed through vinylated and alkyd type condensates to the use of epoxy esters based on maleinized fatty acids. Since epoxy resin-based systems exhibit such good performance as metal primers, solubilized epoxy vehicles play a major role in both anodic and cathodic systems, especially for automotive applications.

Alkyd and acrylic systems have been developed for electrodepositing and have found application in industrial systems. Anodic alkyd systems are based on high acid value resins, particularly trimellitic anhydride derived and may be drying oil modified. They may hence be autoxidative/stoving, or if non-drying oil-containing or oil-free, cured by co-emulsified or soluble melamine/formaldehyde resin. Acrylic systems have been proposed for both anodic and cathodic formulations, in the former case by inclusion of higher than normal concentrations, of, for example, acrylic acid [67, 107], or in the latter case by the inclusion of copolymerized amino monomer, such as dimethylaminoethylmethacrylate [108], or by including a glycidyl monomer which can be subsequently reacted with an amine [109].

Cathodic epoxy systems may be either primary or secondary amino functional. Reacting epoxy resins with amines or quaternary amine salts can be used to give adducts with terminal secondary, tertiary, or quaternary amines or their salts. The relatively weak nature of these amines can result in poor dispersibility of these systems, and it is more desirable to incorporate primary amine groups. This can be difficult to achieve. However, useful methods have been developed; these include reacting excesses of di-primary amine with epoxy resin [110], or blocking primary amine groups on molecules with ketones, before reaction with epoxides through other functionalities [111]. These are illustrated in Figs 2.28 and 2.29.

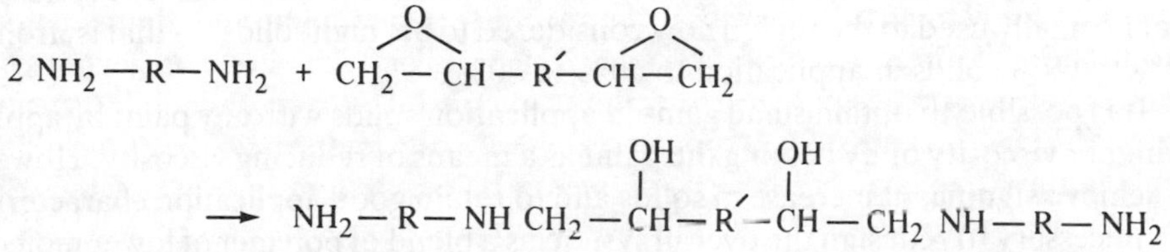

$$2\ NH_2-R-NH_2\ +\ CH_2-CH-R'-CH-CH_2$$

$$\longrightarrow\ NH_2-R-NH\ CH_2-CH-R'-CH-CH_2\ NH-R-NH_2$$

Fig. 2.28 — Reaction of epoxy resin with excess diamine.

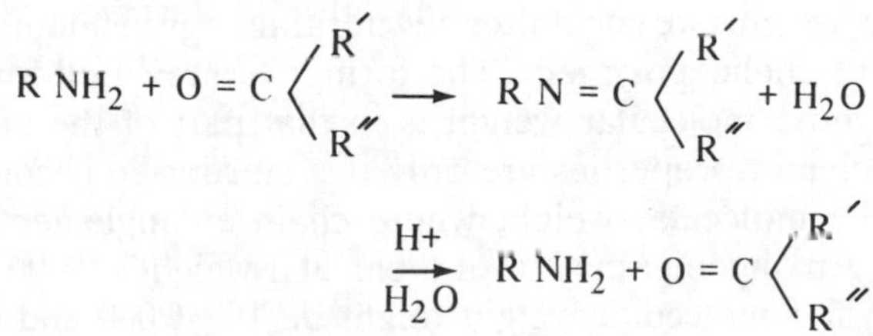

$$R\ NH_2 + O = C \bigg\langle {{R'}\atop{R''}}\ \longrightarrow\ R\ N = C \bigg\langle {{R'}\atop{R''}}\ +\ H_2O$$

$$\xrightarrow[H_2O]{H+}\ R\ NH_2 + O = C \bigg\langle {{R'}\atop{R''}}$$

Fig. 2.29 — Ketimine formation and hydrolysis.

Because of their extremely high film alkalinity these systems crosslink sluggishly with melamine and phenolic crosslinkers, but can be effectively crosslinked with blocked isocyanates, which are designed to be stable at bath temperatures but which unblock at reasonable stoving temperatures [112, 113]. Acrylic cationic systems may also be cured with these curing agents for one-coat systems where good colour requirements are more important than corrosion resistance.

2.18 HIGH SOLIDS COATINGS

An anti-pollution approach which reduces the actual organic solvent content of the coating system is to formulate to higher solids. Economy in application may be gained in that less stoving energy goes into evaporation of solvent; however, the approach may not necessarily lead to overall economy if more expensive materials are needed to formulate to higher solids, as can indeed be the case.

A number of approaches to achieving high solids are possible, and powder coatings are of course one kind of 100% solids system. Reactive diluents (e.g. unsaturated monomers) may be used in place of solvent, as is the practice with unsaturated polyesters and with radiation-curing polymers. The addition of reactive diluent to autoxidatively drying alkyds in place of solvent is also possible where the diluent is again a vinyl monomer, which in this case copolymerizes into the structure through the free radical nature of the autoxidative drying process [9]. As well as the more usual polyfunctional monomers, newer materials are becoming available including unsaturated melamine resins specially formulated as alkyd diluents.

What are most usually referred to as high solids systems are formulations previously described where design criteria have been revised to raise the solids from that normally used to the region now considered to be 'high solids' — that is, from say 40% to 60% solids at application to 60% to 80%.

It is possible to obtain small gains in application solids with any paint by applying at higher viscosity or by heating the paint as a means of reducing viscosity. However, to achieve significant increase in solids and to retain good application characteristics it is necessary to redesign the overall system as a blend of polymer of lower molecular weight than normal with diluent, the result of lower molecular weight being that less diluent is required to achieve application viscosity. To design true high solids systems it has been necessary to devise new techniques for formulating polymers. Low molecular weight demands crosslinking systems in order to achieve final film properties, and lowered molecular weight has the consequence that special care is necessary with the concentration of reactive functional groups and their distribution.

With high solids systems we encounter new terminology including the concepts of 'oligomeric' and 'telechelic' polymers. The former term is used for low molecular weight polymers whose molecular weight is in that part of the molecular weight spectrum where polymer properties are only just starting to become apparent; in particular below the molecular weight where chain entanglement is significantly affecting viscosity. An oligomer may be of molecular weight 1–5000 where a normal thermosetting polymer molecular weight might be 10–40 000 and a thermoplastic polymer 80–100 000 [114]. Telechelic refers to polymer molecules possessing two functional terminal groups and the synthesis of these for high solids systems has been described [115].

To achieve low viscosity/high solids, good understanding is necessary of the relation between molecular weight and viscosity for the paint components and for the interaction between them (oligomer, crosslinker, solvent, *and* pigment). The theoretical basis for this is becoming well developed [116], though it is acknowledged that further research is needed. Molecular weight and molecular weight distribution both require careful control.

In conventional coatings relatively few functional groups need react to yield crosslinked films. In high solids systems a substantially larger number of groups must react to reach the same final molecular weight. This increase in functional group content does, however, tend to increase viscosity since these groups are generally polar, this in part offsetting the viscosity decrease derived from molecular weight reduction. The formulator must nevertheless ensure that each oligomer molecule has at least two functional groups attached.

Both oligomeric acrylic and polyester resin formulations are available for high solids coatings. In the case of acrylic resins additional problems are experienced in preparation over normal formulations. High chain transfer agent concentrations may be necessary, leading to residual odour problems, difficult to tolerate for some applications, The use of high solids in preparation which may be desirable or indeed essential, results in high chain transfer to polymer, leading to undesirably broad molecular weight distribution and consequent adverse effects on rheology. Polyester formulation is relatively simpler, using the same ingredients as for higher molecular weight polyester, and useful formulations have been disclosed. It is also relatively simple with polyesters to ensure difunctionality. Alkyd formulations of novel and controlled structure have also been proposed [17] using the properties of Cardura E (Shell Chemicals). These formulations for acrylic and polyester are hydroxy functional, and hexamethoxymethyl melamine resins have found the most usual application as crosslinkers; the reasons for their suitablity have been discussed [117]. With corresponding higher concentrations of functional groups in the oligomer part of the system, amino resin may constitute as much as 50% or more of the composition, compared with 10–40% in more conventional finishes.

Other systems may be formulated in the high solids region; for example, polyester/isocyanate adduct systems, and for thesc the same design criteria apply to the polyester, as described above [118]. Equally, epoxy/polyamine systems may be considered high solids when low molecular weight liquid di-epoxides are used; here possibilities existing of using monoepoxy compounds as reactive diluents.

2.19 RADIATION-CURING POLYMERS

Radiation cure includes electron beam cure (EBC), ultraviolet (UV), and infrared (IR), and all thcse methods are practised commercially. Reduced energy consumption, improved environment protection, and, in particular, suitability for wood, paper, and plastic substrates have contributed to recent growth. For UV cure photoinitiators are required, and for IR cure thermal initiators, in both cases normally to initiate chain addition polymeiization. For EBC no added initiator is required, the electron beam producing radicals in the polymerizing layer through its own high energy. The systems employed are normally high solids, and typically the resin system is a blend of prepolymer and reactive diluent. Unsaturated polyesters are typical prepolymers and are used in combination with suitable vinyl or acrylic monomer diluents. The systems and mechanisms for UV, IR, and EBC have been reviewed [119, 120].

In addition to the unsaturated polyesters described earlier in this chapter, epoxy and polyurethane resins are also used. Epoxy acrylates are prepared by the reaction of bisphenol epoxy resins with acrylic acid. Polyurethane resins for photocuring are obtained by the complete reaction of di-functional isocyanates or polyfunctional isocyanate adducts with hydroxy ethyl acrylate.

There is also some use made of polyether and polyester acrylates, prepared by the esterification of hydroxy functional polyether or polyester with acrylic acid. In the preparation of all acrylate resins careful control of inhibitor levels is necessary to prevent polymerization, and final dilution with monomer may be necessary to retain fluidity. A major advantage of this type of prepolymer is low viscosity.

Monomers are necessary as reactive diluents especially for viscosity control, and apart from their viscosity and solvating ability other factors needing consideration include their volatility and photo-response. Crosslinking mechanisms are intricate, and for any resin/monomer system there will be an optimum resin/monomer ratio for a given dosage of radiation to attain the required film characteristics. Acrylates give the fastest cure speed, and polyfunctionality is normally necessary; hence such monomers as hexanediol diacrylate and trimethylol propane triacrylate find considerable application. A broad range of such multifunctional monomers are now available commercially. The order of reactivity for unsaturated groups in radiation-curing systems is acrylate>methacrylate>allyl>vinyl.

UV/visible photo-initiators for unsaturated systems will be free radical generators, for example aromatic ketone; ketone/amine combinations are also used [121]. Curing reactions are those described earlier for acrylic polymerization.

The cationic cure of epoxy resins using photo-initiators which release Brönstead acids on irradiation is a modern development in this field; systems using this method are based mainly on cycloaliphatic epoxy resins and aliphatic diepoxides; bisphenol epoxides degrade the curing efficiency of this system.

2.20   POWDER-COATING COMPOSITIONS

Powder coatings possess advantages over conventional coatings in that no polluting solvent loss occurs on application; and, owing to the use of electrostatic spray, little material is lost. Electrostatic spraying is also employed with solvent-borne coatings and is a method particularly suited to automation. The formulation of powder-coating resin compositions [122] presents special limitations due to the solvent-free nature of the product; most powders are produced by comminuting an extruded melt which means that the final curing composition has to withstand this melting process. (Pigmentation is also achieved in the course of the extrusion.) The ingredients, and particularly the final composition, have to be solid and glassy at room temperature in order to grind and for the powder to stay free-flowing on storage, and this requires that the blend must not soften or sinter below 40°C. In formulation and production additional factors such as particle size control, melt flow characteristics, and resistivity need careful attention.

Epoxy compositions are widely used and are based on bisphenol resins cured with amines (dicyandiamide derivatives in particular) or phenolic crosslinkers. Optionally, saturated polyester may be mixed with the epoxy. Dimerized fatty acid modification may also be carried out to improve film levelling and adhesion, and epoxy resins containing branded species derived from epoxidized novolacs may be introduced to increase hardness. Dicarboxylic acid anhydrides and acidic polyesters are also used to cure epoxy powders, and these compositions have better performance than amine cured epoxies, being non-yellowing.

Polyester powder coatings may be hydroxy functional and use melamine/formaldehyde or masked isocyanate curing agents [94], or may be formulated from acidic polyesters cured with added epoxy resin or triglycidyl isocyanurate; the latter offers excellent durability and colour stability. Caprolactam is used as a blocking agent for masked isocyanates for curing powders, though release of this agent from the curing film may be troublesome. Typical curing agents for powder coatings are shown in Fig. 2.30.

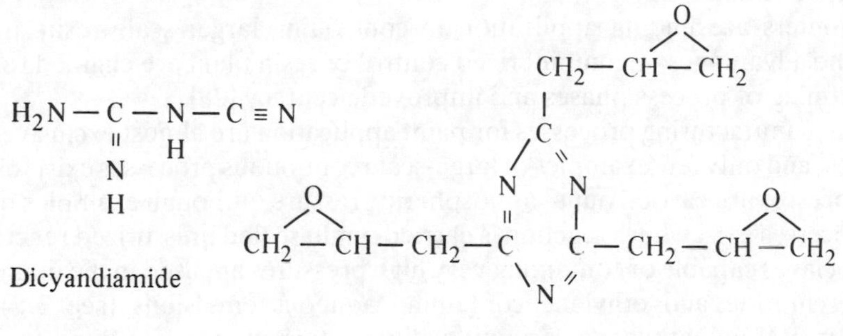

$$H_2N - \underset{\underset{H}{\overset{\|}{\underset{N}{N}}}}{\overset{}{C}} - \underset{H}{\overset{}{N}} - C \equiv N$$

Dicyandiamide

Tri glycidyl isocyanurate

Fig. 2.30 — Typical curing agents for powder coatings.

Acrylic systems are available but are only moderate in performance, and they have shelf-life problems.

Thermoplastic powders are also used containing, for example, nylon 11 or nylon 12, polyvinyl chloride, cellulose acetate and butyrate, or polyethylene. Redispersed powders may be applied from water as the so-called slurry coatings or aqueous powder suspensions, and may also be electrodeposited (electropowder coatings).

2.21 RESIN MANUFACTURE

In the preceding sections of this chapter, the manufacturing methods normally used for each resin type have been indicated. This brief section summarizes these, highlights certain general points, and indicates future trends.

Resin processes divide mainly between high temperature (over 180°C) stepwise polymerization processes and lower temperature (80–150°C) chain addition polymerization processes. The former are mildly endothermic, whereas the latter are generally more fiercely exothermic. Most resins are made in melt or solution and are finally used in liquid solution form. Many disperse products are now also being produced, either made by emulsion or dispersion polymerization, or by post-emulsification of prepared polymer.

Resin manufacturing plant has moved from the simple portable stirred heated pot, satisfactory only for oleoresinous vehicles and chain addition polymers such as alkyds, to the almost universal use of fixed plant. The reactor of this plant will have efficient agitation, be fitted with means of heating and cooling with sophisticated control, be equipped with vapour-condensing systems capable of azeotropic distillation, and will normally be attached to a thinning and cooling vessel. The plant may be multipurpose, but it is more usual to segregate high temperature, low temperature, and aqueous emulsion resins to different plants, one reason being the need of the latter two types of product for feed vessels for monomers. Emulsion reactors, because of the inherent sensitivity of emulsions to shear, often need more carefully designed agitation systems and require other special facilities such as aqueous feed preparation vessels.

Computers are finding application in controlling larger scale resin and paint plant. The advantages of computerized control of resin plant are claimed to include stricter timing of process phases and improved accuracy [28].

Resin manufacturing processes for paint application are almost exclusively batch processes, and only few examples of large-scale continuous processes exist [69]. Most resin processes are carried out at atmospheric pressure, although examples of acrylic resin processes exist where reaction is carried out in sealed pressurized reactors [50], and autoclaves capable of containing very high pressures are used in the manufacture of vinyl chloride and ethylene containing aqueous emulsions (see 2.8). Resin reactors may frequently have vacuum facilities attached, both for fume control and for use in vacuum distillation.

BIBLIOGRAPHY

Margerison, D. & East, G. C., *Introduction to polymer chemistry*, Pergamon Press, 1976.
O.C.C.A. Australia, *Surface coatings*, Volume 1, Raw Materials and their Usage, Chapman & Hall, 1983.
Solomon, D. H., *The chemistry of organic film formers*, Robert E. Krieger Publishing, 1977.
Paul, S., *Surface coatings, science and technology*, Wiley, 1985.

REFERENCES

[1] Ennor, K. S. & Oxley, J., *J. Oil Col. Chem. Assoc.* **50** 577 (1967).
[2] Lower, E. S. *Manufacturing Chemist*, **52** (12) 50 (1981).
[3] Bayer, A. G., German Patent 2,742,584 (1979).
[4] Pacific Vegetable Oil Corporation, United States Patent 3,278,567 (1966).
[5] American Cyanamid, United States Patent 2,350,583 (1944); The Maytag Company, United States Patent 2,575,529 (1951).
[6] Wexler, H., *Chemical Reviews* **64** 591 (1964).
[7] Khan, N. A. *Can. J. Chem.* **37** 1029 (1959).
[8] Barrett, K. E. J. & Lambourne, R., *J. Oil Col. Chem. Assoc.* **49** 443 (1966).
[9] Larson, D. B. & Emmons, W. D., *J. Coatings Technol.* **55** (702) 49 (1983); Kuzma, E. J. & Levine, E., *J. Coatings Technol.* **56** (710) 45 (1984).
[10] Love, D. J., *Polymers, Paint, Colour J.* **175** 436 (1985).

[11] Kraft, W. M., *American Paint J.* **41** (28) 96 (1957).

[12] Heydon Newport Chemical Corporation, United States Patent 2,973,331 (1961).

[13] Rooney, J. F., *Official Digest* **36** (475 Part 2) 32 (1964).

[14] Shell Chemicals, *Cardura E*10, Technical Manual CA 1.1.

[15] Herzberg, S., *J. Paint Technol.* **41** 222 (1969).

[16] Shell Chemicals, *The manufacture of 'Cardura' resins*. Technical Manual CA 2.1; Shell Chemicals, *A review of technical information of Cardura E*10, Technical Note TN-R.83.01.

[17] Holmberg, K. & Johansson, J., *Proc. 8th International Conference Organic Coatings Science and Technology*, Athens, 255 (1982).

[18] Baker, A. S., *Paint and Resin* **52** (2) 37 (1982); Imperial Chemical Industries, British Patent 1,370,914 (1972).

[19] Bobalek, E. G., Moore, E. R., Levy, S. S., & Lee C. C., *J. App. Polym. Sci.* **8** 625 (1964).

[20] Imperial Chemical Industries, British Patent, 1,242,054 (1971).

[21] Imperial Chemical Industries, British Patent, 1,594,123 (1978).

[22] Pond, P. S. & Monk C. J. H., *J. Oil Col. Chem. Assoc.* **53** 876 (1970).

[23] Tysall, L. A., *Calculation techniques in the formulation of alkyd and related resins*, Paint Research Association, 1982.

[24] Patton, T. C. *Alkyd resin technology (formulating techniques and allied calculations)*, Interscience, 1962.

[25] Bernardo, J. J. & Bruins, P., *J. Paint Technol.* **40** 558 (1968).

[26] Weiderhorn, N. M., *American Paint J.* **41** (2) 106 (1956).

[27] Nelen, P. J. C., *17th FATIPEC Congress* 283 (1984).

[28] Anon., *Polymers Paint Colour J.* **173** 786 (1983).

[29] Shackleford, W. E. & Glaser, D. W., *J. Paint Technol.* **38** 293 (1966).

[30] Turpin, E. T., *J. Paint Technol.* **47** (602) 40 (1975).

[31] AMOCO Chemicals Corporation, *How to process better coating resins with AMOCO IPA and TMA*.

[32] AMOCO Chemicals Corporation, *Processing unsaturated polyesters based on AMOCO isophthalic acid (IPA)*.

[33] Imperial Chemical Industries, United States Patent, 3,050,533 (1962).

[34] Fradet, A. & Maréchal, E., *Advances in Polymer Science* **43** 51 (1982).

[35] Price, J. G., *Proc. 3rd International Conference Organic Coatings Science and Technology*, Athens 464 (1977).

[36] Price, J. G., *Silicone resins as modifiers for waterborne coatings*, Dow Corning, England, 1978.

[37] California Research Corp., Unites States Patent 2,904,533 (1959).

[38] Standard Oil, United States Patent, 3,252,941.

[39] Imperial Chemical Industries, British Patent, 784,611 (1957).

[40] Farbenfabriken Bayer, British Patent 810,222 (1956).

[41] Brown, W. H. & Miranda T. J., *Official Digest* **36** (475 Part 2), 92 (1964).

[42] Klein, D. H., *J. Paint Technol.* **42**, 335 (1970).

[43] O'Driscoll, K. F., *Pure and Applied Chemistry* **53** 617 (1981).

[44] Scott, G. E. & Senogles, E., *J. Macromol. Sci. — Revs. Macromol. Chem.* **C9** 49 (1973).

[45] Ham, G. E., *Copolymerisation*, High Polymers *XVIII*, Interscience, 1964.
[46] Wittmer, P., *Makromol. Chem.*, Suppl. 3, 129 (1979).
[47] Greenley, R. Z., *J. Macromol. Sci — Chem.* **A14** 455 (1980).
[48] Greenley, R. Z., *J. Macromol. Sci — Chem.* **A14** 427 (1980).
[49] DuPont de Nemours E.I., British Patent 763,158 (1956).
[50] DuPont de Nemours E.I., British Patent 807,895 (1959).
[51] Zimmt, W. S., *Ind. Eng. Chem. Prod. Res. Dev.* **18** 91 (1979).
[52] Taylor, J. R. & Price, T. I., *J. Oil Col. Chem. Assoc.* **50** 139 (1967).
[53] Christenson, R. M. & Hart, D. P., *Official Digest* **33** 684 (1961).
[54] Vogel, H. A. & Bittle H. G., *Official Digest* **33** 699 (1961).
[55] PPG Industries, British Patent 1,556,025 (1975).
[56] PPG Industries, British Patent 1,559,284 (1976).
[57] Murdock, J. D. & Segall, G. M., *Official Digest* **33** 709 (1961).
[58] Yanagihara, T., *Progress in Organic Coatings* **11** 205 (1983).
[59] Thompson, M. W., In: *Polymer colloids*, Buscall, R., Corner, T., & Stageman, J. F. (eds), Elsevier Applied Science, 1985.
[60] Imperial Chemical Industries/Dulux Australia, United States Patent 4,322,328 (1982); Bromley, C.W.A., Colloids & Surfaces **17(1)** (1986).
[61] Blackley, D. C., *Emulsion polymerisation*, Applied Science, 1975.
[62] Llewellyn, I. & Pearce, M. F., *J. Oil Col. Chem. Assoc.* **49** 1032 (1966).
[63] Bondy, C., *J. Oil Col. Chem. Assoc.* **49** 1045 (1966).
[64] Various papers in *Science and technology of polymer colloids*, Volume 1 and 2, Poehlein, G. W., Ottewill, R. H., & Goodwin J. W. (eds) Martinus Nijhoff, 1983.
[65] Imperial Chemical Industries, United States Patent 4,403,003 (1983).
[66] Backhouse, A. J., *J. Coatings Technol.* **54** (693) 83 (1982).
[67] Westrenen, W. J. van & Nieuwenhuis, W. H. M., *J. Oil Col. Chem. Assoc.* **54** 747 (1971).
[68] Gordon, P. G., Davies, M. A. S., & Waters, J. A., *J. Oil Col. Chem. Assoc.* **67** 197 (1984).
[69] Geddes, K., *Chemistry and Industry*, 21 March 1983, 223.
[70] Hoechst, British Patent 1,541,891 (1976).
[71] Rohm and Haas, British Patent 1,072,894 (1967).
[72] Walbridge, D. J., In: *Science and technology of polymer colloids* Volume 1, Poehlein, G. W., Ottewill, R. H., & Goodwin J. W. (eds), Martinus Nijhoff, 1983.
[73] Walbridge, D. J., In: *Dispersion polymerisation in organic media*, Barrett, K. E. J. (ed.), Wiley, 1975.
[74] Waite, F. A., *J. Oil Col. Chem. Assoc.* **54** 342 (1971).
[75] Rempp, P. F. & Franta, E., *Advances in polymer Science* **58** 1 (1984).
[76] Barrett, K. E. J. & Thompson, M. W., In: *Dispersion polymerisation in organic media*, Barrett, K. E. J. (ed.), Wiley, 1975.
[77] Imperial Chemical Industries, British Patent 1,157,630 (1969).
[78] Parekh, G. G., *J. Coatings Technol.* **51** (658) 101 (1979).
[79] Zuylen, J. van., *J. Oil Col. Chem. Assoc.* **52** 861 (1969).
[80] Koral, J. N. & Petropoulos, J. C., *J. Paint Technol.* **38** 600 (1966).
[81] Bauer, D. R. & Dickie, R. A., J. Poly. Sci. Poly. Phys. Ed. **18** 1997 (1980).

[82] Bauer, D. R. & Dickie, R. A., J. Poly. Sci. Poly. Phys. Ed. **18** 2015 (1980).

[83] Holmberg, K., Proc. 10th International Conference Organic Coatings Coatings Science and Technology, Athens, 71 (1984).

[84] Fischer, R. F., J. Poly. Sci. **44** 155 (1960).

[85] Shechter, L. & Wynstra, J., Ind. and Eng. Chem. **48** 86 (1956); Shechter, L., Wynstra, J., & Kurkjy, R. P., Ind. and Eng. Chem. **48** 94 (1956).

[86] Lee, H. & Neville, K., Handbook of epoxy resins, McGraw-Hill Book Company, 1967; Shell Chemical Company, Epikote resins for paint, RES:64:4.

[87] Somerville, G. R. & Parry M. L., J. Paint Technol. **42** 42 (1970).

[88] Belanter, W. L. & Schulte, S. A., Modern Plastics, Nov. 1959, 154; Enikolo-pyan, N. S., Markevitch, M. A., Sakhunenko, L. S., Rogovina, S. Z., & Oshmyan, V. G., J. Poly. Sci. Poly. Chem. **20** 1231 (1982).

[89] Moss, N. S., Polymers Paint Colour J. **174** 265 (1984).

[90] Saunders, J. M., Rubber Chem. and Technol. **32** 337 (1959).

[91] Frisch, K. C. & Rumao, L. P., J. Macromol. Sci.-Revs. Macromol. Chem. **C5** 103 (1970).

[92] Wicks, Z. W., Progress in Organic Coatings **3** 73 (1975).

[93] Kordomenos, P. I., Dervan, A. H., & Kresta, J., J. Coatings Technol. **54** (687) 43 (1982).

[94] McBride, P., J. Oil Col. Chem. Assoc. **65** 257 (1982).

[95] Blank, W. J. & Hensley, W. L., J. Paint Technol. **46** (593) 46 (1974).

[96] Krishnamurti, K., Progress in Organic Coatings **11** 167 (1983).

[97] Lomax, J. & Swift, G., J. Coatings Technol. **50** (643) 49 (1978).

[98] Moss, N. S. & Demmer, C. G., J. Oil Col. Chem. Assoc. **65** 249 (1982).

[99] Robinson, P. V., J. Coatings Technol. **53** (674) 23 (1981).

[100] Hill, L. W. & Wicks, Z. W., Progress in Organic Coatings **8** 161 (1980).

[101] Grant, P. M., J. Coatings Technol. **53** (677) 33 (1981).

[102] Schenck, H. U., Spoor, H., & Marx, M., Progress in Organic Coatings **7** 1 (1979).

[103] Cook, B. A., Ness, N. M., & Palluel, A. L. L. In: Industrial electrochemical processes, Khun, A. T. (ed.), Elsevier, 1971.

[104] Wismer, M., Pierce, R. E., Bosso, J. F., Christerson, R. M., Jerabek, R. D., & Zwack, R. R., J. Coatings Tech. **54** (688) 35 (1982).

[105] Kordomenos, P. I. & Nordstrom, J. D., J. Coatings Technol **54** (686) 33 (1982).

[106] Imperial Chemical Industries, European Patent 15655; Imperial Chemical Industries, European Patent 109760; Doroszkowski, A., Colloids & Surfaces **17** (1), 13 (1986).

[107] PPG Industries, United States Patent 3,947,339 (1976).

[108] BASF., United States Patent 3,455,806 (1969).

[109] SCM., United States Patent 3,975,251 (1976).

[110] Imperial Chemical Industries, British Patent 2,050,381 (1981).

[111] PPG Industries, United States Patent 4,017,438 (1973).

[112] PPG Industries, United States Patent 3,799,854 (1974).

[113] PPG Industries, United States Patent 4,101,486 (1978).

[114] Gibson, D. & Leary, B., J. Coatings Technol. **49** 53 (1977); Noren, G. K., Polymer News **10** 39 (1984).

[115] Athey, R. D. *Progress in Organic Coatings* **7** 289 (1979).

[116] Hill, L. W. & Wicks, Z. W., *Progress in Organic Coatings* **10** 55 (1982).

[117] Blank, W. J., *J. Coatings Technol.* **54** (687) 26 (1982).

[118] Potter, T. A., Schmelzer, H. G., & Baker, R. D., *Progress in Organic Coatings* **12** 321 (1984).

[119] Roffey, C. G., *Photopolymerisation of surface coatings*, Wiley, 1962; O'Hara, K., *Polymers Paint Colour J.* **175** 254 (1985).

[120] Dowbenko, R., Friedlander, C., Gruber, G., Pruenal, P., & Wismer, M., *Progress in Organic Coatings* **11** 71 (1983).

[121] Green, P. N., *Polymers Paint Colour J.* **175** 246 (1985); Hageman, H. J., *Progress in Organic Coatings* **13** 123 (1985).

[122] Harris, S. T., *The technology of powder coatings*, Portcullis Press, 1976.

3

Pigments for paint

Miss J. F. Rolinson

3.1 INTRODUCTION

This chapter about pigments for paint is designed for a newcomer to the paint industry. The subject matter is condensed and provides a simple background to enable the aspiring paint technologist to gain a basic understanding of the role of pigments in paint. More detailed information is available in suppliers' literature and in the many books and articles already published on the subject.

Pigments are first defined and their required qualities are then listed. Their classification is described and the terminology used in the paint industry explained. The methods of manufacture of pigments are mentioned and reference is made to concerns about toxicity and the environment, and the effect of these on the paint industry. The properties of pigments are then enumerated and some basic rules are proposed for use in the selection of pigments for paint.

Lastly, some of the more important pigments are briefly described according to chemical types, being grouped together in colour and special families — White, Black, Brown, Green, Yellow, Orange, Red, Violet, and Special Effects — Aluminium and Pearlescent, Extender and Corrosion–inhibiting Pigments.

3.2 DEFINITION

Pigments may be used in paint to provide, in conjunction with the film former:

(1) aesthetic appeal — colour and opacity or transparency and special effects.
(2) surface protection — weathering and corrosion resistance.
(3) auxiliary properties — film reinforcement, hardness, fire retardance, anti-condensation, non-skid surfaces, and so forth.

A pigment may be defined as a material available as solid particles largely insoluble in the film former, and in the solvents or diluents used in the paints, and capable of being dispersed in the paint constituents to give maximum benefits in terms of the required paint properties.

3.3 REQUIRED QUALITIES OF PIGMENTS

3.3.1 Appearance

Under the general heading of pigments as defined above, there are two classes — true pigments and extender pigments. The former are used, say, for the whiteness, colour, or opacity they impart, whereas extenders may be added to paints to modify the gloss or sheen, having in this instance little effect on colour and opacity. At higher concentrations extenders do affect colour and opacity, as well as other important physical properties of the paint film. Extender pigments are also used to modify such properties as the rheology and settling characteristics of the wet paint and the flow out of the film after application. They are also added to paint to provide economies.

The first and foremost requirement of a pigment is to provide colour to opaque or transparent films, and so high tinting strength is an important attribute.

In many instances the substrate needs to be obscured, and so pigments which provide opacity are necessary. Other finishes involve the use of coloured pigments which are transparent or semi-transparent.

3.3.2 Resistance

It is important that the pigments used in a particular paint are resistant to attack by the other paint constituents, by water, and by the solvents and chemicals to which the paint film is exposed. If the film former requires heating to produce the crosslinking, then the pigment must be thermally stable.

3.3.3 Durability

Pigments are required to be durable in terms of colour retention, that is, neither fading nor darkening. Additionally, pigments must not detract from, indeed may enhance, the binder in providing a cohesive, hard, elastic film with good adhesion to the substrate. Thus a tough durable abrasion-resistant film, which protects the coated surface and increases its life expectancy, is ensured. The durability of a clear film when pigmented is improved many-fold on exterior exposure by the inclusion of pigments. The paint film needs to be thick enough to give maximum protection, again a property markedly improved by correct pigmentation.

Corrosion resistance provided by paint coatings is a particularly important feature, and corrosion-inhibiting pigments and techniques of anti-corrosive paint design form a technology of their own.

3.4 PIGMENT CLASSIFICATION

Pigments are available which, when wetted with binder, are white, coloured, and colourless or 'extender' materials. Coloured pigments have been traditionally classified as inorganic and organic, but advances in technology are making this division in terms of pigment properties increasingly diffuse.

Sophisticated surface treatments for pigments, together with modern manufacturing and processing techniques, are generating pigments which increasingly need to be treated as individual special materials, rather than generic types.

However, it is easier to begin with simple classifications. Thus we have:

Inorganic pigments — These include all the whites and extenders and a useful range of colours, both synthetic and naturally made.

Organic pigments — In present times these are generally synthetic.

The broad properties are shown in Table 3.1.

Table 3.1 — General properties of inorganic and organic pigments.

Pigment property	Inorganic	Organic
Colour	Sometimes dull	Usually bright
Opacity	Normally high	Relatively low
Colour strength	Usually low	Normally high
Fastness to solvents — 'bleed resistance'	Good	Varies widely from good to bad
Resistance to chemicals	Varies	Varies
Heat resistance	Mostly good	Varies
Durability	Usually good	Varies
Price	Relatively inexpensive	Varies, but some are very expensive

Further classification of inorganic and organic pigments into chemical types is shown in Table 3.2, and typical properties in Table 3.3.

3.5 PIGMENT NOMENCLATURE

Paint technologists identify groups of pigments by using the traditional terms for chemical types shown in Table 3.2. They identify the pigment types by using 'common names' e.g. 'Prussian Blue', which have grown up in the industry, and they use the *Colour Index* number, for the chemical type and the commercial names for actual suppliers' grades.

The *Colour Index* [1] is an important international coding system because it is

Table 3.2 — Pigments for paint — white, coloured, and extenders.

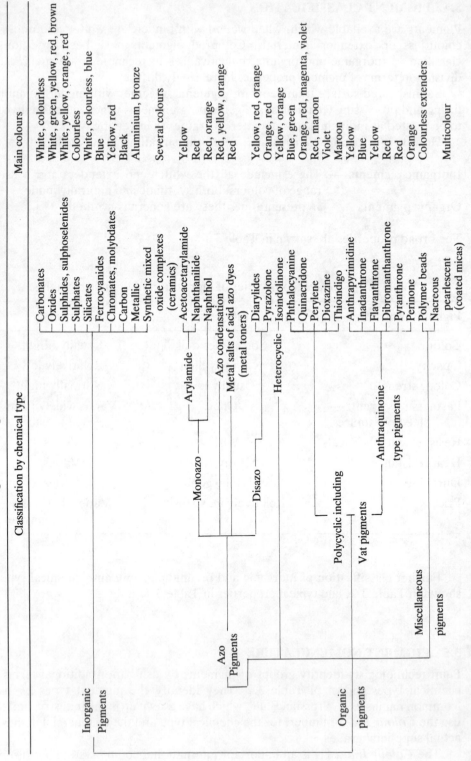

Table 3.3 — General properties of pigment chemical types

Property in paint	Coated rutile titanium dioxide	Iron oxides	Lead chromates	Prussian blue potassium ferro ferricyanide	Carbon black	Mono azo	Dis azo	Condensation azo	Metal azo toners	Phthalo-cyanine	Polycyclic incl. Vat and quinacridone
Colour	E	G	E	G	E	E	E	E	E	E	E
Opacity	E	E (except transparent versions)	E	E	E	F	G to F	G to P to P	G to F	P	G to F
Tinting strength	E	F–P	F	E	E	E	E	G	E	E	G
Heat stability	E	E (except yellow)	G	G	E	F	G	E	G	E	G
Fastness to solvent:											
a) Solubility	E	E	G to F	E	E	P	F	E	G	E	G
b) Bloom	E	E	P		E	P	E	E	E	E	E
c) Overspray	E	E	G	E	E	P	F	E	E	E	G–E
Resistance to:											
Acids	E	E	E	F	E	E	E	E	P	E	E
Alkalis	E	E	P	P	E	E	E	E	P	E	E
Chemical reagents	E	E	G	P	E	E	E	E	P	G	G
Light fastness in:											
Full shade	E	E	G	G	E	E	F	G	F	E	G to E
Medium tints	E	E	G	F	E	F	P	F	P	E	G to E
Pale shades	E	E	P	P	E	P	P	P	P	G	G to E
Dispersibility	E (except transparent versions)	E (except transparent versions)	F*	F	F	G	G	G	G	F	F
Flow properties	E	E	E	G	G	F	P	F	F	P	F
Flocculation resistance	G	G (except red)	E	G	P	G	G	G	G	FPG	GF
Price	Low	Low	Low	Low	Low	Low	Medium	Medium	Medium	Medium	High

E — Excellent   G — Good   F — Fair   P — Poor   *'Easily dispersible' types are available logical and comprehensive, and enables clear identification of the pigment type to be determined.

The *Colour Index* (third edition, second revision 1982) is published by the Society of Dyers and Colourists in association with the American Association of Textile Chemists and Colorists. This very useful reference book gives general information about pigments, classifying them under:

(1) *Colour Index* group name, colour type and number
(2) Chemical constitution
(3) Commercial or trade name

Thus:

(1) *Colour Index* group name, colour type and number

The materials available are grouped together according to whether they are pigments or solvent dyes, and into colour groups. The number refers to all the materials of the same chemical constitution in the designated colour family.

For example:

CI Pigment Yellow 3 is a yellow pigment, and the entry distinguishes the group from, say, CI Solvent Yellow 3, which is a dye.

(2) Chemical constitution

The chemical constitution for each pigment type is given a *Colour Index* Constitution Number. Thus, in the above example CI Pigment Yellow 3 has the CI Constitution Number 11710.

All the yellow pigments which have their basic chemical construction resulting from coupling 4-chloro-2-nitroaniline onto 2-chloro-acetoacetanilide are given the number 11710.

(3) Commercial or trade name

The pigments available on the market within each colour index classification are listed under their commercial or trade names.

The common name for Pigment Yellow 3 is 'Arylamide Yellow 10G'. The 'G' in this name denotes a green-toned yellow and '10' indicates the pigment is 10 units or 'traces' greener then 'Arylamide Yellow G' which is Pigment Yellow 1. These yellows are also called 'Hansa®† Yellows'.

The *Colour Index* lists over 700 colour index pigment types available as over 5000 different pigments for all uses, not only paints. (See Table 3.4). Other uses are for printing inks, plastics, rubber, cement, and paper.

---

† ® Registered trade mark of Hoechst A.G., Germany.

Table 3.4 — Numbers of chemical types of pigments listed in the *Colour Index*.

Colour Index pigment	No. of different chemical types listed
White	33
Black	32
Brown	26
Blue	66
Green	36
Yellow	163
Orange	54
Red	244
Violet	49
Metal	6

Of all these pigments, excluding extenders, only about 30 account for 80% of the total manufacture and consumption. Some of the leading pigments, so far as quantity is concerned, are listed in Table 3.5.

Table 3.5 — Leading pigments.

Pigment	Type
CI Pigment White 6	Titanium dioxide
CI Pigment Red 101	Red iron oxide
CI Pigment Yellow 42	Yellow iron oxide
CI Pigment Blue 15, 15:1, 15:2, 15:3, 15:4	Copper phthalocyanine
CI Pigment Red 49:1	Monoazo, Lithol Red (barium salt of acid azo dye) used mainly in printing inks
CI Pigment Red 57:1	Monoazo, Lithol Rubine toner (calcium salt of acid azo dye) used mainly in printing inks

3.6 FURTHER GROUPS OF PIGMENTS — TERMS USED IN THE PAINT INDUSTRY

A large group of modern coloured pigments derives from dyestuff technology, as used in colouring fibres and fabrics. These, by their nature, need to be soluble at some stage of the dyeing process. The technology could be often adapted to prepare insoluble coloured powders suitable for use as paint pigments. According to their derivation, pigments are referred to frequently and sometimes loosely by paint technologists as:

Hansa®1 — see under monoazo yellows (3.11.6).

3.6.1 Pigment toners — Metal toners

Pigment toners are those materials which, in the pure form, are insoluble in water and most paint resins, solvents, and diluents, as opposed to dyes — which are soluble. The term is used in America to describe all coloured pigments except lakes, but in Europe it is usually reserved for a particular group called the 'metal toners'. The metal toners are made by reacting certain soluble, acidic azo dyes with bases to give the metal salts which are insoluble in water, such as the barium, calcium, or manganese toners. The lightfastness is partly determined by the metal ion, which also influences the colour thus:

Na   Ba   Ca   Sr   Mn

Improving lightfastness   ———————→
Orange   ←———————→   Bordeaux

A typical metal toner is Pigment Red 48:2, a calcium salt of 2-chloro-4-toluidine-5-sulphonic acid coupled with beta hydroxy-naphthoic acid. The common name for this pigment is 'Calcium Red 2B toner'.

Some toners are rosinated during manufacture, and the presence of the rosin during precipitation yields smaller particles, giving higher strength and transparency. (See section 3.6.4.)

3.6.2 Lithols®2

Lithol is the historical name for the metal toner class of pigments. They are mainly used in the printing ink industry, and the name is said to be derived from the term 'lithography'. The use of lithols in paint is limited to the cheaper, low-durability systems because of the poor heat- and lightfastness properties that the lithols possess.

®1 Registered name of Hoechst AG., Germany
®2 Registered name of BASF Germany

An example is Pigment Red 49:1 which is Lithol Red, the barium salt of 2-naphthylamine-1-sulphonic acid coupled with 2-naphthol.

3.6.3   Pigment lakes

Many soluble acid or basic dyes cannot be made into insoluble simple metal toners, but they can be precipitated or adsorbed on to an inorganic base such as alumina with barium sulphate ('Blanc Fixe'), which acts as a 'carrier', to give a pigment.

Other lakes involve the use of hydroxy anthraquinone dyestuffs complexed with metals such as aluminium and calcium. Examples are:

Pigment Red 60:1.

'Pigment Scarlet 3B Lake' which has a complicated formula, but can be described as the barium salt of the mono azo dye 'Mordant Red 9' extended with aluminium hydroxide and zinc oxide.

Pigment Red 83.

'Alizarine Madder Lake', 'Alizarine Lake', or 'Alizarine Red B' is dihydroxy anthraquinone complexed with calcium and alumina, with a small percentage of sulphonated castor oil (Turkey red oil) and aluminium and calcium phosphates included.

By contrast and to illustrate the looseness of terminology in the pigment industry, one may quote Pigment Red 53:1. This is a printing ink pigment known as 'Red Lake C' or 'Lake Red C Toner'. It is the barium salt of the mono azoacid dye, and in the strict sense of the definitions given above, is a metal toner, not a true lake.

3.6.4   Rosinated pigments

Rosination is the treatment of the reactants during manufacture of the pigment, with rosin (abietic acid) or an alkaline earth metal rosinate, obtained from boiling rosin with, say, sodium hydroxide. It is most widely used with the metal toner pigments. The inclusion of the sodium rosinate in the process during the precipitation of the metal salt, favours the production of smaller particles and yields a brighter pigment of higher strength and increased transparency compared with non-rosinated versions. Rosination is often used as a method of improving the dispersibility of pigments.

Modified rosins, such as tetrahydroabietic acid, with higher melting points and greater resistance to oxidation, provide better pigments in terms of mechanical stability and paint storage stability.

Rosinated pigments are widely used in the printing ink industry, where high transparency is essential.

3.6.5   Naphthol, Naphtol AS®1, and Naphthanilide pigments

These are all red mono azo pigments.

Simple mono azo reds

The simple red pigments are monoazo materials based on 2-naphthol (beta naph-thol) coupled with primary aromatic amines.

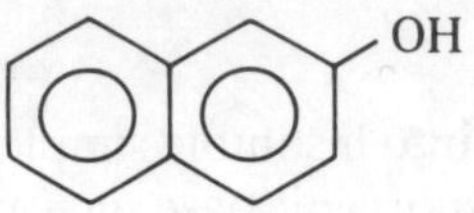

OH

2-naphthol

(See section 3.11.8 for examples.)

Naphthol and Naphtol AS® pigments

Where 2-naphthol in the simple azos above is replaced by the group 3-hydroxy-2-naphthoic acid or substituted derivatives of it, a series of red Napthol and Napthol AS pigments is generated.

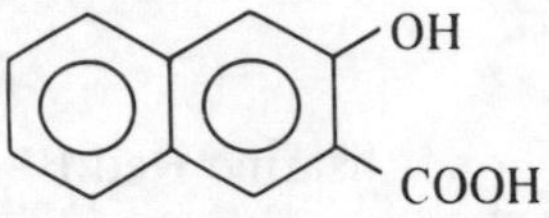

OH

COOH

3-hydroxy-2-naphthoic acid
(BON acid)

Certain of these pigments are sometimes called 'BON Arylamides' because the arylamide is produced by reacting BON acid with an amine. (See section 3.6.6).
For example:

Pigment Red 2 is a Naphthol Red

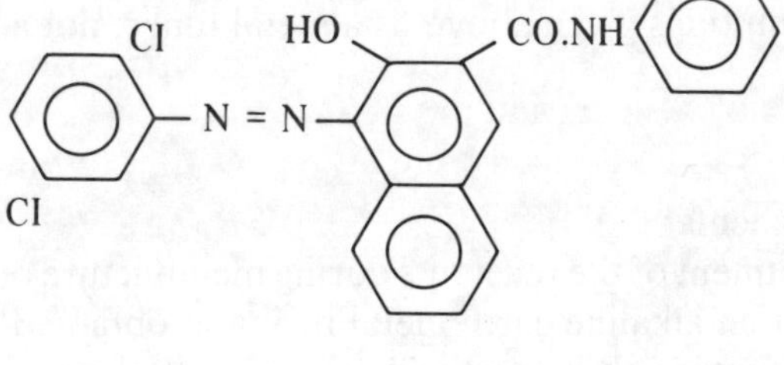

CI
HO
CO.NH

— N = N—

CI

2,5-dichloroaniline coupled with 3-hydroxy-2-naphthanilide.

Naphthanilide pigments

The term Naphthanilide reds derives from the presence of the 3-hydroxy-2-naphtha-nilide group or substituted variants of it, in the coupling with the primary aromatic amine.

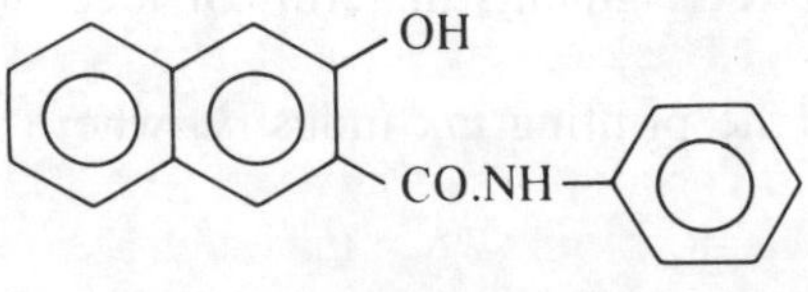

OH

CO.NH—

3-hydroxy-2-naphthanilide

Pigment Red 2, shown as the previous pigment example above, is therefore known as a Naphthol pigment, often called a 'BON arylamide' and also referred to as a Naphthanilide pigment.

All these terms are widely used in the pigment and paint industry and contribute to illustrate the illusive or disconnected logic in the naming of pigment groups. This loose terminology lends greater value to the *Colour Index* nomenclature and numbering systems, [1].

3.6.6 BON pigments

These are usually azo-based red and maroon pigments and metal toners where 3-hydroxy-2-naphthoic acid, known as **B**eta-**O**xy **N**aphthoic acid, or BON acid is used as part of the chemical building block of the pigment.

3-hydroxy-2-naphthoic acid —
'BON' acid

Thus we have 'BON arylamide' and 'BON toner' pigments, and so forth. (See section 3.6.5).

'BON' is a loose term and is frowned upon by some experts. However, many pigment and paint technologists continue to use it in their work.

3.6.7 Vat pigments

The term 'Vat' comes from the dye industry where cloth was coloured in containers called vats. The vat dyes are converted into pigmentary form using a series of complicated and difficult chemical and physico-chemical processes.

Many vat pigments can be derived from anthraquinone, giving rise to Indanthrone Blue, Flavanthrone and Anthrapyrimidine Yellows, Pyranthrone Red, Perinone Oranges, and Anthraquinone Red. Dibromanthanthrone Red is also classified as an anthraquinone pigment. The Thioindigo and Perylene pigments are also included as vat pigments.

These complex polycyclic pigments are expensive and are used in high-quality industrial finishes and motor car paints. Their use is justified by their good heat, chemical, and solvent resistance and good durability in full and pale shades. One or two of these pigments, however, particularly the Flavanthrone Yellow at high concentrations, darken on exposure to light.

3.6.8 Ceramic pigments (synthetic mixed phase oxides)

Ceramic pigments, so-called because they were developed for colouring ceramic glazes in the china and porcelain industry, are mixed phase pigments based on metal oxides where a foreign ion is incorporated in certain crystal lattices. The spinel–$MgAl_2O_4$, or the rutile–TiO_2 lattices give different colours when ions such as cobalt, nickel, chromium, iron, or copper replace some of the metal ions in the basic crystalline structures.

Many blue ceramic pigments have the basic spinel structure, being mixed cobalt and aluminium oxides. Brown ceramic pigments have the rutile structure with titanium, chromium, and antimony. The well-known nickel titanates have the rutile structure with nickel and antimony in the crystal lattice; they are coloured yellow.

The ceramic pigments are made by calcination reactions at extremely high temperatures. They are therefore outstanding for thermal stability and have excellent light, weather, and chemical resistance. Some of these pigments have brilliant pure shades of colour approaching that of organic pigments.

Unlike most inorganic coloured pigments, they do not give high opacity. They have low tinting strength, and some give low gloss at normal pigment volume concentrations. They are generally used in coil coatings and some automotive paints.

3.6.9 Flushed pigments

Flushing is the process whereby pigment is transferred from an existing suspension in water to a second liquid with which water is immiscible, for example, oil, resin, or paint solvent, using a mechanical mixing device. The effectiveness of the transfer depends on the affinity of the pigment to water, or the oil, resin or solvent. Successful flushing can produce finer dispersions than the normal dispersion techniques such as bead milling, thus yielding pastes of higher tinting strength with brighter colour, but lower opacity. This is because the pigments retain the original fine particle size that was achieved during manufacture and do not become aggregated into larger 'clusters' by the drying process.

3.7 MANUFACTURE OF PIGMENTS

Many pigments are obtained from natural sources; others are made using by-products from other industrial processes. Several are the result of reacting various reagents with mineral ores, but the vast majority are produced synthetically from coal tar or petroleum distillates.

Inorganic pigments which occur naturally and are mined and then processed for use in paint include natural iron oxides, and the extender pigments calcium carbonate and china clay.

The synthetic pigments include both inorganic and organic products.

3.7.1 Synthetic inorganic pigments

These pigments are produced by simple chemical reactions or by calcination at very high temperatures, or both. Thus lead chromate is prepared by a relatively simple precipitation, albeit under rigorously controlled conditions. Zinc oxide can be prepared by the controlled oxidation of zinc metal vapour; ultramarine blue, a complex sulphide, is produced by calcination alone and titanium dioxide can be prepared first by precipitation, followed by a calcination process. Certain improvements in properties can be obtained by such calcination, i.e. to change or modify the crystal structure.

   The ceramic pigments mentioned earlier are obtained by fusing materials together at high temperatures — 1000°C.

   The synthetically-produced inorganic pigments are manufactured in relatively short periods in batch or continuous processes, taking, on average, one week to complete, including perhaps crystallization, calcination, and other processes such as washing, filtering, drying, milling, and classification. Additives to improve the manufacturing procedures and also to modify the final pigment properties are included at various stages in the overall production process.

3.7.2 Synthetic organic pigments

These pigments, on the whole, take longer to manufacture, particularly the more complex ones. Some high-performance pigments may take many weeks to synthesize. They are made in various well-defined stages, and materials from each stage may be sold between pigment manufacturers to achieve such benenfits as cost savings derived from economies of scale and greater utilization of specialist equipment.

   The materials at each stage are given the terms:

(1) 'Primaries'
(2) 'Substituted primaries'
(3) 'Intermediates'
(4) 'Crudes'

   The primaries are now obtained largely from petroleum. The important primaries are the aromatic hydrocarbons:

Primaries

Benzene

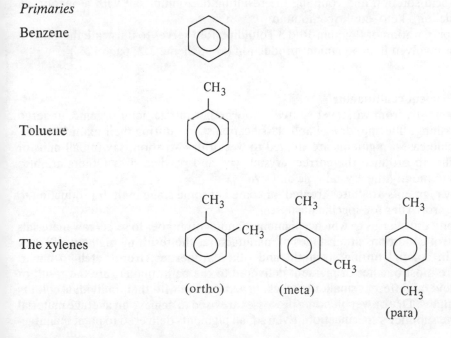

Toluene

The xylenes

(ortho)   (meta)

(para)

Naphthalene

Anthracene

The primaries are converted into 'substituted primaries' and then into 'intermediates' by a variety of well-known classical organic chemical reactions. There are hundreds of intermediates involved in pigment production, and it is mainly these that are traded between pigment suppliers.

Schematically, typical production processes can be as shown in Scheme 3.1 (opposite).

The production of azo pigments may be taken as an example of the reaction of substituted primaries and intermediates to give the crude pigment.

Azo pigment manufacture

The azo colours account for about 80% of all manufactured organic pigments. They are easily prepared from readily available intermediates and yield a wide range of colours, but chiefly yellows, oranges, and reds. The azo pigments are characterized by the presence of one or more azo groups (−N=N−) in the molecule.

They are prepared by diazotizing a primary aromatic amine with nitrous acid in mineral acid solution and coupling the resulting diazonium salt with a phenol, an arylamine, or a keto–enolic compound.

The preparation of Pigment Red 3 Toluidine Red serves to illustrate the general reactions involved in azo pigment production (see Scheme 3.2, page 126.)

3.7.3 Pigment conditioning

Most pigments, from whatever source, and whether inorganic or organic, undergo many washing, filtering, drying and classification stages during their manufacture.

In some cases, pigments are treated to wet or dry attrition, say in ball mills or bead mills, to produce the correct crystal type and particle size. Others are also calcined to modify the crystal type and size.

Many pigments are 'after-treated' or coated at some stage in their production to improve properties for specific purposes.

Despite the lengths to which a pigment manufacturer goes to select raw materials and control conditions at all stages of manufacture, the resulting pigment will vary slightly in colour, tinting strength, and other properties, from batch to batch. Therefore most batches of pigment delivered to the paint maker, are the result of mixing several different smaller batches, in accordance with their individual colours and strengths. Dry or wet blending processes are used to achieve an average material within the supplier's specification. Even so, all pigments delivered to paint manufacturers still vary marginally from batch to batch. Those that are made on a so-called 'continuous process' also vary, but identification of their variability is more difficult, simply because there are no discrete batches.

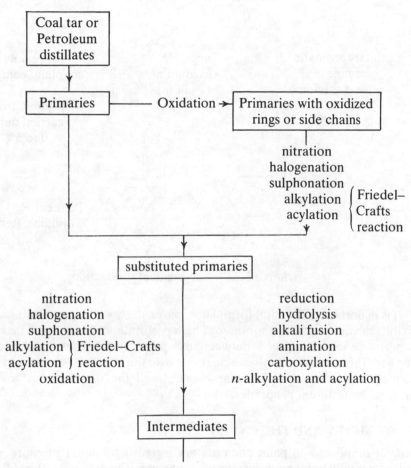

Scheme 3.1. Typical production processes for Organic Pigments.

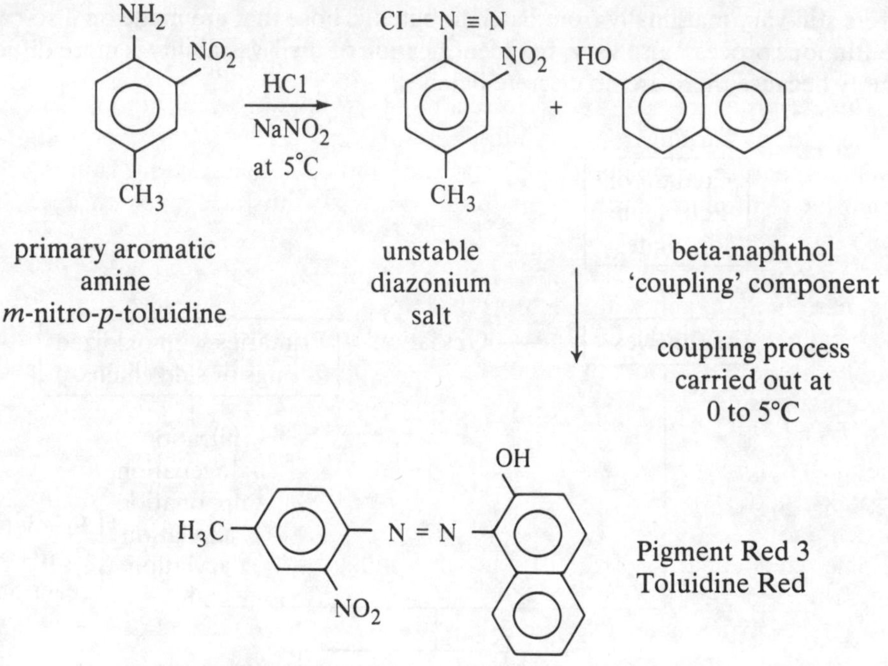

Scheme 3.2 — Azo pigment production.

It is important to the paint formulator, particularly where customers have 'colour mixing schemes', or where continuous high performance is required, that standards of performance are agreed between the paint manufacturer and the pigment supplier. This normally involves agreement over the so-called 'standard batch' which uniquely represents the particular pigment and the specification, and limits of variation for different properties.

3.8   TOXICITY AND THE ENVIRONMENT

Pigment suppliers and paint chemists are increasingly under pressure to provide environmentally 'safe' materials for people to use. The definition of 'safe' is difficult to derive in circumstances where human health and protection of the environment are concerned. Consequently what is considered 'safe', and by contrast, what is considered harmful to people, animals, plants, and the environment, is constantly changing.

In the last few years two main groups of pigments have been highlighted in some European countries. These are the lead and zinc chromates and cadmium pigments. So far only the zinc chromates have given positive evidence of harm to health and are classified as carcinogens. Handling of all these pigments, in accordance with the supplier's recommended instructions, minimizes exposure and the risk of harm to health.

Changes in quality of pigments are sometimes necessary because ingredients or by-products have been identified as suspicious.

The classification as potential carcinogens of the chlorinated bis-phenyls which may occur during the manufacture of phthalocyanine pigments demonstrates this point. Production processes need to be modified to ensure that such by-products either are not generated or that they are rendered 'safe' in some way, as far as life and the environment are concerned.

Concern is expressed over the discharging of waste materials into rivers, lakes, and seas, or by burning into the atmosphere. Any new process may alter the delicate balance of nature, and pigment suppliers and paint manufacturers are finding it more expensive to produce 'harmless' materials which are satisfactory to customers.

3.9 PROPERTIES OF PIGMENTS

The basic colour and resistance properties of a pigment are determined by its chemical nature, but the hue, intensity, opacity, tinting strength, and durability are modified by other factors in an involved and complex manner. The main features concerned are:

Described in Section:

Chemical nature	.3.9.1
Crystal structure	.3.9.2
Particle shape	.3.9.3
Particle size and particle size distribution — oil absorption	.3.9.4
Refractive index	.3.9.5
The nature of the pigment surface and coatings on the pigment particle .	.3.9.6
Dispersion characteristics	.3.9.7
The colour and other properties of the paint film former.	.3.9.8

3.9.1 Chemical nature

The colour of a pigment is determined by the selective absorption and reflection of the various wavelengths of visible light, 0.4 to 0.7 μm, which impinge upon it. A blue pigment appears so because it reflects the blue wavelengths in the incident white light and absorbs the other wavelengths; black pigments absorb all of the wavelengths of incident light almost totally; white pigments reflect all the visible wavelengths.

The differing absorption and reflection characteristics of pigments are attributed to the arrangement of the electrons within the molecules of the pigment and their energy and frequency vibration.

Absorption of light by the pigment molecule excites the electrons and involves the promotion of an electron in the molecule from an orbital of lower energy E_1 to an orbital of higher energy E_2. The promoted electron subsequently returns to the lower energy level by a pathway in which the excess energy is dissipated, generally as heat.

The wavelength of the light absorbed is determined by the difference in energy E between the two orbitals concerned:

$$E = E_2 - E_1 = \frac{hc}{\lambda}$$

where h is Planck's constant, c is the velocity and λ the wavelength of the light.

A given molecular species has a limited number of orbitals, each with its own characteristic energy. This means that the energy difference E described above has certain definite values. The pigment molecule, therefore, absorbs light only at particular wavelengths determined by the energy difference E and characteristic of the given molecular species. The other wavelengths are reflected and determine the colour of the pigment.

Further explanation of the phenomenon of colour is given in Chapter 17.

For organic pigments it has been found that certain groups of atoms are essential for colour. These groupings are called 'chromophores', and groups which modify the colour are called 'auxochromes'.

Chromophores: $>C=C<$ $>C=O$ $>C=S$ $>C=NH$
                 $-N=N-$ $-N=C$ $-NO_2$

Auxochromes: $-OH$ $-NH_2$ $-NHR$ $-NR_2$
                $-SO_2H$ $-COOH$ $-NO_2$ $-CH_3$ $-Cl$ $-Br$

The following illustrate the effect of chromophores and auxochromes on the colour of organic molecules:

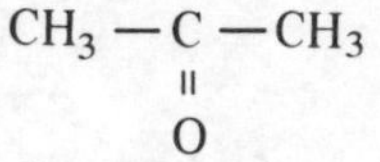

$$CH_3 - \underset{\underset{O}{\|}}{C} - CH_3 \qquad \text{acetone} - \text{colourless}$$

$$CH_3 - \underset{\underset{O}{\|}}{C} - \underset{\underset{O}{\|}}{C} - CH_3 \qquad \text{diacetyl} - \text{yellow}$$

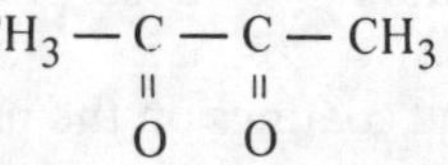

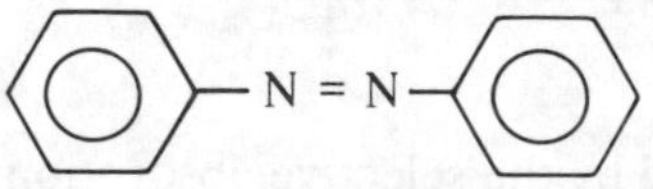

orange

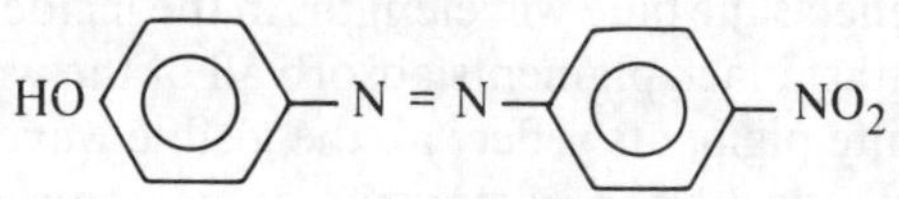

deep yellow

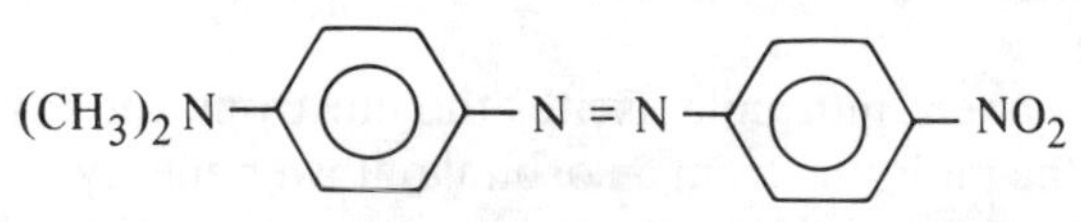

deep red

A very helpful explanantion of colour in organic compounds can be found in Chapter 36 of *Organic chemistry* by Fieser & Fieser, published by Heath & Co in 1950.

The opacifying effect of a pigment is determined principally by its chemical nature. Thus titanium dioxide scatters light, whereas carbon black absorbs light. Both phenomena create an opacifying effect in paint films. Coloured pigments have varying opacifying effects, which are governed by absorption and scattering in the different regions of the visible spectrum.

Inorganic pigments generally have greater hiding power than organic pigments of similar colour, owing to the differences in refractive index in the scattering regions of the visible spectrum (see p. 134).

The chemical nature of pigments also dictates the basic resistance properties to heat, solvents, acids, alkalis, and other chemical reagents.

The simple monoazo pigments (reds, oranges, and yellows) for example, while having good resistance to acids, and good durability in full shades (but not in pale reductions where they fade), are sensitive to heat and solvents. This limits their use to air-drying paints containing weak solvents, and only then in the deeper colours.

The more complex and expensive condensed azo pigments (reds, oranges, browns and yellows) have similar properties to the monoazo in all respects, except that they have excellent resistance to heat and solvents, and so can be used in stoving paints.

Prussian Blue is fast to acids but loses colour in alkaline conditions, changing to brown ferric hydroxide; but Ultramarine Blue — a complex sodium aluminium silicate containing sulphur — is sensitive to acids, losing colour. Phthalocyanine Blues, on the other hand, are stable in most conditions.

3.9.2 Crystal structure

Pigments are basically crystalline or non-crystalline, i.e. amorphous. That is, the atoms within each pigment molecule are arranged in either an orderly or a random fashion. Some materials are capable of existing in several different crystalline forms, and these are known as polymorphic materials.

There are generally different properties associated with the different crystal structures. Some of the crystal forms may be more stable than others, and some crystals are not particularly suitable for use as pigments. The less stable crystals are sometimes modified by the inclusion, say, of a different ion into the lattice to render them more stable.

Examples of polymorphic pigments are the phthalocyanine blues, the quinacridones, the lead chromates and titanium dioxide (see Table 3.6).

Modern techniques of pigment manufacture are enabling pigment producers to control the crystal shape and size, so that particular colour and performance benefits can be achieved.

3.9.3 Particle shape

The primary shape of pigmented particles is determined by the chemical nature, the crystalline structure, or lack of it, and the circumstances in which the pigment is created in nature, or made synthetically. Pigments as primary particles may be:

Spherical
Nodular — irregular spheres
Cubic
Acicular — needle or rod-like
Lamellar — plate-like

These shapes are illustrated in Scheme 3.3.

However, they are usually supplied to the paint chemist as aggregates or

Table 3.6 — Examples of crystal forms for polymorphic materials.

Material	Crystal form	Property
Phthalocyanine blue Pigment Blue 15:1	Stabilized Alpha P. Blue 15:1 & 15:2	Reddish blue
15:2 15:3 15:4	Beta P. Blue 15:3, 15:4	Greenish blue
	Gamma	Not used as a pigments
Quinacridone	Alpha and Delta	Not used as pigment
Pigment Violet 19	Beta	Violet
	Gamma	Red
Lead chromate with lead sulphate	Orthorhombic (including 30–50% lead sulphate and stabilised)	Pale primrose
Pigment Yellow 34	Monoclinic (including 20–45% lead sulphate)	Lemon
	Anatase	(1) Refractive index 2.55 (2) Very white (3) Some tendency to chalk on exterior exposure
Titanium dioxide Pigment White 6	Rutile	(1) Refractive index 2.76. This gives the possibility of higher opacity than with the Anatase form (2) Slightly yellower than the Anatase form (3) More resistance to chalking on exterior exposure
	Brookite	Not used as a pigment

agglomerates. Aggregated particles are difficult to separate because they are firmly held together 'or even joined as a result of, say, sintering during the pigment manufacture or drying processes. Agglomerates are loose associates of primary

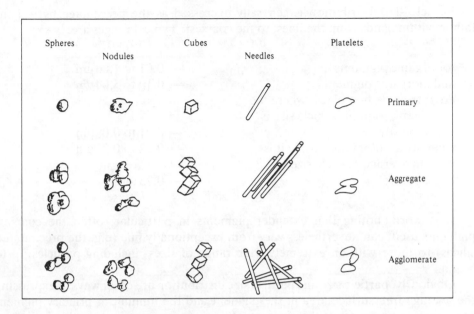

Scheme 3.3 — Particle shapes.

particles which are easily broken down during dispersion of the pigment in the paint ingredients.

Aggregated or agglomerated particles are more easy to handle in paint-making than the primary particles, because the latter can be so small (see p. 132) and therefore very light, that they behave like smoke when disturbed.

Suppliers offer 'non-dusting' grades of pigments, where the shape and size of the agglomerates or aggregates are modified to facilitate easier handling.

The state of aggregation or agglomeration is a key factor in the pigment dispersibility (see p. 138).

The shape and size of the primary pigment particles affect the way they pack in a paint film. Thus, properly dispersed acicular shapes are said to reinforce the film like the fibres in glass fibre reinforced plastics. Lamellar pigments, such as aluminium and mica, tend to form an overlapping laminar structure as in roof tiles, which offer resistance to the passage of water through the paint film.

Particle shape can also influence the shade of colour of a pigment. It has been shown [2] that beta copper phthalocyanine pigments (Pigment Blue 15:3) give different hues, depending on the deviation from isometry to acicularity.

3.9.4 Particle size and particle size distribution

Particle size is normally expressed as the average diameter of the predominant, primary pigment particle assuming sphericity (see Chapter 7).

Particle size distribution is generally expressed as the percentage by weight, falling within bands from the finest to the coarsest. Typical ranges are:

for organic pigments	— 0.01 to 1.00 µm
and inorganic pigments	— 0.10 to 5.00 µm
apart from carbon black, where the distribution of particle size is typically	— 0.01 to 0.08 µm
whereas for titanium dioxide it is	— 0.22 to 0.24 µm
with an average particle size of about	— 0.23 µm

It is worth noting that extender pigments in particular, often the coarsest pigments used, can nevertheless vary from exceptionally fine (e.g. the precipitated silicas) to those with an extremely wide range of sizes, including particles up to 50 µm.

Obviously, particle size and particle size distribution are other ways of expressing the average free surface area of the pigment and the number of primary pigment particles in a unit mass of the pigment. Where pigments are replaced by others of significantly different particle size distribution, simple formulating performance predictions based on the concept of pigment volume concentration and critical pigment volume concentration (see Chapter 1) are not likely to be satisfactory. The well-established parameter of 'oil absorption' (the weight in grams of acid refined linseed oil to just form a paste with 100 grams of pigment) is directly dependent on particle size distribution, although the state of aggregation of the pigment, its packing density, and the ease with which it is wetted by the oil, are important modifying factors.

Aggregation and agglomeration increase the effective particle size of most pigments to at least 20 µm. Such a cluster could contain up to one million fundamental particles.

The particle sizes of pigments have first-order effects on the primary properties of paints in which they are used. For this reason much of the development work by pigment manufacturers has been to approach even nearer to the optimum size required.

For example, the particle size of the pigment affects the opacifying properties. For white pigments, this is because the scattering of light increases as the size decreases until the optimum size for scattering is reached. Further size reduction decreases scattering and increases transparency. For coloured pigments there is also an optimum particle size range for opacity, but both scattering and absorption phenomena are involved.

Titanium dioxide is a very good opacifying pigment. First of all it has a high refractive index (2.76); secondly the optimum particle size for light scattering can be achieved, i.e. half the average wavelength of white light in air — 0.25 µm.

An understanding of the effect of particle size on opacity can be gained by considering a block of ice. As a massive lump, the material is transparent, but crushed to a powder it appears opaque and white. Multiple refractions occur at the interface of ice and air because of the difference in their respective refractive indices.

The particle size also affects the colour hue. This is most dramatically exemplified in the spherically shaped red iron oxide pigments, where the smaller particle sizes (0.09 to 0.12 μm) give yellow toned reds and larger sizes (0.17 to 0.70 μm) give increasingly bluer toned reds.

Generally speaking, the finer particle size pigments below 0.40 μm give brighter, purer shade pigments.

Another pigment property which can be affected by particle size is the tinting strength. Generally for organic pigments, the finer the particle size within its chemical class, the higher the tinting strength, but the extent of this varies for different pigments.

The gloss of a paint film is also affected by particle size, high gloss being generally obtained from pigments — at the right concentration — where the particle size does not exceed 0.30 μm. Larger particle sizes reduce the gloss, and in the extreme cases give matt paints.

The durability of certain pigments, chiefly the lower molecular weight organic materials, is affected by particle size. For example, the arylamide yellows (Pigment Yellows 1 and 3) can be produced in large or small particles, giving the comparative properties listed in Table 3.7.

Table 3.7 — Arylamide Yellow (Pigment Yellows 1 and 3): Properties related to primary particle size.

	Large particle size	Smaller particle size
Tinting strength	lower	higher
Opacity	higher	lower
Lightfastness	better	poorer

3.9.5 Refractive index

The refractive index (RI) of a material is a key to its performance as a pigment. The greater the difference between the refractive index of the pigment and that of the medium in which it is dispersed, the greater the opacifying effect.

Consideration of the refractive indices of some typical extender pigments and white pigments in Table 3.8 shows that titanium dioxide will give higher opacity than zinc oxide, whereas talc and calcium carbonate will be transparent in fully bound surface coatings.

As well as having the highest refractive index, rutile titanium dioxide can also be made at the optimum particle size for light scattering, 0.25 μm; thereby enhanced opacity may be obtained.

For coloured pigments, the refractive index of the pigment in the non-absorbing,

Table 3.8 — Refractive indices of pigments and extenders.

Material	RI	Extender pigments	RI	Opacifying white pigments	RI
Air	1.00	Calcium carbonate	1.58	Lithopone 30% (zinc sulphide/ barium sulphate)	1.84
				Zinc oxide	2.01
				Zinc sulphide	2.37
Water	1.33	China Clay (aluminium silicate)	1.56		
				Titanium dioxide:	
Film formers	1.4 to 1.6	Talc (magnesium silicate	1.55	Anatase	2.55
				Rutile	2.76
		Barytes (barium sulphate)	1.64		

or highly reflecting part of the spectrum, affects the performance as an opacifying material. For example, Pigment Yellow 1, Arylamide Yellow G, gives lower opacity than Pigment Yellow 34 Lead Chromate Yellow; all other factors being approximately equivalent.

Manufacturers are endeavouring to increase the opacifying effects of their organic yellow and red pigments, largely by adjusting the particle sizes.

3.9.6 The nature of the pigment surface and coatings on the pigment particle
The nature of the surface of the pigment particle is a most important feature. Its polarity governs the affinity for alkyds, polyesters, acrylic polymers, and so forth carried in solvents, and also for aqueous solutions and dispersions of film formers. It determines how readily pigments are de-aggregated, and hence affects dispersion and also the ultimate stability of the liquid paint.

The fundamental pigment
Consideration of the nature of the following pigments, before surface modification by additives, will illustrate the radical differences in surface polarity due to the chemical nature of the pigment involved:

Phthalocyanine blue	Pigment Blue 15:3
Phthalocyanine green	Pigment Green 7
Perylene red	Pigment Red 179
Titanium dioxide	Pigment White 6

copper phthalocyanine blue
Pigment Blue 15:3

polychlorophthalocyanine green
Pigment Green 7

Phthalocyanine blue has hydrogen atoms at the perimeter of the planar molecule, whereas the green has chlorine atoms. Clearly the phthalocyanine green pigment will be a more polar pigment than the blue. However, closer study of the pigments shows the difference in polarity to be more pronounced. Honigmann *et al.* [3] showed that phthalocyanine blue pigment is composed of stacks of the square planar molecules piled one on top of another, like a pack of playing cards. The axis of the stack is inclined at an angle to the planar molecules, depending on the crystal modification.

The molecules so arranged form an acicular or needle-like structure. The crystal presents markedly different surfaces — non-polar hydrogen atoms along the length of the needle, with the comparatively polar π bonds of the benzene rings and the nitrogen and copper atoms at the surfaces at each end of the needle. In confirmation of this, beta phthalocyanine blue pigments having longer needles, have been shown to be relatively less polar than the isometric cubic crystals of beta copper phthalocyanine blue. They are also more hydrophobic [2].

In contrast, Kaluza [4] considers the perylene imide Pigment Red 179.

Pigment Red 179
Perylene imide

Here the strongly polar acid imide groups are at the periphery of the molecule, and the number of hydrogen atoms is relatively few compared with phthalocyanine blue. Perylene imide, in fact, proves to be relatively more hydrophilic than the non-polar acicular phthalocyanine blue.

Generally, inorganic pigments are much more polar and hydrophilic than organic pigments because the molecules are ionic.

Pigment White 6 rutile titanium dioxide, for example, has a compact lattice which can be considered as a box with titanium ions at the corners, and oxygen inside and in between. Some of the oxygen atoms carry hydroxyl-OH groups, and these form strong hydrogen bonds or may even be exchanged with acidic or basic hydroxy groups in the paint ingredients [4].

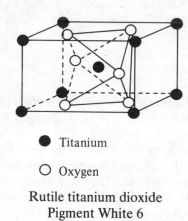

● Titanium

○ Oxygen

Rutile titanium dioxide
Pigment White 6

Surface treatment of pigments
The surfaces of pigments are modified by the manufacturer, either deliberately to alter the performance of the pigment, or as a side effect resulting from the addition of materials to facilitate the manufacture of the pigment itself. Some additives do both. Properties which manufacturers seek to improve by using additives include:

wetting out and dispersion in selected resins and solvents or diluents
flocculation resistance
wet paint stability
paint film durability.

The most notable examples of pigments treated to improve durability, for instance, are the lead chromate and titanium dioxide pigments. Lead chromate yellows and reds are treated with coatings of varying thickness, density, and uniformity, using materials such as alumina, titanium dioxide, silica, and metal phosphates to increase the lightfastness, heat, and chemical resistance.

Titanium dioxide is similarly treated with additives such as alumina, silica, and organic coatings, such as fatty acids and amines, to improve:

Dispersibility in given paint types
Opacity and tinting strength
Durability — gloss retention
           — resistance to chalking
           — colour retention
Protection of the binder.

In the last-named context, the photoactive nature of titanium dioxide, which can promote film polymer breakdown, is masked by the coating on the titanium dioxide crystal. However, titanium dioxide is a good absorber of ultraviolet radiation, and this property affords protection to polymers in the film which would be degraded by ultraviolet radiation. The type and quality of the coating on the crystals is therefore a compromise to provide a balance in protection and destruction of the paint film.

Phthalocyanine blue pigments are a further instance where crystal and surface modification are a major factor in the pigment performance. The alpha form Pigment Blue 15 is unstable, and in solvents such as xylene or at temperatures over 95°C, it reverts to the beta form. The stabilized alpha version is widely used in the paint industry and has its own colour index sub-group number, Pigment Blue 15:1. The stabilization can be achieved by the inclusion of a chlorine atom in the molecule. Pigment Blue 15:2 is the stabilized alpha grade version of phthalocyanine blue which is also modified with surface coatings to give flocculation resistance and improved flow properties (see p. 158).

Rosination, chiefly used in azo pigment preparation either as the rosin acid or sodium rosinate, for example, inhibits crystal growth. This, therefore, favours the production of fine particle size pigments with good dispersibility, leading to high colour strength and good transparency.

Amine treatment, again widely used with azo pigments such as Pigment Yellow 3, arylamide yellow, gives improved dispersibility in given systems such as air-drying alkyd paints carried in white spirit [5]. However, amination may also retard oxidative drying, and once again a balance needs to be struck to achieve a good compromise in pigment performance.

Pigments may also be treated with substituted versions of the parent pigment molecule. Thus, phthalocyanine blue pigments treated with phthalocyanine methylene amines are more stable and have better dispersion and flocculation resistance in solventborne paints.

The whole technology of surface coatings of pigments presents a bewildering array of pigments for the paint technologist, particularly as manufacturers are reluctant to reveal detailed information about their particular grades of pigment. This means that the paint technologist does not know what precisely he or she is buying. Each pigment has to be tested in each paint system. Therefore every pigment and paint type must be considered as a unique system, and so paint design and manufacture becomes a skilled art. This is achievable only by experience, particularly where combinations of pigments, as is usually the case, are involved. The greater diversity of specialized paints now available has been accompanied by proliferation in the number of grades of pigments offered for specific purposes. This militates against rationalization in a multi-paint type factory. The skill of the paint technologist is in arriving at compromises between complexity and simplicity which yet ensure adequate quality for the customer and survival as a business for the paint manufacturer.

3.9.7   Dispersion properties
The dispersion process involves the separation, employing mechanical forces, of the primary pigment particles from their agglomerates or aggregates, the displacement of occluded air and adsorbed water, and the wetting and coating of the pigment surfaces with the dispersion resin. Ideally, each primary particle, having been mechanically separated during dispersion, is also stabilized against flocculation. The stabilization of coated particles of pigment in solventborne paints is often by steric hindrance. The adsorbed molecules of solvated polymer on the surfaces of the pigment act as a physical barrier to re-association.

The polarity of the pigment surface and the affinity for molecular groups in the dispersing resin, together with the nature and concentration of the solvents present, all affect dispersion and subsequent stability.

The advantages built into the pigment by the manufacturer may be lost if the pigment particles are not properly dispersed and stabilized in the paint, so that the dry film subsequently produced contains deflocculated particles distributed evenly throughout the film.

Some pigments are easily dispersed in the paints where they are normally used. Certain azo pigments are rapidly dispersed in white spirit borne alkyds to give maximum colour strength. Others, the arylamide yellows and certain of the reds — cadmium sulphoselenide and molybdated chromes, for example — can easily be overmilled, leading to loss of strength or change in shade, or both. The so-called high-performance pigments such as the vat and quinacridone pigment types are often more difficult to disperse. Longer dispersion times frequently give higher strength, although this is often accompanied by a change in hue, as the particle size of the pigment agglomerates is reduced.

Detailed consideration of the various parameters influencing dispersion is necessary. Hafner [6] and Schafer [7] advise careful millbase design and minimum energy input to break down pigment agglomerates necessary to achieve the gloss and colour strength required. It is pointed out that co-dispersion of widely differing pigment types such as molybdate red chromes, Pigment Red 104, with quinacridones, Pigment Red 122, does not yield the best results. The longer dispersion times needed for the organic quinacridone pigment provide opportunity for the inorganic red chrome to be 'damaged'. Separate dispersion of the two types is advised. Further discussion on dispersion is to be found in Chapter 8.

3.9.8   The colour and other properties of the paint film former
The polymers used in paint vary widely and are fully described in Chapter 2. In particular, the colour of the resin differs from type to type. Films of nitrocellulose and waterborne latex, for example, are almost colourless, whereas oil-modified alkyds are distinctly yellow or brown. The refractive indices also vary, although in a narrow range, 1.4 to 1.6. The net effect on colour and opacity of a pigment dispersed in widely different types of resin is, therefore, also varied.

In addition, different resins and solvents or diluents have different affinities for pigments, owing to differing polarities. Effective separation of the agglomerates, coating of the particles, and subsequent stabilization vary from one resin to another. Hafner [6] describes the dispersibility of two pigments, CI Pigments Violet 19 and 23 (quinacridone and dioxazine violets) in three different resins, two short oil alkyds in xylene — one based on saturated synthetic fatty acids and the other on dehydrated castor oil — and a highly reactive butylated melamine formaldehyde resin in butanol. He shows that colour strength and rheology are markedly affected by changing the resins.

Kaluza [4] discusses the work of many authors who have clearly demonstrated that titanium dioxide, a polar pigment, is dispersed and stabilized more readily in polymers containing carboxyl groups; hydroxyl groups give weaker bonds. He also goes on to describe the importance of the concentration and size of the polymer molecules and their complexity, and the effect of the nature of the solvent on stabilization of the dispersed pigment.

3.10 CHOOSING PIGMENTS

In selecting pigments for paint, the formulator must think about the type and end use of the material in question. Paint is generally considered in terms of the markets served. Thus, coatings are divided into those for buildings — architectural or decorative paints (see Chapter 9), those for motor cars — automotive and automotive refinish paints (see Chapters 10 and 11), those for general industrial purposes (see Chapter 12), and so forth.

3.10.1 Paint type

However, in the first instance, a more useful classification of paints, from the point of view of pigmenting them, is by considering the type of film former used in the paint and its drying mechanism. This is because many, but not all, pigments are affected by the chemical nature of the resin, the solvents, diluents, and the other paint ingredients, and also by heat.

Thus, for example, a naphthol monoazo red pigment such as toluidine red, Colour Index Pigment Red 3, would be acceptable for full colours in air-drying paints based on alkyds carried in white spirit or in emulsion paints, where water is the diluent. Such a pigment would, however, fade in medium and pale reductions. It would also bleed through any succeeding solvent-based paints, because of its poor solvent resistance. This would be particularly noticeable where the overcoating was a white paint; the white topcoat would become pink.

The solvent susceptibility and poor heat stability of this pigment would preclude its being used in stoving alkyd/melamine formaldehyde paints, for example, where the solvents used would cause the pigments to bleed, and the heat used for film curing would cause the pigment to sublime. To produce a red in this type of system, the formulator would better choose a red disazo condensation pigment such as Pigment Red 144, or a red benzimidazolone pigment such as Pigment Red 208. For maximum performance, a blend of two pigments such as a red molybdated lead chromate with a quinacridone magenta may be appropriate, for example Pigment Red 104 with Pigment Red 122.

The type of resin used in the paint is, therefore, a key factor, since it determines the solvents, diluents, and other paint constituents. The method of film formation is a second important factor.

A selection of typical paint types in broad classification is shown in Tables 3.9 and 3.10, together with pigment types which are generally suitable.

3.10.2 Paint performance

The next considerations are the end use and customer requirements, in terms of film durability, resistance properties, and price.

For example, solvent and chemical resistance, with high durability in use, are demanded by the motor car industry. Pigments for such finishes are known as 'high-performance' pigments, and, in the case of organic pigments, are usually more expensive because generally they comprise more complicated organic molecules to stabilize the pigment properties.

Typical pigments used in high-performance automotive paints are described in Chapter 10.

Where special performance such as corrosion resistance is required, then corrosion-inhibiting pigments are necessary. These are normally included in the primers and primer undercoats.

3.10.3 Opacity and transparency

Having narrowed the field of selection by considering the film former and the durability needed, the next considerations are whether a 'solid' colour, special effect, or transparency is required.

A 'solid' colour means an opaque film and the need for opacifying pigments. A special effect film, such as provided by 'flamboyant', 'metallic', or 'pearlescent' paints, demands transparent or semi-transparent pigments, together with aluminium or coated mica flakes.

Where opacity is required, it is obtained most easily in the case of blacks, browns, blues, and greens by the use of carbon black, iron oxides, phthalocyanine blues, and greens respectively. White paints almost invariably use rutile titanium dioxide. Bright reds, oranges, and yellows are the most difficult colour areas in which to obtain opacity. Opaque, but dull, reds and yellows, involve the use of red and yellow oxides. Lead chrome pigments provide the simplest route to opaque bright reds and yellows, but in Europe, Japan and the USA these are increasingly being avoided in surface coatings because of the concern about the toxicity of hexavalent chromium and lead.

The search for bright, opaque, durable red and yellow pigments is reflected in the number of pigments offered by the manufacturers. There are over 240 chemical types of red and over 160 yellows classified in the *Colour Index*, contrasting with just under 70 blues (see Table 3.4).

'Metallic' and 'pearlescent' paints demand the use of transparent or semi-transparent pigments, in addition to the metallic or pearlescent pigments which provide the special effects.

Types of paint	Titanium dioxide	Iron oxides	Lead chromates	Prussian blue	Carbon black	Mono azo	Dis azo	Condensation azo	Metal azo toners (manganese)	Phthalo-cyanine	Polycyclic incl. Vat and quinacridone
Decorative Air-drying emulsion paints carried in water:											
Full shade	E	E	G*	P	G	P	P	P	G to F to P	G	E**
Medium	E	E	G*	P	G	F	F	G to P	G to F	E	E**
Pale	E	E	P*	P	G	P	P	G	P	E	E
Decorative and industrial air-drying long oil length alkyd paints in 'weak solvents' such as white spirit:											
Full	E	E	G*	E	G	P	P	G**	G to F to P	F(B)	E**
Medium	E	E	G*	F	G	F	F	G to P**	G to F	E	E**
Pale	E	E	P*	P	G	P	P	G**	P	E	E
Air-drying medium oil length alkyds. VT and styrenated alkyds stoving in strong solvents	E	E	E	G	G	P	F	F**	FG	E	E**
Industrial refinish and automotive 'lacquers'. Lacquer-drying nitro-cellulose and acrylics, carried in strong solvents:											
Full/Medium	E	E	E	G	G	F	P	P-F-G	PG	G(B)	E**
Pale	E	E	P	P	G	P	P	P-F-G	PG	E	E
Automotive and industrial 'cross-linking' stoving paints alkyd/MF thermosetting acrylic/MF and thermoplastic acrylic:											
Full/medium	E	E	E	G	G	P	P	F	PG	G(B)	G**
Pale	E	E	P	P	G	P	P	P	PG	E	E
Protective paints for industrial plant — chemically cured isocyanate epoxy amine	E	E	G	P	G	P	P	P	P	E	E

(B) Bronzes in full shade
* Not used in Europe in decorative products
** Too expensive for general use

E — Excellent
G — good
F — Fair
P — Poor

Table 3.10 — Simplified guide to pigment selection

Resin and solvent system	Drying mechanism	Depth of colour	Red	Yellow	Green	Blue	Brown	Black	White
AIR-DRYING PAINTS Polyvinyl acetate copolymers; acrylic and styrene acrylic co-polymers dispersed in water (called latices and used in emulsion paints)	Water evaporates and the latex film coalesces	Full	Naphthol Red P. Red 3, 4, 6 / P. Orange Metal toner / P. Red 48, 57	Disazo Yellow P. Yellow 13, 83 / Arylamide yellow / P. Yellow 1, 3, 73, 74					Titanium dioxide P. White 6 / China clay
		Medium	Naphthanalide Reds P. Red 2, 5, 7, 9 112	Arylamide yellow P. Yellow 1, 3, 73 74		Phthalocyanine blue P. Blue 15:1	Red iron oxide P. Red 101 with / P. Black 6,7	Carbon black P. Black	P. White 19
Long to medium oil alkyds in white spirit	Solvent evaporation followed by cross-linking with oxygen from the air	Pale	Red iron oxide P. Red 101 / Quinacridone P. violet 19 / P. Red 122 / Dibromethane-throne reds / P. red 168	Yellow Iron Oxide P. Yellow 42 / Flavanthrone Yellow P. Yellow 112 / Anthrapyrimidine yellow P. Yellow 108 / Isoindolinone yellow P. Yellow 110, 109 / Nickel Azo Yellow P. Green 10	Phthalicyanine green P. Green 7, 36 or Phthalocyanine blue P. Blue 15:1 with Acetoacetary-lamide Yellows P. Yellow 3, 74	Carbon Black			Calcium carbonate P. White
STOVING PAINTS Alkyd/melamine Formaldehyde in xylene or butanol / Thermosetting acrylic Melamine formaldehyde in xylene or butanol	Solvent evaporation followed by film formation under the influence of heat for certain finishes	Full	Chrome red P. Red 104 with quinacridone / P. Violet 19 or with metal toner / P. red 48:4 or Quinacridone PV19, P. Red 122 / Perylene red	Iron oxide yellow P. Yellow 42 / Chrome yellow P. Yellow	Phthalocyanine green	Iron blue P. Blue 27 or	Red iron oxide		

Paint type	Mechanism	Shade	White	Black	Black (with red)	Blue	Green	Yellow	Red
STOVING PAINTS (contd.) Thermoplastic acrylic in xylene or butanol	Ditto		Titanium dioxide P. White 6	Carbon black P. Black 6, 7	PR 101 with Carbon black P. Black 6, 7	Phthalocyanine blue P. Blue 15:1 or Indanthrone blue	P. Green 7, 36 or Phthalocyanine blue P. Blue 15:1 with	[Diarylide yellow P. Yellow 13, 83]	P. Red 123, 179, 190, 224; Thioindigo P. Red 88; Red iron oxide P. Red 101
LACQUER-DRYING PAINTS Nitrocellulose and acrylic lacquers in aromatic hydrocarbons, higher alcohols, esters and ketones	Solvent evaporates, leaving a dry film	Medium				P. Blue 60	chrome yellow P. Yellow 34 or with [Disazo yellow P. Yellow 155] or [Brunswick greens P. Green 15]	Isoindolirone yellow P. Yellow 109, 110; Iron oxide yellow P. Yellow 42; [Benzimidazolone P. Yellow [120] 151, 154]; Nickel azo yellow P. Green 10	Chrome red P. Red 104 with Quinacridone P. Violet 19 or Dis-azo; Condensation Red P. Red 144, 214; Perylene reds P. Red 123, 179; Thioindigo P. Red 88
CHEMICALLY CURED PAINTS Polyester polyurethanes epoxy resins carried in aromatic ketones, esters, alcohols	Chemical crosslinking	Pale						Iron oxide yellow P. Yellow 42; Flaventhrone Yellow P. Yellow 112; Anthrapyrimidine Yellow P. Yellow 108; [Iso indolinone yellow P. Yellow 109 110]	Red iron oxide P Red 101; quionacridones P. Violet 19; P. Violet 122; Dibromanthanthrone P. Red 168; [Disazo condensation red P. Red 144, 214]

P. — Pigment   Brackets [ ] indicate restricted use in high performance paints.

'Flamboyant' finishes and 'transparent' coloured paints are used where the film is required to reveal the substrate, but protect it. For example, flamboyant bicycle frame paints may use a metallic or bright aluminium basecoat with a coloured transparent topcoat. So-called 'high build woodstains' which show the pattern of hard and soft wood or 'grain', also require the use of transparent pigments.

'Coloured transparency' can be achieved by using transparent pigments or small quantities of semi-opaque pigments. Such pigments include:

Transparent pigments — transparent red and yellow iron oxides which are very fine particle size versions of the opaque pigments.

Semi-opaque pigments — organic reds and yellows, greens, and blues, which are brighter and stronger than the iron oxides, are used in smaller proportions, together with non-leafing aluminium pigments to obtain the sparkle and flip/flop effects required in, say, the motor car industry.

3.10.4 Colour and blending of pigments

The simplest paints, in terms of pigmentation, are those where only one pigment is needed to achieve the desired colour. This occurs in high-gloss, alkyd, topcoat black or white paints for doors and window frames, for example, where the single pigment would be carbon black or titanium dioxide.

More complicated pigmentation is required to meet such demands as:

Hues which are intermediate between those obtainable from single-coloured pigments

Lower sheen paints for semi-gloss, eggshell, and matt finishes, where extender pigments may be used to achieve the matting effect

Special rheology of the wet paint, in the container, during application and for film flow-out properties.

Organic and inorganic coloured pigments can be blended to produce 'subtractive' colours. The following are examples:

yellow + blue → green
red + white → pink
black + white → grey
transparent red oxide + aluminium flakes → 'metallic' brown
red + yellow + white → pale orange.

It is a sound principle to keep the paint formulation as simple as possible, using the minimum number of pigments. Ideally, only pigments which have similar resistance and durability properties should be blended in a single paint. This should ensure that the colour hue of the paint film is maintained during the life of the coating, although the total colour may fade or darken. A pale orange made from small quantities of red and yellow with white, for example, will fade to a pale yellow if a toluidine red Pigment Red 3 is used with arylamide yellow Pigment Yellow 74, and titanium dioxide, Pigment White 6. It would be better to use an arylamide red, Pigment Red 112.

The colours produced by blending pigments are often less clean and bright than the single pigment of that colour type. For example, a mixture of a green-toned arylamide yellow 10G Pigment Yellow 3, with a beta copper phthalocyanine blue, Pigment Blue 15:3, will give an acceptable green, but it would be slightly less clean and bright than phthalocyanine green Pigment Green 36, used alone.

To obtain maximum cleanness and brightness it is better to use adjacent shades of colour in colour space.

For example, to produce a bright orange from two pigments, the better way is to use a red-toned yellow with a yellow-toned red, than a green-toned yellow with a blue-toned red. The latter would produce a duller orange. Thus, for example, Pigment Yellow 1 (red-toned) plus Pigment Red 168 (yellow-toned) gives a better orange colour than Pigment Yellow 3 (green-toned) plus Pigment Red 112 (slightly blue-toned). However, the durability properties of these pigments are not equivalent: Pigment Red 168 is much more lightfast than Pigment Yellow 1, and the orange would become less yellow on exposure. A compromise is then necessary, and so Pigment Red 112, which is less lightfast than Pigment Red 168, would be used with the Pigment Yellow 1. This combination would remain orange, although the colour as a whole will fade.

Another consideration in blending pigments is that the wide differences of particle size and in the nature of the surfaces of the pigments may lead to flocculation, giving rise to floating or flooding. A carbon black pigment (particle size $0.01\,\mu m$ to $0.08\,\mu m$) blended with titanium dioxide (particle size $0.25\,\mu m$) often gives a flocculated grey.

The use of special nacreous (or pearlescent) pigments such as synthetically produced modified micas involves the principles of 'additive' colour mixing. For example, with 'interference' pigments:

red + green → 'white'.

Care needs to be taken, therefore, when using these pigments in conjunction with classical inorganic and organic coloured pigments where 'subtractive' colour mixing is involved.

3.10.5 Quantity of pigments in a paint
The amount of pigment used in a paint film is determined by:

(a) Its intensity and strength
(b) The required opacity
(c) The gloss or sheen level
(d) The resistance and durability properties specified

The paint technologist works according to one of two main concepts:

(i) Pigment volume concentration (PVC)
(ii) Pigment to binder ratio by weight or volume (P:B).

Generally automotive paints and industrial finishes are formulated at low pigment volume concentrations, whereas paints for buildings are formulated from 10% *PVC* up to 90% *PVC*, depending on the substrate and the resistance properties expected. Low-quality high *PVC* paints would be made by using a small amount of prime pigment such as anatase titanium dioxide, with a high quantity of an extender pigment such as calcium carbonate, to give 'dry hiding'; such a paint would be used on ceilings.

3.11   NOTES ON THE FAMILIES OF PIGMENTS

In this concluding section some of the more important pigment types are briefly described, and some types are included to provide examples for comparison. Thus one or two pigments, mainly used in the printing ink industry, are included. Inevitably the list is not exhaustive. The pigments are grouped in colour or performance families. Where appropriate, common names used in the industry have been included.

The relative density (*RD*) figures quoted are indicative and not necessarily accurate. They have been included because most manufacturers buy raw materials and pigments by weight, and sell paint by volume.

3.11.1   White pigments

The most important white pigment is titanium dioxide in its rutile crystal form. It is a 'modern' pigment in that it was made available on a commercial scale to the paint formulator only after the second world war.

As the prime white pigment for all types of paint coatings, it has largely displaced:

	Colour Index Pigment White
White lead	1
Zinc oxide	4
Zinc sulphide	7
Lithopone	5
Antimony oxide	11
Titanium dioxide-anatase	6

Its high refractive index, compared with that of the other pigments (see Table 3.11), together with its excellent resistance properties, are the cause of the displacement.

Titanium dioxide

Colour Index — Pigment White 6		*RD*	Rutile 3.8 to 4.2
Formula — Titanium dioxide	TiO_2	*RD*	Anatase 3.7 to 3.9

Table 3.11 — White pigments.

Pigment	RI	Comments
White lead (basic lead carbonate) and associated lead salts such as basic lead sulphate and basic lead silicate.	2.00	Reacts with acidic media. Discolours in sulphurous air. Lead compounds are suspected of being toxic
Zinc oxide	2.01	Reacts with acidic media
Zinc sulphide	2.37	Chalks on exterior exposure
Lithopone 30% (zinc sulphide/barium sulphate)	1.84	
Antimony oxide	2.05	Expensive, but used in some fire-retardant paints
Titanium dioxide-anatase	2.55	Chalks on exterior exposure
Titanium dioxide-rutile	2.76	The most important white pigment

Anatase and rutile crystal forms of the extremely non-reactive titanium dioxide are both used as prime pigments, but the quantity of rutile grade used far exceeds the anatase version in paints. It has a higher refractive index and higher tinting strength, being also less prone to chalking on exterior exposure. The introduction of the rutile grade of titanium dioxide was a major step in the development of the paint industry. It provided the first durable (low chalking) highly opacifying, high tint strength pigment which was also inert to:

(a) Water, dilute alkalis
(b) Organic and inorganic acids (except hot concentrated sulphuric acid and hydrofluoric acid).

It is insoluble in normal paint ingredients and thermally stable, while also having good colour retention, because it is unaffected by hydrogen sulphide and other atmospheric pollutants.

Rutile titanium dioxide at the optimum particle size (0.23–0.25 µm) is produced by calcining precipitated titanium oxide pulp under carefully controlled conditions. The anatase or rutile version is made from titanium-bearing ores such as 'Ilmenite' ($FeTiO_3$) by a sulphate process, which involves dissolving the ore in hot concentrated sulphuric acid. For the rutile version only, a chloride process may be used. This involves the reduction of titanium ore and chlorine to produce titanium tetrachloride, which is then oxidized to produce titanium dioxide.

The production of titanium dioxide normally takes about two weeks from drying the ilmenite ore to the finalised pigment.

Commercially available grades of titanium dioxide are generally made for specific purposes; for example, for high-performance solvent-based automotive paints or for decorative waterborne emulsion paints. Here the manufacturers have developed sophisticated pigment-coating technology involving the use of alumina, silica, various metals and organic compounds, which render the pigment surfaces more hydrophobic and modify the durability in terms of gloss retention and chalk resistance.

Rutile titanium dioxide is used in all types of paint, and considering its excellent properties, is reasonably priced. Nevertheless, the price is high enough to force paint manufacturers to look for ways of economizing on the use of titanium dioxide-using inorganic and organic extender pigments.

White lead
Colour Index — Pigment White 1   RD 6.7
Alternative names — Basic carbonate white lead
                    Cerussa (from naturally occurring cerussite lead
                        carbonate)
                    Flake lead
                    Flake white
Formula   — Basic lead carbonate — $2Pb\,CO_3\,Pb(OH)_2$

White lead darkens on exposure to sulphurous atmospheres, giving black lead sulphide. It reacts with acidic binders, but it is this property, however — the ability to form soaps with the fatty acid of, say, linseed oil — which gave the pigment its reputation for providing tough, elastic, durable paint films. It was mainly used in primers for wood and in undercoats.

The advent, in Europe, of grave concern about the toxicity of lead compounds has reduced the use of white lead pigment, particularly in decorative house paints, to virtually nothing.

It can be made by a wet process. In the first place, molten pig lead is poured into water, giving metallic 'spattered' lead with a high surface area. This is converted to lead oxide or hydroxide and dissolved in lead acetate. The resulting liquor is clarified and treated with carbon dioxide, yielding the basic lead carbonate .

Zinc oxide
Colour Index — Pigment White 4   *RD 5.6*
Formula   — Zinc oxide ZnO

Zinc oxide is a reactive pigment producing zinc soaps in oleoresinous paints and thereby conferring hardness and abrasion and moisture resistance on such systems. This reactivity with acidic resins is a drawback where, say, high acid value, oil-modified alkyds are concerned, since the zinc oxide produces thickening, which may be unacceptably severe.

Zinc oxide has the reputation of having mildew resistance properties and also having fungicidal and anti-fouling activity in marine paints. It is a good ultraviolet radiation absorber and, as such, reduces the breakdown of the resin of the film former. Zinc oxide is generally used in small proportions with other pigments such as titanium dioxide to enhance the general paint properties and performance.

It is available in several particle shapes and sizes including:

Nodular
Acicular
Lamellar
Colloidal

and several degrees of purity called 'Seals'. 'White Seal' is the purest and whitest. There also exist 'Green' and 'Red' Seals.

There are two major processes for making zinc oxide; the 'direct' process using zinc ore (i.e. zinc sulphide) and the 'indirect' process using zinc metal. In both processes the oxide is formed by burning zinc vapour in air — 'fume' process.

Zinc sulphide
Colour Index — Pigment White 7   *RD* 4.0
Formula   — Zinc sulphide ZnS

Zinc sulphide in the cubic sphalerate structure stabilized by traces of manganese or cobalt, gives the next strongest white pigment after titanium dioxide, having good whiteness, opacity and tinting strength. The sphalerate lattice is more light-stable than the hexagonal wurtzite form, which darkens under the action of ultraviolet radiation — metallic zinc being produced from the photochemical breakdown of the ZnS wurtzite lattice.

The pigment has a high degree of chemical inertness, but a major defect is that it chalks very heavily on exterior exposure. It can be made by the interaction of zinc chloride with hot barium sulphide solution to give the insoluble pigment, ZnS.

Lithopones
Colour Index — Pigment White 5   *RD* 4.3 (30% version)
Formula   — Zinc sulphide/barium sulphate — ZnS 30% / $BaSO_4$ 70%
                                               ZnS 60% / $BaSO_4$ 40%

Lithopone is available as two main types containing 30% or 60% zinc sulphide.

The lithopones were available commercially before zinc sulphide, because of difficulties in developing a satisfactory production process for the latter. They have all the benefits of that pigment except that they are correspondingly weaker, depending on the barium sulphate content. They also exhibit the poor resistance to chalking, typical of zinc sulphides. 30% lithopone is prepared by co-precipitation from aqueous zinc sulphate with hot barium sulphide solution.

Antimony oxide
Colour Index — Pigment White 11   *RD* 5.7
Formula   — Antimony oxide Sb_2O_3

This pigment's major use in the 1920s was in conjunction with anatase titanium dioxide to reduce chalking. Of itself, it is inert and intermediate in opacifying effect, with a refractive index of 2.0.

It is made from the ore 'Stibnite' SbS_3, by heating this with scrap iron. The resulting antimony metal is vapourized in an oxidizing atmosphere to give the crystalline pigment.

Antimony oxide is used today in the production of fire-retardant paints.

3.11.2 Black pigments

The most important black pigments are the carbon blacks. The different chemical types of black pigment available includes:

	Colour Index
Inorganic	Pigment Black
Carbon black	6, 7, 9
Graphite — has anti-corrosive properties	10
Black iron oxide	11
Black micaceous iron oxide	Not listed
Copper chromite	28
Organic	
Aniline black	1
Anthraquinone black	20

Carbon black
Colour Index — Pigment Black 6, 7, 9   *RD* 1.8
Formula   — Carbon

Carbon blacks are known as high, medium, or low colour blacks, depending on the intensity of blackness — or jetness — they demonstrate. A 'blackness' index is quoted by suppliers; the scales are arbitrary and depend on the manufacturer. Examples are 'S-value', 'Nigrometer Index', 'M-value'. An internationally recognized classifying system for pigment carbon blacks (not rubber blacks) is arranged in the order of jetness as a function of particle size and manufacturing process. In this system the following codes are in use:

    HCC — High Colour Channel
    MCC — Medium Colour Channel
    RCC Regular Colour Channel
    HCF — High Colour Furnace
    RCF — Regular Colour Furnace
    LCF — Low Colour Furnace

A further commonly used measure is to determine the tinting strength. This property of carbon blacks is generally related to an 'Industry Reference Black' —

IRB, which is indicated with 100. Blacks with lower figures have lower tinting strength, and vice versa.

There are many carbon blacks available, and the names of the various types give a clue to the original source or to the method of manufacture (see Table 3.12).

Table 3.12 — Various types of carbon black available.

Carbon black pigment	Process
Channel Impingement Gas	Made by burning natural gas in an insufficient amount of air for complete combustion, and allowing the resulting carbon black to impinge on cooled metal cylinders or channels.
Furnace Acetylene Thermal	Made by 'cracking' acetylene, gas, or oil in a furnace at high temperatures — 1300°C, with a controlled quantity of air to liberate elemental carbon.
Lamp Vegetable	Were originally made by burning vegetable oil, e.g. colza and rape seed oil, in a limited supply of air. Now the oil fraction of petroleum or coal tar is used. The colour black is deposited on a cool surface.
Wood Animal { Bone Ivory Drop	Made by burning wood, bones, or ivory in a limited air supply. Yields low carbon blacks. In the case of Wood Black, 50% to 70% carbon with calcium carbonate, and for Animal Black, 10% to 20% carbon with calcium phosphate. This group is not now widely used.

Of these carbon blacks, the Furnace Blacks are produced in the greatest quantities, for the biggest consumer — the rubber industry.

Pure Channel Blacks are no longer made, because of the air pollution caused by their manufacture and poor economics associated with their production. The term 'Channel' continues to be used, however, to denote a high colour black, because the blacks produced by this process gave the most intense 'blackness' or 'jetness' known. Impingement or Gas Blacks are still in production.

The Gas Furnace, and Lamp-types of carbon black give high opacity and tinting strength. They are non-bleeding, insoluble in paint ingredients, and are unaffected by alkalis, acids, and temperatures up to approximately 300°C, short-term. They absorb ultraviolet radiation and so give protection to the film former. They retain their colour in all conditions, giving excellent durability.

The particle size ranges of typical carbon blacks are:

Black	*Typical particle size range*
Channel/Gas	0.010–0.025 μm
Furnace	0.01–0.08 μm
Lamp	0.05–0.10 μm
Animal	0.10—0.50 μm

The finer particle size high-colour blacks (0.015 μm) are used in high-quality systems such as paint for motor cars, where high jetness is required. The medium colour blacks, which are coarser (0.03 μm) are used for intermediate quality paints, and the coarser particle sizes (0.05 μm) are used in decorative house paints. The better tinting blacks range from 0.05–0.095 μm.

There are two disadvantages in working with carbon black pigments. One is in paints such as air-drying alkyd paints containing metal soaps as the catalyst for drying, where the paint loses drying potential during storage, necessitating the addition of compensating quantities of driers. It is often said that the carbon black gradually absorbs the driers from solution during storage of the wet paint, but conclusive evidence for this is lacking. The other disadvantage is the propensity of some grades of carbon black to flocculate in mixtures with other pigments, particularly titanium dioxide.

Carbon blacks are often difficult to wet out and disperse in paints. These drawbacks are associated with the small particle size, which gives a high surface area, and the nature of the materials held within the pigment surface, e.g. oxygen and organic groups such as carboxyl, phenolic, hydroxyl, quinone, hydroquinone, and lactone.

A property particularly associated with carbon blacks is 'structure', which refers to the association of particles of carbon black into more or less chain-like or coiled groups resembling clusters of grapes. The forces between the particles vary, from weak physical attraction to chemical bonding. The degree of structure affects dispersibility, viscosity, jetness, and gloss.

The price of carbon black pigments is related to the high volumes produced for the rubber industry, giving economies of scale. Speciality blacks are produced in smaller quantities and so are more expensive.

Gas blacks have an extremely low yield because of the fine particles resulting from incomplete combustion of gas and gassified oil. As a general rule, the finer the particles size, the higher the cost.

Graphites
Colour Index — Pigment Black 10   *RD* 2.2
Formula   — Carbon crystallized in a hexagonal system

Graphite — used in 'lead' pencils — is a soft pigment consisting of plate-like particles. It is inert and offers resistance to the penetration of water. It has been used, therefore, together with other pigments, in anti-corrosive paints.

Compared with other black pigments, graphite gives low-intensity blacks, has low tinting strength, and low oil absorption. Also, the particle shape and 'slippery' nature of the particles lead to high spreading rates.

Graphite is available in varying degrees of purity from natural sources, but a synthetic version can be obtained by heating anthracite in an electric furnace at high temperature.

Black iron oxide
Colour Index — Pigment Black 11   *RD* 4.6
Formula   — Iron oxide Fe_3O_4 or
                        Ferroso-ferric oxide $FeO\ Fe_2O_3$

Black iron oxide occurs in nature as the mineral 'Magnetite'. This, after crushing and pulverizing, gives a coarse particle size pigment which can be used in paint. Iron oxide black pigments are now more often available to the paint technologist as synthetic types. They are made as a by-product in the reduction of nitrobenzene to aniline when excess iron is used. They are also made by the controlled oxidation of another industrial by-product, ferrous sulphate, in the presence of iron.

Black iron oxide is a cheap, relatively inert and solvent-resistant pigment with good lightfastness. It is unstable at high temperatures and is a low colour black with poor tinting strength, particle size 0.20 µm. However, it does not give rise to floating as do carbon blacks, and so is used with white pigments to give grey colours in the cheaper paints, and also in primers.

Black micaceous iron oxide
Colour Index — Not listed   *RD* 4.5
Formula   — Lamella-shaped (plate-like) particles of natural iron oxide Fe_2O_3

Micaceous iron oxide, so called because its presence in paint films gives a spangle, as does mica, occurs naturally in the ore known as 'flake haematite'; it is also called 'shiny ore'. It gives a dark grey glistening paint. It is an inert, inexpensive material with good heat, solvent, and lightfastness. It also has good electrical resistivity and absorbs ultraviolet radiation, thus affording protection for the polymer in the paint binder.

All these properties make it an excellent pigment from the point of view of durability. The plate-like particles orient themselves parallel to the substrate, thereby reinforcing the paint film and impeding the passage of oxygen and moisture through the film. It is used in heavy-duty coatings for structural steelwork. Care must be taken in making paints containing micaceous iron oxide because the platelets can be broken and rendered ineffective by excessive dispersion.

Copper chromite
Colour Index — Pigment Black 28   *RD* 5.3
Formula   — Copper chromite $CuCr_2O_4$ in the spinel crystal form, often modi-
                        fied by the inclusion of iron oxide Fe_2O_3 or manganese oxide MnO

This ceramic pigment is used especially for high-performance coatings which need to be cured at high temperatures such as 250°C. It has very good lightfastness, excellent hiding power, and good dispersibility. Manufacture is by high-temperature calcination of mixed oxides of copper and chromium.

Aniline black
Colour Index — Pigment Black 1   *RD* 2.0
Formula   — Azine — complex molecule involving benzenoid rings joined to the pyrazine ring, which is the chromophore.

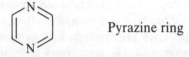

Pyrazine ring

Aniline black pigment is prepared by oxidizing aniline in the presence of a catalyst such as copper or vanadium and chromic acid. The chromate is said to be retained within the pigment molecule.

It has a high oil absorption, leading to matt effects, and is used for specialist finishes, where the low scattering and high absorption of light, a special feature of aniline black, give high opacity and very deep blacks.

The good lightfastness and high heat, chemical, and solvent resistance render the aniline blacks useful pigments, but they are less cost effective than the carbon blacks, because of the relatively lower tinting strength. A typical paint use is in the matt cellulosic or stoving enamels used on high-quality laboratory instruments and machines.

Anthraquinone black
Colour Index — Pigment Black 20   *RD* 1.9

This black pigment reflects infrared wavelengths. It is used, therefore, in specialist paints for military purposes.

3.11.3 Brown pigments
The major brown pigments are the natural and synthetic iron oxides.

	Colour Index
Inorganic	Pigment Brown
Iron oxides natural and synthetic	6, 7
Calcium Plumbate (See page 191)	10
Metal complexes	33
Organic	
Mono and disazo	23,25

Natural and synthetic brown iron oxides
Colour Index — Natural   : Pigment Brown 7   *RD* 3.7
                  Synthetic : Pigment Brown 6   *RD* 4.8
Formula   — Ferric oxide Fe_2O_3

Naturally-derived oxides are known as Burnt Sienna, Burnt Umber and Vandyke Brown, and all are made by heating various naturally-occurring ores which contain iron oxide with other minerals. They are transparent and have low tinting strength.

Synthetic brown iron oxides are usually mixtures of red, black, and yellow iron oxides, dry-ground together, and the properties are typical of each of the components.

The brown iron oxides are not used as widely in the paint industry as the yellow and red oxides described in sections 3.11.6 and 3.11.8 respectively. This is because the colour brown is easily obtained by using combinations of other pigments including black.

Metal complex brown
Colour Index — Pigment Brown 33   *RD* 5.0
Formula   — Zinc iron chromite (Zn Fe) $(FeCr)_2O_4$ in the spinel structure

Zinc iron chromite brown is made by calcining zinc, iron, and chromium oxides at high temperature. Some pigments are modified by the inclusion of oxides of nickel, aluminium, silicon, or titanium in the crystal lattice.

This ceramic pigment shows all the advantages necessary for high-performance paints, typical of the ceramic materials, and is used mostly in the 'coil coating' market where high temperatures are used to cure the paint films.

Monoazo brown
Benzimidazolone
Colour Index — Pigment Brown 25   *RD* 1.6
Formula   —

See under disazo condensation brown below.

Disazo condensation brown
Colour Index — Pigment Brown 23   *RD* 1.5
Typical formula

These two pigments, pigment Brown 25 and 23 are not widely used in paint because they are expensive, and brown colours can normally be obtained using other pigments or blends of pigments; they have good light stability and resistance properties. The main applications are in metallic and camouflage paints.

3.11.4   Blue pigments
This group of pigments contains three major inorganic types and two major organic types:

	Colour Index
Inorganic	Pigment Blue
Iron blues or Prussian blues	27
Ultramarine blue	29
Cobalt blues	36
Organic	
Phthalocyanine blue	15, 16
Indanthrone blues	60, 64

Of all the blues the phthalocyanines are the most important group owing to the colour, strength, and excellent resistance and durability properties they exhibit.

Iron blue pigments — Prussian blues
Colour Index — Pigment Blue 27   *RD* 1.3–1.8
Typical formula $KFe.Fe(CN)_6$ — Potassium ferric-ferro cyanide
                          Potassium ferro-ferricyanide   synonyms

Iron blues, also known as Prussian, Milori, Hamburg, Chinese, or Bronze blues, are potassium, sodium, or ammonium ferric–ferrocyanide. They are low-cost pigments with good durability and lightfastness in mass tone and deep tints. They have high tint strength and good resistance to solvents. Their drawbacks are low opacity, poor alkali resistance-turning to brown ferric hydroxide, even in the presence of weak alkalis, and also bronzing. They may also exhibit poor 'in-can' colour stability in the wet paint. This is thought to be due to reduction effects of the ferric–ferrocyanide back to the white ferro ferrocyanide by the oxidizing paint resins. Colour is usually restored after the paint is applied, owing to natural oxidation processes. Strong light may also cause loss of colour for pale tints.

The particle size is usually small, and the pigments form hard aggregates which give dispersion problems.

Iron blues are used in general industrial finishes to make intense full shade blues. Pigment manufacturers blend iron blues with lead chromate yellows to make lead chrome greens, called 'Brunswick Greens'.

Iron blues are made in their simpler form by reacting aqueous sodium ferrocyanide with ferrous sulphate in the presence of ammonium sulphate to produce white ammonium ferro–ferrocyanide. This is oxidized to the blue ammonium ferric–ferrocyanide at specific pH values.

Ultramarine blue

Colour Index — Pigment Blue 29   *RD* 2.3

General formula — Polysulphide of a complex alumino silicate $Na_7 Al_6 Si_6 O_{24} S_2$

This pigment is the earliest known blue pigment. The naturally derived material is known to have been used in the region around Afghanistan during the 6th century AD.

It can be made nowadays by calcining aluminium silicate, sodium carbonate, sulphur, and carbon at temperatures above 800°C. It is not an expensive pigment.

Ultramarine blue has a clear reddish blue colour and good heat and alkali resistance. However, it has very poor acid resistance, low opacity, and low tinting strength. It gives poor exterior durability in paint films because it fades, especially in pale tints and in industrial atmospheres. It is used in general-purpose industrial paints for interior use where acidity is not encountered.

Cobalt Blues

Colour Index — Pigment Blue 36   *RD* 4.4

Formula   — Cobalt chromium aluminate — a mixed phase cobalt and aluminium oxide in the spinel crystal form

These blues, which are used widely in the ceramic industry, are relatively expensive as far as paint is concerned, because they are weak colours.

They are transparent and bright blue with low tinting strength, but are very lightfast and resistant to heat, solvents, and chemicals. They are made by a mixed phase solid state reaction at very high temperatures.

They are used where high stoving temperatures and excellent durability characteristics are demanded, for example, in high-durability finishes for the coil coating market.

Phthalocyanine blue pigments
Colour Index — Pigment Blue 15, 16   *RD* 1.6, 1.5
Typical formula — Copper phthalocyanine

Phthalocyanine blues are the most important blues for the paint technologist. They are polymorphic and are available in many pigment forms which are derived from the different crystal forms:

Alpha copper phthalocyanine   — red-toned
Beta copper phthalocyanine   — green-toned
Gamma copper phthalocyanine — does not exist as a pigment
Epsilon copper phthalocyanine — slightly redder than alpha
Beta metal-free phthalocyanine — greener than beta.

The principal forms are the alpha and beta versions. The alpha crystal is less stable than the beta crystal. In the presence of aromatic hydrocarbons and some esters, or at high temperatures, it reverts to the beta form.

Therefore a class of blues known as alpha-stabilized types has been developed. This stabilization can be achieved by introducing one or two chlorine atoms into the crystal structure. The resulting crystal growth gives lower strengths and greener shades.

The very small particle size of phthalocyanine blues gives them a tendency to flocculate in mixtures with other pigments. Another group of blues is therefore available, generally described as flocculation-resistant grades. This resistance is achieved by coating the pigment particles to prevent them clustering in flocks. Materials used in coating the blues include:

(a) Sulphonated copper phthalocyanine
(b) Long chain amine derivatives of copper phthalocyanine
(c) Aluminium benzoate
(d) Acidic resins, e.g. rosin.

Many other methods of pigment stabilization are used, resulting in a multitude of blue pigments available to the paint chemist.

The sub-classification of phthalocyanine blues in the *Colour Index* takes account of the different stabilizations:

PB 15   alpha   unstabilized
PB 15:1 alpha   stabilized
PB 15:2 alpha   stabilized, flocculation-resistant
PB 15:3 beta   form
PB 15:4 beta   flocculation-resistant
PB 15:6 epsilon   form
PB 16   beta   metal-free

Phthalocyanine blue pigments are the nearest approach to an ideal organic pigment commercially available. They are produced in large quantities mainly for the printing ink industry and are reasonably priced. They are characterized by their bright hues and high tinting strength combined with excellent fastness to chemicals, solvents, light, ultraviolet radiation, and weathering, as well as having good heat resistance.

Used alone, they are very dark and transparent in mass tone, giving bronzing. It is usual to combine them with other pigments, particularly titanium dioxide and yellow pigments, to give strong blues and greens.

Although all copper phthalocyanines show good durability, the metal-free phthalocyanine blues have better lightfastness in pale tints than the copper phthalocyanines, and so they are particularly useful for pale metallic finishes used in the automotive business.

Copper phthalocyanine blues are made by reacting phthalic anhydride and urea, or phthalonitrile, with a copper salt. The resulting 'crude copper phthalocyanine' is turned into a pigment, using one of two main conditioning processes.

 (i) Acid pasting — the crude material is dissolved in concentrated sulphuric acid
                and the resulting paste in them 'drowned' in large volumes of
                water.
(ii) Salt grinding — the crude material is dry-ground with salt in ballmills.

The alpha pigment stabilized by chlorine alone is produced by acid pasting or salt grinding in the absence of solvent.

The beta pigment is obtained from the salt grinding process when a small amount of solvent is present.

Phthalocyanine blues take one or two weeks to make. Many pigment suppliers buy the crude phthalocyanine from other manufacturers and process the material, using their own specially developed techniques to produce their own grades.

Indanthrone blues
Vat blue
Colour Index — Pigment Blue 60, 64   *RD* 1.6
Formula

Pigment Blue 60

Indanthrone blues are expensive pigments, but give outstandingly good durability and resistance to solvents, chemicals, and heat. They are less likely to flocculate or show bronzing than copper phthalocyanine blues. Pigment Blue 60 is unchlorinated; the chlorinated version, Pigment Blue 64, is redder and slightly less heat stable.

They are used in high-performance paints and, in particular, in very pale metallic colours, and redder shade pale blues, where the resistance to fading in pale colours makes them unique.

Manufacture is by oxidizing anthracene to anthraquinone then converting this to the 2-amino anthraquinone. Fusion with potassium hydroxide and potassium nitrate yields the indanthrone molecule.

3.11.5   Green pigments
The pigments in this group comprise:

	Colour Index
Inorganic	Pigment Green
Chrome greens — Brunswick greens	15
Chromium oxide	17
Hydrated chromium oxide — Guignet's green	18
Organic	
Phthalocyanine green	7, 36, 37
Nickel azo yellow — Greengold	10

Like the phthalocyanine blues, the phthalocyanine greens form a distinctive class of pigments showing outstanding properties compared with other pigments in the respective colour families.

Chrome greens — Brunswick greens

Colour Index — Pigment Green 15   *RD* 3.0–5.0
Formula   — Lead chromate mixed with iron blue — potassium ferric ferrocyanide

The Brunswick greens are prepared by co-precipitating, dry-blending, or mixing the aqeous pastes of lead chromate yellow and iron blue pigments. They are sometimes known as Milori greens.

They are available in a wide range of clean bright colours, but they inherit the strengths and weaknesses of the parent pigments. Hence they are comparatively cheap high-opacity materials with high tinting strength. The durability depends on the content of lead chromate which darkens on exterior exposure, or loses yellow colour (blueing) owing to the attack of atmospheric sulphur dioxide. Their heat resistance is good, but they are affected like the iron blues by alkalis and by the reduction of the ferric–ferrocyanide during storage of the wet paint where oxidizing resins are used.

A major disadvantage of these pigments is their propensity to floating and flooding, due to unequal wetting of the different particle sizes of the lead chromate and iron blue components. Their use is further discouraged in Europe by concern about the toxicity of lead chromate.

Chromium oxide green

Colour Index — Pigment Green 17   *RD* 5.2
Formula   — Chromium sesquioxide Cr_2O_3

Chromium oxide green must not be confused with chrome greens (see above) or hydrated chromium oxide called Guignet's green (see below).

It is a grey–green colour of low tinting strength but good opacity, good heat resistance, excellent lightfastness, and extremely good resistance to chemicals.

It is also substantially non-toxic (the chromium is in the trivalent state; toxicity is associated only with hexavalent chromium).

The feature which makes this pigment important is its ability to reflect infrared radiation. It is therefore used extensively in formulating military camouflage coatings.

Manufacture is by reduction with sulphur of hexavalent chromium compounds such as sodium or potassium di-chromate.

Hydrated chromium oxide — Guignet's green
Colour Index — Pigment Green 18   *RD* 2.9 to 3.7
Formula   — Hydrated chromium oxide $Cr_2O(OH)_4$

This is different from chromium oxide green, Cr_2O_3 (see above). Owing to its water of hydration, it is less heat-resistant and slightly less acid-resistant. It has a brighter, cleaner colour and is almost transparent, having a smaller particle size; but, like chromium oxide, it is very fast to light.

It can be made by fusing sodium dichromate with boric acid and then adding water. It is more expensive than chromium oxide.

Phthalocyanine green pigments
Colour Index — Pigment Green 7, 36, 37   *RD* 2.2 to 2.8
Formula   — Halogenated copper phthalocyanine

Pigment Green 36
polychlorobromo phthalocyanine green

Phthalocyanine green pigments are available in several shades, depending on the degree of chlorination and bromination; the shades range from blue-toned green chlorinated phthalocyanine to yellowish-green chlor-brominated products.

They are similar in performance to their blue counterparts except that they do not bronze in full shade, nor are they affected by solvents, and they do not re-crystallize like the blues; also, they are fairly flocculation-resistant in polar vehicles even without surface treatment.

They are used in high-performance paints and are relatively low-priced, considering their properties.

Manufacture is by treating copper phthalocyanine with pure chlorine and pure bromine.

Nickel azo yellow — Greengold
Colour Index — Pigment Green 10   *RD* 1.6
Formula   — Nickel complex of p-chloroaniline coupled with 2,4 dihydroxy-
                 quinoline

N —O—Ni—O— N

HO   N = N   N = N   OH

Cl   Cl

Greengold pigments are dull greenish yellows with excellent lightfastness in mass tone and tints, which appear quite yellow. They have good heat stability but bleed slightly in some solvents, and are sensitive to acids.

They have relatively low tint strength and are fairly transparent, being used in pale decorative house paints, stoving enamels, and some automotive paints, particularly metallics.

3.11.6   Yellow pigments

Yellow pigments are among the most important to the paint technologist because of their use in yellow, orange, brown, and green paints. They are the second largest group of chemical types — 163 — the largest being the reds with 244.

	Colour Index
Inorganic yellows	Pigment Yellow
Lead chromates	34
Cadmium yellows	37
Yellow iron oxides (including transparent yellow iron oxides)	42, 43
Mixed crystal metal complexes, e.g. nickel with titanium	53
Barium chromate	31
Zinc chromate/tetroxy chromate	36
Strontium chromate	32

Zinc chromate/tetroxy chromate and Strontium chromate: see under corrosion-inhibiting pigments (section 3.11.12)

Organic yellow pigments
Monazo   — Acetoacetanilide
                 Acetoacetarylide
                 Acetoacetarylamide 1,3,10,73,74,97,105,111
                 Arylamide
                 Hansa® yellows
                — Benzimidazolone   120, 151, 154, 156, 175

Disazo   — Dairylide yellow ⎫
                               Arylamide disazo ⎬ 12, 13, 14, 17, 55, 83
                                 yellow   ⎪
                               Benzidine yellow ⎭

Complex metal azo — Copper azo methine   117, 129
pigments   — Nickel azo yellow,
                    known as
                    'Greengold' (see
                    page 163)

Heterocyclic Yellow — Isoindolinone yellow   109, 110, 139

Vat yellow pigments — Flavanthrone   24
based on   — Anthrapyrimidine   108
anthraquinone

Lead chromate yellow pigments
Colour Index — Pigment yellow 34   *RD* 5.5 to 5.9
Formula — Lead chromate and lead chromate/lead sulphate
          $PbCr O_4/Pb SO_4$

The more important yellow lead chromate pigments available are as follows:

Primrose chromes — Greenish, yellow lead sulphochromates (45% to 55% lead
                    sulphate) monoclinic crystal
Lemon chromes   — Yellow lead sulphochromates (20% to 40% lead sulphate
                    monoclinic crystal
Middle chromes   — Reddish yellow chromate monoclinic crystal

All lead chrome pigments are bright in shade and of good opacity. They possess
excellent fastness to heat and solvents, but they are sensitive to alkali and darken on
exposure to light and ultraviolet radiation; also they fade in reduction on exposure in
industrial atmospheres. To counteract the darkening effect, manufacturers surface-
treat their lead chromate pigments with such materials as alumina or silica, and
various metals.

The other drawback which exists at the present time with lead chromate pigments
is the suspicion that they are toxic, in that they contain both lead and hexavalent
chromium. Lead chromate pigment producers sell 'suppressed' versions, in which
they have controlled the amount of 'soluble lead' in their pigments.

Lead chromate pigments are not expensive and are therefore very useful to the
paint technologist. They are made by co-precipitation from sodium dichromate and
lead nitrate and, where necessary, sodium sulphate. The conditions need to be
controlled very carefully to produce the correct crystal form and particle size. The
proportions of the reactants, their strengths, the rate of addition, the reaction

temperature, pH, and the after treatment all contribute to the final pigment properties. The total production process takes about a week.

Lead chromate pigments are used in automotive and industrial coatings, and in some countries in paints for the domestic decorative market.

Toxicity
The evidence so far is that, provided that good standards of industrial hygiene are practised, no hazard to health exists in using lead chromate pigments. However, various countries in Western Europe and North America do not use lead chromate pigments in decorative paints for the home, and more especially on toys or surfaces which might be chewed by children.

Cadmium yellow pigments
Colour Index — Pigment Yellow 37   *RD* 4.7
Formula   — Cadmium sulphide Cd S

Cadmium sulphide occurs in nature in the mineral Greenockite. It is made commercially by the action of hydrogen or sodium sulphide on a cadmium salt. The colour of the pigments can be modified by the incorporation of selenium or zinc to give redder or greener shades respectively. Calcining increases the brightness and strength.

The cadmium yellows are of good opacity and non-bleeding, with excellent heat resistance. They have good alkali resistance and good lightfastness in full shades, but are poor for colour retention in reduced shades. They lose gloss on exposure and discolour in industrial atmospheres.

They are relatively expensive, and because of this and suspicions about toxicity, they are being used in decreasing amounts in Europe. They would be used in general industrial products for inside use.

Yellow iron oxide pigments
Colour Index — Natural: Pigment Yellow 43   *RD* 3.1
          — Synthetic: Pigment Yellow 42   *RD* 4.1
Formula   — Hydrated ferric oxide — $Fe_2O_3\ H_2O$

The natural yellow iron oxides exist as the ore Limonite, and are called Raw Sienna and Yellow Ochre. They are available in various shades from bright to dark yellows.

The synthetic grades have, to a large extent, replaced the natural materials in modern paint coatings. Manufacture is by using by-products from major industries such as the steel and titanium dioxide producers, as well as the iron from the scrap metal industry — old cars, etc.

Yellow iron oxides are needle-shaped. The colour of the pigments depends on the dimensions of the 'needles'. Long thin needles are greenish-yellow and purer in colour, while shorter or fatter ones are more red-toned and duller.

Dispersion techniques can break the needles and consequently change the colour. Care must therefore be exercised in dispersing these pigments.

The needle-shape structure, and in particular the longer needles, give high oil absorption figures because of the way they pack, giving high pore volume.

The outstanding properties of iron oxide pigments are their excellent weather resistance, lightfastness, and chemical inertness. They do not react with acids, alkalis, and normal paint ingredients, and so do not bleed.

They have a high refractive index, and at the optimum particle size for maximum light scattering, give high opacity.

Absorbing ultraviolet light, they give a measure of protection to the polymer in the paint film, retarding film breakdown.

The yellow iron oxides are less heat-stable than the red versions, losing their combined water to give the red iron oxide.

Transparent yellow and red iron oxides
Colour Index — Pigment Yellow 42   *RD* 3.6
                Pigment Red 101   *RD* 3.9
Formulae   — Yellow hydrated ferric oxide $Fe_2O_3xH_2O$
                Red ferric oxide Fe_2O_3

The transparent iron oxides are normally made by synthetic means from, say, ferrous sulphate ('Copperas'). The fine particle size necessary to give transparency in paint films is achieved by:

(1) Controlled precipitation under special conditions and
(2) Attrition in high-energy mills.

They are relatively expensive and are used for automotive, metallic paints and high-quality decorative wood coatings, where the transparency and good colouring properties, in addition to excellent resistance, make them very important. The absorption of ultraviolet light these pigments display also affords some protection for the film former, and so enhances paint film durability.

These pigments are difficult to disperse, requiring careful techniques to ensure the best compromise of gloss, transparency, and colour in the paint film.

Mixed crystal metal complex yellow pigments
Colour Index — Pigment Yellow 53   *RD* 4.6
Formula   — A mixed phase titanium, antimony, nickel oxide $TiO_2/Sb_2O_3/NiO$

These pigments are known as 'nickel titanates' in the paint industry. They are weak in colour strength, but have good resistance to light, ultraviolet radiation, heat, solvents, and chemicals. They do not give high opacity or gloss and are not easy to disperse. They are being promoted as replacements, together with high-performance organic yellow pigments, for lead chromate pigments, and are used for automotive paints. They are also used in coil coating and camouflage paints.

Barium chromate
Colour Index — Pigment Yellow 31   *RD* 4.0
Formula   — $BaCrO_4$

Barium chromate is a dull greenish yellow pigment of very low tinting strength. It has been used with lead chromate pigments to suppress the soluble lead and so render the paints safer in use. Its use in this context has been declining with the introduction of 'low' or 'controlled' soluble lead versions of lead chromate pigments.

Monoazo yellow pigments

Acetoacetanilide   ⎫ *Arylamide or*
Acetoacetarylide   ⎬ *Hansa® yellows*
Acetoacetarylamide ⎭

Colour Index   — Pigment Yellows 1, 3, 10, 73, 74, 97, 105, 111   *RD* 1.4
Typical formula — Pigment Yellow 1
              2-nitro-*p*-toluidine coupled with acetoacetanilide

The name 'Hansa®' is a trade name under which Pigment Yellow 1 and Pigment Yellow 3 were introduced to the market at the beginning of this century. The name 'Hansa' was extended to include other pigments of the same general type, but more recent 'fashions' in the industry have prevailed, and the term 'acetoacetanilide', 'acetoacetarylamide', or equally 'acetoacetarylide' are now in common use. The common name is 'Arylamide Yellow'.

The acetoacetanilide pigments are all yellow. They have a high tinting strength and good chemical resistance in full colour, showing good durability, but in pale tints they fade. They do not give high opacity and are not fast to heat and solvents. This limits their use to air-drying paints with weak solvents such as decorative house paints, although some more complex pigments do have wider application, e.g. Pigment Yellow 97.

Monoazo, benzimidazolone yellow pigments

Colour Index   — Pigment Yellow 120, 151, 154, 156, 175   *RD* 1.5
Typical formula

Pigment Yellow 120

This group of pigments are monoazo pigments into which the cyclic carbonamide group — the benzimidazolone molecule — has been introduced. (See also section 3.11.8 under Benzimidazolone reds). This confers solvent and lightfastness on the monoazo pigments and improves the heat stability, making the benzimidazolones suitable for high-quality stoving industrial finishes, including automotive finishes.

Disazo yellow pigments
Diarylide yellow, including Benzidine yellows — see note below
Colour Index — Pigment Yellow 12, 13, 14, 17, 55, 83, 155   *RD* 1.5
Typical formula

Pigment Yellow 12

The diarylide yellow pigments, in general, have better tint strength, solvent resistance, and heat stability than the simple monoazo acetoacetanilide pigments, but they have relatively poor lightfastness. They are mainly used in inks.

Note: The term 'Benzidine Yellow' has been used to describe the 'Disazo' yellow family of pigments. However, they are derived from dichlorbenzidine rather than from benzidine. The term 'Benzidine Yellow' is now considered inappropriate owing to the carcinogenic effects that are today associated with some benzidine dyes. The term 'Di-arylide' yellow is now favoured.

Copper-azo-methine yellow
Colour Index — Pigment Yellow 117, 129   *RD* 1.7
General formula

The copper-azo-methines are a dull yellowish green in colour, but they have excellent resistance and durability properties. They are highly transparent and so find their main use in automotive metallic paints. They are relatively expensive.

Isoindolinone yellow pigments
Colour Index — Pigment Yellow 109, 110, 139   *RD* 1.8
Typical formula

Tetrachloroisoindolinone

Isoindolinone pigments have excellent chemical and solvent fastness and good heat stability, accompanied by very good lightfastness. Certain of these pigments are used in some high-quality automotive and industrial paints including metallic finishes, where their semi-transparency is an advantage. Methods of manufacture quoted include the reaction of 1,4-diamenobenzene in o-dichlorbenzene with 3,3,4,4,5,6,7-hexachloroisoindolin-1-one in o-dichlorbenzene.

Anthraquinone yellows (vat yellows)
Colour Index — Pigment Yellow 24, 108   *RD* 1.5
Formule

Flavanthrone yellow 24   Anthrapyrimidine yellow 108

These polycyclic pigments have very good lightfastness, especially in tints, although the flavanthrones have a tendency to slight darkening at high concentrations. They have low opacity and are generally used in automotive and refinish metallic paints.

These complex pigments which were first available as vat dyes, are obtained by condensation of anthraquinone derivatives which are then chemically and physically modified to convert them into pigments. The whole pigment-making process can take many weeks.

3.11.7   Orange pigments
Generally, orange colours are achieved by using appropriate blends of reds and yellows. However, interest in the organic orange pigments has increased in recent years where technologists are seeking pigmentations free from lead chromate/molybdates. Sometimes a pigment classified as '*Colour Index* Pigment Orange' is a very yellow-toned red.

	Colour Index Pigment Orange
Inorganic pigments	
Chrome orange	21
'Molybdate orange' — see scarlet chromes, page 174	
Organic pigments	
Monoazo — { Naphthol Orange	5, 38
{ Dinitroaniline Orange	
— Benzimidazolone Orange	36, 60, 62
Disazo — Dianisidine	16
— Pyrazolone	13, 34
Heterocyclic Orange — Isoindolinone	42, 61

Vat orange pigments
based on anthraquinone — Perinone   43

Chrome orange
Colour Index — Pigment Orange 21   *RD* 6.8
Formula   — Basic lead chromate
              $xPbCrO_4\, yPbO$, or
              $xPbCrO_4\, yPb(OH)_2$
              with some lead sulphate $PbSO_4$
              in some grades.

The orange chrome pigments are not so widely used in the paint industry as the yellow and red chromes.

Monoazo orange pigments
Naphthol orange
Di-nitro aniline orange
'Naphtol AS®' orange
Colour Index   — Pigment Orange 5, 38   *RD* 1.5
Typical formula

Pigment Orange 5

Dinitroaniline orange pigments, being simple azo materials, have poor heat and solvent fastness, but good lightfastness in full shades. They are used in decorative solventborne and waterborne paints for buildings.

Many suppliers offer Pigment Orange 5 under the colour name 'Red' because the pigment, although classified as 'Orange' can, in practice, be a yellow-toned red.

Monoazo, benzimidazolone orange pigments
Colour Index   — Pigment Orange 36, 60, 62   *RD* 1.6
Typical formula

Pigment Orange 36

The addition of the cyclic imidazole group into the structure of the azo pigment molecule modifies the properties in the same way as for the red benzimidazolone pigments. The pigments are bright in hue with very high tint strength and are usually transparent, or nearly so. They have excellent heat and solvent resistance, together with good lightfastness. They are used in high-quality industrial and automotive paints, where the more opaque versions are being considered as replacements for lead chromate molybdate orange (Pigment Red 104) which is under suspicion on toxicity grounds.

Disazo orange pigments
Dianisidine orange
Colour Index — Pigment Orange 16   RD 1.4
Formula   — o-dianisidine coupled with 2 moles of acetoacetanilide

Dianisidine Orange has good heat resistance compared with other orange pigments in its class, but its poor lightfastness limits its use to full colour 'interior only' paints for buildings and cheaper industrial products. It is prepared by coupling tetrazotised 3,3' di-methoxybenzidine with acetoacetanilide.

Disazo orange pigments
Pyrazolone orange
Benzidine orange — see note on page 168
Colour Index — Pigment Orange 13, 34   RD 1.4
Formula

Pigment Orange 13

The pyrazolone orange pigments have a bright, clean colour and good tinting strength. The medium-to-poor solvent resistance and lightfastness restrict the use of these pigments to full orange colours in decorative paints for buildings and simple paints for industrial purposes. They are made by coupling tetrazotised 3,3'-dichloro-benzidine with 3-methyl-1-phenyl-pyrazol-5-one, or similar substituted pyrazolones.

Isoindolinone orange pigments
Colour Index   — Pigment Orange 42, 61   *RD* 1.5
General formula

Tetrachloroisoindolinone

These pigments have the same properties and uses as their yellow counterparts.

Vat orange pigment
Perinone
Colour Index — Pigment Orange 43   *RD* 1.6
Formula

Perinone orange has a transparent, clean colour with excellent colourfastness properties when subjected to chemicals, most solvents, and heat. It has high colour strength, and the good durability allows its use in pale tints for high-quality industrial coatings. It is derived from anthraquinone.

3.11.8   Red and maroon pigments

The red pigments comprise the largest group of chemical types available — 244 are classified in the *Colour Index*. To a large extent, this reflects the inadequacy of the red pigments to meet the demands of the paint technologist. The inorganic red pigments, based on cadmium sulphoselenide and lead chromate, are under a cloud of suspicion because of claims that these materials are toxic. The iron oxides do not give bright reds. This leaves the organic red compounds, which are very numerous and widely priced, and usually have been developed to meet specific requirements in paint.

Inorganic	*Colour Index* Pigment Red
Cadmium reds	108
Scarlet chromes — Molybdate oranges	104
Red iron oxides — for transparent red iron oxide (see page 166)	101, 102

Organic red pigments

Monoazo — Simple reds based on beta naphthol	
— Paranitraniline red	1
— Toluidine red	3
— Chlornitraniline	4
— Naphthol and naphthanilide reds	2, 5, 7, 8, 9, 10, 12, 13, 14, 18, 22, 23, 31, 112, 119, 146, 170, 175, 184
— Benzimidazolone red	175, 185, 208
Disazo — Pyrazolone	38
— Condensation red	144, 166, 214

Metal precipitated azo pigments

— Lithol reds	49, 57
— Calcium, barium and manganese toners, including metal BON red and Maroon toners	48, 57, 58, 63

Vat Red pigments

— Thioindigo reds	88
— Perylene reds	123, 149, 179, 190

of the anthraquinone type

— Anthraquinone red	177
— Pyranthrone	216
— Dibromanthanthrone	168
Quinacridone reds and maroons	122, 192, 202, 206

Cadmium red pigments
Colour Index — Pigment Red 108   *RD* 5.1–5.7
Formula   — CdS/CdSe, Cadmium sulpho selenide

Cadmium sulphoselenides (Wurzite lattice) are bright opaque pigments with good lightfastness in full colours and good resistance to solvents. They are heat-resistant to 600°C.

For inorganic materials they are relatively expensive. They have low tinting strength and are sensitive to acids. They also discolour in industrial atmospheres.

They are soft pigments and can be 'bruised' by incorrect or extended dispersion. Therefore care is needed in milling them, for they can become bluer and 'dirtier'. They are used in some industrial finishes.

They are made by reacting a cadmium salt with sodium sulphide which contains dissolved selenium. The resulting precipitate, after washing and filtration, is calcined at 500°C–700°C to produce the final pigment. It is the calcining stage which controls colour hue, intensity, and hiding power.

Manufacturers take a great deal of trouble to reduce the amount of soluble cadmium salts, down to levels that are 'safe'.

Scarlet chromes
Molybdate oranges
Colour Index — Pigment Red 104   *RD* 5.6
Formula   — Lead chromate–lead molybdate.
                 The mixed tetragonal crystals often also contain lead sulphate

The scarlet chromes are available in a wide range of reds, from light orange to intense scarlet. They possess high brilliance and tinting strength. They give good opacity and have good resistance to heat, solvents, and chemicals except alkalis. They have good durability, but care is needed in dispersing these pigments not to damage the pigment surfaces, which most suppliers treat to protect the core pigment from darkening on exposure.

Scarlet chromes are used in conjunction with high-durability organic maroon and violet pigments, e.g. Pigment Red 179 and Pigment Violet 19, to produce bright red colours for automotive paints, and for many industrial paints.

Manufacture and toxicity
The method of manufacture and comments on toxicity mentioned for lead chromate yellow pigments on pages 164 apply equally to the scarlet chromes.

Red iron oxides
Colour Index — Natural   : Pigment Red 102   *RD* 4.4
               — Synthetic : Pigment Red 101   *RD* 5.0
Formula   — Natural   : Aluminium silicate/ferric oxide
                              $Al_2O_3.2SiO_2.2H_2O/Fe_2O_3$
               — Synthetic : Ferric oxide Fe_2O_3

Natural red iron oxide is available in the earth's crust as 'Haematite'. It can also be made from other naturally occurring iron compounds by heating the oxides to remove the water of hydration or by oxidizing other iron compounds, to give the anhydrous brownish red colour.

Burnt Sienna (red) made by calcining raw sienna (yellow) is a well-known artists' colour.

Synthetic red iron oxide pigments are very widely used because they are relatively less expensive than other reds, have excellent opacity, light and solvent fastness and chemical resistance, and are insoluble in most surface coatings.

They are not available as pure bright shades, and they have relatively low tinting strength. In wet paint these pigments show a tendency to settle.

The synthetic red iron oxides are generally made by calcining the synthetic yellow oxides produced by a variety of techniques.

Careful control during production and calcination produces spherical particles in a range of sizes ($0.09\,\mu m$ to $0.70\,\mu m$), and these give quite different shades, varying from yellow-toned to blue-toned reds. Care needs to be taken in dispersing red iron oxides because any alteration in the particle size will affect the colour, tinting strength, and opacity (see page 131).

Monoazo red pigments
Toluidine red
Colour Index — Pigment Red 3   *RD* 1.4
Formule

Toluidine reds are simple mono-azo pigments based on beta-naphthol. They are bright red pigments with poor fastness to heat and solvent. In full shade the durability is acceptable, but in reductions with other pigments the toluidine reds fade. Such pigments find their use in decorative and in low-price industrial paints.

They are made by coupling beta-naphthol with the diazonium compound, made from the reaction of a primary aromatic amine, with nitrous acid in the presence of mineral acid.

The process takes about a day.

Monoazo red pigments
(1) 'Para Red' Paranitraniline red
Colour Index — Pigment Red 1   *RD* 1.5
Formula

(2) Chlornitraniline reds

'Chlorinated para red' — Pigment Red 4   *RD* 1.5
'Para chlor red'   — Pigment Red 6   *RD* 1.5
Formulae

Pigment Red 4   Pigment Red 6

These three pigments based on beta-naphthol — Pigment Reds 1, 4 and 6 — have been mentioned because they are interesting in that 'Para red', Pigment Red 1, is the earliest — 1885. Pigment Red 4 is well known in the printing ink industry, and Pigment Red 6 is used in some decorative paints.

The whole group is not of so much significance to the paint technologist because these relatively simple molecules yield pigments with moderate lightfastness and poor solvent fastness. The introduction of a metal group by way of the acidic azo dyes to produce the 'Lithols' and 'Metal toners', improves solvent resistance. The growth in use has therefore been with the latter pigments, particularly in the ink industry.

Monoazo red pigments
Naphthol red pigments
Naphthanilide reds (including 'Naphtol AS®' and BON reds)

Colour Index   — Pigment Red 2, 5, 7, 8, 9, 10, 12, 13,   *RD* 1.5
                           14, 18, 22, 23, 31, 112,
                           119, 146, 170, 175, 184

Typical formula

Pigment Red 112

The Naphthol red pigments are all suitable for decorative solvent and water-based paints and some industrial finishes. As a class, they have a wide range of colours, from very yellow-toned through blue-toned to maroon shades. They have good colour strength and moderate fastness to heat, light, and solvents. They are very acid- and alkali-resistant.

The performance of individual pigments in the group depends on the specific chemical make-up, as well as the methods of manufacture (see page 126 for a typical manufacturing route). The most stable pigment is that stabilized with the carbonamide ($CONH_2$) group, Pigment Red 170.

Monoazo red pigments
Benzimidazolone red
Colour Index — Pigment Red 175, 185, 208   *RD* 1.4
Formula

Pigment Red 175

The monoazo pigments can be modified by the inclusion of the benzimidazolone group as a building block in the pigment architecture. It is regarded as a carbonamide group in cyclical form:

Benzimidazole

This group confers greater solvent resistance and improves the heat-resistance and lightfastness of the pigment molecule, compared with the simpler monoazo counterparts. The Benzimidazole red pigments are relatively transparent and are used in high-quality pale and metallic paints for industrial finishes. They are made by coupling diazotized 4-chloro-2-nitroaniline to benzimidazolone.

Disazo red pigment
Pyrazolone
Colour Index - Pigment Red 38   *RD* 1.5
Formula

This precise pigment is not so important to the paint technologist. It has been included as an example of this type of molecule. This class of pigment is used, however, in some industrial paints because it has good colour and lightfastness in full shades and good resistance to many solvents, excepting xylene. The disadvantages of the pigment are low opacity, poor lightfastness in pale shades, and the relatively high cost.

Disazo condensation red pigments

Colour Index — Pigment Red 144, 166, 214   *RD* 1.5

Typical formula

The disazo condensation reds have good solvent fastness, chemical resistance, and heat stability, with good lightfastness. They are relatively expensive, however, and although having fairly high tinting strength, the overall balance of properties does not yet encourage technologists to include these pigments in other than high-quality industrial paints. The lightfastness, for example, while good, is not high enough for automotive paints.

The disazo condensation pigments are made by condensing carboxylic acid derivatives of beta-naphthol with a diamine.

Metal precipitated red azo pigments
Lithol® reds

Colour Index   - Pigment Red 49:1, 57:1   *RD* 1.6

Typical formula

Pigment Red 57:1

The two pigments Pigment Red 49:1 and 57:1 are the metal salts of acid azo dyes. They are known variously as:

Pigment Red 49:1 — 'Barium lithol red'
Pigment Red 57:1 — 'Lithol rubine', 'Calcium 4B toner', 'Rubine 4B toner',
                    and '4B toner'.

The term 'Rubine' means blue shade red. For explanation of the terms 'Lithol' and 'Toner' see page 118.

They have good heat- and solvent-resistance, but the poor chemical stability and moderate lightfastness make these pigments of less value to the paint formulator, except where interior full shade inexpensive coatings are involved.

The greatest use of these materials is in the printing industry, where they are very important for inks.

Metal precipitated red azo pigments
Metal toners (for explanation of the term 'toner' please see page 118)
Colour Index   — Pigment Red 48:1, 48:2, 48:4, 57:1,   *RD* 1.7
                        58:4, 63:1, 63:2.
Typical formula

Pigment Red 48:1

Terminology
(1) Pigment Reds 48:1, 48:2 and 48:4 are called 'Red 2B toners', being the barium, calcium, and manganese salts derived from '2B' acid and 'BON' acid,

    viz: '2B' acid: 1-amino-3-chloro-4-methyl benzene sulphonic acid
          also known as: 2-chloro-4-toluidine-5-sulphonic acid
          and 'BON' acid: 3-hydroxy-2-naphthoic acid
          also known as: beta-hydroxy-naphthoic acid.

    These pigments are also called 'Permanent red 2B pigments' and 'Metal 2B toners' (e.g. PR 48:2 'Calcium 2B toner').

    Pigment Red 48:2 and 48:4 are also called 'rubine toners'; (rubine means blue-shade red).
(2) Pigment Red 57:1, the calcium salt of '4B' acid, 4-toluidine-5-sulphonic acid coupled with BON acid, appears under various names (see above).
(3) Pigment Red 58:4, the manganese salt of 4-chloro-aniline-3-sulphonic acid coupled with 'BON' acid in known as 'Rubine G toner' and 'Red B toner'.
(4) Pigment Red 63:1 and 63:2 are the calcium and manganese salts respectively of Tobias and 'BON' acids coupled. Tobias acid is 2-naphthylamine-1-sulphonic acid. They are called 'BON maroons' and 'Claret toners'.

The metal toner pigments are strong colours covering a wide range of red shades from yellow to blue-toned reds. They have good heat and solvent resistance and medium lightfastness. The opacity they contribute is poor, and they have poor alkali resistance. They are often used in blends with inorganic red pigments such as scarlet chromes to give low-cost red paints in the general industrial business.

Vat red pigments
Thioindigo red
Colour Index — Pigment Red 88   *RD* 1.8
Formula

Thioindigo red has an intense maroon colour and is heat-, solvent-, and lightfast, with excellent resistance to acids and alkalis. It is used in high-quality industrial and automotive finishes where the relatively high price is tolerated. It is made by condensation of substituted phenylthioglycollic acids. Contamination by sulphur has been a problem and has been thought to be the cause of colour loss in extreme weathering situations — cycles of high UV exposure, followed by high humidity.

Vat red pigments
Perylene reds
Colour Index — Pigment Red 123, 149, 179, 190   *RD* 1.5
Formula

Pigment Red 179

Perylene red pigments include bright, yellowish reds with good opacity, a range of bright transparent reds, and also a group that are maroon or dirty red in colour. They have good resistance to chemicals, solvents, and heat, and have excellent lightfastness in pale tints; they are used particularly in automotive paints where their transparency enables pure colours and metallic coatings to be formulated by using them in conjunction with quinacridone and iron oxide pigments, together with aluminium flakes where necessary. Perylene reds are synthesized by first oxidizing acenaphthene to give naphthalic anhydride. This is converted by using ammonia to give naphthalimide, and thence by fusion in alkali to give the tetracarboxylic di-imide. Methylation or other such treatments produce the range of perylene pigments.

Vat Red Pigments of the anthraquinone type
(1) Anthraquinone Red
    Colour Index — Pigment Red 177   *RD* 1.5

(2) Pyranthrone Red
    Colour Index — Pigment Red 216   *RD* 1.9

(3) Dibromanthanthrone
    Colour Index — Pigment Red 168   *RD* 1.5

Typical formula

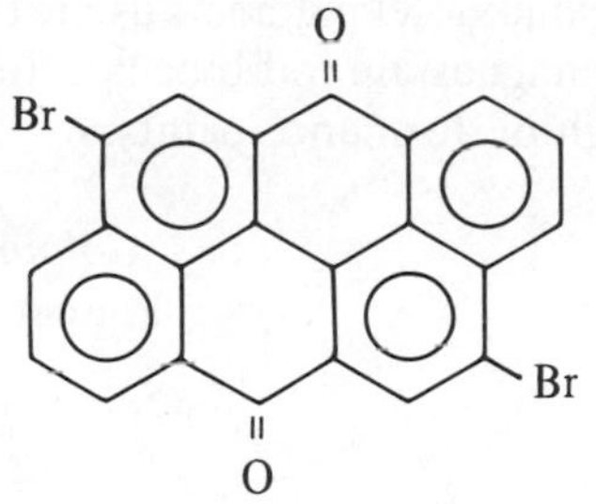

Pigment Red 168

The most important red pigment in this group, and typical of it, is Dibromanthanthrone red, Pigment Red 168.

It is a medium transparent, yellow-toned red pigment with excellent resistance to heat, solvents, acids, alkalis, and soaps, and excellent durability even at low concentrations in the paint film. It is used in high-quality automotive finishes and decorative house paints. It is expensive, because the pigment's molecule is complex and difficult to make.

It can be obtained by cyclization of dinaphthalene dicarboxylic acid in sulphuric acid and bromination, followed by chemical and physical modifications to produce the pigment.

A normal production schedule can take many weeks.

Quinacridone red pigments
Colour Index   — Pigment Red 122, 192, 202, 206   *RD* 1.6
                    — Pigment Violet 19 (very red violet colour)   *RD* 1.6
Typical formula

Pigment Red 122

Quinacridone pigments have exceptional colour brightness with excellent light, heat, chemical, and solvent resistance. However, many are somewhat difficult to disperse and are relatively low in tinting strength. Being expensive, they are used in high-quality industrial and automotive finishes. Their manufacture is either by cyclization of 2,5-diarylaminoterephthalic acid or oxidation of dihydroquinacridone. The two important crystal forms of Pigment Violet 19, beta and gamma, are obtained by ballmilling the crude material with sodium chloride for the beta, and by oxidizing 6,13-dihydroquinacridone with sodium metanitrobenzene sulphonate in the presence of pyridine for the gamma.

3.11.9   Violet pigments
Inorganic
There are no significant inorganic violet pigments used at the present time in Europe. Pigment Violet 48, a 'cobalt magnesium red-blue borate' is described in the *Colour Index* [1] as being used in high-performance paints.

	Colour Index Pigment Violet
Organic	
Dioxazine violets	23, 27
Quinacridone violets	19

Dioxazine violet pigments
Carbazole violet
Colour Index — Pigment Violet 23, 27   *RD* 1.5
Formula

Pigment Violet 23

Dioxazine Violets are very expensive, intense, bright pigments, varying from very red to very blue shades of violet. They have high tinting strength and good resistance to solvents, acids, and alkalis. The excellent lightfastness they display, even in pale tints, makes them suitable for tinting phthalocyanine blues towards the red side, and tinting white paints to make them appear 'whiter'. In combination with aluminium flakes, however, the lightfastness is reduced.

Dioxazine violet is made by the condensation of amino ethyl carbazole with chloranil in dichlorbenzene, followed by ring closure with benzene sulphonyl chloride. The crude material thus obtained is converted into pigmentary form by acid pasting with concentrated sulphuric acid or grinding the crude material in a ballmill.

Quinacridone violet pigment
Colour Index — Pigment Violet 19   **RD** 1.5
Formula

The properties are as described for Quinacridone Reds (see page 182).

3.11.10   Special effect pigments
There are three major types of special effect pigments:

	Colour Index Pigment metal
Aluminium flake pigments	1
Pearlescent pigments	Not listed
Fluorescent pigments	Not listed

Aluminium flake pigments
Colour Index — Pigment metal 1   *RD* 2.6

Aluminium flake pigments are of two kinds:

Leafing
Non-leafing

The two types differ mainly in the nature of the lubricant on the surface of the flakes, but they give very different effects in paint.

The leafing variety in a correctly formulated paint, is carried during the drying process to the paint film surface, and gives the appearance almost of a continuous film of metal.

The non-leafing variety remains within the film and produces a 'polychrome' effect, when used in conjunction with semi-transparent coloured pigments.

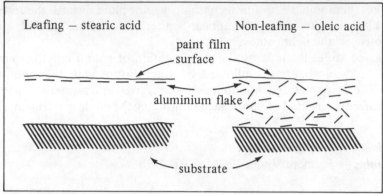

Aluminium Flake Pigments in a paint film (cross-section)

Aluminium flakes are produced by ballmilling finely divided aluminium powder. The aluminium varies in purity from 99.70% to 99.99% aluminium. The purer grades give better resistance to attack by acids. The particle size of the flakes generally varies from $1.2\,\mu m$ to $80\,\mu m$ diameter, and the thickness from $0.03\,\mu m$ to $0.30\,\mu m$.

The particle size and size distribution affect the colour of the paint, the sparkle effect, the opacity, and the 'flip-flop' effect in 'metallic finishes' where the non-leafing material is used.

Sieving and filtering processes are used to provide the different grades for the paint industry.

Leafing aluminium pastes
The leafing grades are produced when stearic acid is used as the milling lubricant in the production of the flakes. They are used to provide paints with the appearance of a bright aluminium surface. Factors which can interfere with the leafing phenomenon are:

(1) Damage to the stearic acid film by, say, excessive dispersion.
(2) Chemical desorption of the stearic acid in more highly polar media than the white spirit in which they are carried.
(3) Reaction with lead soaps such as lead octoates used as driers.

Leafing is best achieved with resins of low acid value and aromatic or aliphatic solvents containing no oxygen, i.e. not alcohols, ketones, or esters.

Non-leafing aluminium pastes
Non-leafing aluminium flakes can be obtained by using lauric or oleic acids in the milling process, as opposed to stearic acid used to make the leafing variety.

Pearlescent pigments
Nacreous pigments
Colour Index — Not listed   *RD* 1.8 to 7.7

Pearlescence is found in a variety of materials, natural and synthetic, including:

(1) Natural pearlescent platelets composed of crystals of guanine with hypoxan-
    thine, commonly called 'fish silver' — being derived from the scales of fish
(2) Basic lead carbonate platelets — $Pb_3(OH)_2(CO_3)_2$
(3) Bismuth oxychloride — Bi O Cl
(4) Mica flakes coated with titanium, iron, or chrome oxides

It is the last-named group that is chiefly used in the paint industry to give pearl and
iridescent effects.

The platelets themselves are of high refractive index (1.8 to 2.5) and transparent.
They partly reflect and partly transmit incident light.

The pearlescent lustre, which is essentially white, comes from multiple reflections caused by the oriented platelets as light passes through the film. The simultaneous reflections, at different levels in the film, create a sense of depth and give rise
to the typical pearlescent lustre.

The iridescent or 'interference' pigments are those where, at a particular optical
thickness, the interference occurring between the partly reflected and partly transmitted light rays, cancelling out some wavelengths and reinforcing others, gives rise
to a play of colours or 'iridescence', as well as the pearly lustre [8].

To work with intereference pigments the paint technologist needs to remember
that:

(1) Traditional pigments achieve colour by partial absorption of incident light. This
    gives the possibility of 'subtractive' colour mixing: blue + yellow = green
(2) Interference pigments give colour by the interference of reflected light rays: blue
    reflecting + yellow reflecting = white reflecting

Fluorescent pigments

Fluorescent pigments are made by incorporating fluorescent dyes into resins. The
dyes themselves, such as rhodamine or the aminonaphthalimides, fluoresce by
absorbing radiation at particular frequencies in the ultraviolet and visible regions,
and re-emitting the energy at lower frequencies, i.e. longer wavelengths. This, added
to the normal reflected light rays, enhances the reflection, and the materials appear
to glow in normal daylight.

The fluorescent pigments for paint are usually 3-4 μm average particle size, and
by their nature are of very poor opacity. They are quite expensive. Their poor
durability in paint, i.e. rapid fading, confines the use of these materials to short-term
interior applications.

3.11.11 Extender pigments

Extender pigments were given this name because such materials were used simply to
'extend' the effect of expensive white and coloured pigments, so cheapening the
paint formulation.

However, it is now recognized that extender pigments contribute much more to paint properties. Careful selection of the type and quantity can affect such features as:

Consistency (rheology)
Flow and levelling   Water permeability
Pigment settlement on storage   Opacity
Film strength   Gloss

Extenders in the dry form normally appear white, but as their refractive index (1.4 to 1.7) is much less than, say, titanium dioxide (2.7) and is close to that of the film formers (about 1.6) they are transparent and colourless when fully wetted with binder.

Extender pigments are available from natural sources or by synthesis.
The main extenders used in paint are:

		Colour Index Pigment White
(1)	Calcium carbonate-natural	18
	Calcium carbonate-synthetic ('precipitated')	
(2)	Aluminium silicate	19
	China clay	
	Calcined clay	
(3)	Magnesium silicate–talc	26
(4)	Barium sulphate, Barytes — natural barium sulphate	22
	Blanc fixe — synthetic barium sulphate	
(5)	Silica	27
(6)	Organic	Not listed

Calcium carbonate
Chalk
Whiting
Paris white
Colour Index — Pigment White 18   RD 2.7
Formula   — $CaCO_3$

Calcium carbonate is available to the paint technologist from natural or synthetic sources, in a wide range of particle sizes from 50 μm down to 1 μm. The natural material is obtained from chalk or limestone. The synthetic grades, 'precipitated calcium carbonate', are often made by using the by-products of other industrial processes. They are also made by heating limestone, reacting the resulting calcium oxide with water to give calcium hydroxide, and then forming the carbonate, for example by reaction with carbon dioxide.

Calcium carbonate reacts with acids, but in other respects is stable. It is used widely in water- and solvent-based paints for interior and exterior surfaces.

The oil absorption figures are comparatively low for all the grades. Certain grades are claimed to give anti-settling properties in wet paints and sag resistance in paint films.

Aluminium silicate
China clay
Kaolin
Calcined clay
Colour Index — Pigment White 19   *RD* 2.6
General formula — hydrated aluminium silicate — $Al_2O_3.2SiO_2.2H_2O$
                        calcined clay — Al_2O_3

China clay is inert and non-reactive. It is obtained from natural sources where it exists as plate-like particles from 0.5 to 50 μm. After extraction from the ground using high pressure water, it needs only milling and classification to provide the extender pigments which can be used as suspending or flatting agents, and as rheology modifiers providing thixotropy. A major use is in waterborne, matt decorative house paints. For some applications, fine particle size aluminium silicates, particularly the calcined versions (0.2 μm to 0.4 μm), can be used to partly replace the more expensive prime pigment, titanium dioxide, without undue loss of opacity.

In paints formulated above critical pigment volume concentration, the good colour of china clay becomes a significant property.

Magnesium silicate
Talc
Asbestine
Colour Index — Pigment White 26   *RD* 2.7
Typical formula — hydrated magnesium silicate $3Mg\ 0.4SiO_2.H_2O$

Talc is the name given to a wide number of naturally occurring minerals with a variety of chemical compositions containing magnesia and silica in various proportions, with impurities such as calcium and aluminium.

Rocks containing the talc are crushed and then milled in ballmills or fluid-energy mills to give lamella or fibrous (asbestine) extender pigments which are then classified.

The inert and hydrophobic nature of talc and the availability, particularly in the plate-like form, make it a useful extender for improving such paint properties as:

Pigment settling in the wet paint
Paint rheology — brushing, flow   Water and humidity resistance
Film toughness   Sanding properties of the film

Talc is used in decorative solventborne and waterborne house paints and in some industrial undercoats and primers.

Barium sulphate
Natural   : barytes
Synthetic : blanc fixe
Colour Index — barytes   : Pigment White 22   *RD* 4.5
                 blanc fixe : Pigment White 21   *RD* 4.4
Formula   — BaSO$_4$

Barium sulphate is stable to light and heat, and extremely insoluble in water. The natural form — barytes — is obtained as the mineral 'Heavy spar'. The synthetic version — blanc fixe — is made by reacting available barium compounds with sulphuric acid or soluble sulphate salts.

The high refractive index, 1.64, compared with other extender pigments where the refractive index is 1.55 to 1.58, gives this material weak pigmenting properties.

Barium sulphate is mainly used in primers and undercoats.

Silica
Colour Index — Pigment White 27   *RD* 2.3 to 2.7
Formula   — SiO$_2$

Silica is available from both natural and synthetic sources. The natural grades are:

diatomaceous silica — Kieselguhr
'amorphous' silica (actually cryptocrystalline)
crystalline silica — Quartz. Can cause 'silicosis' and is therefore a health hazard.

The synthetic grades include precipitated silica which may have particle sizes of extremely small, even colloidal, dimensions. The silicas are used as flatting aids to modify the gloss or sheen level and to improve sanding properties. They are also used as additives to modify other properties, which are listed in Chapter 5.

Organic extender pigments [9, 10, 11]
Colour Index — not listed   *RD* Varies

Recent years have seen the growth in use of organic polymer particles for water-based emulsion paints, to supplement the titanium dioxide and with the intention to provide cheaper paints with similar properties.

The polymer particles may be composed of, for example, a crosslinked polyester--styrene resin, and are generally spheres. Many versions have sealed pores in them which may contain air, water, titanium dioxide, or other inert materials, and others have only one void containing water. In the latter case the water is said to diffuse out of the centre, during film formation, leaving only air — like a micro tennis ball, but with thicker walls. The encapsulated air serves as a light-scattering site. The particle sizes vary from 0.2 μm to 50 μm (see Table 3.13).

The prices of these 'polymer extenders' reflect the complicated emulsion processing technology used in the production of the polymer beads at the correct particle size, particularly those containing vesicles or a single void. However, the prices need to be considerably lower than those for titanium dioxide to justify their use.

Table 3.13 — Typical predominant particle size of pigments and organic extenders.

Material	(μm)
Titanium dioxide	0.20–0.30
'Fine particle size' inorganic extenders	0.20–0.40
Solid polymer beads	0.2–0.5
Vesiculated polymer beads	10.0–25.0
Hollow polymer spheres	0.5–0.6
Latex emulsion film former (included for comparison)	0.1–2.0

3.11.12   Corrosion-inhibiting pigments

Corrosion-inhibiting pigments are effective in helping to prevent the rusting of surfaces by:

(1) helping to provide a physical barrier to the passage of the water and oxygen necessary for corrosion;
(2) being sacrificially destroyed as an anode, thus protecting the anodic sites that would have pitted;
(3) providing soluble passivating ions to protect the metal; or
(4) producing insoluble films which prevent active corrosion.

The main corrosion-inhibiting pigments are:

	Colour Index Pigment
Red lead	Red 105
Basic lead silico chromate	Not listed
Zinc chromate	Yellow 36
Strontium chromate	Yellow 32
Calcium, strontium, and zinc molybdates	Not listed
Calcium plumbate	Brown 10
Zinc phosphate	White 32
Zinc dust	{ Metal 6 { Black 16
Barium chromate†	White 31
Micaceous iron oxide (see page 153)	Not listed

---

† Barium chromate is not widely used as a key corrosion-inhibiting pigment, although such claims are made for most of the partly soluble chromates, especially in the protection of aluminium alloys. It is described under yellow pigments on page 166.

Red lead
Colour Index — Pigment Red 105   *RD* 8.9
Formula   — Pb_3O_4

Red lead is produced by a 'furnace' or 'fume' process. In the furnace process, litharge PbO is calcined in air; in the 'fume' process, molten lead is atomized into an oxidizing flame, and the resulting vapour is cooled.

Red lead is used mainly in primer paints for metal protection. It reacts with acidic groups in the resin to give lead soaps, which are believed to passivate iron and steel surfaces.

Basic lead silico chromate
Oncor
Colour Index — not listed   *RD* 4.0
Formula   — core SiO_2
                coating $PbSiO_3$. 3PbO
                $PbCrO_4$.PbO

Basic lead silico chromate is an orange composite pigment with silica as the core. It is made from litharge, silica, and chromic acid. Calcination and milling yield the final product, and particle sizes of $2 \mu m$ to $5 \mu m$ are used in anti-corrosive paints.

The pigment is easy to disperse, and the lead silicate–lead chromate coating is firmly bound to the silica core. It has relatively low tinting strength and is widely used with other colour pigments for metal protection in the structural steel industry and for automotive paints. The finer grade has found use in electrocoat paints. It is relatively inexpensive.

Zinc and strontium chromates
(1) Zinc chromate*
    Basic zinc chromate
    Zinc tetroxy chromate
Colour Index   — Pigment Yellow 36   *RD* 3.3 to 3.9
General formulae — zinc chromate $ZnCrO_4$
                basic zinc chromate K_2CrO_4.3ZnCrO_4$.$Zn(OH)_2$
                zinc tetroxy chromate $ZnCrO_4$ 4Zn(OH)_2$

(2) Strontium chromate
Colour Index   — Pigment Yellow 32   *RD* 3.9
Formula   — $SrCrO_4$

The zinc chromates are slightly soluble in water to varying degrees. The liberated

*Classified as a carcinogen see page 126.

chromate ions are thought to passivate the metal surfaces, producing a protective film at the anodes which inhibits the anodic reaction. They are used for iron and steel protection and also for some aluminium structures.

The more expensive strontium chromates are thought to act in the same way and are often found in paints for aluminium surfaces.

Calcium, strontium, and zinc molybdates
Colour Index — not listed   *RD*
Formula   — $CaMoO_4$ 4.2
                 — $SrMoO_4$ 4.6
                 — $ZnMoO_4$ 4.5

These three white pigments act as corrosion inhibitors by passivating the anode. So far, the materials are classified as non-toxic. They are made by normal precipitation processes, but care needs to be taken to exclude impurities such as chloride and sulphate ions from the final product, for these ions tend to stimulate corrosion.

Calcium plumbate
Colour Index — Pigment Brown 10   *RD* 5.7
Formula   — Ca_2PbO_4

Calcium plumbate is a white powder prepared by heating calcium hydroxide with lead monoxide in air. It is, in principle, insoluble in water, but it slowly yields calcium hydroxide when suspended in cold water. It is a strong oxidizing agent and reacts with fatty acids (say in linseed oil) to give lead and calcium soaps. These give adhesion properties to the paint film and also confer toughness. It finds uses in paint for zinc and galvanized iron surfaces.

Its use as a corrosion-inhibiting pigment is thought to be associated with:

(1) The capacity to oxidize soluble iron compounds which might be formed in anodic areas, to form an insoluble film of iron compounds at the anode, thus neutralizing that element of the 'corrosion cell' and preventing further corrosion.
(2) The generation of calcium carbonate at the cathodic area of the corrosion cell.

Zinc phosphate
Colour Index — Pigment White 32   *RD* 3.3
Formula   — $Zn_3(PO)_44H_2O$

This material, at present regarded as non-toxic, is found to give, in addition to corrosion inhibition:

Good durability
Excellent intercoat adhesion
Excellent flowout of brushmarks in primers

It is useful in aqueous and non-aqueous systems and is used instead of the toxic corrosion-inhibiting pigments. It is white but has low reducing strength.

The corrosion inhibition is attributed to its reaction with ammonium sulphate produced in industrial atmospheres, and also complex formation with hetero-acids to inhibit corrosion.

Zinc dust
Colour Index — Pigment Metal 6   RD 7.0
                       Pigment Black 16

Zinc dust is a fine bluish-grey powder which is a reactive pigment. It gives zinc soaps with drying oils and zincates with alkalis. It gives corrosion resistance by sacrificial electrochemical reaction, in preference to the steel substrate.

Zinc dust absorbs ultraviolet radiation and so protects film formers in exterior coatings; it is non-toxic. 'Zinc-rich' paints are used for steel constructions, even where there will be subsequent welding.

It is made by vapourizing zinc metal in an air– and oxygen-free atmosphere and collecting the fine particle metal dust from cooled surfaces.

ACKNOWLEDGEMENTS

The guidance and help of colleagues in the following industrial firms is gratefully acknowledged:

Bayer (Inorganic & Organic Divisions); BASF; Blythe Colours; Ciba-Geigy; Degussa; Hoechst; ICI (Organics and Paints Divisions); Sandoz; Tioxide International.

3.12 REFERENCES

[1] *Colour Index — pigments and solvent dyes*, 3rd edition, Society of Dyers and Colourists (1982).
[2] Sappok, R., *JOCCA* **61** 299–308 (1978).
[3] Honigmann, B., Lenné, H. U., & Schrödel, R., *Zeit. für Kristallographie* **122** Nr314 (1965).
[4] Kaluza, U., Physical/chemical fundamentals of pigment processing for paints and printing inks, *Lack u. Chemie*, Edition (1981).
[5] Merkle, K., & Herbst, W., *Farbe u. Lack* **74** Nr 12 (1968).
[6] Hafner, O., *JOCCA* **57** 268–273 (1974).
[7] Schafer, H., *Farbe u. Lack* **77** 1081–1090 (1971).
[8] Baumer, W., *Farbe u. Lack*, **79** 530–536 (1973).
                       Farbe u. Lack **79** 638–645 (1973).
                       Farbe u. Lack **79** 747–755 (1973).
[9] Chalmers, J. and Woodbridge, R. J., In: European supplement to *Polymer, Paint & Colour Journal* 5 Oct. 1983.

[10] Andrew, R. W. & Lestarquit B., *Additives* June 27 1984.
[11] Goldsborough, K., Simpson, L. A., & Tunstall, D. F. *Prog. in Organic Coatings* **10** 35–53 (1982).

BIBLIOGRAPHY

Payne, L. F. *Organic coating technology* John Wiley & Sons (1961).

Tawn, A. R. H. *Paint technology manuals* Part 6, Pigments, dyestuffs and lakes OCCA/Chapman & Hall (1966).

Patton Temple, C. *Pigment handbook*, I, II, & III John Wiley & Sons (1973).

Oil & Colour Chemists Association Australia *Surface coatings* Vol 1 Raw materials and their usage, Vol 2 Paints and their applications Chapman & Hall (1983) (1984).

Fieser & Fieser *Organic chemistry* Heath & Company (1950).

Finch, R. *Colour chemistry* ICI Publications for Schools (1972).

Solomon & Hawthorne *Chemistry of pigments and fillers* (1983).

Lewis, Peter A. *Pigments in paint, American Paint and Coatings Journal* Mar–Nov (1984).

Anon *Glossary of terms used in the paint industry* British Standards 2015 1965 revised (1973).

Moilliet, J. & Plant, D. A. *Surface treatment of pigments JOCCA* **52** 289–305 (1969).

Patton, T. C. *Paint flow and pigment dispersion,* John Wiley & Sons (1979).

Turner, G. P. A. *Introduction to paint chemistry and principles of paint technology,* Chapman & Hall (1980).

Parker, D. H. *Principles of surface coating technology,* John Wiley and Sons, (1965).

Rechman, H. 'The effect of particle size & particle charge on interactions between pigments and other substrates as exemplified by titanium dioxide' *7th Fatipec Congress Vichy* (1974).

Goldsmidt, P., Hantschke, B., & Fock, G. F. *Lacke und Farben* Glazurit Handbook (1984).

Anon, *Paint applications of microspheres, Polymers Paint Colour Journal,* 27 Nov 1985 **175** (4156) 834–5.

Dickson, E. M. & Griffiths, J. B. *Pigments and extenders filling a colourful market Industrial Minerals* May 1985 (1985).

Parfitt, G. D. and Sing K. S. W., *'Characterisation of Powder Surfaces',* Academic Press (1976).

4

Solvents, thinners, and diluents

R. Lambourne

4.1 INTRODUCTION

With the notable exception of water, all solvents, thinners, and diluents used in surface coatings are organic compounds of low molecular weight. There are two main types of these materials, hydrocarbons and oxygenated compounds. Hydrocarbons include both aliphatic and aromatic types. The range of oxygenated solvents is much broader in the chemical sense, embracing ethers, ketones, esters, ether-alcohols, and simple alcohols. Less common in use are chlorinated hydrocarbons and nitro-paraffins.

The purpose of these low molecular weight materials is implied by their names. Thus a solvent dissolves the binder in a paint. Thinners and diluents may be added to paints to reduce the viscosity of the paint to meet application requirements. These statements do not, however, go far enough to describe the complex requirements for solvents, thinners, and diluents to achieve their practical purpose. This purpose is to enable a paint system to be applied in the first instance, according to a preselected method of application; to control the flow of wet paint on the substrate and to achieve a satisfactory, smooth, even thin film, which dries in a predetermined time. They are rarely used singly, since the dual requirements of solvency and evaporation rate usually call for properties of the solvent that cannot be obtained from the use of one solvent alone.

In the foregoing we have been largely concerned with the use of solvents or solvent mixtures that dissolve the binder. There is a growing class of binders (see Chapter 2) that are not in solution, but nevertheless do require the use of a diluent, which is a non-solvent for the binder, for application. Water is thus a diluent for aqueous latex paints, and aliphatic hydrocarbons are examples of diluents for some non-aqueous polymer dispersions. Even these materials may require a latent solvent or coalescing agent to ensure that the dispersed binder particles coalesce to form a continuous film on evaporation of the diluent.

Factors that affect the choice of solvents and the use of solvent mixtures are their solvency, viscosity, boiling point (or range), evaporation rate, flash point, chemical nature, odour, toxicity and cost. Solvency is not an independent property of the liquid since it depends on the nature of the binder or film former. Odour, toxicity, and cost have become of increasing importance in recent years.

4.2 THE MARKET FOR SOLVENTS IN THE PAINT INDUSTRY

Solvents are not produced for the paint industry alone. Nevertheless, the paint industry represents the largest outlet for organic solvents. The second largest sector is probably that for metal cleaning, i.e. principally degreasing, but this market is heavily dependent on chlorinated solvents. Other major industrial users of solvents are the adhesives and pharmaceutical industries.

The production of solvents has changed dramatically during the last two decades because of two main factors: (i) environmental considerations that have led to considerable legislation particularly in the USA, and (ii) the economic recession. Solvent consumption dropped by almost 20% in 1975 and has taken almost ten years to recover, but with a different balance of solvents.

The main technical changes have been the decrease in the usage of aromatic solvents as a result of environmental legislation, notably the introduction of rule 66 in Los Angeles in 1967 and subsequently rules 442 and 443. This legislation was introduced to limit the emission of ultraviolet radiation absorbing materials into the atmosphere in an attempt to eliminate or minimize the well known Los Angeles photochemical smog. It is important to note that legislation does not prohibit the use of solvents, only their emission into the atmosphere. This has meant that some users have not discarded aromatics and other atmospheric pollutants, but have had to seek alternative means of dealing with them. Two methods are commonly used: the introduction of solvent recovery plant enabling the re-use of solvents, for example the recovery of toluene and xylene in large printing works; and the use of 'after burners' on the fume extraction plant of industrial paint users. Both of these methods must be able to justify the capital expenditure. The savings resulting from re-cycling are obvious. In the case of the installation of after burners, the cost is recovered at least to some extent by utilizing the energy obtained by the combustion of the solvent.

Reliable data on the production and use of hydrocarbons are not available, largely because most hydrocarbons are produced for fuel and the solvent fractions represent only a minor proportion of the total. Data on oxygenated solvents are more reliable, but in some cases the solvent may be used as a raw material in chemical synthesis as well as a solvent, e.g. acetone, methanol. Ball [1] has recently reviewed solvent markets and trends. Acknowledging the difficulties of obtaining reliable data, he estimates that for a range of commonly used solvents, excluding hydrocarbons, about 90% of the Western European usage can be accounted for within the geographical area of the EEC countries, plus Spain. Solvent consumption in Western Europe and the USA are compared in Table 4.1. Ten solvents commonly used in the paint industry are listed. It is interesting to note that although there continued to be a small increase in the use of organic solvents in Western Europe, overall the USA market shows a decline. This is due to the effects of the legislation referred to previously which has led to the development of alternative technology.

Table 4.1
Solvent consumption in Western Europe[†] and the USA — '000 tons [1]

Solvent	Western Europe		USA	
	1974	1978	1974	1978
Methyl ethyl ketone	140	130	238	280
Methyl isobutyl ketone	65	50	86	70
Ethyl acetate	165	173	70	81
n/Isobutyl acetates	90	123	32	44
Perchlorethylene	350	375	292	315
Trichlorethylene	270	255	157	115
Isopropanol	250	250	427	360
Acetone	238	244	325	335
Methanol	250	265	270	240
Glycol ethers	142	150	121	140
Totals	1960	2015	2018	1980

†EEC and Spain.

The use of aliphatic hydrocarbons was estimated to be at the rate of about two million tons per year in Western Europe in 1978. This compares with about 2.5 million tons in the USA for the same period.

It is to be expected that the decline in the use of organic solvents will continue in the foreseeable future for a variety of reasons, owing to the effect of anti-pollution regulations, odour, and toxicity, quite apart from economic considerations.

4.3 SOLVENT POWER OR SOLVENCY

The dissolution of a polymer or a resin in a liquid is governed by the magnitude of the intermolecular forces that exist between the molecules of the liquid and the molecules of the polymer or resin. It is these same forces that govern the miscibility of liquids and the attraction between colloidal particles suspended in a liquid. Intermolecular forces have such an important bearing on so many of the factors that control practical paint properties that it is worth considering how they arise. There are three kinds of intermolecular force: dispersion or London forces, polar forces, and hydrogen bonding. Dispersion forces arise as a result of molecular perturbation such that at any instant there is a transient but finite dipolar effect leading to an attraction between the molecules. In the case of hydrocarbons this is the only source of molecular attraction. Polar forces originate from the interaction between permanent dipoles (Keesom effect) or between dipoles and induced dipoles (Debye effect). Hydrogen bonds, largely responsible for the high anomalous boiling point of water, are well known to the organic chemist. When a hydrogen atom is attached to an electronegative atom such as oxygen or nitrogen, there is a shift in electron density from the hydrogen to the electronegative atom. Thus the hydrogen acquires a small net positive charge and the electronegative atom becomes slightly negatively charged. The hydrogen can then interact with another electronegative atom within a second molecule to form a so-called hydrogen bond. These forces are responsible for such fundamental properties as the latent heat of vaporization, boiling point, surface tension, miscibility, and of course solvency.

4.3.1 Solubility parameters

The concept of solubility parameters was introduced by Hildebrand [2] who was concerned with its application to mixtures of non-polar liquids. It was derived from considerations of the energy of association of molecules in the liquid phase, in terms of 'cohesive energy density', which is the ratio of the energy required to vaporize 1 cm^3 of liquid to its molar volume. The solubility parameter is the square root of the cohesive energy density of the liquid,

$$\delta = \sqrt{\frac{\Delta E_v}{V_m}}$$

where ΔE_v=energy of vaporization and V_m=molar volume. For liquids having similar values of δ, it is to be expected that they should be miscible in all proportions. Where δ is significantly different for two liquids it is unlikely that they will be miscible. Whilst this was an important step forward in an attempt to put solvency on a sound footing, it was of little immediate help to the paint technologists of the day who were dealing with a wide range of solvents both polar and non-polar and almost invariably used as mixtures. Moreover, since volatilization of polymers without degradation is impossible, the availability of data on polymers was not forthcoming at that time. Burrell [3] made the next major step forward in recognizing that a single parameter as defined by Hildebrand was insufficient to characterize the range of solvents in use in the paint industry. He did, however, recognize the potential of the solubility parameter concept and carried out some of the early experiments to determine solubility parameters of polymers by swelling. He saw the need to classify the hydrogen bonding capacity of solvents and proposed a system in which solvents were classified into one of three categories, poorly, moderately, and strongly hydrogen bonding. In doing so he established an empirical approach to the determination of the solubility parameters of polymers using a 'solvent spectrum'. A typical solvent spectrum is shown in Table 4.2. The solubility parameters of liquids can be calculated from experimentally determined energies of vaporization and their molar volumes. Solubility parameters of polymers can be determined empirically by contacting them with a range of solvents (e.g. the 'spectrum' in Table 4.2) and observing whether or not dissolution occurs. The solubility parameter is taken as the centre value of the range of δ values for solvents which appear to dissolve the polymer. An alternative method is to use a lightly crosslinked form of a particular polymer and to identify the solvent that is capable of causing the maximum swelling. The δ value for the polymer is then equated with that of the solvent.

Small [4] proposed a method for calculating the δ value for a polymer from the simple equation,

Table 4.2

A solvent 'spectrum' for determining the δ values of polymers

Group I (Weakly H-bonded)		Group II (Moderately H-bonded)		Group III (Strongly H-bonded)	
δ		δ		δ	
6.9	Low odour mineral spirits	7.4	Diethyl ether	9.5	2-ethyl hexanol
7.3	n-hexane	8.0	Methylamyl acetate	10.3	n-Octanol
8.2	Cyclo-Hexane	8.5	Butyl acetate	10.9	n-Amyl alcohol
8.5	Dipentene	8.9	Butyl 'Carbitol'	11.4	n-Butanol
8.9	Toluene	9.3	Dibutyl phthalate	11.9	n-Propanol
9.2	Benzene	9.9	'Cellosolve'	12.7	Ethanol
9.5	Tetralin	10.4	Cyclo-pentanone	14.5	Methanol
10.0	Nitrobenzene	10.8	Methyl 'Cellosolve'		
10.7	1-Nitropropane	12.1	2,3-butylene carbonate		
11.1	Nitroethane	13.3	Propylene carbonate		
11.9	Acetonitrile	14.7	Ethylene carbonate		
12.7	Nitromethane				

Note: Several of these solvents are highly toxic. They will not be used in practical situations as solvents for binders. The purpose of the 'spectrum' is to determine δ values for polymers only.

$$\delta = \frac{\rho \Sigma G}{M}$$

where ρ is the density of the polymer and M its molecular weight. G is the 'molar attraction constant' for the constituent parts of the molecule which are summated. Although this is only in most cases a crude approximation it serves to give a value which is a guide to the choice of the range of solvents that one might select for either of the experimental methods mentioned above. Small's approach is not satisfactory for many of the polymers that are currently used in the paint industry. In the case of those polymers that are of low molecular weight and are converted at a later stage in the film forming process into higher molecular weight products, the reactive group in the molecule may have a disproportionate effect on the solubility of the polymer. With high molecular weight polymers there has been a growth in the use of amphipathic polymers for many purposes. These may be block or graft copolymers in which different parts of the molecules have very different solubility characteristics. The implications of these structural considerations, e.g. in the context of non-aqueous dispersion polymerization (see Chapter 2), illustrate this point. Molar attraction constants (G values) are given in Table 4.3.

The need to use solvent mixtures arises for a number of reasons. We have already mentioned the need to control evaporation rate. On the other hand, a suitable single solvent may not be available. If available it may be too expensive or too toxic, or it may be particularly malodorous. It is therefore useful to predict as accurately as possible the solvency of mixtures having known individual δ values. A reasonable approximation is to calculate the δ value for the mixture simply from the volume fraction of the constituent solvents. For example, a polyester resin that is readily

Table 4.3

Molar attraction constants at 25°C according to Small [4]

Group	G	Group	G
$- CH_3$	214	H	80–100
$- CH_2 -$	133	O (ether or hydroxyl)	70
$- CH -$	28	CO (carbonyl)	275
$- C -$	–93	COO (ester)	310
$CH_2 =$	190	CN (nitrile)	410
$- CH =$	111	Cl (mean)	260
$> C =$	19	Cl (single)	270
$Ch \equiv C -$	285	Cl (twinned as in $> CCl_2$)	260
$- C \equiv C -$	222	Cl (triple as in $- CCl_3$)	250
Phenyl	735		
Phenylene	658		
Naphthyl	1146		
5 membered ring	105–115		
6 membered ring	95–105		
Conjugation	20–30		

Example: Epikote 1004

Equivalent molecular mass of repeating unit,
$\qquad C_{18}H_{20}O_3 = 284;\qquad \rho = 1.15\ g/cm^3$

ΣG;
2 ether O	@	70 =	140
1 hydroxyl O	@	70 =	70
1 hydroxyl H	@	100 =	100
2 phenylene	@	658 =	1316
2 methyl	@	214 =	428
			2348
1 tert C	@ —93	—93
		ΣG =	2255

$$\delta = \frac{\rho \Sigma G}{M} = \frac{1.15 \times 2255}{284} = 9.13$$

Experimentally determined values of δ for Epikote 1004 range from 8.5 to 13.3, give a mean value of 10.9.

soluble in ethyl acetate ($\delta=9.1$, Group II) is not soluble in either n-butanol ($\delta=11.4$, Group III) or xylene ($\delta=8.8$, Group I). However, it is soluble in a mixture of xylene and butanol in the ratio 4/1, by volume. The average δ value of the mixture is 9.3.

Interesting as these observations may be it is clear that the single value Hildebrand solubility parameter, even when used in conjunction with a broadly defined hydrogen bonding capacity, falls short of what is required to define solvent power in the paint industry. Nevertheless it did have sufficient promise as a method for predicting solubility for subsequent workers in this field to attempt to refine the concept and to improve its validity

4.3.2 Three-dimensional solubility parameters

All subsequent developments in this field have been based upon the resolution of the single 'solubility parameter' into three constituent parts related to the three types of inter molecular forces described in 4.3, namely dispersion forces, polar forces, and hydrogen bonding. Crowley, Teague, & Lowe [5] were the first to put forward a scheme which used a solubility parameter δ (equivalent to that of Hildebrand), the dipole moment of the molecule, μ, and a hydrogen bonding parameter, γ. The hydrogen bonding parameter, γ, was derived from the spectroscopic data published by Gordy [6]. Although an improvement on the original Hildebrand solubility parameter as applied by Burrell to paint polymer and resin systems, the Crowley method has not gained the acceptance within the industry as have two subsequent treatments, due to Hansen [7] and Nelson, Hemwall, & Edwards [8].

4.3.2.1 *The Hansen solubility parameter system*
Hansen considered that the cohesive energy density of a solvent results from the summation of energies of volatilization from all of the intermolecular attractions present in the liquid,

$$\frac{\Delta E_t^v}{V_m} = \frac{\Delta E_d^v}{V_m} + \frac{\Delta E_p^v}{V_m} + \frac{\Delta E_h^v}{V_m}$$

where ΔE^v, subscripts T, d, p, and h respectively represent the energies per mole of solvent, and energy contributions arising from dispersion, polar, and hydrogen bonding respectively. V_m is the molar volume. Alternatively, this may be written in terms of the solubility parameter δ, in the form

$$\delta_s^2 = \delta_d^2 + \delta_p^2 + \delta_h^2$$

where δ_s, δ_d, δ_p, and δ_h are the solubility parameters corresponding to the solvent, dispersion, polar, and hydrogen bonding respectively. Hansen used a homomorph concept to estimate the value of δ_d for a given solvent. This involves the assumption that a hydrocarbon having approximately the same size and shape as the solvent molecule will have the same energy of volatilization at the same reduced temperature (the ratio of the absolute temperature to the critical temperature) as the solvent, i.e. the dispersion force contributions are likely to be the same.

The difference between the energies of volatilization of the homomorph and the solvent will thus give a measure of the contributions from polar and hydrogen bonding forces. In solubility parameter terms this means,

$$\delta^d \text{ for solvent} \quad = \delta \text{ for homomorph}$$

therefore

$$\delta_s^2 - \delta_d^2 = \delta_p^2 + \delta_h^2$$

Hansen determined the values of δ_p and δ_h initially by an empirical method involving a large number of solubility experiments for several polymers and then selecting values of δ_p and δ_h which gave the best 'volumes' of solubility for the system. Later he used another method of calculating δ_p and δ_h independently and found generally good agreement with his previous empirical values. Because the concept uses three numerical values it is convenient to plot them on three axes, hence the term 'volume' of solubility which is simply a reference to the volume on a three-dimensional plot representing the three parameters, into which the parameters of a given solvent would have to fall if the solvent is to dissolve a given polymer or resin. In Hansen's case this would be an oblate spheroid. However, for convenience the scale for the dispersion contribution is doubled to convert the volume of solubility

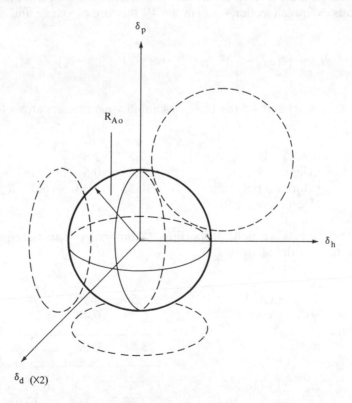

Fig. 4.1 — Diagrammatic representation of a polymer solubility plot based on Hansen's parameter system. (The dotted circles represent projections of the 'sphere of solubility').

into a sphere. In this form the sphere of solubility of a polymer or resin can be defined in terms of its position relative to the three axes, i.e. by the coordinates of its centre and its radius, R_p. An alternative way of treating solubility data is to use two-dimensional plots, where one is in effect using the projection of the sphere of solubility onto one or other of the two-dimensional axes as indicated in the diagram (Fig. 4.1). The third parameter defines the circle which should enclose solvents for the polymer or resin in question. There is, however, no sound theoretical basis for the assumption that all solvents for a given polymer will fall within the sphere of solubility, particularly as the sphere is only arbitrarily defined and without resorting to the artifice of doubling the scale for one of the parameters would not be a sphere at all. Since the sphere is determined empirically the concept does have a certain simplicity which makes it easy to apply to practical systems, and many workers have adopted Hansen's method because of this.

The distance, R_s, between the three-dimensional coordinates of the solubility parameters of the solvent and the coordinates of the centre of the sphere of solubility of the polymer determines whether or not the polymer will be soluble in the solvent. That is, if $R_p > R_s$ the polymer will be soluble in the solvent because the coordinates for the solvent fall within the sphere of solubility for the polymer.

As an illustration of this point it is worthwhile seeing how the Hansen method compares with the Burrell method for predicting the solubility of the polyester of our previous example (section 4.3.1) in the 4/1 mixture of xylene and n-butanol.

$$R_s = [4(\delta_{do} - \delta_d)^2 + (\delta_{po} - \delta_p)^2 + (\delta_{ho} - \delta_h)^2]^{\frac{1}{2}}$$

From Tables 4.4 and 4.5 the Hansen solubility parameters are as follows:

	δ_d	δ_p	δ_h	
n-butanol	7.8	2.8	7.7	
xylene	8.7	0.5	1.5	
saturated polyester	10.53	7.30	6.0	$R_p = 8.2$
(Desmophen 850)				

Since the solvents are in the proportion 1/4, butanol/xylene, the equivalent solubility parameters for the mixture are:

	δ_d	δ_p	δ_h
20% butanol	1.56	0.56	1.54
80% xylene	6.96	0.40	1.20
therefore			
mixture	8.52	0.96	2.74

$$R_s = [\ 4(10.53 - 8.52)^2 + (7.30 - 0.96)^2 + (6 - 2.74)^2\ ]^{\frac{1}{2}}$$
$$= [\qquad 16.16 \qquad + \qquad 40.20 \qquad + \qquad 10.63 \qquad]^{\frac{1}{2}}$$
$$= (66.99)^{\frac{1}{2}} = 8.18$$

Table 4.4

Hansen's solubility parameters for a selection of solvents

Solvent	δ_d	δ_p	δ_h
n-hexane	7.3	0	0
Cyclohexane	8.2	0	0.1
Toluene	8.8	0.7	1.0
Xylene	8.7	0.5	1.5
Acetone	7.6	5.1	3.4
Methyl ethyl ketone	7.8	4.4	2.5
Methyl isobutyl ketone	7.5	3.0	2.0
Ethyl acetate	7.7	2.6	3.5
n-Butyl acetate	7.7	1.8	3.1
Ethanol	7.7	4.3	9.5
n-Propanol	7.8	3.3	8.5
n-Butanol	7.8	2.8	7.7
Ethoxyethanol	7.9	4.5	7.0
Tetrahydrofuran	8.2	2.8	3.9
Ethylene glycol	8.3	5.4	12.7
Glycerol	8.5	5.9	14.3
Dichloromethane	8.9	3.1	3.0
Carbon tetrachloride	8.7	0	0.3
Water	7.6	7.8	20.7

Table 4.5

Hansen's solubility parameters for a selection of polymers

Polymer type	Trade name	Supplier	Computed parameters			
			δ_d	δ_p	δ_h	R_p
Short oil alkyd (34% O.L.)	Plexal C34	Polyplex	904	4.50	2.40	5.20
Long oil alkyd (66% O.L.)	Plexal P65	Polyplex	9.98	1.68	2.23	6.70
Epoxy	Epikote 1001	Shell	9.95	5.88	5.61	6.20
Saturated polyester	Desmophen	Bayer	10.53	7.30	6.00	8.20
Polyamide	Versamid 930	General Mills	8.52	−0.94	7.28	4.70
Isocyanate (Phenol blocked)	Suprasec F5100	ICI	9.87	6.43	6.39	5.70
Urea/formaldehyde	Plastopal H	BASF	10.17	4.05	7.31	6.20
Melamine/formaldehyde	Cymel 300	American Cyanamid	9.95	4.17	5.20	7.20
Poly(styrene)	Polystyrene LG	BASF	10.40	2.81	2.10	6.20
Poly(methyl methacrylate)	Perspex	ICI	9.11	5.14	3.67	4.20
Poly(vinyl chloride)	Vipla KR	Montecatini	8.91	3.68	4.08	1.70
Poly(vinyl acetate)	Mowilith 50	Hoechst	10.23	5.51	4.72	6.70
Nitrocellulose	H23 (½ sec)	Hagedorn	7.53	7.20	4.32	5.60

Since R_s is less than R_p this predicts that the polyester will be soluble in this solvent mixture. But since the values are almost equal it suggests that the polyester is only just soluble. It can be argued, therefore, that Hansen's method can give more information than the Burrell method, since the latter does not give any indication of the influence of degree of disparity between the δ values for polymer and solvent. A selection of the Hansen δ values for solvents is given in Table 4.4, and a selection of δ and R_p values for polymers in Table 4.5. A more extensive range of values will be found in reference [9].

4.3.2.2 *The Nelson, Hemwall, and Edwards solubility parameter system*
Nelson and co-workers developed their system almost contemporaneously with Hansen, but quite independently. They proposed a system which uses the Hildebrand solubility parameter δ, a fractional polarity term, and a 'net hydrogen bond accepting index', θ_A. The latter allows for differences in hydrogen bonding characteristics of the various solvent types. It thus attempts to quantify what Burrell had recognized as an important factor (cf. his 'solvent spectrum').

$\theta_A = K\gamma$, where γ is the spectroscopic value for hydrogen bonding [6]. The coefficient K has three values: -1 for simple alcohols, 0 for glycol–ethers, and $+1$ for all other solvents.

The system is commonly used in the form of 'solubility maps' an example of which is shown in Fig. 4.2. This illustrates the application of the method to Epikote 1004, an epoxy resin (ex Shell Chemicals plc). A and B on the map represent the solubility parameters of xylene and *n*-butanol respectively, neither of which is a solvent for the resin. However, the line joining these two points intersects the region of solubility at the two points C and D, so that it would be expected that Epikote 1004 will be dissolved by mixtures of xylene and *n*-butanol for all compositions lying between C and D. Since the solubility parameters combine linearly, the composition of the mixtures at C and D can be calculated. They turn out to be 85% xylene, 15% *n*-butanol at C, and 25% xylene and 75% *n*-butanol at D. These blends now have to be checked against the fractional polarity contours on the map. Neither mixture may have a fractional polarity of greater than 0.05 if they are to be true solvents. The fractional polarity for xylene is 0.001 and for *n*-butanol is 0.096. Thus the average fractional polarities for the two mixtures are

Blend C 85/15   xylene/butanol $= 0.00085 + 0.0144 = 0.01525$

Blend D 25/75   xylene/butanol $= 0.00025 + 0.072 \ \ = 0.07225$

These results indicate that Blend C would be a solvent for the resin, but that Blend D would not. Clearly one would only have to calculate the proportions of these two solvents to give a value of fractional polarity equal to or less than 0.05 in order to achieve solubility. A number of solubility maps of this sort have been published, and for more detailed information the reader is directed to reference [10].

Both the Hansen method and the Nelson, Hemwall, and Edwards method lend themselves readily to computerization, e.g. in Rocklin & Edwards [11].

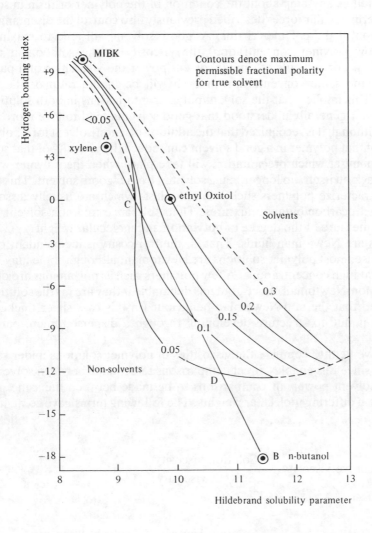

Fig. 4.2   Solubility map for Epikote 1004 (solids > c. 30% W).

4.4 SOLVENT EFFECTS ON VISCOSITY

Establishing that a given solvent or mixture of solvents is capable of dissolving the resinous or polymeric components of a paint is only the first step in solvent selection. The solubility parameter concept may aid in choosing solvents for a given system, but it does not tell us anything about the condition of the polymer or resin in solution; i.e., the intermolecular forces described previously also control the disentanglement and extension of the polymer chains. A 'good' solvent will tend to maximize (at relatively low polymer concentration) the viscosity of the solution at a given concentration. For solvents of similar solvent power the viscosities of a particular polymer at the same concentration in each will be in the ratio of the solvent viscosities. This implies that the solvent/polymer interactions in both solutions are equal. It is not generally understood that good solvents tend to give more viscous solutions, although it is recognized that the addition of a non-solvent for a polymer to a solution of that polymer in a good solvent can reduce the viscosity of that solution. There is a point at which precipitation will take place when the polymer will have collapsed back on itself, no longer being solvated by the 'good' solvent. This effect is used to characterize polymers such as aminoplasts which are usually dissolved in aromatic hydrocarbon/butanol mixtures. Their tolerance to a non-solvent such as hexane is a measure of the degree of butylation and molecular weight.

Solvents are Newtonian fluids. That is, their viscosity is independent of sheer rate. Likewise, most polymer solutions are Newtonian, although some may be non-Newtonian at high concentrations. A few polymers such as polyamides are designed to produce non-Newtonian effects, but it is debatable if they are in true solution, and it is believed that the non-Newtonian behaviour (in this case sheer thickening or thixotropy) is due to structure developing through a dispersed micro-particulate phase.

It is convenient to compare the viscosities of polymer solutions under standard conditions, since this enables direct comparisons to be made between solvents with respect to solvent power, or comparisons to be made between the same polymer type, but with differing molecular weights. The following terms have come into use:

$$\text{Relative viscosity} = \frac{\text{Solution viscosity}}{\text{Solvent viscosity}}$$

$$\text{Specific viscosity} = \frac{\text{Solution viscosity} - \text{Solvent viscosity}}{\text{Solvent viscosity}}$$

$$= \text{Relative viscosity} - 1$$

$$\text{Reduced viscosity} = \frac{\text{Specific viscosity}}{\text{Concentration (g per decilitre)}}$$

Relative viscosity and specific viscosity are dimensionless quantities; reduced viscosity has the dimension decilitres per gram. If reduced viscosities are measured for a range of dilute solutions, e.g. <1% concentration, and they are plotted against concentration, a quantity known as the intrinsic viscosity of the polymer can be obtained by extrapolation to zero concentration. The intrinsic viscosity of a polymer is related to its molecular size and thus its molecular weight.

The foregoing is applicable to linear flexible homopolymers or random copolymers in solution. With block or graft copolymers it is possible for the polymer to be dissolved by a solvent for one component block or graft, the remaining component remaining unsolvated and collapsed into a tight coil. In these cases the polymer is likely to be in a colloidal state, in micellar solution. Such a solvent is described as a 'hemi-solvent' for the polymer. The effects of solvent on polymer configuration will be described in more detail in connection with the use of polymer stabilizers for dispersion (see Chapter 8).

4.5 EVAPORATION OF SOLVENTS FROM COATINGS

In arriving at a suitable solvent blend for a surface coating, one has to consider the rate of evaporation of the solvent from the paint in relation to the conditions of application and cure. For example, thinners supplied for motor refinishing may be formulated for rapid or slow evaporation according to the ambient conditions. Also, the rate of solvent loss from a thin film governs the propensity of the coating to flow, so that it may be important to use a solvent blend of two or more solvents. This enables a rapid increase in viscosity to occur by the evaporation of a volatile component, and flow to be controlled by a less volatile one.

Evaporation takes place in essentially two stages. Initially the solvent loss is dependent on the vapour pressure of the solvent and is not markedly affected by the presence of dissolved polymer. As the polymer film is formed, solvent is retained within the film and is lost subsequently by a diffusion-controlled, slow process. This latter stage may be dominant when as much as 20% of the solvent is retained in the film.

The initial rates of evaporation can be predicted from an application of Raoult's Law, from a knowledge of activity coefficients of the components of a mixture of solvents and its composition. The measurement of evaporation rates is done very simply by a gravimetric method. Shell have developed an automatic evaporometer which is now accepted as an ASTM Standard, No. D 3539.76 [12].

Two methods are commonly used, evaporation from a thin film of the liquid, or from a filter paper, using equal volumes of liquids in each case. The results obtained are comparable for all except the most volatile of solvents. The filter paper method is preferred by many since the evaporating surface remains constant throughout the determination. Newman & Nunn [13] quote relative evaporation rates for some common solvents and tabulate their order of retention. It is common to use n-butyl acetate as a standard for evaporation and to quote all other solvents in comparison with butyl acetate, taken as unity. It is interesting to note from Table 4.6 that the retention of a given solvent is not related to its relative initial evaporation rate or its molar volume. However, within a homologous series the expected order of retention seems to apply, e.g. acetone, methyl ethyl ketone, and methyl isobutyl ketone.

Table 4.6
Order of solvent retention (data from reference [13])

Solvent	Relative evaporation rate	Molar volume (cm³/mole)	
Methanol	3.2	40	
Acetone	7.8	73	
2 Methoxy ethanol	0.51	79	
Methyl ethyl ketone	4.6	90	
Ethyl acetate	0.35	97	
2 Ethoxy ethanol	4.3	97	
n-Heptane	3.3	146	Increasing order of retention
2 n-Butoxy ethanol	0.076	130	
n-Butyl acetate	1.0	132	
2 Methoxyethyl acetate	0.35	117	
2 Ethoxyethyl acetate	0.19	135	
Toluene	2.3	106	
Methyl isobutyl ketone	1.4	124	
Isobutyl acetate	1.7	133	
Cyclohexane	4.5	108	
Methyl cyclohexane	3.5	126	
Cyclohexanone	0.25	103	
Methyl cyclohexanone	0.18	122	

4.6 FLASHPOINT

The flashpoint of a solvent mixture is important to those concerned in the manufacture, handling, storage, transport, and use of solvent-containing products. A considerable amount of legislation has been enacted to control the use of flammable liquids because of the fire hazards they represent, and standard methods have been established for the determination of flashpoint.

The flashpoint of a liquid may be defined as the lowest temperature at which the liquid, in contact with air, is capable of being ignited by a spark or flame under specified conditions. Two methods of determining flashpoint are in common use, the Abel method (Institute of Petroleum Test Method 170) and the Pensky–Marten closed cup method (ASTM D93). Each calls for a specific piece of equipment designed for the test. The apparatus and its method of use is described in each of the test methods referred to.

There have been many attempts to predict the flashpoints of paints containing solvent mixtures [14, 15]. They are based broadly on the assumption that the flashpoint is related inversely to the vapour pressure of the solvent and its activity coefficient. These predictive approaches have gained little support, however, since to meet most regulations related to transportation and storage an experimentally determined value by a specified method is almost always required. In practice it is necessary to know whether a product has a flashpoint above a specified limit to satisfy regulations regarding shipment. For example in the UK a product is not classed as 'highly flammable' unless it has a flashpoint below 32°C and is combustible. Flashpoints of a number of common solvent constituents of paints are given in Table 4.7.

Table 4.7
Boiling points, flashpoints, and *TLV*s for some common solvents

Solvent	BP °C	FP °C	TLV (ppm)
KETONES			
Acetone	57	−17	750
Methyl ethyl ketone	80	−4	200
Methyl isobutyl ketone	116	13	100
Diacetone alcohol		54[†]	50
Di isobutyl ketone		47	
ESTERS			
Ethyl acetate	77	−3	400
Isobutyl acetate		16	
n-Butyl acetate	127	25	150
2 Ethoxyethyl acetate	156	52[†]	50
2 butoxyethyl acetate	1	115[†]	
ALCOHOLS			
Methanol	65	10	200
Ethanol	78	12	1000
n-Propanol	97	22)	200
Isopropanol	82	10	400
n-butanol	118	33	50
Isobutanol	108	27	50
Sec butanol	100	14	
HYDROCARBONS			
Toluene	111	5	100
Xylene	138 144	24	100
White spirit (LAWS)	155–195	39	500
SBP 6	138–165	28	—

FP by Abel method, except [†] which were by the Pensky–Marten closed cup method.
TLV quoted is time weighted average (TWA).

As a general guide the flashpoint of a mixture of solvents can be assumed to be that of the solvent with the lowest flashpoint. However, this is not always the case, and lower values are sometimes obtained experimentally.

4.7 TOXICITY AND ENVIRONMENTAL POLLUTION

In recent years it has become evident that many of the solvents in common use represent a health hazard, and legislation has been introduced to control their use. In addition, even with solvents of low toxicity their odour may be unacceptable both in the working environment and in the vicinity of manufacturing and user plants. As has been mentioned previously, solvent emission into the atmosphere can be controlled by the use of afterburners on extraction systems or by recycling.

Studies of the toxicity of solvents, as with any chemical compound, cover a comprehensive range of standard test procedures, including animal testing for toxic effects arising from skin contact with the liquid, exposure to vapour at a range of well-defined concentrations, and by ingestion. Data have also been collected over many years on the effects of exposure of workers in most industries to the chemicals which they use. These data, which are constantly being revised as more information is obtained, are used to define acceptable working conditions for those exposed to the solvents in their daily work.

The most commonly applied limit on solvent concentration in the atmosphere is the 'threshold-limit value' (*TLV*). This is usually applied as a time weighted average value which takes into account the period that an individual may be exposed during his or her working day, e.g. eight hours. Thus the *TLV* gives the average concentration (ppm) that should not be exceeded within the working day. Such exposure is regarded as being insufficient to be a health hazard. Clearly the concentration may be exceeded for a short period if most of the time an individual is working in an atmosphere in which the concentration of the solvent is less than the *TLV*. An exception to this is if a 'ceiling' value for a given material has been defined. In this case the concentration of the material must not exceed the 'ceiling' value if a health hazard is to be avoided. Fortunately, few solvents come into this category. Some *TLV*s are given in Table 4.7.

ACKNOWLEDGEMENT
Acknowledgement is due to Mr. P. Kershaw and the Shell Chemical Company for permission to produce Figs. 4.1 and 4.2 from the *Shell Bulletin* cited.

REFERENCES

[1] Ball, T. M. D., *Polymer Paint Colour J.* **20** 595 (1960).
[2] Hildebrand, J. H. & Scott, R. (i) *The solubility of non-electrolytes* 3rd ed, Rheinhold, N.Y. (1950); (ii) *Regular solutions,* Prentice Hall, N. J. (1962)
[3] Burrell, H., *Off. Digest* **27** No. 369, 726 (1955).
[4] Small, P. A., *J. App. Chem.* **3** 71 (1953).
[5] Crowley, J. D., Teague, G. S. & Lowe, J. W., *J. Paint Tech.* **38** No. 496, 269 (1966); **39** No. 504, 19 (1967).
[6] Gordy, W., *J. Chem. Physics* **7** 93 (1939); **8** 170 (1940); **9** 204 (1945).
[7] Hansen, C. M., *J. Paint Tech.* **39** No. 505, 104 (1967); **39** No. 511, 505 (1967).
[8] Nelson, R. C., Hemwall, R. W. & Edwards, G. D., *J. Paint Tech.* **42** No. 550, 636 (1970).

[9] Kirk Othmer, *Encyclopaedia of chemical technology*, 2nd ed., Supplement volume, 889 (1971).

[10] Shell Chemicals Technical Bulletin, *Solubility parameters*, ICS(X) 78/1.

[11] Rocklin A. L. & Edwards, G. D., *J. Coating Tech.* **48** No. 620, 68 (1976).

[12] Shell Chemicals Technical Bulletin, *Evaporation of organic solvents from surface coatings*, ICS/77/4.

[13] Newman, D. J. & Nunn, C. J., *Progress in Organic Coatings* **3** 221 (1975).

[14] Walsham, J. G. In: *Solvents, theory and practice*, R. W. Tess, ed., Advances in Chemistry Series, 124 (1973).

[15] Shell technical Bulletin, *Predicting the flashpoint of solvent mixtures*, ICS/77/5.

5

Additives for paint

R. A. Jeffs and W. Jones

5.1 INTRODUCTION

In this chapter an additive is counted as one of those substances included in a paint formulation at a low level which, nevertheless, has a marked effect on the properties of the paint. Generally we do not include ingredients that we regard as part of the fundamental formulation. For example, textured coatings and fire-retardant paints use special formulating ingredients. Although these are used in relatively small quantities one does not usually convert a standard paint to products in these classes by a simple addition, therefore they remain outside the scope of this chapter.

The brief remarks made under each heading will, we hope, prove helpful, but it must be remembered that:

(a) Any listing of this type can never be complete or up-to-date with new developments.
(b) Sometimes an additive is very specific to a particular formulation, being effective in some types of paint and valueless in others.
(c) The use of additives very frequently gives rise to undesirable secondary effects, especially if used unwisely or to excess.
(d) Care must be taken that additives are not the cause of 'spiral formulation' where, for example, a surfactant is used to aid surface wetting but gives rise to foaming which is corrected by silicone solution, thus causing cissing, in turn overcome by a surfactant.
(e) In any formulation which has been developed over a long period, perhaps at the hands of several formulators, expediency may have caused the inclusion of a variety of additives. When further troubles arise it may be sounder practice to *take out* additives rather than seek yet others.

The present authors have often stripped-back a formulation to its known essentials, and only returned additives to the formulation as an up-to-date

review has found them truly necessary. Some had become a permanent feature, although originally added to overcome the eccentricity of a particular batch.
(f)   Additives are invariably related to curing paints' shortcomings. The description of paint faults is often imprecise. Make sure that the fault an additive claims to overcome is the one you are interested in. Foaming and biological protection are two areas with pitfalls for the unwary.

Some semblance of order is attempted in the sub-headings. The classes of product are in alphabetical order.

In an effort to make the entries helpful to the chemist in trouble we list, after each section, raw materials and products mentioned in the text or others that may be useful. The numbers that follow in brackets refer to the listing at the end of this chapter, where you will find the names of suppliers and, where relevant, their UK agent. This is not a comprehensive list of additive suppliers — in venturing to assemble it we are aware that we risk losing some friends. Further information and detailed addresses are available from trade publications such as *Polymers Paint Colour Year Book* (published by Fuel and Metallurgical Journals Ltd, Redhill, Surrey).

Many of the product names listed are registered trade names.

5.2   ANTI-CORROSIVE PIGMENT ENHANCERS

Various proprietary materials are offered to enhance the corrosion protection properties afforded by conventional anticorrosive pigments. Some of these are tannic acid derived (Kelate). Albarex is a treated extender used to part-replace true anticorrosive pigments such as zinc phosphate. Alcophor 827, described as a zinc salt of an organic nitrogen compound, also augments prime anticorrosive pigments. Ferrophos is recommended to part-substitute zinc dust in zinc-rich primers.

Alcophor 827	[27]	Albarex	[44/15]
Kelate	[45/34]	Ferrophos	[42]

5.3   ANTIFOAMS

Latex paints are stabilized with surfactants and colloids which, unfortunately, also help to stabilize air introduced during manufacture or during application, and thus form a stable foam. Non-aqueous paints (indeed any liquid other than a pure one) may also show bubbling. The antifoams on the market may be directed to a particular class of paint or offered for general use. Sometimes two antifoam additions are made, one at an early stage of manufacture and the other just prior to filling-out.

Usually anitfoams are of high surface activity and good mobility whilst not being actually soluble in the foaming liquid. Commonly they work by lowering the surface tension in the neighbourhood of the bubbles, causing them to coalesce to larger, less stable bubbles which then break. At their simplest, these additives may be solutions of single substances such as pine oil, dibutyl phosphate, or short chain (C_6–C_{10}) alcohols. On the other hand, they may be complex undisclosed compositions comprising mineral or silicone oils carried on fine-particle silica in the presence of surfactants. Many work by providing an element of incompatibility, and thus create centres from which bubble collapse can start. The slow wetting-out or emulsification of antifoam agents is the reason why they so often lose their effectiveness during prolonged processing or on long storage.

Testing a candidate for effectiveness can be difficult, for rarely do the shaking or stirring tests often resorted to, give more than an indication of what happens in real production conditions. As so often, an additive that works excellently in one composition will be worthless in another. Nevertheless the following are given as a few examples of the host on the market:

Foamaster(Nopco)NDW	mineral oil based	[18]
Byk 069	minerals oils, alcohols, soaps	[8/39]
Byk 031	mineral oil based: stable to over 100°C	[8/39]
Bevaloid GS32	silicone oil based	[5]
Perenol E2	non-silicone for epoxy systems	[27]
Defoamer 1512M	mineral oil, silica, silicone	[28]
Defoamer L409	silicone based: for stoving alkyds to latices	[20]
Defoamer L413	silicone/silica based, for aqueous preparations	[20]
EFKA range		[22/26]
Schwegofoam range		[52/12]

5.4 ANTISETTLING AGENTS

The prevention of settling of pigment when a solvent based paint is stored is often a matter of compromise. Good dispersing agents and deflocculants aid gloss and opacity but are not favourable for settling. Controlling the final rheological properties of the paint is also a careful balance between resisting the settlement and harming the gloss. As an additive to the dispersion stage the well-established but nonetheless valuable soya-lecithin can have profound antisettling effects. It is possible, however, to exceed the optimum level and induce very heavy vehicle separation.

To impart a small measure of thixotropy, proprietary products abound. Some years ago it was not uncommon to include a fraction of a percent of aluminium stearate. This proved difficult on plant scale. More handleable are the stearate-coated calcium carbonates (e.g. Winnofil and Sturcal) where calcium carbonate is a permissible extender. Another family used to prevent settling are the modified hydrogenated castor oils. These are usually purchased as gels in solvent and are chosen to suit the paint and the milling temperature if 'seeding' on storage is to be avoided (Thixomen, Crayvallac, Thixotrol, MPA).

The modified montomorillonite clays, e.g. Bentone 34 and Perchem 44, also impart some structure, but they need to be activated with a polar solvent such as an alcohol. Depending on how effectively this is done, some variation in structure can result. Byk Anti-terra 203 is claimed to be more reliable as a gellant. Recently grades of montmorillonite-based structuring agents have reached the market which are easier to use and require no polar solvent (Bentone SD1 and SD2, and Easigel).

It should be borne in mind that often antisettling aids are used with each other to enhance the effect. The fine-particle pyrogenic silicas (e.g. Aerosil and Cab-o-sil) are useful on their own or with, for example, Byk Anti-terra 203.

Since so many of the above additives affect viscosity at low shear rates (thus holding pigments in suspension despite gravitational forces) they are also used for the other low-shear effects, principally antisagging, flowcontrol, and colour flotation.

In latex paints, structure is so commonly formulated into the system that settling is not a problem. For other types of aqueous paints the choice of colloid is important, and judicious use of bentonite or pyrogenic silicas can be added to the pigmentation, or recourse may be made to cellulosic or synthetic polymeric thickeners. Guidance on antisettling may be sought from suppliers of:

Aluminium stearate	[21]	Bentones	[55]
Soya lecithin	[10/61] [9]	Perchem	[43]
Winnofil	[30]	Byk Anti-terra	[8/39]
Sturcal	[57]	Aerosil	[17]
Thixomen	[30]	Bentonite	[55]
Crayvallac	[14]	Easigel	[43]
Thixotrol	[41]	EFKA range	[22/26]
MPA	[41]		

5.5 ANTISKINNING AGENTS

The driers in autoxidative air-drying paints, of course, are essential for the proper balance of surface and through-drying characteristics. Unfortunately they may also cause the formation of a skin on the surface of stored paint. Originally a percent or two of pine oil or dipentine was used to alleviate the problem, with phenolic antioxidants such as guaiacol (less than 0.1%) held in reserve for the more stubborn cases. They have largely been displaced by the more easily used oximes. Butyraldoxime, and especially methyl ethyl ketoxime, bought as such, or under one of the proprietary names, are now widely used at about 0.2% on the paint. Cyclohexanone oxime is a powder and finds use because of its mild odour. Being volatile the oximes are lost from the film at an early stage and therefore do not significantly retard the drying. This volatility can, however, be a disadvantage in that a container, once opened, may now skin on being re-lidded and stored.

Some antioxidants in certain systems can cause loss of drying potential on storage, so do check drying as well as skinning performance.

Butyraldoxime	[53/40]	Dipentine	[37] [59]
Methyl ethyl ketoxime	[53/40] [6/11]	Guaiacol	[7] [49]
Cyclohexanone oxime	[53/40] [4]	Exkins 1, 2, and 3	[53/40]
Pine oil	[37] [59]		

5.6 CAN-CORROSION INHIBITORS

Aqueous paints of all types, even when packaged in lacquered tinplate containers, tend to cause corrosion, especially at internal seams and handle studs, etc.

Sodium nitrite and sodium benzoate, often in conjunction with each other at about 1% levels, are long-established inhibitors deriving from car antifreeze experience. A possible disadvantage of using salts of this type is their adverse effect on the water sensitivity of the paint film.

Ser-Ad FA179 at less than 0.3% is claimed to be an effective inhibitor of this type of corrosion, and has the advantage of insolubilizing during film formation.

Ser-Ad FA179   [53/40]   Borchicor Antirust D   [6/11]

5.7 DEHYDRATORS/ANTIGASSING ADDITIVES

With some paints it is important to keep the moisture level low for storage stability reasons. Moisture curing polyurethane paints can only use pigments from which adsorbed moisture has been largely removed, if crosslinking and gassing in the can on storage is to be avoided.

Additive TI is a monomeric isocyanate which reacts with water avidly and is therefore used to scavenge water from pigments at the dispersion stage. Additive OF (a triethyl orthoformate) is added to the paint to control the effects of residual moisture during storage.

Another group of products in which water can cause problems on storage are the aluminium paints and zinc-dust primers where reaction of the metal with moisture liberates hydrogen, causing pressure in a closed can. Sylosiv A1 and ZN1 are commonly used here, although they will not cope if grossly water-contaminated solvents or binders have been used.

Additive TI   [4]   Sylosiv A1   [25]
Additive OF   [4]   Sylosiv ZN1   [25]

5.8 DISPERSION AIDS

One of the main aims of the paint formulator is the optimization of the dispersion of pigments in the binder. Wherever possible he will do this without recourse to additives, but some binders are poor wetters, and some pigments are difficult to wet (see Chapters 6, 7 and 8). If long dispersion times, poor hiding, inferior colour development, and unsatisfactory gloss are to be avoided, then dispersion aids may well be required. The choice of dispersants, in the broadest sense, is very wide, and they are often specific to the binders, pigments, and solvents in the formulation. In fact the earliest ones used were the metal soaps, normally added last in the preparation of paint as through-driers, which were found to improve dispersion markedly if added to the millbase. Calcium and zinc octoates still find use as additives to the dispersion stage and should not be overlooked.

Additive manufacturers often publish charts showing which dispersant to use to best effect. Some are suitable for both water and solvent-carried systems. The offers may be anionic, cationic, non-ionic, or amphoteric. Usually they are characterized by possessing anchor groups which are attracted to the pigment surface and are necessarily rather specific to particular groups of pigments, and commonly a polymeric, or at any rate, higher molecular weight component that is compatible with the binder and solvents in use.

Few general rules on their use can be given, except that it is usual to incorporate a wetting aid at an early stage so that it has the best opportunity to meet up with the pigment surface and not have to compete with — or worse, to displace — other liquid components of the composition. Secondly, it is common to make later liquid additions to the millbase carefully, and with stirring if the dispersing aids themselves are not to be displaced from the pigment surface.

Of the variety of these products on the market, one range, the Solsperse 'hyperdispersants' highlights how effective, but equally how specific, such additives can be. This range usually needs first a bridging molecule chosen to be of similar nature to the pigment being dispersed, which then attracts strongly to a stabilizing molecule suited to the polarity of the solvent. When this is achieved, pigment loading of a millbase can be manyfold greater than dispersing in conventional binders. A short selection from the products on the market include:

Byk 104S	(organic with silicone)	for water and solvent systems	[8/39]
Disperbyk 163	(cationic surfactant)	for solvent systems	[8/39]
Lactimon	(anionic surfactant)	for solvent systems	[8/39]
Solsperse range	(polymeric surfactants)	for solvent systems	[31]
Centrol 3FDB	(soya lecithin based)	for solvent systems	[10/61]
Colorol E	(modified soya lecithin)	for emulsion paints	[15]
Dispex A40	(ammonium polyacrylate)	for aqueous paints	[3]
Tamol 731	(di isobutylene-maleic dispersant)	for emulsion paints	[50]

5.9 DRIERS

Although it might be argued that the use of driers is so fundamentally a part of the paint formulation that it ought not to be considered an additive, it is, by convention, included here.

The oxidation and polymerization reactions that occur as a paint dries will take place without a catalyst, but are speeded greatly by the presence of certain metal organic compounds. The catalytic activity depends on the ability of the metal cation to be readily oxidized from a stable lower valency to a less stable higher valency. All driers are therefore multivalent. In practice, usually a mixture of metal driers is used. Whereas inorganic salts, say cobalt nitrate, can be used to some effect in an autoxidative aqueous paint, by far the greatest number of driers are metal compounds of organic acids ('soaps'), which will emulsify into aqueous systems.

In earlier days lead oxide (PbO), for example, was reacted with linseed oil fatty acids or rosin to make respectively the so-called lead linoleate and lead rosinate. These worked but were unstable on storage, and of uncertain metal content. Most paint manufacturers buy-in carefully controlled, stable metal driers, usually based on octoic acid (of which the most readily available isomer is 2-ethyl hexanoic acid). The acid radical provides solubility; the metal cation is the all-important part of the drier. The literature shows that over 40 transition metals have been examined, and ten or so showed worthwhile activity.

Notes on driers
Cobalt
By far, the most powerful drier. Used alone it gives a pronounced surface dry, leaving the underfilm mobile, which gives rise to wrinkling in thick films. It is used on its own occasionally, e.g. in aluminium air-dry paints (where very thin films are applied and where Pb would dull the aluminium), and in some stoving paints.

Lead
Traditionally the most commonly used metal, ever since pigments were seen to aid drying. A 'through' drier. Still used in combination with other metals unless toxicity or sulphide staining of the dry film rules it out.

Manganese
The second most powerful drier, but not in the class of cobalt. Surface and through-drying is contributed. Its limitation is its colour; manganous, light brown goes predominantly manganic, dark brown. With most driers there is no benefit in adding excessive quantities — with manganese, drying markedly falls off beyond an optimum level.

Iron
Rather an old-fashioned drier of poor colour. Was used in cheap stoving enamels in dark colours. No real use in air-drying paints.

Zinc
Used as the 'soap' or added as zinc oxide pigment in drier quantities. It slows the initial surface dry but accelerates the through dry. The resulting film is harder when zinc is used.

Calcium
Not a prime drier, in that little effect can be seen when used on its own. As an auxiliary drier to a lead/cobalt drier system it helps prevent precipitation of the lead on storage, and drying at low temperature is improved. Some years ago when 'blooming' of dry alkyd films was a problem the presence of calcium drier helped.

Cerium
Promoted some years ago as a substitute for lead (through drier). Although effective, the pronounced yellowing of the film discouraged its use.

Vanadium
Comments rather as cerium; discoloration of the film and loss of dry on storage. No real usage.

Barium
An auxiliary drier, as is calcium. No general legislation on toxic hazard at the moment but the level of barium allowed in paints for toys, etc. is strictly controlled. Finds use in some lead-free drier combinations, but concern exists for its long-term future.

Zirconium
Finding wide favour as a through-drier to replace lead, although the chemistry of its activity is unrelated. Sold also in combination with cobalt to make lead drier substitution easy. Toxicity clearance seems satisfactory.

Aluminium
Comes up for a review every few years as a rather unusual non-toxic through-drier for a paint system that is lead-free. It is not a straightforward substitution for lead — generally the vehicle needs to be tailor-made to suit this drier. Good colour, good through-drying, and hard films are the potential benefits.

Drier combinations
Most resin manufacturers will provide recommended drier combinations and levels to use with their products. Technical service is also available from the drier manufacturers themselves.

For drier manufacturers, see listing at end of chapter, numbers: [5/11], [21], [39], 53].

5.10 ELECTRICAL PROPERTIES

It may be necessary to modify the electrical properties of a liquid paint for two main reasons.

(1) Antistatic additives may be added to hydrocarbon solvents to improve conductivity and therefore help avoid electrostatic build-up and the associated spark and fire risk when storing and transporting.
(2) To lower the electrical resistance of paints based on non-polar solvents, to enable them to be applied satisfactorily by electrostatic spray.

For the first use, it is not uncommon for some companies to make an addition automatically of, say, Antistatic Additive ASA3 at 0.3% to all incoming hydrocarbon solvents.

For electrostatic spraying the formulator will choose his solvents for optimum resistivity. If conductivity needs increasing, then minor additions of products such as Byk E80, Ransprep, or Lankrostat may be made.

Oxygenated solvents (e.g. the glycol ethers), although often desirable for their solvency characteristics being polar, show too high conductivity for good electrostatic spraying and good 'wrap-around'. Exxates 600 and 700 have been introduced, which, whilst being oxygenated solvents, nevertheless show singularly high resistivity. This makes it easier to obtain optimum solvent balance in thinners for electrostatic spraying.

Antistatic Additive ASA3	[54]	Ransprep	[48]
Byk ES80	[8/39]	Lankrostat	[38]
Exxates 600 & 700	[23]		

5.11 FLASH CORROSION INHIBITORS

Where aqueous paints find use directly onto ferrous metals, e.g. waterborne primers, rust staining may mar the film, usually in the form of a scatter of light brown spots. This so-called 'flash corrosion' may be overcome by the use of inhibitors. Generally, the same additives listed from protection against in-can corrosion find use here but at slightly higher additions.

Prestantil 448   [36/15]   Flash rust inhibitor 179   [1/35]

5.12   FLOATING AND FLOODING ADDITIVES

Most coloured paints change their colour slightly during drying, owing to the migration of some of the pigment to the surface. If it always took place in a regular manner there would be little problem.

Floating is the term usually reserved for colour striations or mottled effects, whereas flooding is the marked development of colour which, although at first uniform, is lightened again if a part dried area is disturbed (e.g. by lapping-in by brush). Because of the high viscosity of the setting paint-film the area does not then re-develop its colour. It follows that both effects may manifest themselves when two or more pigments are presented in a formulation. Flooding is often considered to be a severe form of floating. Both can be alleviated by dispersing the pigments intimately together, so that any flocculation that takes place is co-flocculation. Fine particle-size extenders such as aluminium oxide and modified precipitated calcium carbonates included at the dispersion stage are often of value. Disperbyk 160 is claimed to even out the electrical charges of organic pigment surfaces, although the level required to do so is greater than that normally expected of an additive. At the other end of the scale, introducing some controlled flocculation, e.g. with Anti-terra P, may prove a remedy.

Corrective additions to a tinted paint include the various silicone-containing products, e.g. Dow Corning PA3 or non-silicone additives such as Henkel's Product 963.

Suiting the additive to the formulation is particularly relevant in this area. In checking the control of floating, the authors have found that observing the lack of Bénard cell formation on small pools of paint dropped on to a glass panel, especially useful in judging both effectiveness and the right level of addition of these additives.

Aluminium oxide [17]   PA3   [19/34]
Disperbyk 160   [8/39]   Product 963 [27]
Anti-terra P   [8/39]

5.13   IN-CAN PRESERVATIVES

Latex and other aqueous paints are particularly prone to spoilage by microorganisms. Good housekeeping is vital to prevent major infection occurring at the production unit, but preservatives will still be needed to prevent deterioration of colloids, etc. The evil-smelling breakdown products of glue-size and casein are now rarely met, but bacterial infection can still cause gassing in the can, and enzyme breakdown of cellulosic thickeners will cause loss of viscosity.

Ideally, the biocide is added to the first charge of water, thus ensuring its presence from the earliest stage. It follows that in-can preservatives are readily water-soluble.

Although organo-mercurials and organo-tin compounds are effective and still find use, there has been a marked swing to metal-free organic biocides, for low mammalian toxicity is becoming increasingly important. End use will determine the importance of this and other properties. Where the product finds household use, then low irritancy and low sensitization will also rate high; pale colours will need a biocide that does not cause discoloration.

Generally, useful products available include the benzisothiazolinone derivatives, e.g. members of the Proxel range, substituted oxazolidines e.g. Nuosept 95, and proprietary blends of 'heterocylics' with or without formaldehyde release agents (e.g. Mergal K6N and Mergal K7).

Proxel range	[32]	Mergals	[29]
Nuosept 95	[21]	Preventol	[4/60]

5.14   IN-FILM PRESERVATIVES

Given the right environment, most classes of paint film will support mould growth, causing, typically, black stains at the junction of walls and ceiling in bathrooms. Similar growth may be seen on the outside of buildings, and the role of greenish algae in defacing exterior surfaces is increasingly recognized. Biocides incorporated into paints can help to keep the films free from such growth, but do not expect too much from them if the surface is obliterated by nutrients. The present authors recall the difficulties of keeping painted walls clean when they were heavily coated with syrup, in a Barbados sugar refinery.

Biocides are not usually added as an afterthought; neither are they planned into the formulation, but generally the level used is below 5% and therefore they are often considered as additives. Unlike biocides, used as in-can preservatives for emulsion paints which need to be highly soluble, those used for film protection need to be almost insoluble if they are not to be leached quickly from the film and the effectiveness lost with condensation or the first showers of rain.

Proprietary film preservatives are often blends of biocides to give best protection against a broad spectrum of fungi and algae. Testing for fungicidal and algicidal efficacy is a specialist activity, but fortunately many of the suppliers offer an evaluation service to help choose type and level of biocide to suit a particular paint composition and the environment it must meet.

Suppliers include:

Acticide APA	[58]	Mergal 588	[29]
Parmetol A23	[56]	Proxel range	[32]
Nopcocide N40	[18]		

5.15   INSECTICIDAL ADDITIVES

The control of household flies by the use of an insecticide in a coating seems to be worthwhile. Two main problems arise. Firstly, to render the active ingredient available to a settled insect the level of additive needs to be high and is formulated into a flat paint. Secondly, and more fundamental, is the desirability of the end effect. Where flying insects are a problem the public prefer them on-the-wing rather than dropping lifeless into their food. There are special applications where such paints find use, e.g. in ships' holds to kill cockroaches. Nor must one overlook the insecticides used in penetrating compositions for the protection of wood. Organo-metallic compounds (e.g. tributyl tin oxide, zinc octoate, and copper naphthenate) have found use in woodcare compositions.

Chlorinated aromatic compounds, e.g. 6-chloro epoxy hydroxy naphthalene and 1-dichloro 2,2' bis (p. chlorophenyl) ethane, our records show, have been offered as insecticides for addition to paint.

| Zinc octoate | [21] [39] | Tributyl tin oxide | [2] |
| Copper naphthenate | [16] | Priem Insecticide | [46] |

5.16 OPTICAL WHITENERS

These materials absorb ultraviolet wavelengths and re-emit the energy in the visible waveband. If they are chosen to emit in the blue–violet region they can give a boost to the colour of whites, overcoming any tendency to yellowness. Although widely used in the paper and detergent industries they have not found a significant outlet in the surface coating industry because of their short-lived effectiveness and the cost premium incurred. An example of a product of this type is Uvitex OB.

The use of optical whiteners is not to be confused with the 'improvement' of a white by the addition of, say, C1 Pigment Violet 23 to counter yellowness. This it does, but the resulting undesirable greyness often rules out this use.

Uvitex OB   [13]

5.17 REODORANTS

Most paints smell when drying. The mild odour of latex paints is quite acceptable to most people and receives little attention, but that from solventborne systems is more noticeable. It stems, in the first place, from the evaporation of the solvents used, but in oxidative paints a secondary and more persistent odour from the drying reactions follows. Still sought is an additive that will truly remove these odours.

Generally, care can be taken in the initial formulation to avoid particularly offensive-smelling components, e.g. by choosing lower odour solvents. However, the odour from both sources can be modified or masked by the use of industrial perfumes, such as find use in polishes and household sprays. It is the writers' experience that the majority of people prefer the original paint odour to the 'muddy' combination-odour that so often results from the use of reodorants in paints.

Reodorant suppliers are listed at the end of the chapter under numbers [7], [24], [47], [62].

5.18 ULTRAVIOLET ABSORBERS

Many pigments fade and many binders degrade owing to the effect of incident radiation, especially ultraviolet. The use of a coat of varnish was shown to slow down the fading of a fugitive paint many years ago, but unfortunately unpigmented varnish films themselves degrade quickly. Recently ultraviolet absorbers (akin to those used in sun-screen cosmetic creams) have been shown to improve the performance of such a protective clearcoat. A further advancement in this technology has been achieved by combining a light-stabilizer with the UV absorber. The resultant protection is achieved in two stages. The UV absorber (about 1%) converts the undesirable short wavelengths to heat energy, and the light-stablizer (a sterically hindered amine) captures the free radicals generated that would cause film degradation. This technology has made possible the use of 'base coat plus clear coat' automotive finishing systems and overcomes the early problems of under-film chalking, delamination, and cracking of the top clear coat.

Tinuvin 900 — UV absorber   [13]
Tinuvin 901 — hindered amine   [13]
Sanduvor 3206 — UV absorber   [51]
Sanduvor 3046 — hindered amine   [51]

5.19   ADDITIVE SUPPLIERS

All addresses are in the UK unless otherwise specified.

[1] Acima Chemical Industries Ltd, PO Box CH9470, Buchs SG (Switzerland).
[2] Albright & Wilson Ltd, PO Box 3, Oldbury, Warley, W. Midlands B68 ONN.
[3] Allied Colloids Ltd, PO Box 38, Low Moor, Bradford.
[4] Bayer (UK) Ltd, Bayer House, Strawberry Hill, Newbury, Berks.
[5] Bevaloid Ltd, Flemingate, Beverley, Yorkshire.
[6] Borchers (Gerb) AG, Elsabethstrasse 14, Postfach 1812, D-4000, Dusseldorf 1, W. Germany.
[7] Bush Boake Allen Ltd, Blackhorse Lane, London E17 5QP.
[8] Byk Chemie GmbH, Abelstrasse 14, D-4230 Wesel, W. Germany.
[9] Capricorn Chemicals Ltd, 1 Sugar House Lane, Stratford, London E15 2QN.
[10] Central Soya, 1825 North Lawrence Avenue, Chicago 39, Ill, USA.
[11] Chemitrade Ltd, Berkeley Square House, Berkeley Square, London W1X 5LA.
[12] Chemacord Ltd, 43 Upper Wickham Lane, Welling, Kent DA16 3AJ.
[13] Ciba–Geigy, Industrial Chemicals Division, Trafford Park, Manchester M17 1WT.
[14] Cray Valley Products Ltd, Farnborough, Kent BR6 7EA.
[15] Croxton & Garry Ltd, Curtis Road, Dorking, Surrey RH4 1XA.
[16] Cuprinol Ltd, Adderwell, Frome, Somerset BA11 1NL.
[17] Degussa Ltd, Stanley Green Trading Estate, Cheadle Hulme, Cheshire SK8 6RW.
[18] Diamond Shamrock Process Chemicals Ltd, 147 Kirkstall Road, Leeds LS3 1JN.
[19] Dow Corning Ltd, Reading Bridge House, Reading, Berks, RG1 8PN.
[20] Drew Ameroid International, PO Box 6, 4930 AA Geertrudenberg, Netherlands.
[21] Durham Chemicals Ltd, Birtley, Chester-le-Street, Durham DH3 1QX.
[22] EFTA Chemicals BV, PO Box 358, 2180 AJ Hillegom, Netherlands.

[23] Esso Chemical Ltd, Arundel Towers, Portland Terrance, Southampton SO9 2GW.

[24] Fritzsche Dodge & Alcott, Finedon Road, Lindhurst Estate, Links Road, Wellingborough, Northants.

[25] W. R. Grace Ltd, Northdale House, North Circular Road, London NW10 7UH.

[26] Haeffner, H. & Co. Ltd, Nursery Industrial Estate, Station Road, Chepstow, Gwent NP6 5PB.

[27] Henkel Chemicals GmbH, Dusseldorf, W. Germany, (UK) Merit House, The Hyde, Edgware Road, London NW 9.

[28] Hercules Ltd, 20 Red Lion Street, London WC1R 4PB.

[29] Hoechst (UK) Ltd, Hoechst House, 50 Salisbury Road, Hounslow, Middx.

[30] ICI plc, Mond Division, The Heath, Runcorn, Cheshire.

[31] ICI plc, Organics Division, Hexagon House, Blackley, Manchester.

[32] ICI plc, Speciality Chemicals, Cleeve Road, Leatherhead, Surrey.

[33] Industrial Perfumes Ltd, Commerce House, Commerce Road, Brentford, Middx.

[34] K&K Greeff Industrial Chemicals Ltd, Suffolk House, George Street, Croydon CR9 3QL.

[35] KMZ Chemicals, Claire House, Bridge Street, Leatherhead, Surrey K22 8BZ.

[36] Fratelli Lamberti, S.p.A, 21041 Albizzate, Italy.

[37] Langley Smith & Co Ltd, Langley House, 19–21 Christopher Street, Finsbury Square, London EC2.

[38] Lankro Chemicals Ltd, PO Box 1, Eccles, Manchester M30 0BH.

[39] Manchem Ltd, Ashton New Road, Clayton, Manchester M11 4AT.

[40] ML Chemicals (UK) Ltd, 2 National Trading Estate, Norman Avenue, Hazlegrove, Stockport, Cheshire.

[41] NL Chemicals (UK) Ltd, St Ann's House, Wilmslow, Cheshire SK9 1HG.

[42] Occidental Chemical Corporation, 10889 Wiltshire Boulevard, Los Angeles California 90024, USA.

[43] Perchem Ltd, West Road, Temple Fields, Harlow, Essex.

[44] Pluss Stauffer, CH-L665 Oftrigen, Switzerland.

[45] PRB SA, Avenue de Brogneville 12, B1150, Brussels, Belgium.

[46] Priem Wilhelm GmbH & Co, D-4800 Bielefield 1, Osningstrasse 12, W. Germany.

[47] Proprietary Perfumes International Ltd, Wilesborough Road, Ashford, Kent TN24 0LT.

[48] Ransburg (UK) Ltd, Ransburg House, Hamm Moor Lane, Weybridge, Surrey K15 2RH.

[49] Rhone Poulenc (UK) Ltd, Hutton House, 161/166 Fleet Street, London EC4.

[50] Rohm & Haas (UK) Ltd, Lennig House, 2 Masons Avenue, Croyden, CR9 3NB.

[51] Sandoz Products Ltd, PO Box Horsforth No. 4, Calverley Lane, Horsforth, Leeds LS18 4RP.

[52] Schwegmann, Bernd, KG, Buchenwag 1, D5300 Bonn 3, PO Box 300 860, W. Germany.

[53] Servo BV, PO Box 1, 7490 AA Delden, Netherlands.

[54] Shell Chemicals (UK) Ltd, 1 Northumberland Avenue, Trafalgar Square, London WC2 5LA.

[55] Steetley Minerals Ltd, PO Box 2, Gateford Hill, Worksop, Nottinghamshire S81 8AF.

[56] Sterling Industrial, Capeltown, Sheffield S30 4YP.

[57] John E. Sturge Ltd, Lifford Chemical Works, Lifford Lane, Birmingham B30 3JW.

[58] Thor Chemicals Ltd, Ramsgate Road, Margate, Kent CT9 4JY.

[59] (The) White Sea & Baltic Co Ltd, Concordia Works, Parkinson Approach, Garforth, Leeds, LS25 2HR.

[60] Whitfield Chemicals & Polymers Ltd, 23 Albert Street, Newcastle-under-Lyme, Staffs ST5 1JP.

[61] Wilson & Mansfield, 48 Gresham Street, London EC2.

[62] Wymouth Lehr, 158 City Road, London EC1V 2PA.

6

The physical chemistry of dispersion

A. Doroszkowski

6.1 INTRODUCTION

The physical chemistry of dispersion may range from producing a polymer dispersion, such as latex, where the particle dispersion is formed *in situ*, to pigment suspension.

In preparing a latex, there is a strong element of polymerization kinetics which does not apply to pigment dispersion. On the other hand, in pigment dispersion there may be certain interfacial considerations which do not apply to latex formation. However, from a colloidal point of view, once the dispersion is produced, the physical chemistry of the dispersion is the same, whether polymer or pigment.

In this chapter the emphasis is on examining the physical chemistry of the dispersion where the disperse phase is preformed, i.e. from a pigment standpoint, although the same considerations apply to polymer particles if they constitute the disperse phase.

Our objective is therefore to look at the physical chemistry involved in forming a dispersion; to examine the factors influencing dispersion stability; and to consider how they can be measured or assessed.

For simplicity, we can define a paint as a colloidal dispersion of a pigment (the 'disperse' phase) in a polymer solution (the 'continuous' phase). Emulsion paints have both the polymer and pigment as the disperse phase. While the chemical structure of the polymer is important in determining the properties of a paint, the state of dispersion of the pigment in the polymer is no less important.

In practice it is fair to say that most of the problems that arise in a paint come back to the state of the pigment dispersion. The state of pigment dispersion can affect the:

(1) optical properties, e.g. colour,   [1]
(2) flow properties,   [2]
(3) durability,   [3]
(4) opacity,   [4]

(5) gloss,   [5]
(6) storage stability.   [6]

To produce a 'good' dispersion of colloidal particles from a dry powder we have to go through a number of processes which can be subdivided, arbitrarily, as separate operations, since in practice they may occur simultaneously. They are:

(1) immersion and wetting of the pigment;
(2) distribution and colloidal stabilization of the pigment.

These simple titles may mask many complex processes but are well worth consideration, as is the 'simple' practical question "Is it better to disperse TiO_2 pigment in an alkyd solution using an aromatic or aliphatic solvent?"

6.2 IMMERSION AND WETTING OF THE PIGMENT

Suppose we consider a single solid particle of pigment which will become immersed in a liquid. This can be represented schematically as a three-stage process:

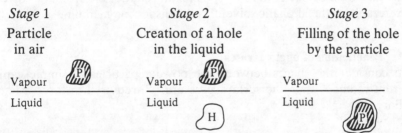

Stage 1	*Stage* 2	*Stage* 3
Particle in air	Creation of a hole in the liquid	Filling of the hole by the particle

If the particle has an area A, and γ_{SV} is the free energy per unit area, then the total energy of the three stages can be written as:

Stage 1: The surface energy of the particle is simply $A \, \gamma_{SV}$.
Stage 2: The energy of the particle in the vapour plus the work done to create a hole in the liquid identical to the volume and area of the pigment is $A \, \gamma_{SV} + A \, \gamma_{LV}$.
Stage 3: The work done to plug the hole in the liquid with the particle is $A \, \gamma_{LS} - A \, \gamma_{SV} - A \, \gamma_{LV}$.

Then the total energy change on immersing a particle is the sum of stages $2 + 3 - 1$:

i.e. $A \, \gamma_{LS} - A \, \gamma_{SV}$ (6.1)

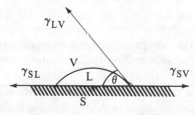

From Young's equation we have $\gamma_{SV} = \gamma_{LS} + \gamma_{LV} \cos \theta$:

The energy change on substitution in equation (6.1) is

$$- A \, \gamma_{LV} \cos \theta \qquad (6.2)$$

If we examine Young's equation in terms of wetting, or contact angle (θ), that is:

$$\cos \theta = \frac{\gamma_{SV} - \gamma_{SL}}{\gamma_{LV}}, \qquad (6.3)$$

Then, provided that $\theta < 90°$, a decrease in γ_{LV} will reduce θ and improve wetting, hence an aliphatic hydrocarbon is preferable to an aromatic solvent, since $\gamma_{LV}(\text{aliphatic}) < \gamma_{LV}(\text{aromatic})$.

If, however, we add a surface active agent, and it adsorbs at the air interface it will reduce γ_{LV}, and if it adsorbs on the particle surface it will decrease γ_{SL}. Both these effects will lead to better wetting.

However, if $\theta = 0$, as in a high-energy surface such as one might expect of a TiO_2 surface, it would be better to have γ_{LV} maximized, i.e. the aromatic solvent should be preferable to the aliphatic solvent as the dispersion medium.

6.2.1   Penetration of agglomerates

If we consider the spaces between the powder particles as simple capillaries of apparent radius r, then the surface pressure (P) required to force a liquid into a capillary is:

$$P = \frac{- 2\gamma_{LV} \cos \theta}{r} \qquad (6.4)$$

Force $= - 2\pi r \, \gamma_{LV} \cos \theta$
Area $= \pi r^2$

Hence penetration will occur spontaneously (ignoring gravitational effects) only if $\theta < 90°$ and if the pressure within the capillaries does not build up to counter the ingress of liquid.

Thus, to enhance liquid penetration of agglomerates it is desirable to maximise γ_{LV} and decrease θ. But since changes in γ_{LV} go hand in hand with θ this is difficult to realize. The addition of surface active agents will tend to decrease both γ_{LV} and θ, especially in aqueous media, hence the assessment of which is the dominant effect is best obtained by trial.

The argument used above is rather a simplistic one which uses the surface energy of a solid in equilibrium with the vapour of the liquid. Heertjes & Witroet [7] have examined the wetting of agglomerates and have shown that only when $\theta = 0$ can complete wetting of the powder aglomerate be obtained.

The rate of liquid penetration into an agglomerate was derived by Washburn [8] as:

$$\frac{dl}{dt} = \frac{r\gamma_{LV} \cos \theta}{4\eta} \tag{6.5}$$

where dl/dt is the rate of penetration of the liquid of viscosity and η in a capillary of radius r and length l.

In a packed bed of powder it is customary to employ an 'effective pore radius' or a 'tortuosity factor'. Thus $r/4$ can be replaced by a factor K, which is assumed to be constant for a particular packing of particles. Then equation (6.5) becomes

$$l^2 = \frac{Kt\, \gamma_{LV} \cos \theta}{\eta} \tag{6.5a}$$

Thus by inspection of equation (6.5a) we can see that to facilitate penetration of the powder we want to:

(a) maximize $\gamma_{LV} \cos \theta$
(b) minimize the viscosity (η)
(a) have K as large as possible, e.g. loosely packed agglomerates of pigment.

The Washburn equation describes a system in which the walls of the tube are covered with a duplex film (i.e. one where the surface energy of the film is the same as that of the surface of the bulk material). Good [9] generalized the Washburn equation to cover the case where the surface is free from adsorbed vapour. That is,

$$l^2 = Kt \left[\frac{\gamma_{LV} \cos \theta + \Pi_e - \Pi_{(t=0)}}{\eta} \right] \tag{6.5b}$$

where Π_e is the spreading pressure of the adsorbed film that is in equilibrium with the saturated vapour, and $\Pi_{(t=0)}$ is the spreading pressure for the film that exists at zero time.

If we consider the question of whether it is better to disperse TiO_2 in aliphatic or aromatic hydrocarbon, then while it is difficult to assign a specific surface energy to the TiO_2, because of its varied surface coating, which consists of mixed hydroxides of alumina, silica and titania, it is nevertheless a high surface energy material and can be likened to a water surface for simplicity [10], therefore $\cos \theta = 1$ for both the aromatic and aliphatic hydrocarbons, unless one of the liquids is autophobic [11]. Hence, in order to maximize $\gamma_{LV} \cos \theta$ it is better to use the aromatic hydrocarbon.

Crowl [12] has demonstrated the effect of $\gamma_{Lv} \cos \theta$ on grind time, as shown in Table 6.1.

Table 6.1 — Adhesion tension ($\gamma_{LV} \cos \theta$) and rate of milling: Laboratory ballmilling of rutile titanium dioxide

Medium	Adhesion tension ($\gamma_{LV} \cos \theta$)	Time (hr) to reach	
		Hegman 3†	Hegman 5†
5% isomerized rubber	21.9	0.9	2.0
5% alkyd	16.3	1.0	1.7
10% alkyd	14.4	1.7	2.5
10% isomerized rubber	12.3	2.0	4.0

† Hegman readings on the 0–8 scale, where 0 = coarse or poor dispersion, 8 = fine or best dispersion.

6.3 DEAGGLOMERATION (MECHANICAL BREAKDOWN OF AGGLOMERATES)

Pigmentary titanium dioxide exists in powder form as loose agglomerates of about 30 to 50 μm in diameter. The surface coating of the pigment, by the manufacturer, has a large effect on reducing the cohesive forces of the powder and thus assists in the disintegration (or deaggregation) process [13]. It is difficult to define the 'grind' or dispersion stage where the loose agglomerates are broken down into finer particles after all the available surface has been wetted out, and the process is generally treated empirically, as described in Chapter 8.

The surface coating applied by the pigment manufacturer, for example onto TiO_2, is proprietary information and probably not completely understood, though there are many publications attempting to describe it [14, 15]. The chemical analysis of the pigment and surface coating is generally in terms of 'equivalent to' Al_2O_3, SiO_2, ZnO, and TiO_2 etc. even though the coating consists of mixed hydroxides which are probably more correctly described by $M_x(OH)_y$, where M can be a mixture of Al, Si, Ti, Zr, etc. (as well as organic treatments with polyols, to facilitate stages of paint manufacture). The main purpose of the surface coating is to deactivate the surface of the rutile pigment, which would otherwise accelerate the degradation of the resin on weathering. The pigment coating is also there to aid pigment dispersion by the paint manufacturer. The 'grinding stage' in millbase manufacture is not a comminution stage but a dispersion stage of the pigment to the primary particle size as made by the pigment manufacturer. Some of the 'primary' particles consist of sinters of TiO_2 crystals produced during the 'surface coating' stage in pigment manufacture and remain intact *per se* on completion of the 'grind stage', as can be shown by particle size analysis before and after incorporation of the pigment in a paint (see Fig. 6.1); that is, the median particle size by count ($d_{gc} = 0.16$ μm) and distribution ($\sigma = 1.52$) of the TiO_2 before 'grinding' was found to be the same as that obtained by sedimentation analysis after 'grinding' in a ballmill (using the appropriate Hatch–Choate equation to convert particle size by mass to that of size by count;

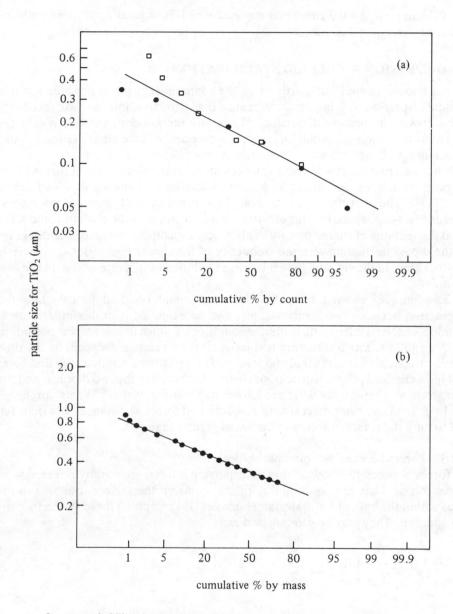

Log — probability plots of particle size
(a) by electron microscopy, prior to milling; d_{gc}=0.16 micron
     σ=1.52, therefore d_{dm}=0.3
     □ counting all particles (even when obvious aggregates)
     ● counting only single crystals (provided more than half the perimeter was visible)
(b) by x-ray sedimentation, after milling; d_{gm}=0.36, =1.5

Fig. 6.1 — Particle size analysis of TiO$_2$
(Log-probability plots)

$d_{gc} = 0.16 \ \mu m = d_{gm} = 0.3 \ \mu m$ when $\sigma = 1.52$ which is in good agreement with the measured size $d_{gm} = 0.36 \ \mu m$, $\sigma = 1.5$).

6.4 DISPERSION — COLLOID STABILIZATION

It is not enough to 'wet out' particles by the continuous phase to produce a stable colloidal dispersion. It is important to realize that attractive, interparticle forces are always present in pigment dispersions. These are the London, van der Waalls (or surface) forces. These attraction forces are a consequence of the attractive interatomic forces amongst the atoms which constitute all particles.

Polar materials exert electrostatic forces on other dipoles (Keesom forces [16]), and polar molecules can attract non-polar molecules by inducing dipoles (Debye forces [17]). The attraction between non-polar atoms or molecules was not understood until it was realized that the electron cloud surrounding the nucleus could show local fluctuations of charge density. This produces a dipole moment, the direction of the dipole fluctuating with the frequency of the charge fluctuations. If there is another atom in the vicinity, then it becomes polarized and it interacts with the first atom.

London [18] showed how this treatment could be used to calculate the interaction between two similar atoms, and he extended it to dissimilar atoms. London's analysis is based on instantaneous dipoles which fluctuate over periods of 10^{-15} to 10^{-16}s. If two atoms are further apart than a certain distance, by the time the electric field from one dipole has reached and polarized another, the first atom will have changed. There will be poor correlation between the two dipoles, and the two atoms will experience what are known as 'retarded' van der Waals' forces.

In general, we can expect strong non-retarded forces at distances less than 100 Å (10 nm) and retarded forces at distances greater than this.

6.4.1 Forces between macroscopic bodies

The forces between two bodies, due to dispersion effects, are usually referred to as surface forces. They are not only due to the atoms on the surface, but also to the atoms within the bulk of the material. Hamaker [19] computed these forces by pairwise addition. They may be summarized as:

$$V_A = \frac{-A}{48\pi} \left[\frac{1}{d^2} + \frac{1}{(d+\delta)^2} - \frac{2}{(d+\frac{\delta}{2})^2} \right] \tag{6.6}$$

if

$$d \gg \delta \text{ then } V_A = \frac{-\delta^2 A}{32\pi d^4} \tag{6.6a}$$

$$d < \delta \qquad V_A = \frac{-A}{48\pi} \left[\frac{1}{d^2} - \frac{7}{\delta^2} \right] \tag{6.6b}$$

$$d \ll \delta \qquad V_A = \frac{-A}{48\pi d^2} \tag{6.6c}$$

where V_A is the attraction energy for two plates of thickness δ at a distance of separation of $2d$ from each other, and A is the Hamaker constant for the substance comprising the plates.

Hamaker [20] also showed that the attraction energy (V_A) between two like spheres of radius a separated by a surface to surface distance d is: −

$$V_A = \frac{-A}{6}\left(\frac{2}{S^2 - 4} + \frac{2}{S^2} + \ln\left\{\frac{S^2 - 4}{S^2}\right\}\right) \tag{6.7}$$

where
$$S = 2 + \frac{d}{a}$$

If $d \ll a$ then $V_A \simeq \dfrac{-Aa}{12d}$, \hfill (6.7a)

and for spheres of radii a_1, a_2:

$$V_A \simeq \frac{-Aa_1a_2}{6d(a_1 + a_2)} \tag{6.7b}$$

We can compute these attraction forces in terms of particle size, medium, and distance of separation, and express the result as an energy of attraction (V_A). However, the Hamaker constant A is the value *in vacuo*, and if one is considering the attraction energy between two particles in a fluid, it has to be modified according to the environment. Hamaker showed that particles of substance A_1 in a medium of substance A_2, had a net Hamaker constant of $A = A_1 + A_2 - 2\sqrt{(A_1A_2)}$. Thus if we have two different materials — material 1 and material 2 in medium 3 — the net Hamaker A_{132} is modified according to

$$A_{132} = (A_{11}^{\frac{1}{2}} - A_{33}^{\frac{1}{2}})(A_{22}^{\frac{1}{2}} - A_{33}^{\frac{1}{2}}) \tag{6.8}$$

For a list of Hamaker constant values for common substances see Visser [21]. Gregory [22] showed how the Hamaker constant could be estimated from simple experimental measurements.

The point to note is that these forces are always attractive; they may be modified to varying extents depending on one's wish for accuracy, e.g. Vold's correction due to adsorbed layers [23] or Casimir & Polders correction due to retardation of the attraction forces [24, 25]. However, simple examination of the published values of the Hamaker constants reveals that they may vary by an order of magnitude and therefore rarely justify laborious correction for varying subtle effects such as that of Vold [23] and Vincent [26].

Although there is always an attraction force between two like particles, Visser [27] points out that in certain circumstances it is possible to have a negative Hamaker constant in a three-component system such as when $A_{11} < A_{33} < A_{22}$ or $A_{11} > A_{33} > A_{22}$ where A_{33} is the individual Hamaker constant of the medium. It is important to note that there is still an attractive potential between like particles of A_{11} in A_{33} as well as A_{22} particles attracting each other in medium A_{33}, but there will be a repulsion between A_{11} particles with A_{22} particles. A suggested example of this type of non-association is thought to be exemplified by PTFE and graphite particles in water.

To obtain a 'colloidal dispersion' one must somehow overcome the omnipresent attraction energy by generating some kind of repulsion energy (V_R) between the particles, such that when the repulsion energy and attraction energy are added, there is still a significant net repulsion energy. Since the attraction and repulsion energies are dependent on interparticle distance, it is important to know something of particle spacing, especially in paint systems which, in conventional colloidal terms, are considered to be 'concentrated'. It can be shown that in a hexagonally close-packed array of spheres, the ratio of surface-to-surface separation (S) to centre to centre separation (C) is related to total volume by

$$\frac{S}{C} = 1 - \left(\frac{PVC}{0.74}\right)^{\frac{1}{3}}$$ (6.9)

where PVC is pigment volume concentration (%).

The surface-to-surface distance in relation to pigment volume concentration is given in Table 6.2, where the distance S is expressed in terms of sphere diameter.

Table 6.2 — Relationship between pigment volume concentration (PVC) and surface-to-surface spacing (S)

PVC	S†
5	1.43d
10	0.95d
15	0.71d
20	0.55d
25	0.44d
30	0.35d
35	0.28d
40	0.23d
45	0.18d

† S is expressed in terms of particle diameter (d)

Thus for any particle size of mono disperse spheres the average separation is equal to one particle diameter at a PVC of 9.25%. Therefore the average interparticle distance for TiO_2 at this concentration is about 2000 Å (200 nm), and for smaller particles at the same volume concentration the distance is proportionally smaller.

If we compare repulsion energies and attraction energies, and remember that if particles are in Brownian motion, which at room temperature have an average translational energy of $\frac{3}{2}kT$, then to stabilize them we must have an energy barrier (V_{max}) which is significantly greater than this (see Fig. 6.2).

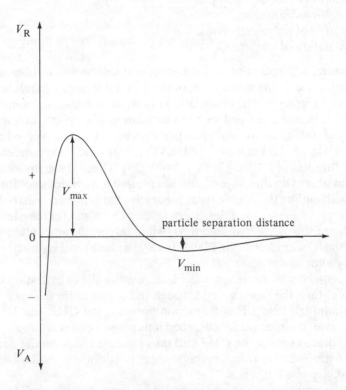

Fig. 6.2 — Typical net energy — distance of separation curve, showing 'secondary minimum' (V_{min} flocculation).

Depending on the nature of the net energy curve we can have a 'secondary minimum' or an energy trough which may bring about flocculation which is weak. This is called 'weak flocculation' (V_{min}) or 'secondary minimum flocculation' and is used to differentiate it from the stronger flocculation which occurs on closer approach of particles sometimes called 'primary flocculation'. It is worth noting that sometimes in the literature [28] the terminology is different from that used in the paint industry, e.g. the term 'Coagulation' in the literature is used to mean flocculation, when normally that term is reserved by industrial chemists for the irreversible association of latex or emulsion particles.

6.4.2 Dispersion stabilisation by charge
To produce a stable colloid dispersion, a source of repulsion energy must be induced onto particles so that within the interparticle distances (at the desired particle concentration, i.e. *PVC*) there is sufficient net energy to prevent flocculation.

This repulsion energy may arise from coulombic forces as described by the

DLVO theory or from 'steric stabilisation'. Charge may be generated at a surface in different ways:

(1) Preferential adsorption of ions.
(2) Dissociation of surface groups.
(3) Isomorphic substitution.
(4) Adsorption of polyelectrolytes.
(5) Accumulation of electrons.

Preferential adsorption of ions is the most common way of obtaining a charged particle surface, e.g. the adsorption of Ag^+ or I^- ion on Ag halide sols, or ionic surfactants on pigments. The dissociation of surface groups is a common feature with latexes which become charged by the dissociation of sulphate or carboxyl groups. The theory of stabilization of colloidal particles by charge has evolved over the years. It culminated in what is known as the DLVO theory, and is expounded by Verwey & Overbeek (the VO of DLVO) in their book [29]. There are many excellent reviews [30, 31] and papers on the subject, and the present author does not intend to go into the theory. In outline the double layer theory at a flat surface consists of an innermost adsorbed layer of ions, called the 'Stern layer'. The plane (in the Stern layer) going through the centre of the hydrated ions (when no specific adsorption takes place) is known as the 'Outer Helmholtz Plane' (or Stern plane) and is directly related to the hydration radius of the adsorbed ions.

If some specific adsorption takes place (usually the dehydration of the ion is a prerequisite) then the plane going through the centre of these ions is known as the 'Inner Helmholtz Plane'. The distinction between the OHP and IHP is generally necessary since, the specifically adsorbed ions allow a closer approach to the surface than the hydrated ions at the OHP and thus increase the potential decay.

Beyond the Stern (or compact) layer there is a diffuse layer, known as the 'Gouy (or Gouy–Chapman)' layer.

When the charge in the Stern layer, diffuse layer, and the surface is summed the total = 0 because of electroneutrality. The potentials just at the solid surface (ψ_0), at the IHP (ψ_s) and OHP (ψ_d) are, as Lyklema [30] points out, abstractions of reality, and are not the zeta potential, which is the experimentally measureable potential defined as that occurring at the 'slipping plane' in the ionic atmosphere around the particle.

However, while the zeta potential and the potential at the Helmholtz plane are not the same, for simple systems such as micelles, surfactant monolayers on particles, etc., there is considerable evidence to equate them [32].

6.4.3   Zeta potentials
Zeta potentials can be estimated from experimentally determined electrophoretic mobilities of particles. The equation used to convert observed mobilities into zeta potentials depends on the ratio of particle radius (a) to the thickness of the double layer $\left(\dfrac{1}{\kappa}\right)$, e.g. for 10^{-3} M aqueous solution at 25°C with a 1:1 electrolyte, $\dfrac{1}{\kappa} = 1.10^{-6}$ cm.

Values for other electrolyte types or other concentrations modify κ on a simple proportionality basis, since for an aqueous solution of a symmetrical electrolyte the double layer thickness is $\dfrac{1}{\kappa} = \dfrac{3 \times 10^{-8}}{z\, c^{\frac{1}{2}}}$ cm, where z is the valency, and c is expressed as molarity.

If $\kappa a > 200$ then the mobility (μ) is related by the Smoluchowski equation to the zeta potential (ζ) by $\mu = \dfrac{\varepsilon \zeta}{4\pi\eta}$ where ε is the permittivity of the medium and η = viscosity.

If the mobility is measured in μm s^{-1}/volt. cm^{-1} then $\zeta = 12.8\mu$ mV$_{\text{per mobility unit}}$ for aqueous solutions.

When $\kappa a < 0.1$ then the Huckel equation $\mu = \dfrac{e\zeta}{6\pi\eta}$ applies.

For $200 < \kappa a < 0.1$ it is necessary to use the computations of O'Brien & White [33] which are an update of the earlier Wiersema *et al.* [34] computations relating mobility to zeta potential.

The conversion of electrophoretic mobility to zeta potential is based on the assumption that the particles are approximately spherical. If they are not, then unless κa is large everywhere, there is obviously doubt as to the value of the zeta potential obtained from the mobility calculation.

If the particle consists of a floccule comprising small spheres and if κa (for a small sphere) is large, the Smoluchowski equation applies.

Hence the zeta potential has frequently a small uncertainty attached to its value, and many workers prefer just to quote the experimentally determined electrophoretic mobility.

6.4.4 Measurement of electrophoretic mobility

The electrophoretic mobility of small particles can be measured by microelectrophoresis, where the time taken for small particles to traverse a known distance is measured, or alternatively by a moving-boundary method. The microelectrophoresis method has many advantages over the moving-boundary method, and it is the more frequently adopted method, although there are some circumstances which favour the moving-boundary method [35].

An alternative method based on the electrodeposition of particles has been devised by Franklin [36]. While the quoted values appear to be in good agreement with microelectrophoresis measurements, the method is fundamentally unsound because electrodeposition of a dispersion is based on electrocoagulation and does not depend on electrophoresis [37].

If the mobility of charged particles is examined in a microelectrophoresis cell then it will be noticed that there is a whole range of velocities, with some particles even moving in the opposite direction. This effect is due to electro-osmotic flow within the cell. The true electrophoretic mobility can only be determined at the 'stationary levels' where the electro-osmotic flow is balanced by the hydrodynamic flow. The position of the stationary levels depends on the shape of the microelectrophoresis cell, i.e. whether it is circular or rectangular. There are many varieties of microelectrophoretic cell [38], and there are many refinements to enable rapid and accurate measurements to be made, such as the Rank Brothers' microelectrophoresis apparatus [39] equipped with laser illumination, rotating prism, and video camera and monitor. For details on how to measure electrophoretic mobilities the reader is referred to Smith [32], James [40], and Hunter [41].

In extremis one can construct a simple 'flat cell' suitable for viewing under an ordinary microscope, just using a microscope glass slide and two sizes of glass cover slips, as shown in Fig. 6.3.

6.4.5 Application to colloid stability

When two charged surfaces approach each other they start to influence each other electrostatically as soon as the double layers overlap. For surfaces of the same sign the ensuing interaction is repulsion. In a qualitative interpretation a number of points have to be taken into account.

The essence of DLVO theory is that interparticle attraction falls off as an inverse power of interparticle distance and is independent of the electrolyte content of the continuous phase, while the coulombic (or electrostatic) repulsion falls off exponentially with a range equal to the Debye–Hückel thickness $1/\kappa$ of the ionic atmosphere.

The Stern layer does not take a direct role in the interaction, but its indirect role in dictating the value of ψ_d is enormous. In the DLVO theory, the double layers are considered as if they were purely diffuse, but in reality the theory applies to the two diffuse parts of the two interacting double layers. The surface potential ψ_0 should be replaced by Stern potential ψ_d, which in turn is replaced by the zeta potential for non-porous substances.

Typical double-layer thicknesses at varying electrolyte concentrations are given in Table 6.3 calculated from

$$\kappa^2 = \frac{F^2 \Sigma c_i z_i}{\varepsilon_r \varepsilon_0 RT} \tag{6.10}$$

and illustrated in Fig. 6.4

where F = Faraday constant

$RT = 2.5 \times 10^{10}$ ergs at 298°K

$\varepsilon_r \varepsilon_0$ relative permittivity and permittivity in a vacuum

$z_i; c_i$ are the charge and number of all ions in solution.

For an aqueous solution of a symmetrical electrolyte the double-layer thickness is

$$\frac{1}{\kappa} = \frac{3 \times 10^{-8}}{z \, c^{\frac{1}{2}}} \text{ cm}$$

where z is the valency and c is expressed as molarity.

6.4.5.1 *The dynamics of interaction*

Much depends on the timescale of particle interaction, e.g. diffusion relative to adjustments of the double layers on overlap. If the rate of approach is fast, the double layers have little or no time to adjust.

If there is slow approach, then the double layers are continually at equilibrium.

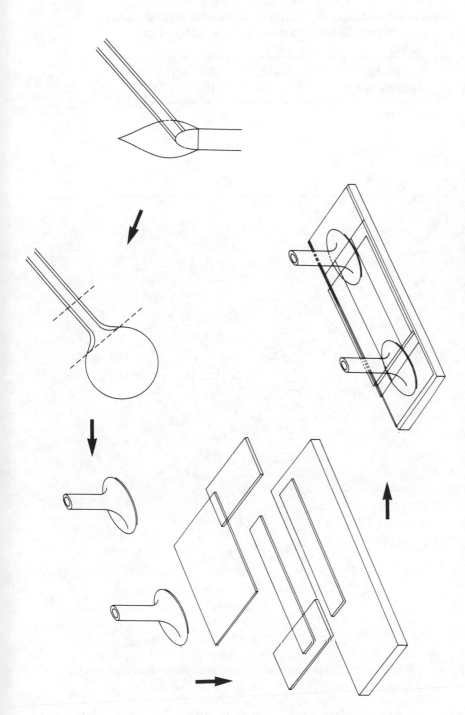

Fig. 6.3 — Construction diagram of a simple 'flat' electrophoretic cell made from a microscope slide and cover slips, cut and glued together with an epoxy adhesive. The electrode compartment is made by blowing a piece of glass tubing to flare the ends so that it can seal the opening to the cell as shown.

Table 6.3 — Effect of electrolyte on double layer thickness
Concentration of 1 : 1 electrolyte in water at 25°C

Molar concentration	double layer thickness $\frac{1}{\kappa}$ in cm $\times 10^{-8}$
10^{-5}	1000
10^{-4}	300
10^{-3}	100
10^{-2}	30
10^{-1}	10

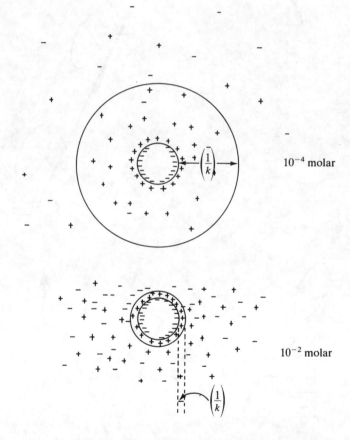

Fig. 6.4 — Schematic effect of electrolyte concentration on double layer thickness $\frac{1}{\kappa}$.

The DLVO approach, in this latter case, is that the energy of repulsion, V_R, can be calculated by reversible thermodynamics as the isothermal reversible work to bring the particle from infinity to distance H.

The underlying idea is that in an equilibrium encounter the potential should remain constant, since it is determined by adsorption of the charge-determining ion, the chemical potentials being fixed by their values in the bulk. This is what is known as 'constant potential' interaction.

'Constant charge' interaction is expected to occur if the interaction proceeds so fast that there is no readjustment of ionic adsorption/desorption, and hence the surface potential, ψ_0, is not constant but rises during overlap.

When all is said and done, there is not a great difference in V_R for the two types of interaction, but the constant charge case tends to give larger V_R than constant potential.

For two small spherical particles of radius a at constant charge the energy of repulsion is given by Verwey & Overbeek [29] as

$$V_R = \varepsilon a \psi_0^2 \left(\frac{e^{-\tau(s-2)}}{s} \right) \gamma \tag{6.11}$$

while for constant potential

$$V_R = \varepsilon a \psi_0^2 \left(\frac{e^{-\tau(s-2)}}{s} \right) \beta \tag{6.12}$$

where

$$\gamma = \frac{1 + \alpha}{1 - \dfrac{e^{-\tau(s-2)}}{2s\tau} \left(\dfrac{\tau - 1}{\tau + 1} + e^{-2\tau} \right)(1 + \alpha)}$$

and

$$\beta = \frac{1 + \alpha}{1 + \left(\dfrac{e^{-\tau(s-2)}}{2s\tau} \right)(1 - e^{-2\tau})(1 + \alpha)}$$

$s = R/a$; R is the centre-to-centre distance between two particles of radius a. Since γ and β are always between 0.6 and 1.0 we may neglect their influence in many cases [29, p. 152]. Hence equation (6.13) is a good approximation for the free energy of electrostatic repulsion. (Overbeek points out that there is no exact equation [30]).

$$V_R = 2\pi\varepsilon_r\varepsilon_0 a \left[\frac{4RT}{zF} \gamma' \right]^2 \ln\left[1 + \exp(-\kappa H) \right] \tag{6.13}$$

$$\simeq 2\pi\varepsilon_r\varepsilon_0 a\left[\frac{4RT}{zF}\gamma'\right]^2 \exp{(-\kappa H)} \tag{6.13a}$$

where $\gamma' = \tanh(zF\psi_0/4RT)$
ε_r = relative permittivity of medium
ε_0 = permittivity of a vacuum
R,T,F,z have their usual meanings

$$\kappa^2 = \frac{F^2\Sigma c_i z_i}{\varepsilon_r\varepsilon_0 RT} \tag{6.13b}$$

For the interaction of two particles of different surface potential (ψ_{01},ψ_{02}) and different radii, a_1 and a_2 (provided that $\kappa H > 10$ and $\psi_0 < 50$ mV) is, according to Hogg et al. [42],

$$V_R = \frac{\varepsilon a_1 a_2(\psi_{01}^2 + \psi_{02}^2)}{4(a_1 + a_2)}B \tag{6.14}$$

where

$$B = \frac{2\psi_{01}\psi_{02}}{(\psi_{01}^2 + \psi_{02}^2)}\ln\left(\frac{1+2\exp - \kappa H}{1-\exp - \kappa H}\right)\ln[1-\exp(-2\kappa H)].$$

Figures 6.5, 6.6, 6.7 illustrate the evaluation of V_{net} ($= V_R + V_A$) in terms of surface-to-surface distance (H) when the particle size and electrolyte content are kept constant, but the surface potential is varied; the surface potential and electrolyte content are kept constant, but the particle size is varied, and when the particle size and surface potential are kept constant, and the electrolyte content is varied.

6.4.5.2 Flocculation by electrolyte

From the evaluation of $V_{net} = V_R + V_A$ shown in Fig. 6.7 it is seen that the addition of an electrolyte will make a charge-stabilized dispersion flocculate. This property was recognized at the end of the nineteenth century when Schulze found that the flocculating power of a counterion is greater, the higher its valency. It was confirmed by Hardy, and is now known as the Schultz–Hardy rule.

We can define the onset of dispersion instability using Overveck's approach [31] as the condition when

$$V_{net} = V_R + V_A = 0 \tag{6.14a}$$

and $\dfrac{\partial V_{net}}{\partial H} = 0$ are satisfied for the same value of H. That is,

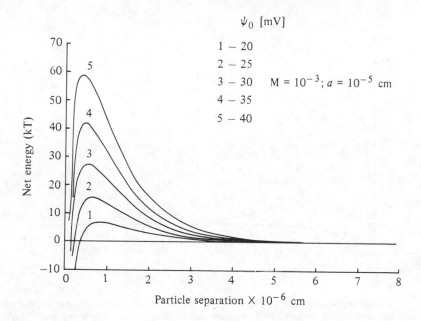

Fig. 6.5 — Net interaction energy ($V_{net} = V_R + V_A$) — particle separation curves, where particle size and electrolyte content are constant and the surface potential is varied (using equations (6.13) and (6.7).

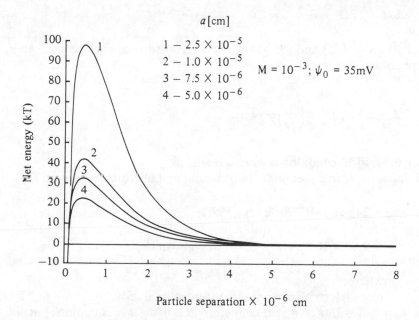

Fig. 6.6 — Net interaction energy ($V_{net} = V_R + V_A$) — particle separation curves, where electrolyte content and surface potential are constant and particle radius is varied (using equations (6.13) and (6.7).

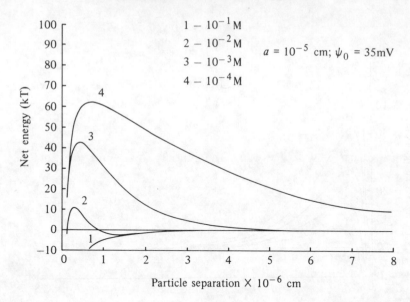

Fig. 6.7 — Net interaction energy ($V_{net} = V_R + V_A$) — particle separation curves, where particle size and surface potential are constant and the electrolyte content is varied (using equations (6.13) and (6.7).

$$\frac{\partial V_{net}}{\partial H} = \frac{\partial V_R}{\partial H} + \frac{\partial V_A}{\partial H} = 0,$$

using equations (6.13a) and (6.7a), where $d = H =$ surface–surface separation distance, we obtain

$$\frac{\partial V_{net}}{\partial H} = -\kappa V_R - \left(\frac{1}{H}\right) V_A = 0,$$

therefore the critical condition is when $\kappa H = 1$.
Substituting κ for H in equation (6.14a), we obtain the critical condition as:

$$\kappa_{crit} = 24\pi \varepsilon_r \varepsilon_0 (4RT/F)^2 \, e^{-1} (\gamma')^2 / A z^2$$

$$= 2.03 \times 10^{-5} (\gamma^1)^2 / z^2 A \qquad \text{in c.g.s. units,}$$

where:
  $\varepsilon_r =$ dielectric constant (78.5) for water at 25°C
$\varepsilon_0 = \varepsilon_{i/4\pi} = 1/4\pi$ ($\varepsilon_i = 1$ and dimensionless in c.g.s.–e.s.u. units, but not in SI units)
  $e = 2.718$
  $R = 8.314 \times 10^7$ erg T^{-1} mol^{-1}
  $T = 298°$K

$F = 2.892 \times 10^{14}$ e.s.u. mol^{-1},

and for water at 25°C we can (for symmetrical z–z electrolyte) write equation (6.13b) as:

$$\kappa = \frac{zM^{\frac{1}{2}}}{3 \times 10^{-8}}\ cm^{-1},$$

and substituting for κ_{crit} we obtain the critical flocculation concentration (M_{crit}) as

$$M_{crit} = \frac{3.7 \times 10^{-25}\ (\gamma')^4}{z^6\ A^2}\ \text{moles/litre.} \qquad (6.14b)$$

If the surface potential (ψ_0) is high ($\psi_0 \geqslant 100$mV) then $\gamma' = \tanh\ (zF\psi_0/4RT) \to 1$. Hence the DLVO theory predicts that the flocculation efficiency of indifferent electrolytes should be inversely proportional to the sixth power of the valency (that is, $1/z^6$; hence the relative concentration to bring about flocculation by monovalent- : divalent : trivalent ions is $\dfrac{1}{1^6} : \dfrac{1}{2^6} : \dfrac{1}{3^6}$ or 729 : 11.4 : 1.

For low surface potentials ($\psi_0 < 25$mV) we can equate the surface potential to the zeta potential and simplify equation (6.14b) to

$$M_{crit} = \frac{3.31 \times 10^{-33}\zeta^4}{A^2 z^2}\ \text{moles/litre}$$

where ζ is in mV since $(4RT/F) = 102.8$mV at 25°C and $\tanh x \approx x$, for low values of x. Hence the ratio of flocculating concentrations for monovalent : divalent : trivalent ions becomes $\dfrac{1}{1^2} : \dfrac{1}{2^2} : \dfrac{1}{3^2}$ or 100 : 25 : 11.

In practice there has been found to be good general agreement with the predictions of the DLVO theory, but since flocculation studies are dependent not only on equilibrium conditions, but also on kinetic factors and specific ion effects, care has to be exercised in using this simplified approach, for even ions of the same valency form a series of varying flocculating effectiveness such as the lyotropic (Hofmeister) series where the concentration to induce flocculation of a dispersion decreases with ion hydration according to $[Li^+] > [Na^+] > [K^+] > [Rb^+] > [Cs^+]$.

In emulsion paints the presence of electrolytes is thus seen to be a very important factor. Pigment millbases formulated on the use of charge alone, e.g. Calgon (sodium hexameta phosphate), are very prone to flocculation; sometimes even by the addition of latex, since the latex may carry a considerable concentration of electrolyte. Such dispersions, once made, are usually further stabilized by the addition of a water-soluble colloid (polymer) such as sodium carboxy methyl cellulose, which may be labelled only as a 'thickener', nevertheless it increases the dispersion stability to the presence of electrolyte by its 'protective colloid' action.

6.4.5.3   *Charge stabilization in media of low dielectric constant*

Charge stabilization is of great importance in media of high dielectric constant, i.e. aqueous solution, but in non-aqueous and particularly in non-polar systems (of low dielectric constant) repulsion between particles by charge is usually of minor importance [43].

There have been attempts to use charge to explain colloid stability in media of low dielectric constant such as *p*-xylene [44].

Osmond [45] points out that the attempted application of double layer theory, with all the corrections necessary to apply it to non-polar non-aqueous media, reduces the DLVO approach to simple coulombic repulsion between two charged spheres in an inert dielectric (thus ignoring the existence of ions in solution).

Examination of Lyklema's calculation [46] of the total interaction between two spherical particles of radius 1000 Å (0.1 μm) (e.g. TiO_2 particles) in a low dielectric medium shown in Fig. 6.8, reveals that if the particles had a zeta potential of 45 mV and were at 9% *PVC* — i.e. average spacing of 2000 Å (0.2 μm) — then the maximum energy barrier to prevent close approach is only about 4 kT. While at 20% *PVC* — average spacing 1000 Å (0.1 nm) — only 2 kT is available to prevent flocculation, which is clearly insufficient.

Thus, because the repulsion energy–distance profile in non-polar media is very 'flat' (see Fig. 6.8); concentrated dispersions (their definition implying small inter-particle distances) cannot be readily stabilized by charge. But in very dilute concentration where interparticle distances are very large, charge may be a source of colloidal stabilization. While charge is not considered to be important in stabilizing concentrated non-aqueous dispersions in media of low dielectric constant, however, it can be a source of flocculation in such systems if two dispersions of opposite charge are mixed.

6.5   STERIC (OR POLYMER) STABILIZATION

An alternative source of repulsion energy to stabilize colloidal particles in both aqueous and non-aqueous (including non-polar) media is what is now known as 'steric' or 'entropic' stabilization. In 1966 Overbeek said 'The theory of protective (entropic) action is still in rather a primitive state' [47]. Since then great progress in the understanding of steric stabilization has been made with many comprehensive reviews on the subject [48–50].

Napper [51] presents the latest 'state-of-knowledge' in a very readable form. The important feature of steric stabilization is that it is not a steric hindrance as normally considered in organic chemistry terms, where molecular configuration prevents a reaction, but a source of energy change due to loss of entropy. It is sometimes known as 'polymer stabilization', but since non-polymeric material can also produce stabilization, this name too is wanting, as are the many alternatives that have been suggested, e.g. non-ionic, entropic.

In its simplest form, steric stabilization can be visualized as that due to a solvated layer of oligomeric or polymeric chains irreversibly attached to a particle surface of uniform concentration producing a 'solvated' sheath of thickness δ.

If two such particles approach each other, so that the solvated sheaths overlap, or

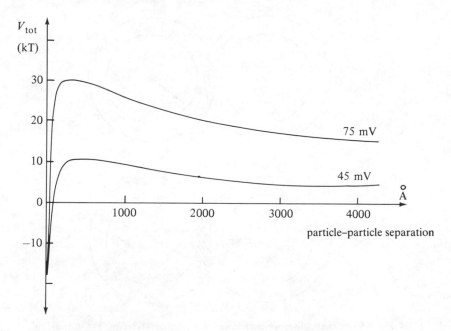

Fig. 6.8 — Net energy — distance plots in non-aqueous medium (from [46]).

redistribute their segment density in the overlap zone, there is a localized increase in polymer concentration, (see Fig. 6.9). This localized increase in polymer concentration leads to the generation of an osmotic pressure from the solvent in the system. Hence the source of the repulsion energy is equivalent to the non-ideal component of the free energy of dilution.

It is possible to estimate the size of the adsorbed layer of polymer around a particle by measuring the viscosity of the dispersion at varying shear rates [52]. From these measurements it is also possible to obtain some idea of the effectiveness of the adsorbed layer by examining the slope in a plot of log (dispersion viscosity) against the square root of the reciprocal shear rate, as suggested by Asbeck and Van Loo [53]. The greater the slope, the larger the degree of flocculation. If the viscosity is independent of shear then there is no detectable flocculation under these specific conditions.

A check on the validity of measuring adsorbed layer thickness was made by synthesizing a series of oligoesters of known chain length which could be adsorbed only through their terminal carboxyl group onto the basic pigment surface. The measured 'barrier thicknesses' were found to be in very good agreement with terminal oligoester adsorption [54], confirming the effectiveness of the method.

From measurements such as these, it is clear that the size of the adsorbed polymer is not all-important as has been suggested by some [55]. Smaller adsorbed layers can be more effective than larger ones [56]. Chain branching, and where it occurs can have a marked influence on the effectiveness of the stabilizing layer [54]. For example, *poly* 12-hydroxy stearic acid, which is used as the soluble moeity in a graft copolymer as described in [57], is very effective.

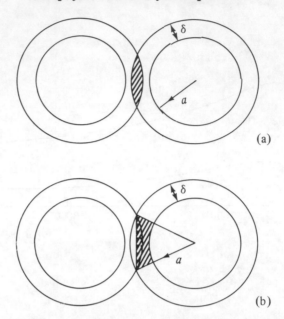

Fig. 6.9 — Schematic representation of steric-barrier overlap: (a) Ottewill [76],
(b) Doroszkowski & Lambourne [69].

The theory of steric stabilization, like most, has undergone evolution. Its origins
are irrelevant, except maybe to historians, whether it is founded in the concept of
'protective colloids' as discussed by Zigmondy [58] with his gold sols in 1901, or the
early attempts to calculate the effect of entropy loss by adsorbed molecules such as
envisaged by Mackor [59]. Nevertheless, steric stabilization was practised by crafts,
including paintmakers, before scientists classified the various stages by nomenclature and explanation. Fischer [60] probably contributed most to the understanding of
steric stabilization, since he envisaged the important influence of the solvent
environment on steric stabilization, which was not appreciated even by later workers
[61].

At present steric stabilization is generally considered to arise from two factors
(which are not necessarily equal):
(1) A volume restriction term (i.e. the configuration of the adsorbed molecules is
reduced by the presence of a restrictive plane V_{VR}).
(2) A mixing or osmotic term (i.e. an interpenetration or compression of adsorbed
layers resulting in an increase in polymer concentration V_M).
These two contributions are regarded as being additive. That is,
$V_{tot} = V_{VR} + V_M$.

It is difficult to evaluate the contribution of these two terms since it entails a
knowledge of the polymer configuration of the adsorbed polymer, and more
importantly the segment density distribution normal to the surface of the particle.

Meier [62] and Hesselink [63, 64] have attempted to calculate this segment
distribution and hence the relative contribution of the two sources of repulsion
energy. However, there is dispute in that by subdividing the repulsion energy in this
way, there is double counting [65].

There have been, unfortunately, only a few attempts to measure the energy of repulsion by direct experiment [66–71], and it would appear that the theoretical and mathematically complex calculations of Meier and Hesselink grossly overestimate the repulsion energy measured by experiment [69]. The more simple estimates of energy of repulsion (derived from purely the 'osmotic' term V_{RM}) appear to be in better accord with practice. Napper, in many investigations [72–74], has shown the importance of the solvent environment in contributing to steric stabilization, and for all intents and purposes, the osmotic contribution to steric stabilization is all that needs to be considered. Therefore the repulsion energy on the interaction of two solvated sheaths of stabilizing molecules may be simply written as

$$V_R = \frac{c^2 RT}{v_1 \rho^2} (\psi_1 - \kappa_1) V, \qquad (6.15)$$

where c = polymer concentration in solvated barrier,
       ρ = polymer density,
       v_1 = partial molar volume of solvent,
       $\psi_1\,\kappa_1$ = Flory's entropy and enthalpy parameters [75],
       V = increase in segment concentration in terms of barrier thickness δ,
       a = particle size and surface-to-surface separation, (h).

Which in the Ottewill & Walker model [76] is

$$V = \left(\delta - \frac{h}{2} \right) \left(3a + 2\delta + \frac{h}{2} \right)$$

and the Doroszkowski & Lambourne model [69]

$$V = \frac{\pi x^2 R_1 (3R_1 - x)(2R_1 - x)\, 4(R_1^3 - r^3) + x R_1 (x - 3R_1)}{2(R_1^3 - r^3) + x R_1 (x - 3R_1)}$$

where $R_1 = a + \delta$; $r = a$; and $x = \delta - h/2$.

One has a more useful relationship if one rewrites equation (6.15) and substitutes $(\psi_1 - \kappa_1)$ by $\psi_1(1 - \theta/T)$ where θ is the theta temperature as defined by Flory [75], i.e. where the second virial coefficient is zero, meaning that the molecules behave in their ideal state and do not 'see or influence each other'.

Thus one can immediately realize that at the theta temperature, $\left(1 - \dfrac{\theta}{T}\right)$ reduces to zero ($\theta = T$), and the repulsion energy disappears, allowing the attractive energy to cause flocculation, as shown by Napper. Furthermore, if $T < \theta$, the repulsion energy not only does not contribute to the repulsion, but it actually becomes an attraction energy.

An important point to bear in mind is the c^2 term, which is the polymer concentration term in the 'steric barrier'. If this is small then flocculation can occur, even when the solvency is better than 'theta', which has been observed in a few instances.

Napper [77] has subdivided steric stabilization into 'entropic stabilization' and 'enthalpic stabilization'. Depending on whether the entropy parameter is positive or negative, e.g. in the case of steric stabilization by polyhydroxy stearic chains in aliphatic hydrocarbon, flocculation is obtained by cooling; and in enthalpic stabilization, flocculation is achieved by heating (as with polyethyleneglycol chains in water). A purist [78] might correctly insist that this is entropic or enthalpic flocculation, and not stabilization. Nevertheless, in simple practice it is a useful distinction.

Another simple way of achieving the theta point is by changing the degree of solvency of the continuous phase, i.e. by the addition of a non-solvent. Napper [79] has illustrated the utility of varying the solvency to achieve incipient flocculation at constant temperature, and has thus again demonstrated the overall importance of the 'osmotic' term in steric stabilisation.

6.6 DEPLETION FLOCCULATION AND STABILIZATION

Steric stabilization arises as a consequence of polymer adsorption onto the surface of particles. The effects of free polymer in solution on colloid stability give rise to what was coined by Napper 'depletion flocculation' [80] and 'depletion stabilization'. (For a review on the topic) see [51, Chapter 17].

The concept of depletion flocculation may be traced back to Asakura & Oosawa [81, 82]. However, it was Li-in-on et al. [83] who showed that at higher concentration of free polymer in solution, the flocculating effect of added polymer disappeared.

There are variations on the theme of depletion flocculation [84–86], but basically the concept arises from consideration of non-adsorbing surfaces and polymer molecules in solution. When surfaces of colloidal particles approach each other to separations less than the diameter of the polymer molecule in solution, the polymer is then effectively excluded from the interparticle space, leaving only solvent. Thermodynamically, this is not favourable, leading to osmosis of the solvent into the polymer solution and thereby drawing the two particles even closer together, i.e. the particles flocculate.

Napper & Feigin [80] extend the depletion flocculation mechanism to depletion stabilization by considering what happens under equilibrium conditions to two inert, flat plates immersed in non-adsorbing polymer solution, and by examining the segmental distribution of a polymer molecule adjacent to the surfaces in the bulk solution from a statistical point of view. They showed that the polymer segment concentration adjacent to the flat, non-adsorbing surface was lower than in the bulk. By estimating the 'depletion-free energy' as a function of plate separation in terms of polymer molecule diameter, they showed that as the two plates were brought together there was first a repulsion as the polymer molecules began to be constrained, followed by an attraction on closer approach of the two surfaces (see Fig. 6.10).

They applied Derjaguin's method of converting flat plate potential energy curves into those for curved surfaces (i.e. sphere–sphere) interactions, by the equivalent of mathematical 'terracing' of a curve as shown in Fig. 6.11.

Vincent et al. studied the flocculating effect of added polymer in sterically stabilised latexes [87] and explained their findings with a semi-quantitative theory that is schematically represented in Fig. 6.12. They showed that the strength of the

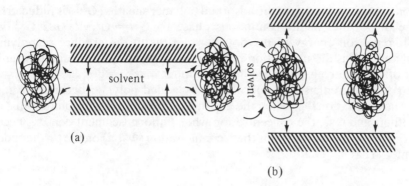

Fig. 6.10 — Schematic representation of depletion flocculation (a) and stabilization (b).

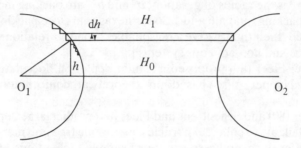

Fig. 6.11 — Derjaguin's flat plate–sphere conversion.

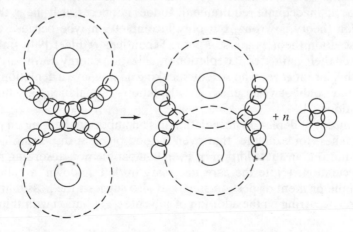

Fig. 6.12 — Vincent *et al.*'s depletion flocculation mechanism (from [88]).

repulsion energy on overlap of the adsorbed polymer sheaths (G^{LL}) is reduced by the presence of polymer in the continuous phase by $G^{LL} = G^{LL,O} - n/2\ G^{PP}$ where $G^{LL,O}$ is repulsion energy of the polymer sheaths in the absence of polymer in solution and G^{PP} is the interaction between two polymer coils in solution; n is the number of polymer coils displaced into solution.

Their theory predicts that flocculation by added polymer occurs between the polymer concentration Φ^{*}_{PS} when the expanded polymer coils in solution just touch one another; and Φ^{**}_{PS}, the concentration when uniform segment density, occurs as the polymer coils contract to their theta configuration ($\Phi_{PS}{}^{**}$) on the further addition of polymer. That is,

$$\Phi^{*}_{PS} = \frac{M}{b^* <S^2>^{3/2} N\rho} \quad \text{and} \quad \Phi^{**}_{PS} = \frac{M}{b^{**} <S^2>_0^{3/2} N\rho}$$

where M is the molecular weight of the polymer of density ρ; N is Avogadro's number; $<S^2>^{\frac{1}{2}}$ is the radius of gyration; b^* and b^{**} are packing factors (8 and 2.52 for cubic close packing, and 5.6 and 1.36 for hexagonal close packing respectively).

When applied, their theory gave a reasonable, broad co-relation with experiment for both aqueous and non-aqueous systems [88].

Scheutjens & Fleer [89] proposed a similar depletion flocculation–stabilization mechanism to Napper, also based on theoretical conformations of polymer molecules.

Both Napper [80] and Scheutjens and Fleer [89] predict large depletion stabilization energies available to colloidal particles on suitable free polymer content, but the two theories differ in several respects. For example, Scheutjens and Fleer predict greater stabilization effects in poor solvency conditions (i.e. theta conditions), while Napper favours good solvency.

One criticism of depletion stabilization theory is that it appears to teach that it is only necessary to have a sufficiently concentrated polymer solution, preferably of high molecular weight, to make a good pigment dispersion. In practice this is clearly found to be an inadequate requirement. It does not necessarily mean that depletion stabilization theory is wrong, but only incomplete, maybe because concentrated polymer solution theories are still poor. Scheutjens & Fleer [90], unlike Napper, have revised their estimates of depletion stabilization energy downwards to be more in line with Vincent *et al.*, who strictly speaking do not have a depletion stabilization energy, but only an absence of depletion flocculation at high polymer concentrations.

The concept of depletion flocculation has an important bearing on paint properties. It teaches, for example, that even if good pigment dispersion is achieved in dilute solution, then the addition of even compatible non-adsorbing polymer may cause flocculation. (Note the care necessary in 'letting-down' a ballmill on completion of the pigment dispersion stage). It also suggests the possibility of pigment flocculation occurring on the addition of solvent, e.g. dilution with thinners prior to spraying.

It also suggests that different methods of paint application which rely on varying degrees of solvent addition may cause different degrees of pigment flocculation, leading to variation in colour development, even when the dispersion is well deflocculated at low polymer content in the continuous phase. Depletion flocculation may also occur with charge-stabilized systems such as latexes on the addition of an 'inert' thickening agent, as shown by Sperry [91] which may produce different types of sedimentation phenomenum.

6.7 ADSORPTION

To obtain good dispersion stability by steric stabilization it is important to fix the stabilizer molecules to the surface of the particle. The more firmly they are held, provided that they are well solvated and free to adopt varying configurations, the better; for example, terminally adsorbed polymeric molecules such as polyhydroxystearic acid, have been found to be excellent stabilizers [57]. If, for example, the adsorbed molecules are attached to the surface of a particle and are able to move about on the surface, then on overlap of the solvated layers the increased concentration could be partly accommodated by a surface redistribution which will result in a lower repulsion energy, than if surface redistribution is not possible. Furthermore, if the adsorption of the stabilizing molecules is weak, then not only could the increase in osmotic pressure be countered by surface redistribution, but also by stabilizer desorption, leaving little, if any, force for approaching particle repulsion.

But, if adsorption of the stabilizer molecules is so strong that the stabiliser molecules are 'nailed flat' to the surface of the particle, effectively producing a pseudo, non-solvated layer, then on the close approach of another similar layer there would be no entropy loss of the polymer chains, and all that would be achieved would be the extension of the particle surface by some small distance. The attraction potential would then be from the new surface, and the particles would have a composite Hamaker constant producing an attraction potential with no source of repulsion energy.

In the design of dispersion stabilizers it is important to have something with which to attach the solvated stabilizing molecule firmly to the surface, preferably at a point, leaving the rest of the molecule to be freely solvated in its best solvent, e.g. chemically bonding the stabilizer to the particle surface either covalently, as with reactive stabilizers such as those described by Thompson *et al.* [92], or by acid–base interaction in non-polar media such as reacting carboxyl terminated fatty chains, e.g. *poly* 12 hydroxy stearic acid with an 'alumina' coated pigment.

Sometimes it is not possible to react a stabilizer to the surface of the particle, and the demand for strong anchoring and maximum solvation can be achieved only by physical adsorption using amphipathic copolymers in hemi-solvents, such as block or graft copolymers, where one portion of the molecule is 'precipitated' onto the surface of the particle (i.e. the anchor component is non-solvated) leaving the solvated portion to produce the stabilizing barrier as described in [57].

6.7.1 Adsorption isotherms

Adsorption isotherms are very useful in obtaining an understanding of what may be happening in a pigment dispersion. For example, it may be expected that when TiO_2 pigment is dispersed in butyl acetate, xylol, and white spirit, using the same dispersant [54], then the attraction energy because of the Hamaker constant values of the respective solvents should be of the order

$$V_A^1 > V_A^2 > V_A^3$$

where V_A^1 = attraction energy in white spirit
V_A^2 = attraction energy in xylol
V_A^3 = attraction energy in butyl acetate

Since the stabilizer is terminally adsorbed (i.e. the barrier thickness is the same), it should give the best stability in butyl acetate solution. However, viscometric measurements indicate that the reverse sequence holds, i.e. the dispersion with the least flocculation was in white spirit, followed by that in xylol, and the worst was in butyl acetate. Inspection of the adsorption isotherms immediately revealed that the surface concentration was the least in butyl acetate (see Fig. 6.13), hence the stabilizing barriers were the weakest in butyl acetate solution.

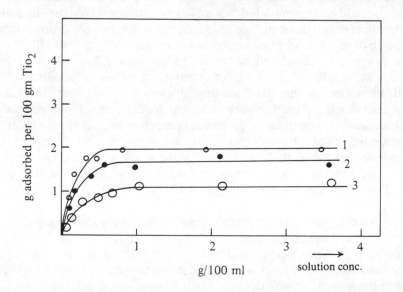

Fig. 6.13 — Adsorption isotherm of dimer 12 hydroxy stearic acid in (1) white spirit; (2) xylol; (3) butyl acetate onto TiO_2 [54].

Adsorption isotherms are most readily measured by placing a fixed amount of pigment in a known amount of polymer solution, in glass jars, along with small (e.g. $\frac{5}{16}$ in.) glass beads as a milling aid, and leaving the jars for at least 24 hours on rollers, so that constant agitation is obtained. Centrifugation will then enable the separation of the continuous phase, so that the amount of material adsorbed may be determined, as described in [54].

An alternative method to obtain an adsorption isotherm is to place the pigment in a chromatographic column and to pass a dilute solution of the adsorbate down the pigment column, as described by Crowl [93].

The advantage of the chromatographic method is that it enables preferential adsorption and adsorption reversibility to be easily determined. The method generally works well with organic pigments; however, with TiO_2 pigments flow rates through the pigment column are in practice very slow, making the method very much less attractive than might appear at first sight.

6.7.2 Free energy of adsoprtion and adsorption isotherms

Estimation of free energy of adsorption ($\triangle G_{ads}$) from adsorption isotherms
Suppose that we examine a particle immersed in a liquid, e.g. water, and we imagine the surface of the particle to consist of a 'mosaic' of adsorption sites where 'S' represents such a site occupied by the solvent. If we introduce a solute molecule, X, into the liquid, then if it were to adsorb on the surface, it will displace the water molecule occupying the site. Let SX represent the adsorption of the solute at a site, and $S(H_2O)$ the water molecule adsorption. We can express this as an equilibrium reaction by:

$$S(H_2O) + X \underset{k_2}{\overset{k_1}{\rightleftharpoons}} SX + H_2O \qquad \text{and} \qquad K = \frac{k_1}{k_2}$$

i.e. K is the partition, or equilibrium, constant for the solute between the surface of the particle and the solvent, and is related to the free energy of adsorption ($\triangle G_{ads}$) by

$$\triangle G_{ads} = -RT \ln K.$$

Using Glasstone's notation [94] where possible we can express K as

$$K = \frac{\text{(mole fraction solute on surface) (mole fraction solvent in bulk)}}{\text{(mole fraction solvent on surface) (mole fraction solute in bulk)}}$$

Suppose the surface has a sites per g adsorbent which are occupied by the solvent molecules in the absence of solute and that x_1, of these are occupied by the solute when dissolved in the solvent, then

$$K = \frac{(x_1/a) \text{ (mole fraction solvent in bulk)}}{\left(\dfrac{a - x_1}{a}\right) \text{ (mole fraction solute in bulk)}}.$$

If we restrict our considerations to low (i.e. initial) adsorption i.e. when $x_1 \ll a$ and have dilute solute concentration, then

$$K \approx \frac{(x_1/a)\,(1)}{(1)\,(N)}$$

hence $\triangle G_{ads} = -RT \ln \left(\dfrac{x_1/a}{N} \right)$

where N is the mole fraction of solute of concentration, 'c' g. moles in solution of a liquid of molecular weight M, and density 'ρ' so that $N = \dfrac{Mc}{1000\rho}$ and hence for aqueous solution $N = c/55.5$.

Let x/m be the number of g moles adsorbed solute per g adsorbent. this is proportional to the fraction of sites occupied by solute, i.e. $x/m = p\dfrac{x_1}{a}$ where p is a proportionality constant.

If the surface is completely covered by a monolayer of solute, then $x_1/a = 1$; and $x/m = p$ = maximum adsorption (or plateau value). Thus for aqueous solution,

$$\triangle G_{ads} = -RT \ln \left(\frac{x/m \times 55.5}{c \times p} \right)$$

At low solution concentration, $\dfrac{x/m}{c}$ = initial slope (IS) of an adsorption isotherm, hence

$$\triangle G_{ads} = -RT \ln \left(\frac{IS \times 55.5}{\text{plateau value[that is, } p]} \right) \tag{6.16}$$

By plotting an adsorption isotherm in terms of g moles adsorbed per g adsorbate against molar equilibrium concentration it is thus possible to estimate the free energy of adsorption in a monolayer or Langmuir-type adsorption isotherm from the initial slope and saturation value.

The usefulness of equation (6.16) is that it gives a 'feel' for the strength of adsorption and especially its 'converse' use, i.e. in estimating the adsorption of a surfactant by particle surfaces such as those of pigment or emulsion droplets, and ensuring that sufficient material is added to satisfy the adsorption requirement of the dispersion. For if the free energy of adsorption is known, and for simple physical adsorption it is of the order of 5 K.cals/g mole, the cross-sectional area of a surfactant molecule is known (obtainable in the literature, or it can be estimated using molecular models). It is thus possible to construct the initial slope and plateau portions of the adsorption isotherm. By joining the initial slope to the plateau level, by drawing a gentle curve in the 'near-saturation' zone (as shown by the broken line in Fig. 6.14), the adsorption isotherm can be completed with a little 'artistic licence', hence the minimum surfactant requirement estimated.

The derivation of $\triangle G_{ads}$ as outlined above is based on a simple model of an adsorption surface composed of adsorption sites which are either occupied by solvent molecules or solute, such as a simple surfactant.

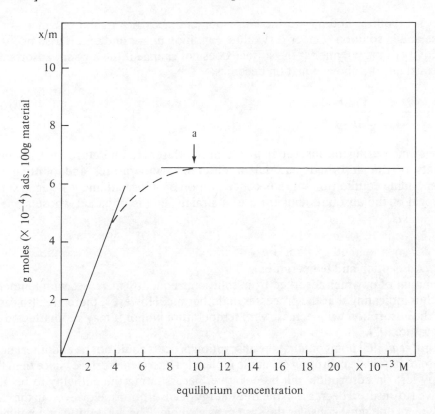

Fig. 6.14 — Prediction of surfactant adsorption using equation (6.16). The construction of the isotherm is based on the physical adsorption of surfactant (aqueous) at 5 K cals. per mole at the rate of 200 Å^2 per molecule onto a powder of density 1.5 g/cm^3, with average particle diameter of 0.5 μm. a is the required surfactant concentration which comprises 6.7×10^{-4} g moles surfactant per 100 g powder in equilibrium with a minimum surfactant solution concentration of 9.5×10^{-3}M.

This is useful in giving an insight into the interpretation of an adsorption isotherm and illustrates the importance of the initial slope, as well as the saturation value (which also applies to polymer adsorption); but the calculation of ΔG_{ads} for a polymer molecule, which may occupy a varying number of sites depending on its conformation, should not be made unless refinements, or constraints to polymer chain configurations, are taken into account. A similar proviso is as made by Koral *et al.* [95] who determined the 'isosteric' heats of polymer adsorption from adsorption isotherms. That is, if the adsorption isotherm is measured at different temperatures, then it is possible to estimate the enthalpy of adsorption (ΔH) using an analogue of the Clausius–Clapeyron equation (Huckel equation):

$$\frac{\Delta H}{R} = \frac{H_s - H_L}{R} = \frac{d \ln(a_L/a_s)}{d (1/T)} \tag{6.17}$$

where H_s and H_L are the partial molar enthalpies of the polymer on the adsorbate

most practical instances one is more interested in the state of equilibrium adsorption in a mixed component system; and with polymer adsorption from solution, the simple principle of first come first adsorbed applies: i.e. adsorption is diffusion controlled [99, 105, 106]. The smaller molecules arrive at a surface first and are adsorbed only to be displaced on the later arrival of the larger molecules which are preferentially adsorbed [107]. This phenomenon of preferential adsorption of the higher molecular weight analogues frequently gives rise to the observed 'maxima' in adsorption isotherms of polymers [56]. In dispersing pigments, e.g. making a millbase, it is important to ensure that the preferentially adsorbed polymer is in sufficient quantity at the very start, since displacement may be very slow.

This is sometimes vividly demonstrated in practice, as for example when pigment is initially dispersed in a polyethylene glycol modified alkyd, and the other polymeric constituents are added afterwards: then good pigment flushing into the aqueous phase occurs immediately, as shown when the freshly prepared oil-based paint adhering to a paint brush is washed in detergent solution (i.e. when the modified PEG alkyd is adsorbed on the pigment surface). However, if the pigment is milled in the other polymeric constituents first, and then the PEG alkyd is added, it then takes about four weeks for the 'brush wash' properties to appear, signalling that the PEG alkyd has displaced the other constituents from the pigment surface [108].

Another example of the slow re-equilibration of adsorbed molecules sometimes manifests itself as a 'colour drift' (a gradual shift in colour on storage), for example when the 'colour' is made by blending various (colour stable) pigment dispersions (tinters), especially if the tinters are based on different resins. One way of solving this colour drift problem is simply to pass the finished paint through the dispersion mill (e.g. a Sussmeyer). The shearing in the mill, as well as the temperature rise, accelerate the re-equilibration of the adsorbed species, just as agitation is found to accelerate adsorption of polymer [106] and thereby stops the colour change.

Stereo regularity is also an important factor in polymer adsorption, as shown by Miyamoto *et al.* [109] who found that isotactic polymethyl methacrylate is more strongly adsorbed from the same solvent than syndiotactic polymer of the same molecular weight. Syndiotactic pMMa, in turn, is more strongly bound than atactic polymer, as shown by TLC development with ethyl acetate [110].

6.7.5 Specific adsorption

Adsorption of polymer from solution onto a surface depends on the nature of the solvent and the 'activity' of the surface.

While in the simple physical adsorption of polymer it can be seen intuitively that higher molecular weight species might be preferentially adsorbed in a homologous series, to the low molecular weight species, specific adsorption may override molecular weight considerations. A typical example is the preferential adsorption of the low molecular weight, more polar species, such as the phthalic half esters present in long oil alkyds. Walbridge *et al.* [111] showed that the adsorption of these materials could be explained in terms of acid–base interactions, where the base was the TiO_2 pigment coating and the metal of the 'driers'. They were able to show that the complex flocculation–deflocculation behaviour depended on the order of addition of drier and dimer fatty acid to a simple white, long oil alkyd-paint. They found that it also depended on the relative acid strength and the formation of irreversible carboxyl–pigment coating bonds, which in turn were a function of the temperature of dispersion. Just as ion exchange resins can bind acids (or bases), so can certain pigment surfaces behave in a similar manner. Solomon et al. [112] studied the effect of surface acid sites on mineral fillers and concluded that they were comparable in strength to, but fewer than, those found in 'cracking' catalysts. The presence of these sites, which were developed by heat, profoundly affected chemical reactions in polymer compounds, particularly in non polar media.

6.8 RATE OF FLOCCULATION

When considering dispersions and flocculation phenomena it is useful to have some idea of the rate of flocculation, at least in order of magnitude terms. Smoluchowski [113] showed that for uniform spheres which have neither attraction nor repulsion forces present and adhere to each other on collision, the rate of decrease in the number of particles can be expressed as: $-dN/dt = KN^2$ where N = number of particles, and K is the rate constant. On integration, this becomes

$$1/N = 1/N_0 + Kt, \tag{6.18}$$

where N_0 is the initial number of particles; i.e. $N = N_0$ at time t_0. If we define a 'half life' time $t_{\frac{1}{2}}$ where the number of particles is one half the original number. i.e. $2N = N_0$ at time $t_{\frac{1}{2}}$ then on substitution in equation (6.18) we have; $2/N_0 = 1/N_0 + Kt_{\frac{1}{2}}$, that is,

$$t_{\frac{1}{2}} = 1/KN_0 . \tag{6.18a}$$

If the rate constant is diffusion controlled, then $K = K_D = 4kT/3\eta$ where k = Boltzman constant; T = temperature; η = viscosity of the medium.

Table 6.4 gives the time taken for different particle sizes at various pigment volumes to reduce their number by one half, according to Smoluchowski's rapid flocculation kinetics based on Brownian motion when the medium is at room

Table 6.4 — The half life of particles of different particle size undergoing rapid flocculation at 25°C in a medium of 1 poise viscosity

Half-life time $(t_{\frac{1}{2}})$ (s)	% pigment volume (e.g. TiO$_2$) ($\sim 0.2\,\mu$m diam.)	Half-life time $(t_{\frac{1}{2}})$ (μs)	% vol (Aerosil) ($\sim 0.014\,\mu$m diam.)
7.6	1	2400	1
1.5	5	480	5
0.76	10	240	10
0.5	15	160	15
0.38	20	120	20

temperature and has a typical paint viscosity (1 poise). If flocculated, a typical gloss paint (TiO_2 at 15% PV) will revert to its flocculated state in half a second after having been subjected to disturbance, e.g. by brushing. Even if the vehicle viscosity is very much greater, the time taken for reflocculation to occur after shear is still very short. For a similar reason when Aerosil (fumed silica) is used as a structuring agent, the quantity required is small compared to TiO_2 volume, and the structure developed through flocculation is almost instantaneous. There have been a number of attempts to verify and improve on Smoluchowski's rapid flocculation kinetics by taking into account interparticular forces and hydrodynamic interactions [114–118]. While these findings suggest that the rate constant should be approximately half of Smoluchowski's value, they do not significantly alter the concept that in practical systems the flocculation of pigment is very fast.

Gedan *et al.* [119] have measured the rapid flocculation of polystyrene particles. While they are in contention with Smoluchowski's approximation that the rate constant for doublet, triplet, and quadruplet formation is the same, nevertheless Smoluchowski's theory is still considered to give a good insight as to the rate of dispersion flocculation and has been well confirmed by experiment [114, 115].

6.8.1 Sedimentation and flocculation

Pigments generally have a density greater than the resin solution constituting the paint, and under the influence of gravity they will tend to settle out, according to Stokes Law. This may be expressed as

$$ d = \left(\frac{18\eta h}{\Delta \rho g t} \right)^{\frac{1}{2}}, \tag{6.19} $$

where a particle of apparent diameter d (Stokes diameter) falls a distance h in a medium of viscosity η for a time t under the influence of gravity g when there is a difference in density of $\Delta \rho$ between the particle and medium.

If well deflocculated, the particle dispersion will sediment according to the above relationship to give a hard, compact deposit which is difficult to redisperse. However, a sedimentation velocity of up to *circa* 10^{-6} cm s^{-1} is generally countered by diffusion and convection, and sedimentation may seem to be absent. If flocculation is present, pigment sedimentation is rapid and the deposited material is soft, voluminous, and readily reincorporated into suspension by stirring.

When sedimentation takes place the volume of the final sediment depends upon the degree of flocculation; the greater the flocculation, the larger the sediment volume. In the extreme, the sedimentation volume may equal the total volume and should not be confused with the deflocculated state when no sedimentation may occur. The sediment volume is not only dependent on the flocculation forces, but it also depends on particle shape and size distribution [120]. Care should be taken to ensure that sedimentation volumes are used only as a relative measure of the degree of flocculation in similar systems.

Plate 6.1 shows how sedimentation tubes may be used to determine the optimum concentration of dispersing agent (Calgon) to produce a well-deflocculated suspension of TiO_2 in water.

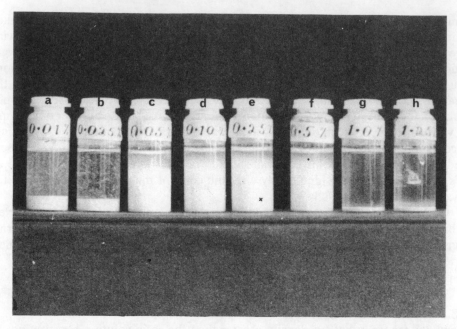

Plate 6.1 — Sedimentation tubes used to determine optimum concentration of Calgon to produce TiO_2 dispersion in water. Tube 'e' is marginally the best — note the lower sedimentation volume of tubes 'g' and 'h', indicating that although flocculated they are not as intensely flocculated as 'a' and 'b'.

A simple alternative way of assessing the degree of pigment flocculation is to spot the dispersion onto filter paper and measure the ratio of pigment stain to solvent stain, in an anlogous manner to Rf values in chromatography; the higher the ratio the more deflocculated the dispersion.

Plate 6.2 illustrates the respective 'spot' testing of the dispersions shown in the sedimentation test. Note how tube and spot tests both show dispersion 'e' to be the best dispersion.

Smith & Pemberton [121] have described a variant of the 'spot' method of assessing flocculation which they claim gives better results.

Generally, in order to assess flocculation by observing sedimentation behaviour or measuring the 'Rf' flocculation value, it is necessary to reduce the viscosity of a paint, by the addition of solvent, otherwise sedimentation is too slow to be of experimental value; though on dilution, care must be exercised, since this may introduce a new source of flocculation due to desorption, or depletion flocculation, or both.

Flocculation is more readily assessed in the liquid state where it can be measured by viscometry [52]. For example, if the viscosity of the continuous phase (η_0) is Newtonian, then if there is no particle–particle interaction, the dispersion viscosity (η) should also be Newtonian. If the dispersion is not Newtonian, then the degree of deviation from Newtonian behaviour is a measure of particle–particle interaction, i.e. of flocculation, provided that the disperse phase volume is $< 30\% +$, otherwise inertial effects may be observed [122]. By plotting $\log \eta$ against the reciprocal square

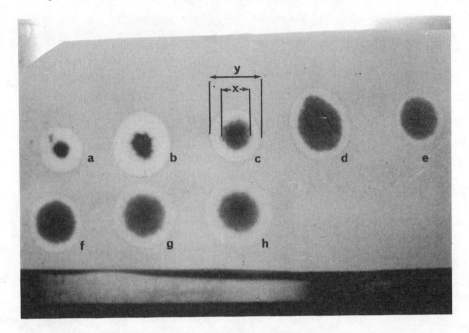

Plate 6.2 — 'Spot' tests on filter paper of the dispersions used in Plate 6.1 to determine optimum Calgon concentration for TiO_2 dispersion. $Rf = x/y$. The larger the value the better the dispersion. Spot 'e' is the best.

root of the shear rate, straight-line plots are obtained. The steeper the gradient the greater the degree of flocculation, as shown in Fig. 6.15.

This method of assessing flocculation has been found in practice to be very reliable and applicable to typical pigment concentrations in paint formulations. The comparison of flocculation by this method is strictly for similar volume concentrations, since the slope in the above plot will increase with disperse phase concentration. However, the effect of dissimilar disperse phase volumes can be estimated, and corrections can be applied to enable comparison.

It is worth noting that the rheological assessment of flocculation is a function of the total interaction of the system which is measured. Therefore, if there is a large difference in particle size between two dispersions, even though the two dispersions are compared at the same disperse phase volume, the dispersion with the smaller particle size and weaker interparticle interactions will appear to be more non-Newtonian (i.e. more flocculated). This is due simply to the interparticular forces being proportional to the first order of particle radius, while the number of particles per unit volume is proportional to the cube of the radius; hence the total 'strength' of a large number of weak bonds amongst small particles is greater than a few strong bonds between larger particles.

There have been attempts to estimate interparticle forces from rheological measurement by determining a 'critical shear rate' [123], but these estimates are very questionable [124].

The degree of pigment flocculation can also be assessed by colour measurement

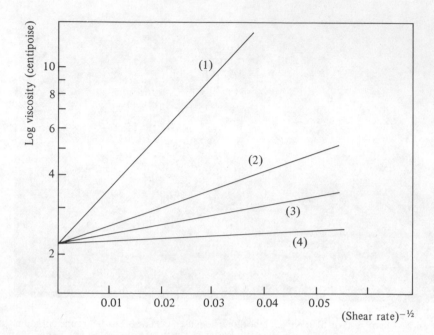

Fig. 6.15 — Log (dispersion) viscosity–(shear rate) $^{-\frac{1}{2}}$ plots as a measure of varying degrees of pigment flocculation (ref 54) (1) most flocculated to (4) least flocculated.

[125], anti-sag properties [111], or by direct visual observation through a microscope, as shown in Plate 6.3. Flocculation in a dry paint is more difficult to assess since microtoming thin sections of paint film for electron micrographs may alter the apparent state of pigment dispersion [126]. Freeze fracture, followed by oxygen etching methods [127], has been developed to overcome this shortcoming; but at concentrations around 12% or more, visual interpretation becomes more difficult.

An alternative approach to the assessment of pigment flocculation has been developed by Balfour & Hird [128] who used the measurement of scatter coefficients determined in the IR region which are more sensitive to large particles. However, none of these methods can measure the state of pigment flocculation directly in both the liquid and solid state. Armitage [129] has described a method based on proton straggling which can measure the state of pigment flocculation directly, irrespective of the nature of the phase (i.e. dry or liquid paint).

In conclusion, the apparently simple question asked at the beginning of this chapter, as to whether it is better to disperse a pigment in aliphatic or aromatic hydrocarbon, is in reality extremely complex, for it contains conflicting considerations, viz:

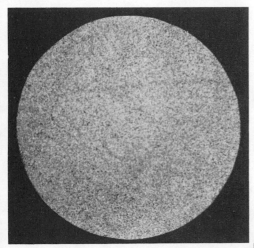

6.3a — A well-dispersed TiO_2. When viewed through a microscope ($\times400$) the particles appear to shimmer, owing to Brownian motion. Although the particles seem to be visible, in fact they are not resolved, and only their diffraction pattern is seen as a dot.

6.3b — Lightly flocculated TiO_2 dispersion. The particles have lost their Brownian motion and are seen as clusters ($\times400$).

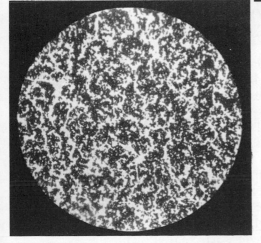

6.3c — Highly flocculated TiO_2 dispersion. The particles are flocculated into large clusters and are reminiscent of the mud-flats of a river bed ($\times400$).

Plate 6.3 — Photomicrographs of TiO_2 pigment dispersion showing varying degress of flocculation.

(1) It is necessary to maximise $\gamma \cos \theta$ for the system, and this may depend on the nature of the substrate and the concentration of alkyd. For TiO_2, which has a high surface energy, it would probably be better to use the aromatic solvent, since $\cos \theta = 1$ for both solvents.

(2) From a consideration of attraction energies it would appear that the aromatic solvent might be the better choice when TiO_2 is the pigment, since this would produce a better match of Hamaker constants.

(3) When considering the nature of the adsorbed layer and its contribution to steric repulsion it would probably be better to use the aliphatic solvent (white spirit) since a greater concentration of resin might be adsorbed, and this may give a more robust dispersion at only a small trade-off in solubility of the solvated layer than using aromatic solvent)toluene).

Finally, consideration (3) will probably outweigh considerations (1) and (2) combined.

REFERENCES

[1] Balfour, J. G., OCCA Conference 39–49 (1977).
[2] Asbeck, W. K., *Off. Digest* **33** 65–83 (1961).
[3] Balfour, J. G., *JOCCA* **60** No. 9 365–76 (1977).
[4] Peacock, J., *XI Fatipec Congress* 193–9 (1972).
[5] Elm, A. C., *Off. Digest* **33** (433) 163.
[6] Patton, T. C., p. 146 In *Paint flow and pigment dispersion* Interscience Publ. (1964).
[7] Heertjes, P. M. & Witvoet, W. C., *Powder Tech.* **3** 339 (1970).
[8] Washburn, E. O., *Phys. Rev.* **17** 374 (1921).
[9] Good, R. J., *Chem. and Ind.* 600 (1971).
[10] Doroszkowski, A., Lambourne, R., & Walton, J., *Colloid & Polymer Sci.* **255** 896–901 (1977).
[11] Hare, E. F. & Zisman, W. A., *J. Phys. Chem.* **59** 335 (1955).
[12] Crowl, V. T., *JOCCA* **55** (5) 388–420 (1972).
[13] Parfitt, G. D., *XIV Fatipec Congress* 107 (1978).
[14] Hughes, W., *Fatipec Congress*, 67–82 (1970).
[15] Howard, P. B. & Parfitt, G. D., *Croatica Chemica Acta* **45** 189–194 (1973) : ibid **50** 15–30 (1977) see also *JOCCA* **54** 356–362 (1971).
[16] Keesom, W. H., *Phys. Zeit* **22** 129 (1921).
[17] Debye, P., *Phys. Zeit* **21** 178 (1920).
[18] London, F., *Zeit Phys.* **63** 245 (1930).
[19] Hamaker, H. C., *Rec. trav chim* **55** 1015 (1936) ibid **56** 3,727(1937).
[20] Hamaker, H. C., *Physica* **4** 1058 (1937).
[21] Visser, J., *Adv. in Colloid and Interface Sci.* **3** 331 (1972).
[22] Gregory, J., *Adv. in Colloid and Interface Sci.* **2** 396 (1970).

[23] Vold, M., *J Coll. Sci.* **16** 1 (1961).

[24] Casimir, H. B. G., and Polder, D., *Phys. Rev.* **73** 360–372 (1948).

[25] Schenkel, J. H. & Kitchener, J. A. (see also Tabor, D.) *Trans. Farad. Soc.* **56** 161 (1960) (see also Tabor, D.) in p. 23–45 in: *Colloid dispersions* Ed. J. W. Goodwin Special Publ./Royal Soc. of Chem.

[26] Osmond, D. W. J. & Vincent, B., & Waite, F. A., *J. Colloid Interface Sci.* **42** (2) 262–269 (1973) also ibid **42** (2) 270–285 (1973).

[27] Visser, J. *Adv. in Colloid and Interface Sci.* **15** 157–69 (1981).

[28] La Mer, V. *J. Colloid & Interface Sci.* **19** 291 (1969).

[29] Verwey, E. J. W., & Overbeek, J.Th, G., *Theory of lyophobic colloids* Elsevier Publ. Co. Inc. (1948)

[30] Lyklema, H. In: *Colloidal dispersions* Chapter 3 J. W. Goodwin (ed.). Special Publ./Royal Soc of Chem.

[31] Overbeck, Th.G., *Adv. Coll. and Interface Sci.* **16** 17–30 (1982).

[32] Smith, A. L. In: *Dispersion of powders in liquids* p. 127, Chapter 3 3rd ed. by G. D. Parfitt, Applied Sci. Publ. (1981).

[33] O'Brien, R. W. & White, L. R., *J. Chem. Soc. Farad* 11 **2** 1607 (1978).

[34] Wiersema, P. H., Loeb, A. L., & Overbeek, Th. G. *J. Coll. Sci.* **22** 78 (1966).

[35] Van Oss, C. J. *J. Coll. Interface Sci.* **21** 117 (1966).

[36] Franklin, M. J. B. *JOCCA* **51** 499–523 (1968).

[37] Beck, F. *Progress in Organic Coatings* **4** 1–60 (1976).

[38] Schenkel, J. H., Kitchener, J. A., *Experientia* **14** 425 (1958) Mackor, E. L., *Rev. Trav. Chim.* **70** 747–62(1951).

[39] Rank Bros. Bottisham, Cambridge UK

[40] James, A. M., In: *Surface and colloid science* R. J. Good & R. R. Stromberg (eds), Plenum Press Chapter 4, vol. 11 121–185 (1979).

[41] Hunter, R., In: *Zeta potential in colloid science* Chapter 4, p. 125–178 Academic Press (1981).

[42] Hogg, R., Healy, T. W., & Fuerstenan, D. W., *Trans. Faraday Soc.* **62** 1638 (1966).

[43] Overbeek, Th. G., *Disc. Farad. Soc.* **42** 10 (1966).

[44] McGowan, D. N. L. & Parfitt, G. D. *Disc. Farad. Soc.* **42** 225 (1966).

[45] Osmond, D. J. W. ibid p. 247.

[46] Lyklema, H., *Adv. Coll. and Interface Sci.* **2** 94 (1968).

[47] Overbeek, Th G., *Disc. Farad. Soc.* **42** 10 (1966).

[48] Vincent, B. *Adv. Colloid and Interface Sci.* **4** 193–277 (1974).

[49] Parfitt, G. D. & Peacock, J. In: *Surface and colloid science* Vol. 10 Chapter 4 Matijevic (ed.) Plenum Press (1978).

[50] Osmond, D. J. W. *Sci. and Tech. of Polymer Colloids* Vol. 2 369 (1983) NATO ASI series Poehlein, Otteweill, & Goodwin (eds).

[51] Napper, D. H. *Polymeric stabilisation of colloidal dispersions* Academic Press (1983).

[52] Doroszkowski, A. & Lambourne, R. *J. Coll. and Interface Sci.* **26** 214–221 (1968).

[53] Asbeck, W. E. & Van Loo, M., *Ind. Eng. Chem.* **46** 1291 (1954).

[54] Doroszkowski, A. & Lambourne, R. *Disc. Farad. Soc.* **65** 253–262 (1978).

[55] Garvey, M. J., *J. Coll. and Interface Sci.* **61** (1) 194–196 (1972).
[56] Doroszkowski, A. & Lambourne, R. *CID Congress Chemie Physique et Applications Protiques* Vol. 2 part 1 73–81 Barcelona (1968).
[57] Walbridge, D. J., In: *Dispersion polymerisation in organic liquids* Chapter 3 Barrett (ed.): Wiley-Interscience Publ. (1974).
[58] Zigmondy, R. *Z. Anal. Chem.* **40** 697 (1901).
[59] Mackor, E. L. *J. Coll. Sci.* **6** 492 (1951).
[60] Fischer, E. W. *Kolloid-Z* **160** 120 (1958).
[61] Clayfield, E. J. & Lumb, E. C. *J. Coll. Sci.* **22** 269 (1966).
[62] Meier, D. J. *J. Phys. Chem.* **71** 1861 (1967).
[63] Hesselink, F. Th. *J. Phys. Chem.* **73** 3488 (1969).
[64] Hesselink, F. Th. *J. Phys. Chem.* **75** 65 (1971).
[65] Evans, R. & Napper, D. H., *Kolloid Z-Z Polymere* **251** 409–414 (1973) ibid 329.
[66] Ottewill, R. H., & Barclay, L., & Harrington, A., *Kolloid-ZuZ Polymers* **250** 655 (1972).
[67] Ottewill, R. H., Cairns, R. J. R., Osmond, D. W. J., & Wagstaff, I. *J. Coll. and Interface Sci.* **54** 45 (1976).
[68] Homola, A. M. & Robertson, A. A. *J. Coll. Interface Sci.* **54** 286 (1976).
[69] Doroszkowski, A. & Lambourne, R. *J. Polymer Sci.* **C34** 253 (1971)
[70] Doroszkowski, A. & Lambourne, R. *J. Coll. and Interface Sci.* **43** 97 (1973).
[71] Klein, J. *Nature* **288** 248 (1980) see *Adv. Coll and Interface Sci.* **16** 101 (1982).
[72] Napper, D. H., *Trans. Farad. Soc.* **64** 1701 (1968).
[73] Napper, D. H., *J. Coll. and Interface Sci.* **32** 106–114 (1970).
[74] Evend, R. & Davison, J. B., Napper, D. H. *J. Polymer Sci.* **B10** 449–453 (1972).
[75] Flory, P. J., In *Principles of polymer chemistry*, p. 523 Cornell Uni. Press.
[76] Ottewill, R. H. & Walker, T. *Kolloid Z* **227** (1/2) 108–116, (1968).
[77] Napper, D. H., *Kolloid Z u Z Polymere* **234** 1149 (1969).
[78] Waite, F. A. & Osmond, D. J. W., Vincent, B., *Kolloid Z u Z Polymere* **253** 676–682 (1975).
[79] Napper, D. H. *Trans. Farad. Soc.* **64** 1701 (1968).
[80] Napper, D. H. & Feigin, R. I. *J Coll. and Interface Sci.* **75** 525 (1980).
[81] Asakara, S. & Oosawa, F. *J. Chem. Phys.* **22** 1255 (1954).
[82] Asakara, S. & Oosawa, F. *J Polymer Sci.* **33** 183 (1958).
[83] Li-in-on, F. K., Vincent, B., & Waite, F. A. *ACS Symposium Series* **9** 165 (1975).
[84] Vrij, A., *Pure Appl. Chem.* **48** 471 (1976).
[85] Joanny, J. F. & De Gennes, P. G., *J Polymer Sci. Polymer Phys.* **17** 1073 (1979).
[86] Pathmamanoharan, C., Hek, Th. de. & Vrij, A., *Colloid Polym. Sci.* **259** 769 (1981).
[87] Vincent, B., *J. Coll. and Interface Sci.* **73** No. 2 (1980).
[88] Clarke, J., Vincent, B., *J. Chem. Soc. Farad. Trans. I* **77** 1831–43 (1981).
[89] Scheutjens, J. M. H. M. & Fleer, G. J., *Adv. Coll. Interface Sci.* **16** 361 (1982).

[90] Fleer, G., Scheutjens, J. H. M. H., & Vincent, B., In *Polymer adsorption and dispersion stability* pp. 245–263 B. Vincent & E. D. Goddard (eds.) A.C.S. Symp. Ser. (1984) **240**.

[91] Sperry, P. R., *J Coll. and Interface Sci.* **99** 97–108 (1984).

[92] Thompson, M. W., Graetz, C., Waite, F. A., Waters, J. A., EP13478.

[93] Crowl, V. T., *JOCCA* **46** 169—205 (1961).

[94] Glasstone, S. *Textbook of physical chemistry* 2nd ed. p. 822 MacMillan & Co. (1956).

[95] Koral, L. & Ullman, R., Eirich, F., *J Phys. Chem* **62** 541 (1958).

[96] Gilliland, E. R. & Gutoff, E. B., *J. Applied Polymer Sci.* **3** (7) 26–42 (1960).

[97] Ellerstein, S. & Ullman, R., *J Polymer Sci.* **55** 161, 129 (1961).

[98] Mizuhara, K., Hara, K., & Imoto, T., *Koll.-Z u Z Polymers* **229** 17–21 (1969).

[99] Howard, G. J. & McConnell, P., *J Phys. Chem.* **71** 2974, 2981–2991 (1967).

[100] Perkel, R. & Ullman, R., *J Poly. Sci.* **54** 127–148 (1961).

[101] Kiselev, A. V., Lygin, V. I. *et al. Colloid J* USSR **30** 291–295 (1968).

[102] Thies, C., *J Phys. Chem.* **70** No. 12 3783–3789 (1966).

[103] Eirich, F. R., *Effects of polymers on dispersion properties Symposium* 125–143 (1983).

[104] Lipatov, Yu. S. & Sergeeva, L. U., *Adsorption of polymers*, Halsted Press (1974) ISBN 0–470–54040–0.

[105] Hobden, J. F. & Jellinek, H. H. G. *J Poly. Sci.* **11** 365 (1953).

[106] Patat Von, Franz & Schliebeuer, C., *Macromol Chem.* **44–6** 643–668 (1961).

[107] Felter, R. E., Moyer, E. S., & Ray, L. N., *J Polymer Sci.* **B7** 529–533 (1969).

[108] Baker, A. S., Jones, G. M., & Nicks, P. F., BP 1370914 (1974).

[109] Miyamoto, T., Tomoshige, S., & Inagaki, H., *Poly J.* **6** No. 6, 564–570 (1974).

[110] Butter, R., Tan, Y., & Challa, Cr., *Polymer* **14** (4) 171–2 (1973).

[111] Walbridge, D. J., Scott, E. I., & Young, C. H. *Vth Conference in Organic Coatings Science and Technology* 526–546 (1979) Athens.

[112] Solomon, D. H., Swift, J. D., & Murphy, A. J., *J. Macromol. Sci. — Chem.* **A5** (3) 587–601 (1971).

[113] Smoluchowski, M. *Z Phys.* **17** 557 (1916); *Z Phys. Chem.* **92** 129 (1917).

[114] Lichtenfeld, J. W., Th. Pathmamanoharan., & Wiersema, P. H., *J Coll and Interface Sci.* **49** 281 (1974).

[115] Lips, A. & Wills, E., *Trans. Farad. Soc.* **69** 1226 (1973).

[116] Overbeek, J. Th. G., In: *Colloid science* Vol. 1 p. 282 Kruyt (ed.), Elsevier (1952).

[117] Spielman, L. A. *J Coll. and Interface Sci.* **33** 562 (1970).

[118] Honig, E. P., Roebersen, G. J., & Wiersema, P. H., *J Coll. and Interface Sci.* **36** 97 (1971).

[119] Gedan, H., Lichtenfeld, H., & Sonntag, H., *Coll. and Polymer Sci.* **260** 1151 (1982).

[120] Vold, M. J., *J Coll. Sci.* **14** 168–174 (1959).

[121] Smith, A. E. & Pemberton, E. W., *Pig. and Resin Tech.* **9** Nos. 11, 8 (1980).

[122] Strivens, T. A. *J. Coll. Interface Sci.* **57** (3) 476–87 (1976).

[123] Albers, W. & Overbeek, J., Th., G., *J Coll. Sci.* **15** 480–502 (1960).

[124] Doroszkowski, A. & Lambourne, R., *J Coll. Sci.* **26** 128–130 (1968).

[125] Kalwza, U., *Pig. Resin Tech.* **9** Nos. 4–7 (1980).
[126] Hornby, M. R. & Murley, D. R., *Progress in Organic Coatings* **3** 261 (1975).
[127] Menold, R., Luttge, B., & Kaiser, W., *Adv. Coll. and Interface Sci.* **5** 281–335 (1976).
[128] Balfour, J. G. & Hird, M. J., *JOCCA* **58** 331, (1975).
[129] Doroszkowski, A. & Armitage, B. H., XVII *Fatipec Congress* **2,** 311–330 (1984).

7

Particle size and size measurement

A. Doroszkowski

7.1 INTRODUCTION

To understand the behaviour and properties of particle suspensions such as pigments in paint it is necessary to know something about the size and the size distribution of the particles. Particulate material may constitute pigment, extenders, emulsion droplets, or latex particles.

The question of 'What is a particle?' and hence 'What size is it?' is not always easy to answer, as shown in Plate 7.1

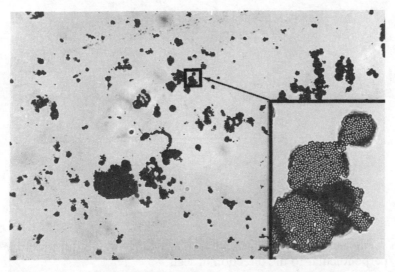

Plate 7.1 — What is a particle? (Insert at × 12 increase magnification).

The answer to the question resides in another question: 'What property is to be investigated?' Particle size measurement is usually a means to an end and rarely justified in itself. In any particle size measurement it is important to use a method most suited to the problem in hand. When surfactant adsorption estimates are required, sizing by surface area should be carried out; if sedimentation considerations are required then particle sizing should be by a volume (mass) method in preference to a count by number method.

When the particles are perfect spheres and are all of the same size, then it is only necessary to know the particle diameter to be able to describe fully the size and size distribution. However, in practice such an occurrence is extremely rare. Particles are frequently irregular in shape and usually have a range of sizes, and what might be an adequate definition of 'particle size' for one set of conditions will be inadequate for another. It is therefore important to use a definition of particle size which is relevant to the property being considered. For example the variable floccule size shown in Plate 7.1 has no relation to the size of the monodisperse–primary–particles constituting the floccules.

7.2 DEFINITIONS

Spherical particles are readily described by their diameter, but non-spherical particles may be measured in many different ways. Some of the ways to express the size of irregular particles are illustrated below [1].

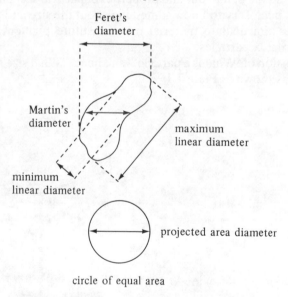

Fig. 7.1 — Non-spherical particle diameter.

Feret's diameter is the distance between two tangents on opposite sides of the particle, perpendicular to the direction of scan.

Martin's diameter is the length of a line parallel to the direction of the scan that divides the particle profile into two equal areas.

Maximum and minimum linear diameters are two obvious linear measurements that may be used. These values can be amalgamated to give a single value in the form of the square root of their products, which is more representative of size than either value alone. However, the process is rather laborious, and frequently special scales consisting of a series of circles of different diameters are placed on photographs (or in the eye-piece of a microscope) where the irregular particles can be equated to a circle of equivalent area (or equivalent perimeter), i.e. the *projected area diameter* is the diameter of a circle having an area equal to the projected area of particle. Typical examples are the Patterson–Cawood graticule where the series of circles is graduated in an arithmetic series, and the Porton graticule which is graduated in a series based on $2^{\frac{1}{2}}$ [2].

7.2.1 Counting requirements

The question of how many particles should be counted to obtain a representative particle size distribution is frequently encountered. Intuitively, one can see that this must depend on the range of sizes. If the particle size distribution is monodisperse then a small number of measurements will suffice. Likewise the more polydisperse, the larger the number of measurements that will be required. Fig. 7.2 is a simple practical demonstration of 'average' particle size measured against number of particles counted; the two types of 'average' being number and weight averages (described later). It is seen that somewhere after counting 350 particles there appears to be little change in the 'average' size from counting 1000 particles.

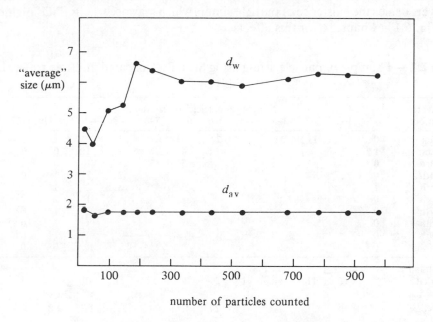

Fig. 7.2 — Average particle size vs number of particles.

Time or fatigue is the usual criterion for limiting the number of particles measured. In practice the American Society of Testing and Materials [3] recommends that not fewer than 25 particles in the modal class should be measured, and that at least 10 particles should be present in each size class.

Sichel [4] devised a technique called 'truncated multiple traversing' to minimize the number of measurements, yet maintaining reliability. The method is an adaptation of 'stratified sampling' [5]. The concept is that at least 10 particles must be observed in every size class that has a significant influence on the size curve. The method is exemplified in Table 7.1. In the first traverse, an area was searched and sufficient measurements were obtained in sizes 2–3, 3–4, 4–5, and 5–6 units. A second traverse of a similar, but not the same area as the first, was made, except that only particles greater than 5–6 and less than 3–4 units were counted. The process was repeated until Sichel's criteria were satisfied by all major size classes, with the result that only 180 particles were counted, yet the reliability was equivalent to the counting of 1250 particles not using this method (ten traverses in each field giving an equivalent count of about 1250 particles, i.e. 125 × 10).

The overriding criteria in particle size counting are that the sample being measured is representative and that the sample preparation technique does not introduce bias. Therefore it is insufficient to measure a large number of particles from, say, one electron micrograph; more accurate results may be obtained by measuring fewer particles but from many different electron micrographs. The particles being counted must represent the total population from which they were obtained, must be dispersed randomly without reference to shape or size, and agglomerates are deflocculated. A frequent requirement for single particle counting is that more than half the particle perimeter is visible before it is counted. This, however, denies the existence of particle sinters which may occur, e.g. TiO_2 pigment (see Fig. 6.1 in Chapter 6 for this effect).

Table 7.1 — Example of particle sizing using Sichel's 'truncated multiple traversing' [4]

Traverse number	Number of particles in size range												
	0–1	1–2	2–3	3–4	4–5	5–6	6–7	7–8	8–9	9–10	10–11	11–12	Totals
1st	0	5	11	34	41	24	5	3	1	0	1	0	125
2nd	1	7					4	2	1	1	0	0	
3rd	0						5	5	3	1	0	0	0
4th	0							2	2	1	0		
5th	1							3	2	1	1		
6th	0								1	2	0		
7th	0								2	1	0		
8th	0								1	1	0		
9th	0									1	0		
10th	0									0	0		
Column total	2	12	11	34	41	24	14	12	11	10	8	1	180
Nos. per traverse	0.2	6	11	34	41	24	4.7	4	2.2	1.3	0.8	0.1	129.3
Nos. per cent (frequency)	0.15	4.64	8.51	26.30	31.71	18.56	3.63	3.09	1.70	1.01	0.62	0.08	100

Total number counted = 180 particles (equivalent to counting 1250 particles)

7.2.2 Average particle size — ways of expressing size

If a particle size plot is made of size against the number of particles in that size (frequency), a particle size distribution graph is obtained, or a histogram, depending on the method of presentation. While this is very informative it is cumbersome to use, hence the need of a simple expression to summarize the range of sizes, i.e. an 'average'. There are many ways of describing an 'average', depending on the property being emphasized (weighting).

Some of the definitions frequently encountered are given in Table 7.2 and their centre of gravity on a curve is represented in Fig. 7.3.

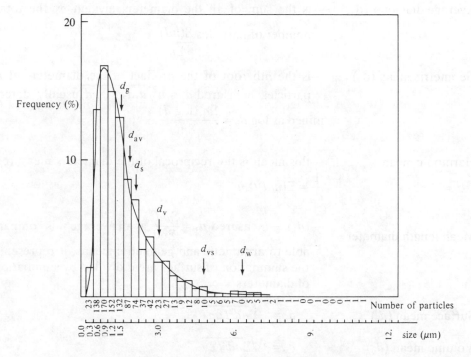

Fig. 7.3 — Effect of definition on weighting of 'average'.

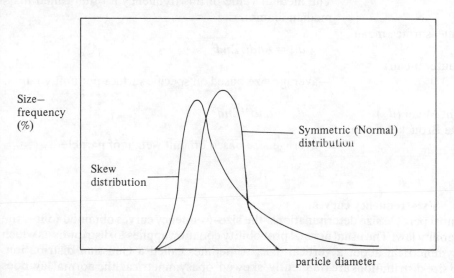

Fig. 7.4 — Distribution curves.

Table 7.2 — Definitions of average diameters

Mode	—is the diameter which occurs most frequently.
Median diameter	—is the diameter for which 50 percent of the particles measured are less than the stated size.
Average diameter (d_{av})	—is the sum of all the diameters divided by the total number of particles $\dfrac{\Sigma(nd)}{\Sigma n}$
Geometric mean (d_g)	—is the nth root of the product of the diameters of n particles measured $d_g = n\sqrt{d_1 d_2 \ldots d_n}$ usually determined as $\log d_g = \dfrac{\Sigma n\,(\log d)}{\Sigma n}$
Harmonic mean	—this mean is the reciprocal of the diameters measured $\dfrac{1}{d_n} = \Sigma(n/d)/\Sigma n$
Mean length diameter	—(d_1) – measured $d_1 = \dfrac{\Sigma nd^2}{\Sigma nd}$ as an average it is comparable to arithmetic and geometric means; it represents the summation of surface areas divided by summation of diameters.
Surface mean (d_s)	$d_s = \sqrt{\overline{\Sigma nd^2/\Sigma n}}$
Volume mean (d_v)	$d_v = \sqrt[3]{\overline{\Sigma nd^2/\Sigma n}}$
	The median value of this frequency is often called mass median diameter.
Volume–surface mean (d_{vs}) (or Sauter mean)	$d_{vs} = \Sigma nd^3/\Sigma nd^2$
	—average size based on specific surface per unit volume.
Weight Mean (d_w) (or De Broucker mean)	$d_w = \Sigma nd^4/\Sigma nd^3$
	—average size based on unit weight of particle.

7.2.3 Size–frequency curves

For most particle size determinations, the size–frequency curves obtained follow the probability law. The usual normal probability equation applies to distributions which are symmetrical about a vertical axis, sometimes called a Gaussian distribution. Since size distributions are frequently 'skewed' or asymmetrical, the normal law does not apply (see Fig. 7.4).

and in solution: a_s and a_L are the corresponding activities of the polymer at the surface and in solution. Under very dilute condition $a_L = c$ and, if it is assumed that the activity of the polymer at the surface does not change if the amount adsorbed is kept constant, the above equation becomes:

$$\frac{\triangle H}{R} = \frac{d \ln c}{d \, (1/T)} \, .$$

(6.17a)

Hence by measuring the adsorption isotherm at different temperatures; taking equal amounts of the adsorbed material at different temperatures and plotting the corresponding equilibrum solution concentration on a logarithmic scale against the reciprocal of the absolute temperature, a straight line is obtained whose slope is equal to $\dfrac{\triangle H}{R}$.

6.7.3 Adsorption and temperature

The amount of polymer adsorbed from solution frequently increases with temperature, thus indicating that the process is endothermic. However, there are also cases when the adsorption will decrease with temperature [96] or it may be unaffected by temperature [97].

Koral *et al.* [95] point out that the adsorption process could not be endothermic in the case of physical adsorption of simple molecules on a clean surface, since then the entropy lost on adsorption will be negative, necessitating the enthalpy to be also negative (to ensure a decrease in the free energy of adsorption). However, in the case of polymers one must consider the system as a whole. The adsorption of a polymer molecule at several sites on a surface requires that several solvent molecules are released from the surface to the solution. The translational entropy of the polymer molecule, along with some of its rotational and vibrational entropy, is lost on adsorption, because of partial restriction to its segmental mobility. Thus the solvent molecules which are desorbed gain their translational entropy, which cumulatively is much larger than that of the polymer molecule. The net result is that there is an overall entropy gain in the system on the adsorption of a polymer molecule which displaces solvent molecules, even if the process is endothermic.

The quality of the polymer solvent also has an effect on the amount of polymer adsorbed by the substrate. Generally the poorer the solvent the greater the amount adsorbed [95], and adsorption can be related to the solubility parameter [98]. However, the adsorption of polymer does not depend on polymer solvent interaction alone, but also on substrate-solvent interaction as well [99]. It is also possible to obtain less polymer adsorption by a substrate on the addition of small amounts of non-solvent to a polymer solution, if the solvent has a strong interaction with the substrate [100].

6.7.4 Rate of adsorption and equilibration

It is difficult to define how quickly adsorption takes place since it depends on many factors, especially concentration [101], but it is generally considered to be fast [102, 103]. For a review of polymer adsorption see Lipatov & Sergeeva [104]. However, in

Fortunately, the asymmetrical curves can be made symmetrical in most cases, if the size is plotted on a logarithmic scale (the frequency remaining linear). Such distributions are known as log-normal distributions. Hatch & Choate [6] showed the importance of this property.

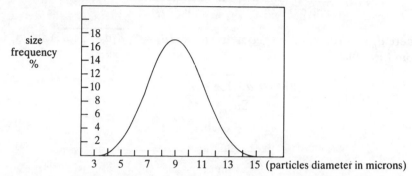

Fig. 7.5 — Typical symmetrical size distribution (Gaussian or Normal law) plot.

The equation of the normal probability curve (Fig. 7.5) as applied to size–frequency distribution is;

$$F(d) = \frac{\Sigma n}{\sigma\sqrt{2\pi}} \exp\left[\frac{-(d-d_{av})^2}{2\sigma^2}\right] \tag{7.1}$$

where $F(d)$ is the frequency of observations of diameter d; n is the total number of observations; d_{av} is the arithmetic average of the observations($\Sigma(nd)/\Sigma n$), and σ is the standard deviation given by $\sigma = \sqrt{\Sigma n[d-d_{av}]^2/\Sigma n}$. *The constants d_{av} and σ completely define the frequency distribution of a series of observations.* Thus if the particle sizes are plotted on an 'arithmetic-probability' grid, the summation curve is a straight line where the mean value (the 50% value) is the simple arithmetic average d_{av}: the standard deviation (σ) is given by $\sigma = 84.1\%$ size $-$ 50% size $= 50\%$ size $-$ 15.9% size (if plotted with negative slope). (Fig. 7.6).

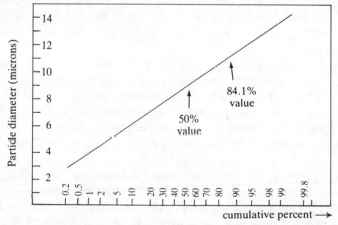

Fig. 7.6 — Gaussian curve (of fig 7.5) plotted on arithmetic-probability chart.

The asymmetrical distribution curve, where particle size is plotted on a linear scale (Fig. 7.3), can be converted into a symmetrical one if the particle diameters are fitted onto a log scale as in Fig. 7.7; i.e. equation (7.1) becomes

$$F(d) = \frac{\Sigma n}{\log(\sigma_g)\sqrt{2\pi}} \exp \frac{-(\log d - \log d_g)^2}{2 \log^2 \sigma_g} \tag{7.2}$$

where d_g now refers to the geometric mean ($d_g = \sqrt[n]{d_1 d_2 \ldots d_n}$) and σ_g is obtained from the equation

$$\log \sigma_g = \sqrt{\frac{\Sigma[n(\log d - \log d_g)^2]}{\Sigma n}} \quad .$$

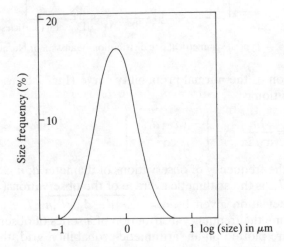

Fig. 7.7 — Transformation of skew distribution (of fig. 7.3) into symmetrical plot using log scale for particle size (μm).

The terms $\log d_{gl}$ and $\log \sigma_g$ are called the log-geometric mean diameter and the log-geometric standard deviation respectively. These two values are very important since they completely define a log-normal size distribution which is typical of a dispersion process [7].

A simple way of plotting a log-normal distribution is to use log-probability graph paper (see Fig. 7.8 cf. arithmetic-probability with a normal distribution), where the particle size is plotted as the ordinate and cumulative per cent by weight (or number) is plotted along the abcissa. The geometric median diameter (d_g) is the 50% value of the distribution and the geometric standard deviation (σ_g) is the 84.1% value divided by the 50% value (or the 15.9% value divided by the 50% value if the plot is with a negative slope).

The geometric standard deviation is always the same in a log normal particle size distribution, whether the sizes are plotted as cumulative percent by count or by weight. However the median values are different and hence care must be exercised to denote if the geometric median value is by weight (d_{gw}) or by count (d_{gc}).

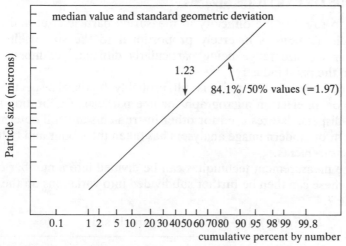

Fig. 7.8 — Plot of size distribution (shown in fig. 7.3 and 7.7) represented on log-probability chart showing derivation.

The Hatch-Choate transformation equations [6] enable us to convert the geometric median by weight to that of the geometric median by count. They also enable us to convert one type of "average" to that of another and hence are most useful in comparing distribution size measurements carried out by one method with that of another (see Table 7.3).

In the transformation of particle size distributions, by number to that of weight, errors may be introduced since the largest and heaviest particles are frequently present in statistically small numbers. Jackson *et al.* [8] have calculated the errors which are likely to be encountered when converting count by number to that of mass, and say what steps have to be taken to ensure that the errors be small.

Table 7.3 — Hatch–Choate transformation equations

To convert from	To		Use
d_{gm}, the geometric median mass	$d_{gc} = \text{antilog}\left(\dfrac{\Sigma n \log d}{\Sigma n}\right)$	geometric median by count	$\log d_{gc} = \log d_{gm} - 6.908 \log^2 \sigma_g$
	$d_{av} = \Sigma nd / \Sigma n$	arithmetic mean	$\log d_{av} = \log d_{gm} - 5.757 \log^2 \sigma_g$
	$d_s = \sqrt{\Sigma nd^2 / \Sigma n}$	surface mean	$\log d_s = \log d_{gm} - 4.605 \log^2 \sigma_g$
	$d_v = \sqrt[3]{\Sigma nd^3 / \Sigma n}$	volume mean	$\log d_v = \log d_{gm} - 3.454 \log^2 \sigma_g$
	$d_{vs} = \Sigma nd^3 / \Sigma nd^2$	volume-surface	$d_{vs} = \log d_{gm} - 1.151 \log^2 \sigma_g$
	$d_w = \Sigma nd^4 / \Sigma nd^3$	weight mean	$\log d_w = \log d_{gm} + 1.151 \log^2 \sigma_g$
d_{gc}, the geometric median by count	d_{av}		$\log d_{av} = \log d_{gc} + 1.151 \log^2 \sigma_g$
	d_s		$\log d_s = \log d_{gc} + 2.303 \log^2 \sigma_g$
	d_v		$\log d_v = \log d_{gc} + 3.454 \log^2 \sigma_g$
	d_{vs}		$\log d_{vs} = \log d_{gc} + 5.757 \log^2 \sigma_g$
	d_{gm}		$\log d_{gm} = \log d_{gc} + 6.908 \log^2 \sigma_g$
	d_w		$\log d_w = \log d_{gc} + 8.023 \log^2 \sigma_g$

7.3 METHODS OF PARTICLE SIZING

It is a truism to say that the difficulty in measuring particle size, and especially particle size distribution, is inversely proportional to the size itself, the sub-micrometre particle size range being particularly difficult, yet most frequently encountered in the paint industry.

The basic method of particle sizing is still, probably, by visual inspection through a microscope, or an electron micrograph for fine particles. Calibration standards such as mono-disperse latexes used for other instruments are still determined this way. The advent of modern image analysers has taken the labour and tedium away from these measurements.

Particle size measurement techniques can be divided into a number of generic methods, and these can then be further subdivided into variations on the theme as shown in Table 7.4.

Table 7.4 — Methods of particle sizing.

	Size range	Size distribution	Average only
1. DIRECT OBSERVATION METHODS			
a) microscopy	> 1000 nm	yes	
b) electron microscopy/SEM–TEM	5 nm – 5 μm	yes	
2. SEDIMENTATION METHODS			
a) natural sedimentation (or creaming)	> 100 nm	yes	
b) enhanced sedimentation			
—centrifugation/ultra-centrifugation	> 50 nm	yes	
—elutriation	5–10 μm (very vari – able)	yes	
—permeability	1–50 μm		yes
3. CHROMATOGRAPHIC METHODS			
a) GPC	> 3 nm	yes	
b) hydrodynamic and size exclusion	30–1500 nm	yes	
c) field flow fractionation	10–1000 nm	yes	
4. APERTURE METHODS			
a) mechanical (sieving)	> 20 μm	yes	
ultrapore (filtration)	0.1–5 μm	yes	
b) electrical–electric zone sensing	0.5–500 μm	yes	
5. OPTICAL (light scattering) METHODS			
a) turbidity	50—300 nm		yes
forward angle ratio	< 500 nm	yes	
maxima–minima approach	> 200 nm	†yes	
dissymmetry	< 200 nm	yes	
polarization ratio	< 500 nm	†yes	
variable wavelength	95–1000 nm	†yes	yes
flow ultra microscope	100 nm–2 μm		yes
b) Fraunhofer diffraction	4–1800 μm	yes	
c) Photon correlation spectroscopy	10 nm–3000 nm	†yes	
6. SURFACE AREA METHODS			
a) gas adsorption via BET theory	< 5 μm		yes
b) macromolecule adsorption—			
dye absorption	5 nm–100 μm		yes
soap titration	5 nm–1000 nm		yes
c) calorimetric measurement			
—Harkins, Jura method			yes

† indication of size distribution

7.3.1   Direct measurement

Microscopy or electron microscopy is probably the definitive sizing technique producing full-size distribution analysis. Monodisperse particles such as latexes or gold sols, used as standards for other techniques, are certified by this means for use as calibrants in other particle sizing methods. There are well established counting procedures such as those described at the beginning of this chapter, and many ways of simplifying the counting procedures such as the use of modern image analysers [9, 10] like the Quantimet 920 [11], Magiscan 2 [12], or the Zeiss Kontron sem-ips. These instruments can be directly coupled to an optical or electron microscope, or they can operate using just photographs of particles to be measured.

Image analysers consist of two components, a unit for converting the optical image into an electrical signal, and a computer which analyses the electric signal to give quantitative image information. The image scanner can enhance the image definition, or contrast, and be made to assess the particle in a variety of ways such as particle area, perimeter, Feret's or Martin's diameter, radius of equivalent disc area or perimeter; in fact the only limitation appears in the 'software' or computer programmes available. Light pens can be used to discard or separate 'touching' particles. Unfortunately, image analysers are very expensive, costing more than some types of electron microscope. Their accuracy is nevertheless dependent on proper, representative sample preparation.

When preparing a sample for direct measurement by microscopy and especially electron microscopy, great care has to be taken not to introduce instrumental artefacts, such as aggregation of the particles during slide preparation.

Soft or liquid particles are particularly difficult to prepare for electron microscope examination. They require crash-cooling techniques followed by freeze fracture [13] to produce replicas suitable for electron microscopy. If crash-cooled at high concentration (e.g. 15%) soft particles may be forced to coalesce even when cooling rates are in excess of 1000°K/sec. [14] to give unrepresentative emulsions (see Plate 7.2) for size analysis. Under these conditions one has to dilute emulsions to at least a few percent before crash-cooling, using Arcton/liquid nitrogen in order to obtain samples suitable for making replicas for electron microscope particle sizing. Even when successfully freeze-fractured, special counting techniques [15] have to be employed to obtain the true size of particles, since not all fractures will be through the diameter of the particle.

7.3.2   Sedimentation methods

Particle sizing by sedimentation will give full particle size distribution analysis whether it is due to natural sedimentation, i.e. under gravity alone, or enhanced sedimentation due to centrifugation. The particle diameter that is measured is 'Stokes diameter', which is the diameter of a sphere of the same density and free-falling speed as the particle being assessed.

Particle size measurements are always made at high dilution, and great care must be taken to ensure that the particles are well dispersed (deflocculated).

The simplest approach to obtain a particle size distribution with readily available, inexpensive equipment with an accuracy of between 2 and 5%, is to use the Andreasen pipette [16] (see Fig. 7.9) which consists of a cylindrical sedimentation

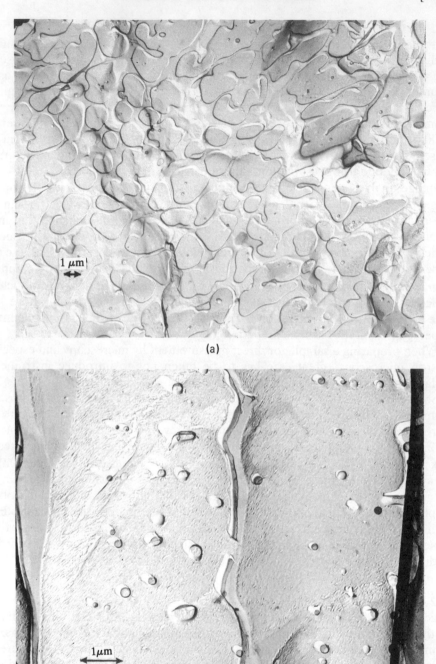

(a)

(b)

Plate 7.2 — Electron micrographs of replicas of freeze-fractured emulsion. (a) Effect of crash-cooling at too high a disperse phase volume with soft particles (15% DPV) — particles have coalesced to form larger particles about $2\,\mu m$ in size. (b) The same particles crash cooled at high dilution ($\frac{1}{2}$% DPV) (note that (b) is $\times 2$ magnification of (a).)

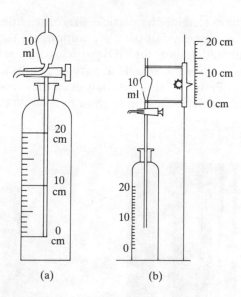

Fig. 7.9 — Schematic representation of Andreasen pipette (a) and variable depth modification.

vessel (about 550 ml) with a 10 ml pipette fitted with a two-way stopcock to enable sampling at a predetermined depth of usually 20 cm. A well-dispersed suspension of not more than 1% disperse phase volume is placed inside the vessel which is maintained at a constant temperature by means of a water bath. Samples of the dispersion are withdrawn slowly at regular time intervals (usually in a 2:1 time interval progression), and their solids per unit volume are determined.

The particle size calculations are based on determining the initial concentration of particles (zero time) and the concentration of the sample after a lapse of time t.

The calculation of particle size is dependent on Stokes' Law, that is,

$$d = \left(\frac{18\eta h}{(\sigma - \rho)gt}\right)^{\frac{1}{2}}.10^4 \tag{7.3}$$

where d is Stokes' diameter in μm

σ = apparent density of particle (g/cm^3) [equal to true density for non-porous particles only]

ρ = density of medium (g/cm^3)

η = absolute viscosity of medium (poise)

h = distance (cm) through which particle falls in time t (secs)

g = acceleration due to gravity (cm/s^2).

One thus obtains a cumulative percent by weight versus particle diameter plot. The Andreasen pipette method is a British Standard test method and is fully described in reference [17]. The particle size range is usually between 1 and 100 μm, but it can be used to determine the particle size distribution of dense sub-micron particles such as TiO_2 pigment if suitable precautions are taken. However, because of the lengthy sedimentation times for TiO_2 particles of about $0.2\ \mu m$ the Andreasen pipette is frequently modified to sample at varying depths in order to speed up the measurements — see Plate 7.3 [18]. An analysis of the errors that can occur in size measurement by sedimentation has been given by Svarovsky & Allen [19].

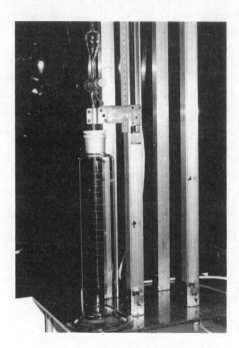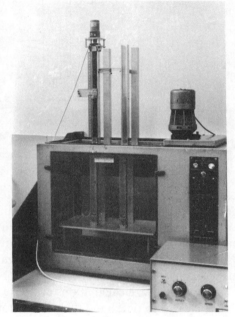

Plate 7.3 — Adjustable height Andreasen pipette, and racking mechanism placed in thermostatted bath ready for use.

A variation on particle size distribution by sedimentation under gravity is that obtained using a hydrometer [20]. In this procedure the suspension density is estimated at the effective distance beneath the suspension surface at the centre of gravity of a floating hydrometer, known as a soil hydrometer, having a density range from 0.995 to 1.038. The hydrometer is placed in a sedimentation cylinder containing a well-dispersed suspension of about 50 g pigment in 200 ml of dispersion medium; the density of the dispersion is read off at time intervals increasing exponentially, along with that of a control cylinder containing the dispersion medium (e.g. Calgon solution); the temperature is maintained to $\pm 0.1°C$. The percent pigment remaining in suspension after t minutes, at the level which the hydrometer measures the density of the suspension, is the ratio of $(R - B)$ to $(R_0 - B)$ multiplied by 100, where

R = hydrometer reading in pigment suspension at time t, B = hydrometer reading in blank solution, and R_0 = initial hydrometer reading in pigment suspension.

Stokes' diameter is determined from equation (7.3), which can be related to the percent pigment remaining in suspension — the method is applicable generally to particles ranging in size from 1 to 15 μm, and even smaller if the pigment density is greater.

The use of a soil hydrometer assumes that the difference in density between the suspension and the suspension medium is proportional to the pigment concentration, and that the hydrometer measures the density at a given level somewhere near the centre of gravity of the hydrometer. The position of this level is somewhat uncertain. Furthermore, the hydrometer sinks as sedimentation progresses, hence the sampling depth increases. Removal and re-insertion of the hydrometer into the suspension also distrubs the settling of the suspension. To overcome these problems one can use small glass divers known as 'Berg's divers' of different densities which are immersed in the suspension [21, 22]. These take up an equilibrium position at the level of equal suspension density, which is assumed to be proportional to concentration. Since the divers are about 1 cm in length the uncertainty in height is small. The disadvantage is that the diver is not usually visible and must be moved laterally to the side of the vessel with a magnet, and might therefore disturb the suspension.

An alternative approach to size determination by simple sedimentation based on Stokes Law is to use a sedimentation balance as described by Oden [23]. Modern techniques employ micro-balances such as those built around the Cahn or Sartorius [24, 25] microbalances, thus enabling automatic recording and size calculation using a simple computer programme. It is claimed that this method is more accurate, automatic, and uses a more rigorous analysis than that applicable to the Andreasen pipette. However, it is also very much more expensive, and like the pipette method is limited by the need to have a reasonable density difference between the particles and suspension medium, as well as a deflocculated dispersion.

When determining particle size by sedimentation the requirement of an analytical method to determine sample concentration is solved, in the Andreasen pipette method, by careful sample removal for analysis by non-volatile assay, and, in the sedimentation balance, by *in situ* weighing. Another convenient method of determining concentration continuously is to measure the attenuation of a beam of radiation passing through the sample. For this approach the attenuation must be proportional to the mass of sample lying in the beam. Visible light absorption can be used for coloured pigments which have negligible light-scattering properties. However, for particles which strongly scatter light, such as TiO_2, X-rays have to be used. Murley [26] describes such an apparatus (shown schematically in Fig. 7.10). A commercially available instrument working on this principle is built by Micromeritics in the form of the 'Sedigraph 5000ET'.

All the aforementioned methods rely on sedimentation due to gravity and hence are time-consuming for small particle sizes. Centrifugation can be applied to speed up sedimentation, as with the Joyce–Loebl Disc Centrifuge; and for finer particle sizes an ultra centrifuge is used [27]. The Joyce–Loebl Disc Centrifuge [28] has a hollow transparent disc-like cylinder which is rotated at a preselected speed. The particles to be measured are injected into the disc, using a buffered line start

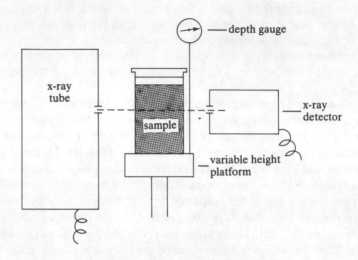

Fig. 7.10 — Schematic representation of static X-ray apparatus [26].

technique to prevent particle streaming and 'settle' through the spin fluid. A modified relationship of Stokes Law is used to determine the particle size:

$$t = 9\eta/2w^2 \ d^2\Delta\rho \ \ln \ (r_2/r_1)$$

where t = time elapsed after sample injection
η = viscosity of spin fluid
w = rotational speed of disc
d = equivalent particle diameter
$\Delta\rho$ = density difference between particle and spin fluid
r_1 = initial radius of spin fluid
r_2 = radius of sampling depth.

A stroboscope is necessary to detect that particle streaming does not occur, which otherwise invalidates the measurement. The early J–L instruments used Perspex for the construction of the centrifuge disc, which greatly restricted the type of spin fluid that could be used. The most recent model, the 'Joyce-Loebl Disc Centrifuge 4' (Plate 7.4) has a solvent-resistant centrifuge disc, and it is claimed by the manufacturers to produce comparative size distribution curves of particles in the range 0.01 to 60 μm (depending on density) by continuously monitoring the progress of particles past a light beam. True particle size distribution curves can be obtained from the instrument by calibration with standard samples, but careful interpretation of the data is required [29].

The problem of light-scattering particles where the attenuation of light per unit weight of sample is strongly particle size dependent, was overcome by Hornby & Tunstall [30] using X-rays, as mentioned earlier. They applied this principle to the design of a disc centrifuge which readily enabled them to measure fine particle size,

Plate 7.4 — Joyce–Loebl disc centrifuge (photo by courtesy of Joyce–Loebl).

and used this method to compare the particle size distribution of TiO$_2$ pigment milled for varying times in different grinding mills [26].

Elutriation [31]
The corollary of particles moving through a fluid is that a fluid moves through the particles, which is the basis of elutriation and permeametry. Elutriation (the converse of sedimentation) is where a fluid (usually a gas) is forced through a powder bed, and can thus be used to determine particle size distributions. The basis of the method is the ability of fluid velocities to support particles smaller than a given size. It is based on the following equations.

(1) for streamline motion $V_m = K_s \left(\dfrac{\rho - \rho_0}{\rho_0} \right) d^2 V^{-1}$

(2) *turbulent motion* $V_m = K_t \left(\dfrac{\rho - \rho_0}{\rho_0} \right)^{\frac{1}{2}} d^{\frac{1}{2}}$

where (c.g.s. units)
    ρ = density of particle
    ρ_0 = density of fluid
    d = particle size (diameter)
    V = fluid velocity
    V_m = maximum particle velocity
    K_s and K_t are constants dependent on particle shape.

Terminal-velocity constants for differently shaped particles are [32]:

Shape	K_s (c.g.s. units)	K_t
Sphere	54.5	24.5
Irregular (quartz)	36.0	50.0

The particles can be fractionated according to size by varying the quantity of air passing through the air jets or by altering the size of the elutriation chamber. One can thus obtain a cumulative size distribution by collecting the various fractions in settling jars. Elutriation methods of particle sizing are nowadays not favoured because the process is generally difficult to govern [33].

Permeametry
The use of gas permeability through a packed powder bed as proposed by Carman [34] was used by Lea & Nurse [35] to build the earliest permeability apparatus for routine particle sizing. Gooden & Smith [36] produced a variation of the Lea & Nurse apparatus; they incorporated a nomograph to simplify particle size calculations, and the instrument was commercialized in the form of the 'Fischer Sub-Sieve Sizer'.

The apparatus provides good results in the 1 to 50 micrometre range, but should not be used beyond this range, because the flow mechanism in beds of sub-micron particles is largely molecular and not streamline, which is the basis of the sub-sieve analyser's size calculation.

Rigden [37] showed how to account for molecular flow and particle clusters which can exhibit a 'natural' porosity at low flow rates in which the constituent particles act as individuals. Pechoukas & Gage [38] and Carman & Malherbe [39] made modifications to the apparatus, as did Hutto & Davies [40] who were able to show that their results were in good agreement with BET surface area measurements.

7.3.3 Chromatographic methods
Gel permeation chromatography (GPC) has become widely adopted as a means of sizing polymer molecules. The principle of sizing is based on treating a polymer molecule as a 'particle' of a certain size, and molecular weight distributions are calibrated on an equivalence to a polystyrene molecular weight standard which corresponds to a given volume. The method is now widely used to determine molecular weights via molecular volume [41, 42], and it is not proposed to discuss this approach, since the coiled molecular chain constituting the 'particle' is not a particle in the context of this chapter. There are many excellent commercial instruments and many articles and text books dealing with the subject. Although GPC instrumentation is expensive, it need not necessarily be so for the determination of molecular weight distributions, since there are inexpensive thin-layer chromatography GPC adaptations for molecular weight determinations [43, 44].

The extension of GPC to larger, solid particles is known as 'size exclusion chromatography' or 'hydrodynamic chromatography'. Although the two types of chromatography are different in principle they can occur simultaneously in a chromatographic column. The two types may be schematically differentiated as in Fig. 7.11.

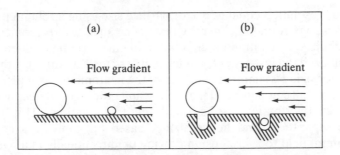

Fig. 7.11 — Schematic representation of (a) hydrodynamic and (b) size exclusion chromatography.

In hydrodynamic chromatography, because larger particles project further from the wall of a capillary they will be subjected to faster flow towards the centre of the channel than smaller particles, whose centres can approach closer to the stiller flow at the channel wall. In size exclusion chromatography, a small particle may find refuge from the velocity gradient in a pore which is unavailable to a larger particle. In both instances the larger particles pass through the column faster than the small particles, and are sized according to their retention time as a parameter of particle size, as in GPC analysis. Although there are many papers dealing with particle size determination by chromatographic means [45–49], this sizing approach is as yet not of general, non-research application to pigments. Micromcritics Corporation, however, have recently marketed an instrument suitable for routine latex analysis in the form of a 'Flow Sizer HDC5600' (Plate 7.5) which provides a full particle size distribution analysis of latex emulsions ranging from 30 to 1500 nm on this principle,

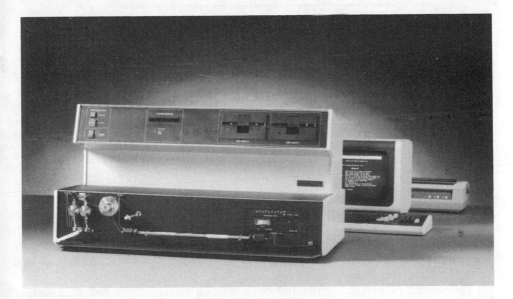

Plate 7.5 — Micromeritics HDC Flowsizer (photo by courtesy of Micromeritics Instrument Corporation).

with a claimed resolution equal to 5% of particle size using a column packed with a cation ion-exchange resin [50], thereby restricting analyses to anionic latexes. The column is stabilized by continuous circulation of eluent (initially eluent with latex), and once stabilized the instrument is quick and easy to use. Although the fractionating principle for sizing is simple, the design of the instrument is complex, requiring a fairly powerful small computer to deconvolute and process the detector signal to give the particle size distribution results. Field flow fractionation [51] is a method for particle size separation, and hence sizing, based on applying a concentration-disturbing lateral field to a suspension flowing unidirectionally in a narrow tube as shown in Fig. 7.12. The applied field may be a thermal gradient [52] or an electrostatic or gravitational field (Sedimentation field flow fractionation). Kirkland *et al.* [53, 54] describe a technique using an apparatus with a changing gravitational field developed by them called TDE–SFFF (Time delayed exponential sedimentation field flow fractionation) which has enabled them to obtain particle size distribution analysis in the range <0.01–1.0 μm of various pigments such as TiO_2, carbon black, and phthalocyanine blue as well as various latexes. Relative to constant-Field SFFF, TDE–SFFF has the advantage of faster analysis time and enhances detection sensitivity while maintaining resolution. Retention time correlates with the logarithm of particle size (the largest particles have the longest retention time) provided that the particles have the same density. In the case of a sample containing several components with different densities it is not possible to attach an accurate particle size scale to retention time. Nevertheless, qualitative variations in the constituents enable the 'fingerprinting' of the sample.

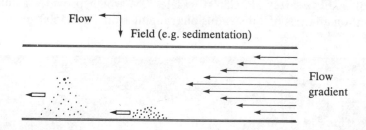

Fig. 7.12 — Principle of field flow fractionation.

7.3.4　Aperture methods
Sieving
Sieving is an old, well-established method of grading particles according to size; however, it is not as simple a method as it might appear. Although micro-sieves with an aperture width down to 5 μm have been developed [55], the method is generally used for particle fractionation from about 20 μm to 125 mm particles using standard woven wire test sieves.

The weight of particles collected between two sieves at the end of a sieving process can be readily determined. However, the nominal aperture width cannot be taken to represent the 'cut size' of the sieve. Sieving is a non-ideal classification process, and cut size has to be measured independently by calibration, using a known material for accurate work.

Sieves used for sieve analysis are standardized. Their nominal aperture widths follow a progression series. They may be woven wire screens as in BS410: 1969, or electroformed micromesh sieves (ASTM E161–607). The aperture width of a standard sieve itself may vary, and it should therefore be characterized by its aperture width distribution.

The particle size classification of a specific sieve depends on a number of factors which are independent of the mesh, e.g. solids loading, shaking action, tendency for solids to agglomerate, amount of near-mesh particles, etc. Leschowski [56] suggests that it would be better to plot sieving weight results against the median aperture width of a sieve rather than its mesh size. A sieve should be characterized by its median aperature, its confidence interval, and the standard deviation as obtained by calibration.

Great care has to be exercised in cleaning fine sieves; one should avoid brushing a fine sieve, e.g. below 200 μm, as this can readily damage the mesh, and it should be cleaned in an ultrasonic bath. Extremely fine micromesh sieves demand special wet sieving techniques as described by Daeschner [57] and Crawley [58].

Principle of sieving
When sieving, a powder is separated into two fractions, one that passes through the apertures in the mesh and a residue which is retained. The process is complicated in that non-spherical particles will only pass through the holes in the sieve when presented in a favourable orientation.

The sieving process may be divided into:

(1) the elimination of particles considerably smaller than the sieve apertures, which occurs fairly rapidly;
(2) the separation of 'near-mesh' particles, which is a gradual process, never reaching final completion.

The general approach to sieving is to eliminate the fines, and define an 'end point' to the test, when the elimination of 'near-mesh' particles has attained a practical limit. Whether sieving is performed in the dry or wet state depends on the characteristics of the material. Very fine particles are more quickly eliminated by wet sieving; it may also reduce the breakage of friable materials, but it may give results different from dry sieving.

There are two alternative recommended methods to define the 'end point' which marks the completion of 'near-mesh' particles:

(1) To sieve until the rate at which particles passing through the mesh is reduced to a specified weight or percentage weight per minute.
(2) To sieve for a specified time.

Method (1) is the better approach, but to simplify the procedure for routine tests with fairly consistent materials, method (2) may be adopted, provided that there is evidence to show that the time selected is adequate.

Both the dry and wet methods of sieving, along with the construction of a simple mechanical vibrator for sieves, are described in BS1796: 1952, (now superseded by BS1796: 1976, but without the design of the simple vibrator).

For a review of the sieving process, along with a discussion of the theory and practice of sieving, the reader is referred to Whitby [59] and Daeschner *et al.* [60].

Arietti *et al.* [61] have extended the sieving approach to particle size distribution analysis in the sub-micrometre size range, using a micro-pore filtration approach which appears to give good agreement with electron microscope sizing.

Electrical resistivity (electric zone — the Coulter principle)

The Coulter principle of particle sizing enables the measurement of particle sizes from 1 μm to a few hundred μm. With great care it is claimed to be able to measure down to 0.4 μm.

It was originally used to count blood cells [62–64]. Modifications were suggested by Kubitschek [65], which enabled the principle to be applied to the measurement of cell volume as well as number. The principle of the system is that a constant electric current (DC) is established between two electrodes placed in two separate containers, (see Fig. 7.13) and linked by a glass sensing zone (orifice). A mercury siphon is made temporarily unbalanced by the application of a vacuum. Closing the tap (T) allows the suspension to be drawn through the aperture, as the mercury returns to its equilibrium position. Electrical contacts triggered by the mercury allow a precise,

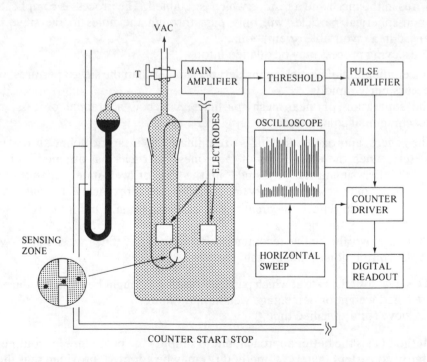

Fig. 7.13 — Schematic representation of the Coulter principle.

reproducible volume of suspension to pass through the orifice, altering the resistance by displacing their own volume of electrolyte, creating voltage pulses essentially proportional to the volume of the particle. These particle-generated pulses are amplified and counted in a 'discriminating' circuit. By systematically changing the sensitivity of the pulse detection (or counting) a cumulative count of particles (pulses) larger than a given size can be made, as the predetermined unit volume of suspension is transferred from one chamber to the other.

Calibration of the instrument is carried out for the specific electrolyte solution being used, usually with standard monodispersed latexes of known diameter supplied by the manufacturer.

The calibration constant (k) is determined from

$$d_c = k t_c^{\frac{1}{3}}$$

where d_c is the known number-volume diameter; t_c is the threshold level for the average of the pulses generated with the known particle size suspension.

Having thus calibrated the instrument with a standard latex, the constant (k) should hold for all other threshold sizes using that specific electrolyte.

Coincidence correction
The measurement of pulses is based on the assumption that each pulse is caused by a single particle. However, two particles may enter the sensing zone together, giving rise to a disproportionate signal. By making the measurements within certain counting rates, the effects of coincidence can be minimized from a statistical point of view, and they may be estimated experimentally by counting increasingly dilute suspension, as described in the Coulter *Users' manual*.

In conclusion, the Coulter principle for particle size distribution has been found to be a very useful, reliable, and an easy method to operate, as testified by over a thousand references on the subject [66], for the determination of particle sizes from about 0.6 μm upwards, and the reader is referred to Allen's review on the subject for further reading [67]. Particle Data Inc [68] manufacture a similar instrument to the Coulter Counter, making use of the electric sensing zone technique; their latest instrument, the Elzone 180, can be upgraded by plugging in additional microprocessing equipment and/or readily complemented with a variety of larger computers for data storage/retrieval. The minimum particle size that can be measured by this method ($\sim 2\%$ orifice diameter) is restricted by the inability to produce a suitable orifice, i.e. sensing zone, smaller than about 12 to 15 μm in diameter, not by the electronics of the instrument.

7.3.5 Optical (light-scattering) methods
A good review of light-scattering theory, as applied to particle size, is given by Kerker [69] and common methods used in size measurement are discussed by Collins *et al.* [70].

The source of scattered light is the re-radiation of light due to oscillating dipoles in polarizable particles induced by the oscillating electric field of a beam of light. The polarizability per unit volume at any position and time can be considered as the sum of a constant portion (giving rise to refraction) and a fluctuating part which produces scattering.

If we limit our considerations to spherical, non-adsorbing, and non-interacting particles, the light-scattering behaviour is mainly determined by two factors; (i) the ratio of particle size to the wavelength of the incident light beam in the medium (d/λ), and (ii) the relative refractive index, $m = n_1/n_2$ where n_1 and n_2 are the refractive indexes of the particles and their suspending medium respectively. In practice a dilute dispersion is illuminated by a narrow, intense beam of monochromatic light, and the intensity of the scattered light is measured at some angle Θ from the incident beam.

The three most common approaches to the measurement of particle size are:

(1) By turbidity (transmission) where Θ is fixed at $180°$ and the light intensity is measured.
(2) The light intensity is measured at some fixed angle (usually $90°$)
(3) The intensity of the scattered light is measured as a function of the angle.

The measurements usually have to be carried out at infinite dilution so that Rayleigh (when $d/\lambda \ll 1$), Rayleigh–Debye (when $(n_1-n_2)\, d/\lambda \ll 1$) or Lorenz–Mie (when $(n_1-n_2)d/\lambda \lesssim 1$) theories can be applied, which are for single scattering centres. For a Rayleigh scatterer the intensity of the scattered light I is angle-independent and given by:

$$I = \frac{16\pi^4 d^6}{x^2\lambda^4}\left(\frac{n_1^2 - n_2^2}{n_1^2 + n_2^2}\right)^2,$$

where x is the distance between the sample and the detector.

Scattering intensity is angle-dependent with a Rayleigh–Debye scatterer, and even more complex angle-dependency occurs with a Lorenz–Mie scatterer. Concentrated solutions produce multiple scattering, and the multiple scattering light theories become indeterminate in practice except on a semi-empirical basis. The advantages and disadvantages of the transmission, dissymmetry, maximum–minimum techniques, forward angle ratio, and polarization ratio, are all discussed by Collins et al. in their review paper already cited [70]. A recent extension of the turbidimetric approach has enabled the particle size distribution to be determined as well [71]. In addition to these methods of particle sizing is a method of variable wavelength patented by Tioxide International Limited [72] which entails the measurement of transmitted light at three different wavelengths, using a spectrophotometer. This enables the measurement of a mean particle size and its standard deviation which is useful in determining the efficiency of grinding. Nobbs [73] recently described a method using a thin-film technique which is a modification of conventional light-scattering measurements, but enables the measurement of apparent particle size to be made at high concentrations. Also of interest is a refractive index measurement of a dispersion method to obtain a particle size diameter [74].

Fraunhofer diffraction

As particle size increases and approaches the wavelength of the light source λ, the amount of light scattered at forward angles increases and becomes very much greater than in other directions. When the particle size d is much greater than λ then Fraunhofer diffraction (FD) theory describes the forward-scattering properties of a particle in a beam of light which can be considered as a limiting case of Lorenz–Mie theory. FD theory shows that the intensity of the scattering angle (diffraction pattern) is proportional to the particle size, and that the size of the scattering angle is inversely proportional to particle size, as shown in Fig. 7.14. A Fourier transform

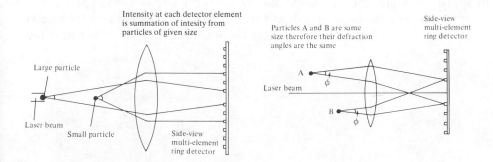

Fig. 7.14 — Diagrammatic representation of sizing by the FD principle.

lens — a lens positioned between the particle field and the detector, such that the detector lies in the focal plane of the lens — is used.

A typical FD instrument used for particle size analysis is shown schematically in Fig. 7.15.

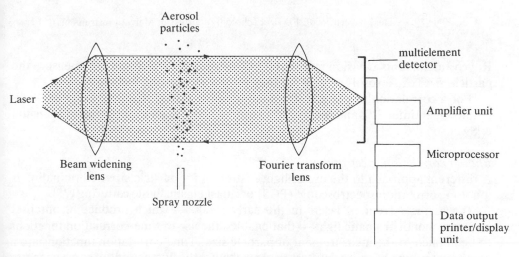

Fig. 7.15 — Schematic representation of typical FD instrument for measuring aerosol particle sizes.

Because the geometry and positioning of the lens in the instrument are so arranged as to meet the requirements of FD theory [75], the diffraction pattern of a moving particle will be stationary when the detector is in the focal plane of the lens.

The detector, consisting of an array of light sensors, analyses the light energy distribution (that of a low-powered laser beam) over a finite area, and a microprocessor computes the particle size distribution.

Instruments of this type, such as Particle Measuring Systems Inc. PDPS 11–C which can be used with different probes, are suitable for accurately measuring particle size distributions from a few micrometres upwards. They are useful in obtaining the particle sizes of aerosols such as those generated when paint is applied by spray-gun applicators. This type of instrument can also be used to determine the particle size distribution in liquids as with the Malvern 3600E Laser Particle Sizer (Plate 7.6). Leeds & Northrup have similar instruments in their Microtrac range.

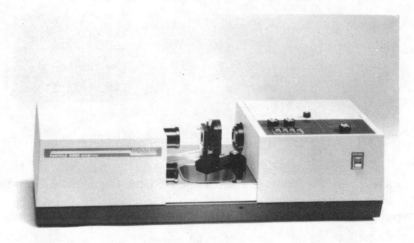

Plate 7.6 — The Malvern 3600E FD sizer (photo by courtesy of Malvern Instruments).

Royco Instruments Division market HIAC instruments which can also measure the particle sizes of Aerosols using what they term a 'shadowing technique'.

For a correlation of particle size analysis using HIAC, Coulter Counter, Sedigraph, Quantimet 720, and Microtrac instruments the reader is referred to Johnston & Swanson [76].

Photon correlation spectroscopy
A different approach to the use of light scattering for particle size determination is 'photon correlation spectroscopy' (PCS) or quasi elastic light scattering [75].

It was the advent of lasers in the early 1960s — which produce an intense, coherent monochromatic light — that enabled the use of time-correlation functions to be applied to the measurement of particle size. Time correlation functions are a way of describing by means of statistical mechanics the fluctuations of some property (in this case the number of photons emitted) on a time basis. This type of analysis requires a coherent, monochromatic incident radiation so that phase relationships in a beam are maintained, and examines the fluctuation of the radiation due to the random motion of the light scattering centres in a small volume which gives information as to the diffusion coefficient of the light-scattering centres.

When we examine scattered light intensity on a time basis, it will be found to fluctuate about an average value, if the light-scattering particles are undergoing random (Brownian) motion. The scattered electric field is a function of particle position and is therefore constantly changing. The intensity (proportional to the square of the electric field) is also fluctuating with time. By measuring these fluctuations it is possible to determine how these fluctuations decay over longer time-averaged periods, using auto-correlation theory to determine a diffusion coefficient for the particle. This in turn can be related to a particle diameter if certain assumptions as to particle shape and the viscosity of the medium are known, through the Stokes–Einstein equation.

A typical experimental set-up is shown in Fig. 7.16.

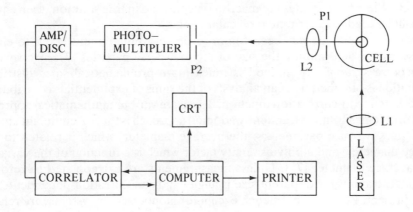

Fig. 7.16—Block diagram of the BI–90 particle sizer. L1, L2 are focusing leneses and P1, P2 are pinholes: CRT is a cathode ray tube.

The intensity of scattered light is an average effect and is a function of the individual fluctuations, just as the pressure of a gas is the average of the individual bombardment of gas molecules on the container wall.

For dilute suspensions with particles smaller than λ (the wavelength of the light beam)

$$< I_s(q) > \; = KNM^2 \, P(\Theta) \, B(C) \tag{7.4}$$

where $< I_s(q) >$ is the time-averaged scattered intensity from the particles,

q is the wave vector amplitude of the scattering fluctuation,

K is an optical constant,

M is the mass of the particle,

N is the number of particles contributing to the scattering,
P is the particle form factor,
Θ is the scattering angle,
B is the concentration factor, and
C is the particle concentration.

The fluctuations due to thermal excitation may be resolved into various frequencies, and at any angle the scattering, due to a particular fluctuation, may be stated as

$$q = \frac{4\pi n}{\lambda} \sin \Theta/2$$

where n = refractive index

The mass of a spherical particle is proportional to d^3, hence the M^2 term in equation (7.4) leads to a d^6 factor in scattering. The particle form factor, $P(\Theta)$, is known for simple shapes and for size less than λ. However, in the limit as Θ goes to zero, the particle form factors tend to unity. Hence angular PCS measurements with extrapolation to zero angle are required for polydisperse samples with large particles. When the particles are macromolecules and in true solution, then equation (7.4) can be used to determine molecular weight.

The PCS technique has been found to be very reliable and equal to electron microscopy for determining the size of monodisperse particles [77]. However, for polydisperse systems the method is much more problematical since distribution information is obtained from an analysis of the sums of exponentials contributing to the measured auto-correlation function. There are various mathematical approaches to this problem. The most frequently adopted approach is that of 'cumulant analysis', which gives two size parameters: the average diameter, which is related to the z-avarage diameter, and a polydispersity factor which is a function of the variance of the z-average diameter [78]. Unless measurements are extrapolated to zero angle and concentration, the apparent size is angle- and concentration-dependent.

At present, PCS measurements, except for monodisperse systems, are relatively insensitive to size distribution. Separation into two peaks appears to be possible only if the size ratio is more than about 2:1, although improvements in this respect are continually being made.

Frequently the instruments display histograms which give size classification along the abscissa; however, the manufacturers are usually careful not to name the ordinate as percent size frequency as might be expected (which it is not), but use terms such as 'percent relative light scatter' or 'particle size distribution (arbitrary units)'.

The particle size distribution is dependent on the value of the second cumulant which is very sensitive to the presence of dust particles. In some instruments the effect of dust particles has been reduced by a subtractive process using a delayed-time baseline approach which effectively ignores larger light scatters than a predetermined size.

In PCS size determinations there is an underlying assumption (as with all light scattering) that all particles are of a homogenous composition, even though their refractive index need not be known (i.e. only that of the continuous phase is required). This means in practice that one cannot measure the average size of a mixture of two different latex particles of different polymers, e.g. a mixture of a polystyrene and an acrylic latex.

There are many commercial instruments on the market for measuring particle size by PCS using the fixed angle (usually 90°) approach, such as Coulter Electronics Nanosizer [79] or their more recent N4 series of instruments which can measure at a number of different angles. Brookhaven Instruments Corp. (USA) produce a B1–90 particle sizer which has an electronic 'dust filter' and is a fixed-angle instrument (Plate 7.7), while they have a B1-200SM 'Automatic Goniometer' with an angular

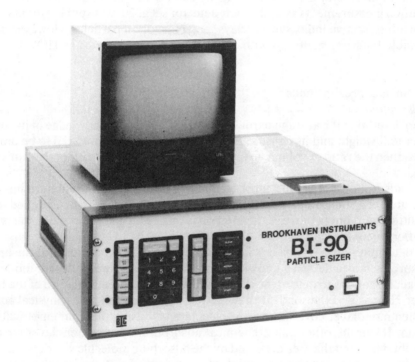

Plate 7.7 — Brookhaven Instruments BI–90 (photo by courtesy of Brookhaven Instruments).

range of 15° to 160° which can be used with their B1–2020 correlator. Malvern Instruments produce a 'System 4600' variable-angle Photon Correlation Particle Analyser and a fixed-angle instrument called the 'Autosizer', and Nicomp Instruments (now part of Hiac/Royco Instruments) also produce a number of models for particle-sizing by PCS, such as their Laser Particle Sizer Model 200D. All these instruments give an 'average' particle diameter and an indication of the particle size distribution.

Ultramicroscopy

A classic method [80] to determine the size of colloidal particles beyond the resolution of an optical microscope was to employ dark-field vision (ultramicroscope) to count the number of particles per unit volume, and by knowing their mass per unit volume their average size was readily estimated.

Derjaguin *et al.* [81] improved on the method by introducing a flow technique to simplify the counting procedure. Since then other workers have extended the procedure [82].

Walsh *et al.* [83] built a flow ultramicroscope capable of sizing particles between 0.1 and 2.0 μm. By detecting and measuring the intensity of laser light pulses using a photo multiplier, they have been able to extend the technique to particle size distribution measurements as well. Their detector set at 20° was suitable for materials with lowish refractive index such as clays, polymer, or oil particles. However, it was not suitable for sizing materials with a high refractive index such as TiO_2.

7.3.6 Surface area methods

Gas adsorption

One can determine the average particle size of a solid from a knowledge of its surface area per unit weight and its density; for if the particles are considered to be uniform spheres, then the ratio of volume to area is $r/3$, where r is the radius of the equivalent sphere.

Gas adsorption [84] is the most frequently used method for determining the surface area of a solid. In principle, gas adsorption techniques can be applied to any gas–solid system, but in practice the method is restricted and is limited to the type of adsorption encountered.

To determine surface area it is necessary to have a suitable value for the area of the adsorbing molecule and a knowledge of the number of molecules for monolayer coverage. Equations to determine these values depend upon the nature of the forces between the gas and the solid. If they are purely non-specific, i.e. physical adsorption, then monolayer coverage can be calculated by using the semi-empirical BET equation. If, on the other hand, chemical adsorption occurs, monolayer coverage can be obtained, but the area occupied by the adsorbing molecule will depend on the lattice spacing of the atoms in the solid. Many gas–solid systems are intermediate in adsorption type and cannot be readily defined, e.g. active carbons or clays.

Classical equipment for surface area measurement used vacuum glassware. Modern commercial equipment now available is generally made of metal and uses electronic gauges instead of traditional mercury manometers and Macleod gauges.

Gas adsorption surface areas are determined from the full adsorption isotherm plot with automatic instruments such as Carlo Erba's Sorptomatic instrument or Micromeritics Digisorb 2600. While automation may reduce other time factors it does not alter the time required to reach equilibrium conditions. To speed up measurements, instruments have been designed to make certain assumptions concerning the nature of the BET equation allowing surface areas to be determined from a single experimental point, such as the Micromeritics 2200 surface area analyser which uses a static method, while Perkin–Elmer's Sorptometer uses a dynamic adsorption (flow) method. Both instruments enable the surface area to be determined in less than an hour, once the sample has been 'conditioned'.

'Conditioning' or sample pretreatment can have a large effect on the measured surface area, and care has to be taken to ensure that the nature of the surface to be measured is not altered by the pretreatment.

One of the advantages of surface area measurement by gas adsorption is that it can indicate the presence of porosity. At liquid nitrogen temperatures, nitrogen condenses in the pores, according to the Kelvin equation; and the shape of the isotherm [85] reflects pore sizes between 1.5 and 30 nm in the solid, and it is most successfully analysed in the type IV isotherm. (Pore sizes in the range 8 nm upwards are usually determined by using mercury porosimetry. Typical commercial porosimeters can operate automatically down to 7.5 nm corresponding to pressures of about 2000 atmospheres.)

Solute adsorption

An alternative way of measuring surface areas is to use the adsorption of a solute from solution such as a fatty acid [86, 87]. While the cross-sectional area of a fatty acid is 20.4 $Å^2$ per molecule in a vertical orientation, it will not necessarily adopt this form of adsorption on all surfaces and in all solvents. It is therefore necessary to establish the nature of the adsorption and the molecules' effective area before using it for surface area determination (unless the measurements are relative ones on the same substrate using the same solvent).

Dyestuff adsorption has also been used to determine surface areas since its concentration is readily and accurately determined colorimetrically [88]. Again, care must be used in assigning an area for molecule occupancy as cautioned by Kipling & Wilson [89] who assigned an area of 102–108 $Å^2$ per molecule for methylene blue. Linge *et al.* [90, 91], however, found that methylene blue tended to absorb at between 69.6 $Å^2$ to 76 $Å^2$ per molecule corresponding approximately to dimer formation.

In measuring surface area by dye adsorption Padday [92] suggests that one should precalibrate the method against some more reliable method such as gas adsorption.

Gregg & Sing [93] suggest that in using a dyestuff approach to the measurement of surface area it must be:

(1) limited to cases where the dyestuff is sufficiently soluble and a clear plateau is obtained in the isotherm; and
(2) the orientation of the molecule must be known and
(3) the number of molecular layers must be known.

Nevertheless, when applicable, surface areas by dyestuff adsorption can be readily determined even without the use of a spectrophotometer to determine dyestuff concentration in solution, which is the usual approach.

For example, to determine the concentration of dyestuff in a mother liquor after adsorption, it is only necessary to make a dye solution weaker in colour than the unknown solution but of known concentration, e.g. prepared by dilution of a standard solution. It is then placed in a test tube and used as a reference colour; by adding a known volume of the unknown colour solution to a second similar tube and diluting the unknown sample with clear solvent (for the dyestuff) till the two tubes appear identical in colour, and recording how much solvent was required to bring the unknown concentration to the colour of the reference, then by simple arithmetic the concentration of the unknown sample can be determined. This makes use of the human eye as a powerful colour comparator, which it is.

'Surface' methods

Calorimetry to determine the heats of immersion also offers a means of measuring surface areas, as for example the Harkins & Jura method [94]. Heats of immersion have been advocated for determination of the specific surface of solids. However, the advantage of a single experimental determination is frequently outweighed by the small quantity of heat evolved and the care and instrumentation required to make a measurement [95].

Surface balance techniques have been used as methods for measuring the particle sizes of aluminium flake [96]. The method can be traced back to Edwards & Mason in 1934 [97].

The flakes are spread on a water surface, and successive expansions and contractions on the Langmuir balance-like equipment are carried out till a constant, compact, but planar area is obtained. From a knowledge of the sample weight and the two-dimensional area occupied, it is possible to calculate the average flat diameter of the particles as well as their thickness. Capes & Coleman [98] have used this approach to determine the size of mica flakes in the size range of 2–30 μm as well as sand particles in the range of 50–200 μm. Sub-micrometre latex particles have also been spread at the air–water interface [99] and the oil–water interface [100] where similar calculations can be made.

7.4   THE BEST METHOD?

The question as to what is the best method for particle sizing cannot be answered unless the question is qualified with a reason for the sizing. Even then, two independent methods are better than one (carried out twice), since in conjunction they will be more informative. For example, if the average TiO_2 pigment particle size is determined from specific surface areas using nitrogen adsorption as well as by sedimentation (or electron microscopy), it will be found that the 'average' values do not necessarily correlate even when all corrections are made. This is due to the surface coating on the pigment, and hence more can be learnt about the nature of the particle, and there is less possibility of being misled from the results of one type of measurement alone.

It is also worthwhile examining particle size distributions from a number and mass point of view. Fig. 7.17, based on HDC measurement by count, gives quite a different impression of size distribution than Fig. 7.18, which is the same distribution on a mass frequency basis.

Fig. 7.19 shows the results of particle size distribution measurements made, using the same sample of latex and plotted on a mass–frequency basis, using electron microscopy (over 4000 particles counted); disc centrifuge; four different PCS instruments; and hydrodynamic chromatography.

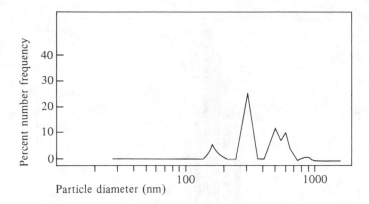

Fig. 7.17 — P.S.D. by count.

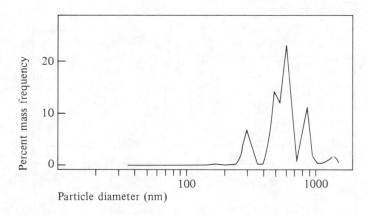

Fig. 7.18 — P.S.D. by mass.

It has not been possible to describe in a single chapter all the 400 or so methods for particle sizing [101]. The author's classification of methods in a 'generic' form may not comply with that of another. However, by using the format chosen it has been possible to give some idea of the diversity of basic techniques available, while not citing all the variations.

The methods described here may be divided into two categories. The first relies on the most recent instrumentation where nearly all the calculations are carried out by microprocessors. The instruments are generally expensive to buy or else require sophisticated instrument-making facilities. In this instance only the underlying principle has been described along with examples of the instruments.

The second category, which enables measurements to be made without resort to a large expenditure of money, has in general been described in more detail to enable one to measure particle sizes on a do-it-yourself principle with readily available

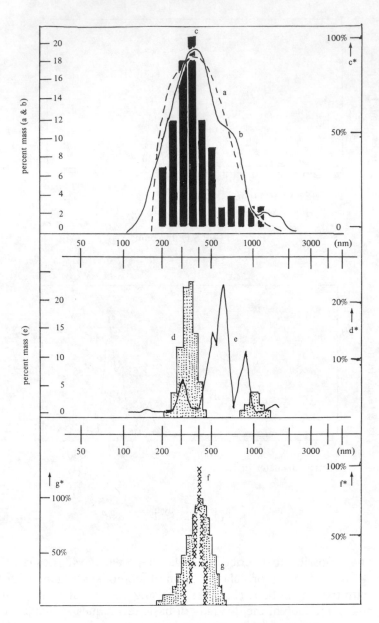

* The y-axis for the four PCS graphs is linear percent, and purely arbitary.
The Nanosizer N4 has a maximum scale of 40% (SDP: Differential Intensity),
whereas the rest have their peaks normalised to 100% ("relative mass" – the
Autosizer does not label the ordinate). Only the Nanosizer correctly identifies
the ordinate as "Size Differential Processor (results): Differenrial Intensity,
which gives an indication of relative mass.

Fig. 7.19 — Comparison of size distribution measurements, on the same sample of latex, using:
(a) Joyce-Lobel disc centrifuge: (a) Electron microscopy (~ 4000 particles counted): (c) PCS,
Brookhaven BI-90: (d) PCS, Nanosizer N4: (e) HDC, Micromeritics HDC 5600: (f) PCS,
Nicomp 200: (g) PCS, Malvern Autosizer.

materials. In general these tend to be more classical methods, and are certainly no less valid than modern automated techniques.

REFERENCES

[1] Kaye, B. H. *Chem. Eng.* **73** 239 (1966).
[2] Delly, J., *The particle analyst* Reticles and Graticules, Vol. 1 Nos. 77 — Ann Arbor Sci. Publ. (1965).
[3] ASTM, E20–68 (1974) section 14.02.
[4] Sichel, H. Sin., Silverman, L., Billings, C., & First, M., *Particle size analysis in industrial hygiene*, New York, NY, Academic Press (1971) p. 248.
[5] Hoel, P. G., *Introduction to mathematics and statistics* 3rd ed. J. Wiley & Sons Inc. New York (1962).
[6] Hatch, T. & Choate, S. P., *J. Franklin Inst.* **207** 369–387 (1929).
[7] Jelinek, Z. K., In: *Particle size analysis* p. 14 Ellis Horwood Ltd (1974). ISBN 85 312 0021.
[8] Jackson, M. R., Iglarsh, H. & Salkowski, M., *Powder Tech.* **3** 317–322 (1969/70).
[9] Swenson, R. A. & Attle, J. R., *Counting, measuring and classifying with image analysis*, Am. Lab. **11** 4(1978).
[10] Attle, J. R., Oneg, D., & Swenson, R. A., *International Lab.* Oct. 1980 pp. 35–48.
[11] Quantimet 920, Cambridge Instruments Ltd Rustat Rd. Cambridge CB1 3QH.
[12] Magiscan 2, Joyce-Loebl, Gateshead NE11 0QW Tyne & Wear.
[13] Menold, R., Luttge, B., & Kaiser, W., *Adv. Coll. & Interface Sci.* **5** 281–335 (1976).
[14] Menold, R., *et al. ibid.* p. 306.
[15] Cruz-Orive, L. M., *J of Microscopy* **131** (3) 265–290 (1983) also Schwartz, H. A. & Saltykov (1958) see Saltykov (1958) *Stereometric metalography* (2nd Ed.) (Moscow: Metallurgizdat).
[16] Andreassen Pipette, *Technico Andreassen pipette apparatus PBW-200-W* technical brochure ex Gallenkamp.
[17] BS3406: Part 2: 1963.
[18] Jelinek, Z. K., p. 77 *loc. cit.*
[19] Svorovsky, L. & Allen, C. J., *Particle size analysis* 442–450 M. J. Groves (ed.) Heyden & Son Ltd (1978) ISBN 085501 1580.
[20] ASTM D3360–06.02.
[21] Berg, S., ASTM Special Tech. Publ. No. 234, p. 143–171 (1958).
[22] Jarrett, B. A. & Heywood, H., In: 'A comparison of methods for particle size analysis' *British J. Applied Sci.* Supplement Vol. 3, p. 21 (1954).
[23] Oden, S., Alexanders *Colloid chemistry* Vol. 1 p. 877–882 Chem. Catalogue Co. NY (1926).
[24] Siebert, P. C., In: *Particle size analysis* Ann Arbor Sci. (1977) Stockham, J. D. & Fochtman, C. G., (eds) p. 52.
[25] Scott, K. J. & Mumford, D., *Powder Tech.* **5** 321–328 1971/2.
[26] Murley, R. D., *XII Fatipec Congress* 377–383 (1974).

[27] Svedburg, T., *Ind. Eng. Chem (Anal. ed.)* **10** 113–127 (1938).

[28] Joyce-Loebl Centrifuge, Vickers Company, Adv. Information brochure 8:83.

[29] Oppenheimer, L. E., *J. Coll. and Interface Sci.* **92**(2) 350–357 (1983).

[30] Hornby, M. R. & Tunstall, D. F., BP, 1,387,442.

[31] Silverman, L., Billings, C. E., First, M. W., *Particle size analysis in industrial hygiene* p. 137–146 Academic Press (1971). ISBN 0-12-643750-5

[32] Dallavalle, J. H., *Micromeritics*, 2nd ed., Isaac Pitman & Son Ltd (1948), p. 20.

[33] Herdan, G., *Small particle statistics* Butterworth 2nd revised ed. (1960) p. 360.

[34] Carmen, P. C., *J. Soc. Chem. Ind.* **57** 225–239 (1938).

[35] Lea, F. M. & Nurse, R. W., *J. Soc. Chem. Ind. London* **58** 277 (1939).

[36] Gooden, E. C. & Smith, C. M., *Ind. Eng. Chem.* (Anal ed.) **12** 497 (1940).

[37] Rigden, P. J., *J. Soc. Chem. Ind.* **66** 130–136 (1947).

[38] Pechukas, A. & Gage, F. W., *Ind. Eng. Chem.* (Anal. ed.) **18** 370–373 (1946) also *J. App. Chem.* (*London*) **1** 105 (1951).

[39] Carmen, P. C. & Malherbe, P. le R., *J. Soc. Chem. Ind.* **69** 139–193 (1950) and *J. Appl. Chem.* **1** 105–108 (1951).

[40] Hutto, F. B. & Davies, D. W., *Off. Dig.* **31** 429 (1959).

[41] *Gel permeation chromatography*, Altgelt, K. H. & Segal, L., (eds.) Marcel Dekker Inc. ISBN 0-8247-1006-1 (1971).

[42] Yau, W. W., Kirkland, J. J., & Bly, D. D., *Modern size exclusion liquid chromatography* J. Wiley & Sons (1979).

[43] Otocka, E. P., Hellman, M. Y., & Muglia, P. M., *Macromolecules* **5** 1273 (1972).

[44] Belenkii, B. G. & Gankina, E. S., *J Chromatog.* **141** 13–90 (1977).

[45] Small, H., Saunders, F. L., & Solc, J., *Adv. Coll. Interface Sci.* **6** 237–266 (1976).

[46] Small, H., *Anal. Chem.* **54** (8) 892A–8A (1982).

[47] McGowan, G. R., *J. Coll. Interface Sci.* **89**(1) 94–106 (1982), also Nagy, D. J., Silebi, C. A., & McHugh, A. J., *J. Appl. Poly Sci.* **26**(5) 1555–1578 (1981).

[48] Giddings, J. C. & Caldwell, K. D., *Anal. Chem.* **56**(12) 2093–9 (1984).

[49] Rudin, A. & Frick, C. D., *Polymer latex* 11', preprints of Conference (item 12), London, May 1985.

[50] Thornton, T. & Maley, R., *ibid.* (item 14).

[51] Giddings, J. C., *J. Chem. Phys.* **49** No. 1 81–85 (1968).

[52] Hovingh, M. E., Thompson, G. H., & Giddings, J. C., *Anal. Chem.* **42** 195–203 (1970).

[53] Kirkland, J. J., Rementer, S. W., & Yau, W. W., *Anal. Chem.* **53** 1730–6 (1981).

[54] Kirkland, J. J. & Yau, W. W., *Anal. Chem.* **55** (13) 2165–70 (1983).

[55] Heidenreich, E., In: *Particle size analysis* Groves (ed.) Heyden (1978). ISBN 035501 1580

[56] Leschowski, K., *Powder Tech.* **24** 115–124 (1979).

[57] Daeschner, H. W., *Powder Tech.* **2** 349 (1968/9).

[58] Crawley, D. F. C., *J. Sci. Instruments Ser.* 2 **1** 576 (1968).

[59] Whitby, K. T., ASTM Special Tech. Publ. No. 234 pp. 3–25 (1958).

[60] DaeschNer, H. W., Seibert, E. E., & Peters, E. D., *ASTM Special Tech. Publ.* No. 234 26–50.

[61] Arietti, R., Tenca, F., & Scarpone, A., *XVI FATIPEC Congress* (1978) 277–300.

[62] Coulter, W. H., USP 2,656,508 (1953).

[63] Coulter, W. H., *Proc. of National Electronic Conf.* 12 1034 (1956).

[64] Morgan, B. B., *Research, London* 10 271 (1957).

[65] Kubitschek, H. E., *Nature* 182 234–5 (1958). See also *Research* 13 128 (1960).

[66] *Industrial Bibliography* Oct. 1982, produced by Coulter Electronics Ltd. Luton UK.

[67] Allen, T., Chapter 13 in *Particle size measurement* 2nd ed. Chapman & Hall, London (1974). SBN 412 13490X.

[68] Particle Data Ltd, Cheltenham, Gloucestershire GL50 3AT.

[69] Kerker, M., *The scattering of light and other electromagnetic radiation* Academic Press (1969).

[70] Collins, E. A., Davidson, J. A. & Daniels, C. A., *J. Paint Tech.* 47 No. 604, 35–56 (1975).

[71] Melik, D. H. & Fogler, H. S., *J. Coll. & Interface Sci.* 92(1) 1983.

[72] Tunstall, D. F., UKP 2,046,898.

[73] Nobbs, J. H., Paint Research Assoc. Progress Report No. 6 33–37 (1984).

[74] Meeten, G., *J Colloid & Interface Sci.* 72 471 (1979).

[75] Weiner, B. B., Ch. 5. *Modern methods of particle size analysis* Barth, H. G. (ed.) Wiley Interscience Publ. (1984) also Plantz, P. E., *ibid.* Chapter 6.

[76] Johnston, P. R. & Swanson, R., *Powder Tech.* 32 119–124 (1982).

[77] Lee, S. P., Tscharnuter, W., & Chu, B., *J. Polym. Sci.* 10 2453 (1972).

[78] Weiner, B. B., Chapter 3 in *Modern methods of particle size analysis* Barth, H. G. (ed.) Wiley Interscience Publ. (1984).

[79] Lines, R. W. & Miller, B. V., *Powder Tech.* 24 91–96 (1979).

[80] Siedentopf, H. & Zsigmondy, R., *Ann. Physic* 10 1 (1903).

[81] Derjaguin, B. V., Vlasenko, G. Ja., Storozhilova, A. I., & Kudrjavteeva, N. M., *J. Coll. Sci.* 17 605–627 (1962).

[82] McFadyen, P. & Smith, A. L., *J. Col. Interface Sci.* 45 573 (1973).

[83] Walsh, D. J., Anderson, J., Parker, A., & Dix, M. J., *Coll. & Polym. Sci.* 259 1003–9 (1981).

[84] Gregg, S. J. & Allen, K. S. W., *Adsorption, surface area and porosity* Academic Press (1967).

[85] Gregg, S. J. & Sing, K. S. W., *ibid.* p. 7.

[86] de Boer, J. H., Houben, G. M. M., Lippens, B. C., & Meijs, J. *Catalysis* 1 1 (1962).

[87] Kipling, J. J. & Wright, E. H. M., *J. Chem. Soc.* 855–860 (1962).

[88] Herz, A. H., Danuer, R. P., & Janusonis, G. A., In: *Adsorption from aqueous solution* Webb, W. W. & Matijevic, E. (eds.) Advances in Chemistry Series 79 (1968) pp. 173–197.

[89] Kipling, J. J., Wilson, R. B., *J. Appl. Chem.* 10 109–113 (1960).

[90] Linge, H. G. & Tyler, R. S., *Proc. Australas Inst. Min. Metall* 271 27–33 (1979).

[91] Barker, N. W. & Linge, H. G., *Hydrometallurgy* 6 (3–4), 311–326 (1981).

[92] Padday, J. F., *Surface Area Determination Proc. Int. Symp.* 1969, 331–340 Everett, D. H. (ed.), Butterworth.

[93] Gregg, S. J. & Sing, K. S., *Adsorption surface area and porosity* Academic Press p. 294 (1967).

[94] Harkins, W. O. & Jura, G., *J. Am. Chem. Soc.* **66** 1362 (1944).

[95] Gregg, S. J., *The surface chemistry of solids* 2nd ed. Chapman & Hall (1961) p. 280.

[96] Edwards, J. D. & Wray, R. I., *Aluminium paint and powder* 3rd ed. (1955) Rheinhold Publ. Corp. p. 18.

[97] Edwards, J. D. & Mason, R. B., *Ind. Eng. Chem. Anal Ed.* **6** 1951 (1934).

[98] Capes, C. E. & Coleman, R. D., *Ind. Eng. Chem. Fundamentals* **12** No. 1 1246 (1973).

[99] Shepard, E. & Tcheurekedjian, N., *J Colloid Interface Sci.* **28** 481 (1968).

[100] Doroszkowski, A. and Lambourne, R., *J Polym. Sci. Part* **C34** 253 (1971).

[101] Scarlet, B., In *Particle size analysis* Stanley-Wood, N. G. & Allen, T. (eds.), Wiley Heyden (1982), p. 219.

ACKNOWLEDGEMENTS

I would like to express my thanks to my colleagues Tony Evans for the computation of the potential energy curves in Chapter 6, and Fred Waite for useful comments on the content of Chapter 7. Not the least, my thanks are also due to my wife Nina for my draft chapters and for improving my English!

8

The industrial paint-making process

F. K. Farkas

8.1 INTRODUCTION

The theory of wetting and the stabilization of pigment particles in paint media is discussed in Chapter 6. It is essential, however, to restate that the purpose of the pigment dispersion process is the wetting and separation of primary pigment particles from aggregates and agglomerates and their subsequent stabilization in suitable paint media, i.e. resin or dispersant solutions, during the process of dispersion.

All stages of this process are important and effect considerably the utilization of pigment, productivity, and the properties of the final product. The process is summarized diagrammatically in Fig. 8.1.

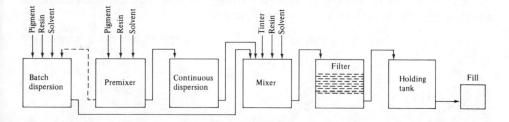

Fig. 8.1 — Flow diagram of the paint-making process.

To prevent reaggregation during and after the dispersion it is important to select the correct ratios of pigments, resins, and solvents. In addition, the second stage of adding further amounts of resin solutions or solvents should be carried out in the dispersion equipment to eliminate the possibility of 'colloidal shock' (flocculation) on the final make up to paint. The process is carried out in various types of milling equipment where shear forces are applied to the pigment aggregates, in order to separate the primary pigment particles. This stage is often called 'grinding'.

Intermolecular forces also play an important part in influencing the wetting of the surface of the pigment and in bringing about some degree of spontaneous dispersion. The maximization of the effect of intermolecular forces within any paint system to achieve rapid and stable dispersion of pigments with the minimum application of shear forces is a desirable objective.

The cleanliness and the strength of the colour, the durability of a number of a shear-sensitive pigments, and the actual time and energy expended during the process are affected by the wetting process. The conditions by which the physical (shearing) forces are applied in the various dispersing machinery have similar influence.

8.2 THE USE OF DISPERSANTS

Wetting plays an important part in the dispersion of pigments and therefore in the production of paints. All pigments have contaminants e.g. air, moisture, and gases, adsorbed onto their surfaces. To wet the particles these contaminants must in most cases be displaced by the dispersing medium. It is essential therefore that the wetting efficiency of the dispersant is strong enough to overcome, or at least reduce, the cohesive forces within the liquid and the surface tension between the solid/liquid interface leading to adhesion of the wetting groups of the dispersant onto the surface of the pigment.

The majority of the vehicles used for paint-making can be considered as dispersants, their wetting efficiency depending on their molecular weight, structure, and the presence of substituent groups, e.g. carboxyl, hydroxyl, amine, and ester. Dispersants are now available especially formulated to be more efficient than the majority of film-forming media in paint, and to a large degree these can be multipurpose in nature, having a wide range of compatibility with a variety of paint systems.

Since the electrical charges in the molecules of liquid and pigment are responsible for the wetting process, the polarity of ingredients, resins, solvents, and solid particles plays an important part in the dispersion stage. Polarity is defined by the shape of the molecule and therefore by the arrangement of the electrical charges and whether they are symmetrical or asymmetrical.

Symmetrical non-polar pigment   Unsymmetrical polar pigment

Thio indigo red Y   Toluidine red

Symmetrical non polar solvent   Unsymmetrical structures have increased
                                        polarity and are soluble in polar solvents

Benzene

Chlorobenzene
Dipole moment 2.25 Debye units.
(Sum of electrical charges ×
distance between them)

Polar solvent
$n\ C_4\ H_9\ OH$ n-butyl alcohol

Dispersant oleic acid: $C_{17}\ H_{35}\ COOH$

| non polar radical | polar reactive group |

Dispersants and paint media adsorb onto pigment surfaces by means of a wetting group (anchor group), leaving the non-polar radical which is soluble in the liquid phase extended. This mode is called 'steric stabilization'. The solvent balance, the compatibility of subsequent resin constituents, and the order of addition are of great importance to maintain stability. If the stabilizing chains partly or wholly collapse as a result of solvency changes, reaggregation and flocculation can take place to the detriment of the final product which, in most cases, is impossible or very expensive to recover.

A large variety of wetting agents (surfactants) are being used in the paint industry to reduce surface tension at the solid/liquid interface. Thus, new surfaces are created onto which the adsorption of paint media becomes easier during the dispersion process. The surfactants may be cationic where the adsorbable ion is positively charged, anionic where the adsorbable ion is negatively charged, or non-ionic, the activity of which may be due to polar and non-polar groups in the molecule.

The wetting agents are salts of organic amines, alkali soaps, sulphated oils, glycol ethers, etc. There are numerous products available with various claims of effectiveness. Some of them, or combinations of them, which may be critical, are standard constituents of waterborne (in particular emulsion) paints to provide ionic stabilization. Suitably designed carboxylated acrylics and hydrolytically stable alkyds, when neutralized with an amine or inorganic base, may also be used to provide a combination of steric/ionic form of stabilization, in particular for the manufacture of tinters where flow is of some importance.

In general, excessive use of surfactants is detrimental to many properties of a paint, hence the determination of the exact amount to give the right degree of dispersion and stabilization is of considerable importance. Prior to this, however, the wetting efficiencies of the various resins and their effect on each of the pigment/extenders present in any paint formulation have to be determined when formulating the dispersion stage to obtain the maximum use of raw materials, machine time, and quality of the end product. Wetting agents should only be used when the paint medium has little or no potential to accomplish the necessary degree of wetting.

8.3 METHODS OF OPTIMIZING MILLBASES FOR DISPERSION

The formulating techniques, applied in the paint industry for the dispersion of pigments, have been based mainly on empirical values, long experience, and time- and material-consuming laboratory evaluations, in order to establish working — but not necessarily optimum — conditions for the process of dispersion.

It is fairly safe to say that at least six laboratory scale ballmill trials have to be carried out on a single pigment single vehicle and single solvent system with variations in pigment volume percent ($PV\%$) and vehicle solids percent ($VS\%$) to obtain some idea of reasonable processability. The same applies to beadmills and high-speed (HSD) type dispersions.

The time spent, including testing of the experimental paint, may be as long as three days, and if the paint is a multipigment and resin system, weeks of experimentation is common. The subsequent semi-technical and plant scale proving trials often require further modifications.

Prior to the pigmentation of a millbase for dispersion, the paint chemist may mix known weights of pigment and resin solution in a can, and when the mix flows, scales up the ingredients to fill the dispersion volume (Dv) of the appropriate milling machine. This technique is called the 'can test' and may or may not result in a satisfactory dispersion. Researchers have long recognized the importance of the process of dispersion and its profound effect on the economics and quality of the subsequent product. No satisfactory dispersion of the pigments/extenders can be achieved with arbitrary millbase formulations. F. K. Daniel [1] developed the 'flow point' technique by which improved millbase formulations were obtained.

The technique consists of the titration of known weights of pigment with resin solutions of varying concentration while the mixture is being agitated with a palette knife. The end point is reached when the mixture begins to flow from the palette knife with a break, then starts flowing again within a second. The results, volume of resin solution used for the known weight of pigment, are then worked out. The concentration of resin solution which gave the lowest value is considered the best wetting medium and is proportionally scaled up to give a millbase for ballmill or beadmill type dispersion as there is little difference in the millbase consistency required for these machines. The complication is that not all premixed millbase will flow satisfactorily to be certain about the end point, e.g. organic pigments and fine particle extenders, in which case a series of ballmill trials are recommended.

Work by Guggenheim [2] was based on a number of production scale experiments with high-speed disc impeller dispersers. It resulted in an empirical expression to determine optimum millbases for this type of dispersion process. Non-volatile (NV) vehicle solids, vehicle viscosity (η), and Gardner–Coleman oil absorption (OA) are related to obtain pigment/vehicle ratios by weight:

$$\frac{W_v}{W_p} = (0.9 + 0.69 \ NV + 0.025\eta) \ \frac{(OA)}{(100)}$$

where W_v = *weight of vehicle*
W_p = *weight of pigment*
NV = *non-volatile fraction of vehicle*
η = *viscosity of vehicle in poises*
OA = Gardner–Coleman oil absorption (lb/100lb)

The above equation is based on a base value of 0.9 OA/100 with upward adjusting factors for the NV resin content 0.69 $NV.OA$/100 and for the viscosity 0.025 $\eta.OA$/100.

In practice a millbase formulation for high-speed dispersion technique using Guggenheim's empirical method could be worked out. For example:

NV of vehicle = 40% by weight, and its viscosity = 5 poises

The OA = 25 for the pigmentation

$$\frac{W_v}{W_p} = (0.9 + 0.69 \times 0.40 + 0.025 \times 5.0) \ \frac{(25)}{(100)}$$

$$= (0.9 + 0.28 + 0.125) \ (0.25) = 1.30 \times 0.25 = 0.325$$

The ingredients of the millbase are therefore:

NV vehicle $(0.40 \times 0.325) = 0.130 = \quad 9.81\%$ by weight
Solvent $\quad (0.60 \times 0.325) = 0.195 = 14.72\%$ by weight
Pigment $\qquad\qquad\qquad = \dfrac{1.000}{1.325} = \dfrac{75.47\%}{100.00\%}$ by weight

There is no doubt that these formulating techniques are useful and help the chemists to do a better job. However, they lack the accuracy essential for the measurement of the order of wetting efficiencies of the vehicle system, an equally important factor required to produce a stable paint. Furthermore, the results obtained are very dependent upon the operator variable which can be as much as $\pm 25\%$ depending on the amount of work put into the mix by hand.

The ratios of pigment/binder obtained in the undispersed state of the pigment do not therefore provide a firm and reproducible base for subsequent formulating, and millbases can become deficient in dispersant during the process of dispersion.

Further separation and stabilization of primary particles from the aggregates becomes increasingly difficult. The dispersion process comes to a halt and the pigment remains in a partly reaggregated state, irrespective of prolonged machine time.

Higher vehicle solids ($VS\%$) than the optimum, in order to induce some flow, will only prolong the time of the dispersion process, (see Fig. 8.2).

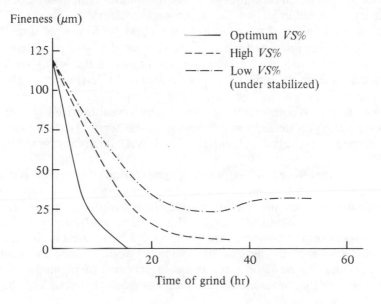

Fig. 8.2 — The effect of $VS\%$ on the rate and fineness of dispersion.

8.4 THE INSTRUMENTAL FORMULATING TECHNIQUE

Wirsching, Haug, & Hamann investigated the oil adsorption characteristics of pigments using a Brabender Plastograph (Torque Rheometer) [3], and found remarkably reproducible values.

This prompted the present author to examine the possibility of an instrumental formulating technique. For over 20 years of research at ICI Paints Division, this technique has been developed and used to formulate optimum millbases for the various dispersion techniques. [4].

From the data of well over 11000 tests carried out on a vast range of pigments, extenders, film-forming media, solvents, and additives, it was found that the relationship and quantitative interactions of these ingredients were specific and accurately measurable.

The criticality of the relationships greatly influences the viscosity, rate of dispersion, and utilization of pigments during the process of dispersion, and determines the order of addition of subsequent make-up ingredients to yield stable products with the highest quality attainable within the limits of that system. Each pigment behaves preferentially in an environment which contains more than one surface-active agent, hence reaches the optimum and proportional amount of the preferred surfactant in the dispersion stage if stabilization is to be maximised.

The degree of dispersibility of the pigment, or alternatively the wetting efficiency of the polymer, surface-active agent solution, is determined instrumentally in the dispersed state of the solid particle (pigment) with a torque/rheometer e.g. Brabender Plastograph. Such an instrument is capable of dispersing pigmented millbases under the standardized conditions and rate of shear to a sufficient degree.

The process of dispersion and the end result cannot be influenced by the operator using a specially developed technique which also simulates temperature of millbases likely to be encountered in plant scale dispersion machinery.

The end result of a set of tests (optimization) may be converted directly into a practical millbase formulation for the dispersion of the said particle(s) for the different types of wet dispersion techniques well known in the art. Alternatively it may be calculated from the individual specific data of $PV\%$ (pigment volume) and $VS\%$ (vehicle solids) previously generated and stored.

Data storage is most conveniently carried out by a computerized system which is capable of producing optimized formulations from the accrued data for any number of pigments, extenders, resins, solvents, and additives which may be required for a product.

It is possible, therefore, to analyse the pigmentation of a colour-matched test panel using a spectrophotometer, and with the ratios obtained to compute the best possible formulation and method of manufacture for the selected resin (film-forming) system. This obviously saves a lot of time and money which would otherwise be spent on the trial and error method still being used in the industry, to obtain a workable but not necessarily optimized formulation. In addition the technique provides the basis of accurate quality control of pigments, extenders, resins, solvents, wetting agents, and additives, provided that standards have been established for comparison.

The final $PV\%$ at the end of the test and the torque curve obtained under specified and standardized conditions will quickly show differences, if any, between batches which are otherwise unmeasurable and can cause manufacturing and/or product problems. The technique also provides the means to monitor the manufacture of resins and dispersants by measuring the effect of rate of stir, feed, and temperature on the wetting efficiency of the product.

As a research tool it is invaluable to develop new dispersants, resins, and pigmentation of the same leading to new products.

The Brabender Plastograph (Plasticorder) (Plate 8.1) is a torque rheometer. The measuring principle is based on the display of the resistance of test material sheared by rotors in the measuring head. The corresponding torque moves the dynamometer from its zero position, and a curve is recorded (torque vs. time). The unit of torque is Newton-metre (N m). The force required to accelerate 1 kg weight to 1 m/second.

Parameters which influence the viscosity of material under test such as temperature and rate of shear can be varied over a wide range, hence test conditions similar to the milling conditions can be simulated. The first Brabender Plastograph was built over 40 years ago for the rubber and food industry and has found wide application since, in particular in the field of thermoplastics, raw material control, and general research. It is a sensitive and robustly built instrument which requires small amounts of materials for the test. The formulating technique consists of the dispersion of a specified volume of pigment in a specified volume of resin solution to provide a stiff paste at around 70% loading by volume of the measuring head. After the dispersion, the millbase is titrated with the same resin solution till the torque drops to a

Plate 8.1 — Brabender Plastograph.

predetermined and standardized value at which the total amount of resin solution used is recorded.

The initial loading data are then used as standard for any subsequent tests employing the same pigment or worked out proportionally if mixtures of pigments are being tested.

8.4.1 Determination of the best-wetting resin solution and optimum millbase formulation for a system

Following the operating technique a range of vehicle solids solutions of the appropriate vehicle(s) or dispersant(s) and solvents should be tested on the pigment or pigment mixture of the system to obtain the wetting efficiency curve(s). This is obtained by plotting the pigment volumes obtained at the end point against the range of vehicle solids tested. This curve is characteristic of the system and shows the best-wetting resin or dispersant solution to be used for the dispersion of the pigment. It also shows the order of wetting efficiencies of the various vehicles in a system, which is the basis of the optimized let-down procedure.

It is advisable to omit the various wetting agents and additives from the standard millbase formulation when evaluating the vehicle system. However, they should be included, and their effect, if any, measured on the optimized millbase formulation. In a very large number of cases they give no further improvement in an optimized system.

The pigment volumes% ($PV\%$) of the millbases which were obtained at the end point on the Plastograph are calculated with the different vehicle solids solutions used for the tests.

Each of the millbases is composed of a fixed amount of pigment and the total volume of vehicle solutions:

$$\frac{w_p}{\rho_p} = V_p \qquad V_v\, \rho_v = w_v \qquad PV\% = \frac{V_p}{V_p + V_v} \cdot 100$$

where:

w_p = weight of pigment
V_p = volume of pigment
w_v = weight of vehicle solution
V_v = volume of vehicle solution
ρ = specific gravity (subscript p for pigment and v for vehicle solution).

Example: A carbon black pigment was tested with an alkyd resin (a) and with a nitrogen resin (n) with the following results:

Total volume of resin solution (ml)			The corresponding $PV\%$		
VS%	Resin (a)	Resin (n)	VS%	Resin (a)	Resin (n)
10	50	43	10	14.25	16.23
15	45	38	15	15.47	17.98
20	43.6	42	20	16.04	16.55
30	57	48	30	12.75	15.00

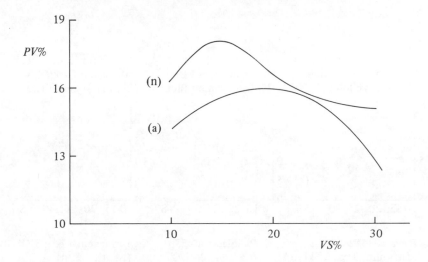

Fig. 8.3 — Wetting efficiency curves of resin (a) and (n) measured using a carbon black pigment.

Resin (n) is a better resin for wetting and therefore should be used for the dispersion of carbon black pigment at 15% VS concentration which gives the highest volume of pigment incorporable/unit volume of millbase and dispersible in the shortest time possible.

The shape of the wetting efficiency curve is different and specific for the different pigments and resin solutions. Curves for several pigment types are shown in Fig. 8.4a and for a single pigment in a range of dispersant solutions in Fig. 8.4b.

The reason why the PV% is worked out instead of its weight becomes apparent when practical millbase formulations are worked out from the test data.

Using the conversion figures it is possible to obtain optimum millbase formulations for all major types of dispersion techniques used in the paint industry by a single calculation. Scale-up is direct from laboratory to plant, and only the optimum formulation is checked on a laboratory scale for the rate and fineness of dispersion prior to implementation.

The wetting efficiency curves of Fig. 8.3 also show that resin (a) is inferior to resin (n) in general wetting properties on the carbon black pigment tested, and although essential for the paint formulation it will not interfere with the stability of the millbase provided that it is added after further amounts of resin (n) are incorporated in the dispersed millbase as a second stage.

In general (with reference to Fig. 8.5a and 8.5b), if lower than the optimum VS% (understabilized region) is used for the dispersion of pigment the millbase may run short of the stabilizing resin and may partly reaggregate during the process. Higher than the optimum VS% (overstabilized region) will, on the other hand, disproportionally extend the time of dispersion to reach the required fineness.

Where compromise is required (e.g. sensitive pigments; formulation is short of solvents) higher than the optimum VS% may be used with the corresponding PV% as determined by the wetting efficiency curve.

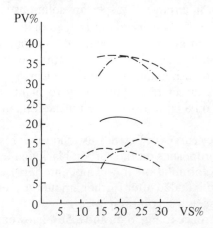

1. Deep Lemon Chrome
    (Chrome Yellow)

2. Monastral Blue FBN
    (Phthalocyanine Blue)

3. Monastral Green GN
    (Phthalocyanine Green)

4. Runa RH472
    (Titanium Dioxide White)

5. Philblack AN550
    (Carbon Black)

6. Permanent Red F3RK70
    (Napthol Red)

7. A composite of the six graphs, illustrated above, to give a direct comparison.

Fig. 8.4a — Wetting efficiency curves for several pigments.

1. A medium oil alkyd in an aliphatic hydrocarbon.

2. MMa polymer in a blend of an ester and an aromatic hydrocarbon.

3. An acrylic copolymer in an aromatic hydrocarbon.

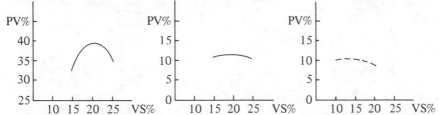

4. A long oil alkyd in a blend of an aliphatic and aromatic hydrocarbon.

5. A 2nd long oil alkyd in a blend of an aliphatic and aromatic hydrocarbon.

6. A 2nd acrylic copolymer in a blend of an aliphatic and aromatic hydrocarbon.

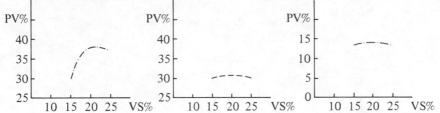

7. A composite of the six graphs, illustrated above, in direct comparison to each other.

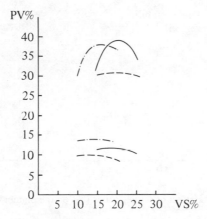

Fig. 8.4b — Wetting efficiency curves for Monolite Red Y in different dispersant solutions.

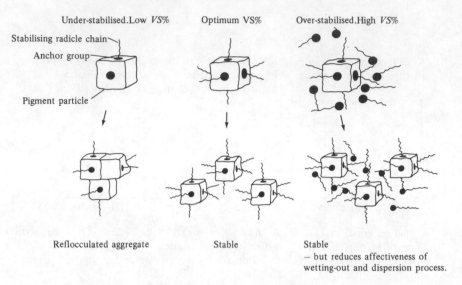

Fig. 8.5a — Particle arrangements.

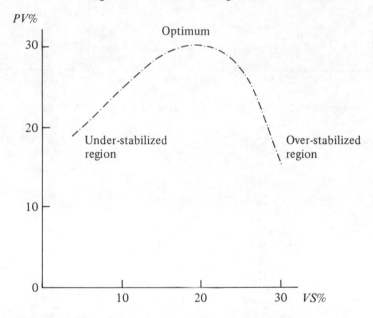

Fig. 8.5b — The effect of vehicle solids (%) on the pigment volume (%).

As an additional and important factor the amount of work done on the mix during the dispersion stage (the area covered from the coherency point) can be used to differentiate between dispersants and resin solutions regarding their wetting and dispersing efficiencies. The less work that is done in Plastograph Units, the better the wetting efficiency of the dispersant or resin solution. A typical Plastograph recording as shown in Fig. 8.6, is obtained after each test, and from it the wetting efficiency value is worked out.

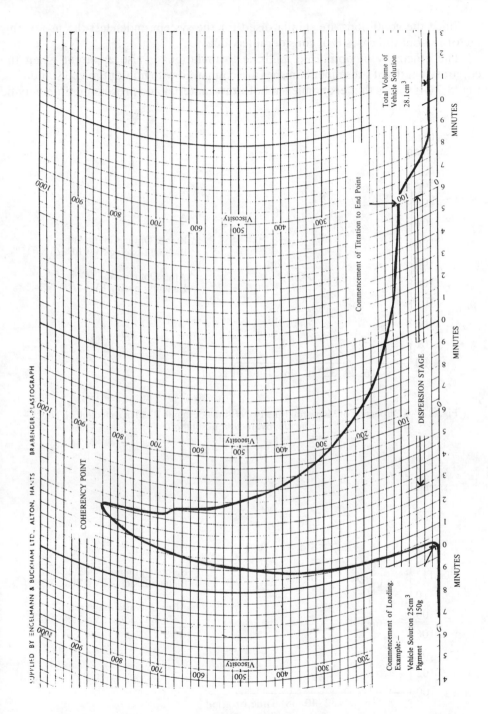

Fig. 8.6 — Plastograph recording.

Since the conditions are standardised for the system under test the use of this factor in doubtful cases is of considerable value.

The efficiency of optimized millbase formulations is also apparent when the colour strengths of subsequent finishes are compared with non-optimized standards. Fig. 8.7 a–f shows that economies in both raw material and processing time may be

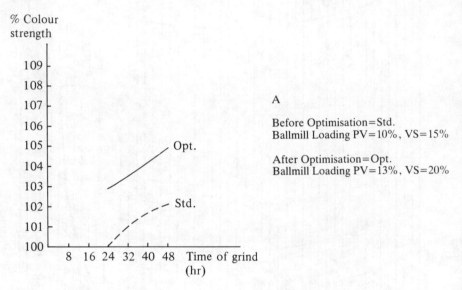

Fig. 8.7a — Dispersion of Monastral Blue FBN (Phthalocyanine Blue) in a medium oil alkyd — aromatic hydrocarbon solution.

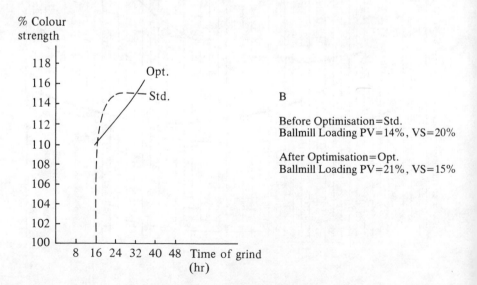

Fig. 8.7b — Dispersion of Monastral Blue FBN (Phthalocyanine Blue) in a long oil alkyd — aliphatic hydrocarbon solution.

obtained. The graphs show the results of colour strength versus time of grind for three pigments dispersed in solutions of:

(1) a medium oil alkyd in an aromatic hydrocarbon
(2) a long oil alkyd in an aliphatic hydrocarbon before and after optimisation.

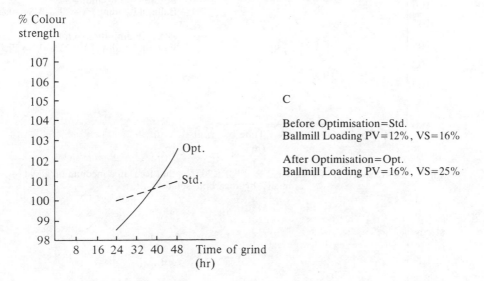

Fig. 8.7c — Dispersion of Monastral Green GN (Phthalocyanine Green) in a medium oil alkyd — aromatic hydrocarbon solution.

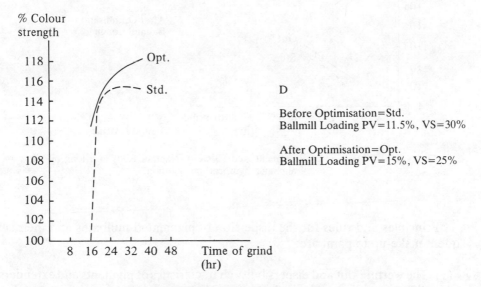

Fig. 8.7d — Dispersion of Monastral Green GN (Phthalocyanine Green) in a long oil alkyd — aliphatic hydrocarbon solution.

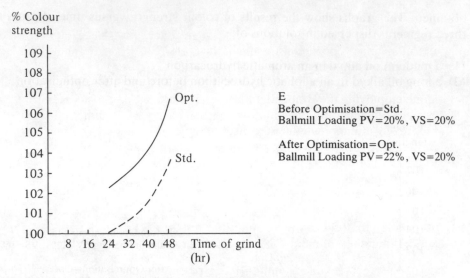

Fig. 8.7e — Dispersion of Permanent Red F3RK70 (Napthol Red) in a medium oil alkyd —
aromatic hydrocarbon solution.

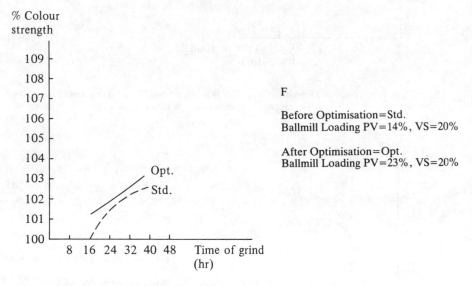

Fig. 8.7f — Dispersion of Permanent Red F3RK70 (Napthol Red) in a long oil alkyd —
aliphatic hydrocarbon solution.

Principles and rules for the dispersion of pigmented millbases and their subsequent make-up to paint are:

(1) The wetting-out and dispersibility characteristics of pigments and extenders are different and specific to the various dispersants, film-forming materials, and additives.

(2) Each pigment will therefore behave preferentially in an environment which contains more than one surface-active agent. Hence it is necessary to use the optimum and proportional amount of the preferred surfactant for each pigment in the dispersion stage if stabilization is to be maximized.

(3) These characteristics can be measured instrumentally using the special technique employing the Brabender Plastograph. The order of dispersibility or the wetting efficiency of the film-forming materials etc. within the limits of a product formulation, defines the dispersion, let-down and make-up stages, and yields the best possible utilization of raw materials and process time.

(4) Even small differences in wetting efficiency between component resins if used in the wrong order in the dispersion and let-down stages can cause long-term stability problems (e.g. colour drift).

Interaction of 'dispersing aids' and additives have a similar effect, and their use is often unnecessary in an optimized system.

(5) Higher than the optimum $VS\%$ may be used in the following cases:

(a) To protect certain pigment surfaces (coatings) from damage brought about by the highly efficient dispersion conditions of the optimum $VS\%$.

(b) To conform to the tight solvent (viscosity) limits of a particular product.

(c) To give more flexibility to plant operations where required.

(6) Never disperse any pigment by any method at more than 5% below its optimum $VS\%$ or level of dispersant, e.g. if optimum $VS\%$ is 20 then 15% is the bottom limit. Otherwise the millbase will run short of stabilizer during dispersion, will reaggregate, and may not reach the fineness of dispersion required irrespective of extended time of grind. In general, 10% higher $VS\%$ than the optimum will extend the time of grind disproportionally.

(7) Never 'second stage' a dispersed millbase with solvents or very high solids resins. This causes partial or total shock and may result in a flocculated base difficult to recover. Instead, the vehicle solids should be increased 8–10% above that of the optimum dispersion $VS\%$ during the process of second stage, preferably with solutions of the appropriate resin added incrementally. This ensures the necessary stabilization of the dispersed millbase for short-term storage and freedom from shock on subsequent make-up.

(8) If a tinter contains a more efficient dispersant than the finish to be tinted, care must be taken. Some pigments in the finish could be destabilized and/or flocculated if more than half the amount of stabilizer required by them is introduced via the tinter. If, however, the tinter is in a less efficient dispersant than the resin system of the finish, partial or total flocculation will occur. The extent of this will depend on the degree of wetting efficiency differences between the tinter resin and the most efficient resin or resin mixtures in the finish. It is assumed that in both cases the media of the tinters are compatible with the film-forming materials of the finish.

(9) It is important to maintain the solvent balance at all stages of manufacture to prevent film-forming materials coming out of solution. Even partial insolubility or incompatibility of components can cause irreversible shock and flocculation. Transparent dispersions are particularly sensitive in this respect.

(10) To obtain maximum benefit from optimized millbase formulations it is to be ensured that the dispersing machinery required for the job is operated at its optimum conditions.

Finishes, and therefore corresponding millbases, can be classified as follows:

(1) Single pigment – single vehicle
(2) Two or more pigments – single vehicle
(3) Single pigment – two or more vehicles
(4) Two or more pigments – two or more vehicles.

In cases (1) and (2) the pigment(s) has little 'choice' and will disperse to its limitations at the optimum vehicle solids (*VS%*).

In cases (3) and (4) the pigment(s) has a choice and therefore must be dispersed in the best resin or resin blend at the vehicle solids which is determined or calculated from the individual optima.

Practical example (1) below shows how to calculate a millbase formulation from *PV%* and *VS%*; it therefore applies to case (1). If the *VS%* is the same for two pigments in a single resin system, case (2), the same calculation is to be used. It is assumed that the *PV%* is also the same. If, however, the *VS%* and *PV%* are different for two or more pigments in the same resin, proceed as practical example (2), and in different resins as practical example (3).

8.4.2 How to use the test data for formulating millbases
Ideally, the data obtained by the systematic optimisation of millbases for dispersion should be filed, and it may be used to formulate different combinations of pigment(s) and resin(s) from the individual values without further tests.

The basic statement of *PV%* and *VS%* defines a millbase formulation from which a unit of 100 by volume should be calculated first before scale-up. This is to be multiplied by the appropriate factor to conform to the different loading requirements of dispersion machinery.

CALCULATIONS
Basic equations
To find weight: $\rho\ V$

$$\text{To find volume: } \frac{w}{\rho}$$

$$\text{To find } \rho\text{: } \frac{w}{v}$$

$$\text{To find } N/V\text{: } \frac{w_r \times R_s}{100}$$

To find w_r and w_s at the $VS\%$ wanted for the weight of dispersion solution required:

$$w_r = \frac{w_{ds} \times VS\%}{R_s}$$

$$w_s = w_{ds} - w_r \text{ therefore } w_r + w_s = w_{ds}$$

where

ρ = density

V = volume

w = weight

w_r = weight of resin

w_s = weight of solvent

w_{ds} = weight of dispersion solution

R_s = resin solids

N/V = non-volatile resin weight

$VS\%$ = vehicle solids %.

To find composite pigment volume ($CPV\%$) at different weight ratios (w):

$$CPV = \frac{T_{vp}}{T_{vp} + V_{ds}} \, 100$$

$$\frac{W_{pr1}}{\rho_1} + \frac{W_{pr2}}{\rho_2} + \frac{W_{pr3}}{\rho_3} + \frac{W_{prx}}{\rho_x} = V_{pr1} + V_{pr2} + V_{pr3} + V_{prx} = T_{vp}$$

where:

CPV = composite pigment volume %

T_{vp} = Total volume of pigment

W_{pr} = Weight ratio of pigment

V_{pr} = Volume ratio of pigment.

To find the proportional amount of dispersant solution (w_{dsr}) and their correspond-ing volume (v_{dsr}) for each pigment from the individual optima.

$$w_{\mathrm{p}} : w_{\mathrm{ds}} = w_{\mathrm{pr1}} : w_{\mathrm{dsr1}} \text{ therefore } w_{\mathrm{dsr1}} = \frac{w_{\mathrm{ds}} \, (w_{\mathrm{pr1}})}{\rho_1} \text{ and } \frac{w_{\mathrm{dsr1}}}{\rho_1} = v_{\mathrm{dsr1}}$$

$$w_{\mathrm{p}} : w_{\mathrm{ds}} = w_{\mathrm{pr2}} : w_{\mathrm{dsr2}} \text{ therefore } w_{\mathrm{dsr2}} = \frac{w_{\mathrm{ds}} \, (w_{\mathrm{pr2}})}{\rho_2} \text{ and } \frac{w_{\mathrm{dsr2}}}{\rho_2} = v_{\mathrm{dsr2}}$$

$$w_{\mathrm{p}} : w_{\mathrm{ds}} = w_{\mathrm{pr3}} : w_{\mathrm{dsr3}} \text{ therefore } w_{\mathrm{dsr3}} = \frac{w_{\mathrm{ds}} \, (w_{\mathrm{pr3}})}{\rho_3} \text{ and } \frac{w_{\mathrm{dsr3}}}{\rho_3} = v_{\mathrm{dsr3}}$$

$$w_{\mathrm{p}} : w_{\mathrm{ds}} = w_{\mathrm{prx}} : w_{\mathrm{dsrx}} \text{ therefore } w_{\mathrm{dsrx}} = \frac{w_{\mathrm{ds}} \, (w_{\mathrm{prx}})}{\rho_x} \text{ and } \frac{w_{\mathrm{dsrx}}}{\rho_x} = v_{\mathrm{dsrx}}$$

therefore

$$w_{\mathrm{ds}} = w_{\mathrm{dsr1}} + w_{\mathrm{dsr2}} + w_{\mathrm{dsr3}} + w_{\mathrm{dsrx}} \text{ and}$$

$$v_{\mathrm{ds}} = v_{\mathrm{dsr1}} + v_{\mathrm{dsr2}} + v_{\mathrm{dsr3}} + v_{\mathrm{dsrx}}$$

where:

w_{p} = weight of pigment and/or extender in individual optimum millbase
w_{ds} = weight of dispersion solution in individual optimum millbase
w_{pr} = weight ratio of pigment
w_{dsr} = weight ratio of optimum $VS\%$ solution
v_{dsr} = volume ratio of optimum $VS\%$ solution
v_{ds} = volume of dispersion solution required.

To calculate composite vehicle solids ($CVS\%$):

$$CVS \qquad = \frac{N/V}{w_{\mathrm{ds}}} \, 100; \quad \text{for example}$$

$$\frac{w_{\mathrm{dsr1}}(VS)}{100} = N/V_1; \quad \frac{w_{\mathrm{dsr2}}(VS)}{100} = N/V_2$$

$$\frac{w_{\mathrm{dsr3}}(VS)}{100} = N/V_3; \quad \frac{w_{\mathrm{dsrx}}(VS)}{100} = N/V_x$$

$$N/V = N/V_1 + N/V_2 + N/V_3 + N/V_x$$

where:

CVS = composite vehicle solids %

N/V = non-volatile resin solids

N/V_x = non-volatile resin solids ratio

VS = optimum individual vehicle solids %.

Calculate weight of resin(s), (w_r) and weight of solvent(s) (w_s) to complete the formulation.

$$w_r = \frac{N/V}{R_s} 100 \text{ and } w_s = w_{ds} - w_r, \text{ for example}$$

$$\frac{N/V_1}{R_s} 100 = w_{r1} \text{ therefore } w_{dsr1} - w_{r1} = w_{s1}$$

$$\frac{N/V_2}{R_s} 100 = w_{r2} \text{ therefore } w_{dsr2} - w_{r2} = w_{s2}$$

$$\frac{N/V_3}{R_s} 100 = w_{r3} \text{ therefore } w_{dsr3} - w_{r3} = w_{s3}$$

$$\frac{N/V_x}{R_s} 100 = w_{rx} \text{ therefore } w_{dsrx} - w_{rx} = w_{sx}$$

$$w_{s1} + w_{s2} + w_{s3} + w_{sx} = w_s.$$

To obtain 100 by volume of millbase, multiply the weight of each ingredient by the factor of

$$\frac{100}{CPV + v_{ds}}$$

PRACTICAL EXAMPLE 1

TiO$_2$ Kronos RN45. This pigment is to be dispersed in a 15% *VS* solution of an aliphatic alkyd resin. The corresponding *PV*% is 43.5%

	Volume units		Density (g/cc)		Weight units
Pigment	43.5	×	4.1	=	178.35
15% *VS* solution of alkyd	56.5	×	0.788	=	44.52
	100.0				222.87

Now calculate the components of 44.52 weight units of 15% *VS* solution of the alkyd resin.

The solids of alkyd = 40%

To find the amount of *N/V* resin in 44.52 of 15% *VS* solution:

$$N/V = \frac{44.52 \times 15}{100} = 6.678.$$

To find the amount of 40% solids alkyd required to give 44.52 15% *VS* solution:

$$\frac{6.678}{40} \, 100 = 16.695 \text{ weight units of } 40\% \text{ alkyd}$$

To find the amount of solvent required:

$$\begin{array}{r} 44.520 \\ -\ 16.695 \\ \hline = 27.825 \end{array} \cdot$$

The millbase therefore consists of:

Pigment	178.35 weight units
40% solids alkyd	16.695 weight units
Solvent	27.825 weight units

Now multiply all weight units with the 'factor' to give the requisite volume of millbase; for example, if a ballmill with a total volume of 1000 is to be used then at 60% loading and charge to voids ratio of 1/1.5 the volume of millbase required will be 240 and the factor is 2.4.

The factor will vary according to conditions and dispersing machinery.

PRACTICAL EXAMPLE 2
A co-grind formulation to be calculated from individual data of TiO_2 pigment and blanc fixe extender at a weight ratio of 20/80. The dispersant is an aliphatic alkyd resin (40% NV) in white spirit.

Step 1. Write individual optimum *PV*% and *VS*% for the pigments required to formulate the millbase.
The following data was obtained using the Brabender Plastograph:—

For TiO_2 $PV = 43.5$ % at $VS = 15\%$ dispersant in white spirit
and Blanc fixe $PV = 38.35\%$ at $VS = 20\%$ dispersant in white spirit.

Step 2. Convert above to actual volume and weight units.

	V	ρ	wt
TiO_2	43.5	4.1	178.35
15% *VS* alkyd solution	$\dfrac{56.5}{100.0}$	0.788	44.52
Blanc Fixe	38.35	4.1	157.24
20% *VS* alkyd solution	$\dfrac{61.65}{100.00}$	0.798	49.20

Step 3. Write determined ratios of the different pigments in millbase and work out
the volume of that pigment mixture.

	wt.%	ρ	V
TiO$_2$	20	4.1	4.88
Blanc Fixe	$\dfrac{80}{100}$	4.1	$\dfrac{19.51}{24.39}$

Step 4. Work out the proportional amount of dispersant solution for each pigment
and their corresponding volume for each pigment.

TiO$_2$

$$178.35 : 44.52 = 20 : x \quad x = \frac{44.52 \times 20}{178.35} = 4.99 \quad 0.77.88 \text{ therefore } \frac{4.99}{0.788} = 6.34$$

with column headings: wt ρ V

Blanc Fixe

$$157.24 : 49.20 = 80 : x \quad x = \frac{49.20 \times 80}{157.24} = 25.05 \quad 0.798 \quad \text{therefore } \frac{25.05}{0.798} = 31.37$$

Step 5. Determine composite pigment volume *CPV%*:

Volume of pigments   24.39 +

Volume of disp.solutions $\dfrac{37.71}{62.10}$ therefore $CPV = \dfrac{24.39}{62.10}\,100 = 39.28\%$

Step 6. Calculate composite vehicle solids *CVS%*

required wt of 15% *VS* alkyd = 4.99 $N/V = \dfrac{4.99 \times 15}{100} = 0.75$

required wt of 20% *VS* alkyd = 25.03 $N/V = \dfrac{25.03 \times 20}{100} = 5.06$

$CVS = \dfrac{5.76}{30.02}\,100 = 19.15\%$ therefore total $N/V = 5.76$

Step 7. Calculate weight of resin (w_r) and weight of solvent (w_s) required to complete the formulation:

$$w_r = \frac{N/V}{R_s} \, 100 \text{ and } w_s = w_{ds} - w_r$$

	w_r		$w_r - w_s = w_{s(required)}$
for 15% alkyd $\dfrac{0.75}{40}\,100 =$	1.86	therefore $4.99 - 1.86 =$	3.13
for 20% alkyd $\dfrac{5.01}{40}\,100 =$	$\dfrac{12.52}{14.38}$	therefore $25.03 - 12.52 =$	12.51
		therefore total $=$	15.64

To obtain 100 by volume of millbase multiply by the factor of

$$\frac{100}{62.10} = 1.61$$

Step 8. Write up the completed millbase formulation:

$$(PV = 39.28\% \; VS = 19.15\%)$$

	V	ρ	wt
TiO_2	7.86	4.10	32.2
Blanc fixe	31.42	4.10	128.8
alkyd (40% NV)	28.44	0.84	23.83
white spirit	32.28	0.78	25.18
	100.00		

PRACTICAL EXAMPLE 3
To find composite vehicle solids $(CVS\%)$ and composite pigment volume $(CPV\%)$ from individual data of $VS\%$ and $PV\%$ to give the proportional amounts of stabiliser for the pigments present in 100 volume units of millbase (co-grind).

Step 1. Write individual optimum $PV\%$ and $VS\%$ for the pigments required to formulate a millbase for an automotive topcoat.

Pigment

Monolite Yellow $PV =$ 8% $VS =$ 20% dispersant melamine resin 1 in xylol

TiO_2 $\qquad\qquad$ $PV =$ 42% $VS =$ 15% dispersant acrylic resin in xylol

Titan Yellow $\qquad$ $PV =$ 36% $VS =$ 25% dispersant melamine resin 2 in xylol

Red oxide $\qquad$ $PV =$ 35% $VS =$ 20% dispersant alkyd resin in xylol

Step 2. Convert above to actual volume and weight units:

	V	ρ	wt
Monolite Yellow	8	1.6	12.8
20% VS Melamine R1	$\dfrac{92}{100}$	0.91	83.72
TiO_2	42	4.1	168
15% VS acrylic resin	$\dfrac{58}{100}$	0.89	51.62
Titan Yellow	36	4.3	154.8
25% VS melamine R2	$\dfrac{64}{100}$	0.92	58.88
Red oxide	35	4.3	150.5
alkyd resin	$\dfrac{65}{100}$	0.90	58.5

Step 3. Determine weight ratios of the four different pigments in the millbase to give the requisite colour of the subsequent finish and work out the overall volume of the pigment mixture. It is assumed that the weight ratios required are as follows:

	$w\%$	ρ	$V\%$
Monolite Yellow	2.85	1.6	1.78
TiO_2	18.36	4.1	4.58
Titan Yellow	46.72	4.3	10.87
Red oxide	32.07	4.3	7.46
	100.00		24.70

Step 4. Work out the proportional amount of dispersant solution for each pigment and their corresponding volume for each pigment:

Monolite Yellow

$$12.8 : 83.72 = \ 2.85 : x \quad x = \frac{\overset{w}{83.72 \times 2.85}}{12.8} = 18.64 \quad \overset{\rho}{0.91} \quad \frac{\overset{V}{18.64}}{0.91} = 20.48$$

TiO_2

$$168 \quad : 51.62 = 18.36 : x \quad x = \frac{51.62 \times 18.36}{168} = 5.64 \quad 0.89 \quad \frac{5.64}{0.89} = 6.34$$

Titan Yellow

$$154.8 : 58.88 = 46.72 : x \quad x = \frac{58.88 \times 46.72}{154.8} = 17.77 \quad 0.92 \quad \frac{17.77}{0.92} = 19.32$$

Red oxide

$$150.5 : 58.5 = 32.07 : x \quad x = \frac{58.5 \times 32.07}{150.5} = 12.47 \quad 0.90 \quad \frac{12.47}{0.90} = 13.86$$

$$\text{Total wt} = 54.22 \qquad\qquad \text{Total vol} = 60.00$$

Step 5. Determine composite pigment volume CPV = volume of pigments 24.70 + volume of disp.soln. $\dfrac{60.00}{84.70}$

$$CPV = \frac{24.70}{84.70} \, 100 = 29.16\%.$$

To obtain 100 by volume of millbase multiply by the factor of $\dfrac{100}{84.70} = 1.181$.

Step 6. Calculate composite vehicle solids $CVS\%$

$$\text{required wt of 20\% } \textit{VS Melamine resin } R_1 \quad = 18.64 \; N/V = \frac{18.64 \times 20}{100} = \; 3.728$$

$$\text{required wt of 15\% } \textit{VS} \text{ Acrylic resin} \qquad = \; 5.64 \; N/V = \frac{5.64 \times 15}{100} = \; 0.846$$

$$\text{required wt of 25\% } \textit{Vs} \text{ Melamine resin } R_2 \quad = 17.77 \; N/V = \frac{17.77 \times 25}{100} = \; 4.442$$

required wt of 20% VS Alkyd resin $= \dfrac{12.47}{54.52} N/V = \dfrac{12.47 \times 20}{100} = 2.494$

$$\text{Total } NV = 11.510$$

$CVS = \dfrac{11.510}{54.22} 100 = 21.11\%.$

Step 7. Calculate weight of resin (w_r) and weight of solvent (w_s) to complete the formulation.

$w_r = \dfrac{N/V}{R_s} 100;\ w_s = w_{ds} - w_r,$ therefore

	w	w_{ds}	$-$	w_r	$=$	w_s

for Melamine resin R_1 $\dfrac{3.728}{67} 100 = 5.56$ and $18.64 - 5.56 = 13.08$

for acrylic resin $\dfrac{0.846}{65.5} 100 = 1.29$ and $5.64 - 1.29 = 4.35$

for Melamine resin R_2 $\dfrac{4.442}{60} 100 = 7.40$ and $17.77 - 7.40 = 10.37$

for alkyd resin $\dfrac{2.494}{50} 100 = 4.99$ and $12.47 - 4.99 = 7.48$

$$\text{Total} = 35.28$$

To obtain 100 by volume of millbase multiply by the factor of $\dfrac{100}{84.70} = 1.181.$

Step 8. Write up the completed millbase formulation.

($PV = 29.82\%\ VS = 21.15\%$)

	V	ρ	w
Monolite Yellow	2.10	1.6	3.37
TiO_2	5.41	4.1	21.68
Titan Yellow	12.84	4.3	55.18
Red oxide	8.81	4.3	37.87
Melamine R_1	6.32	1.04	6.57
Acrylic resin	1.52	1.00	1.52
Alkyd resin	6.20	0.95	5.89
Melamine R_2	8.74	1.00	8.74
Xylol	48.06	0.868	41.67
	100.00		

8.5 METHODS OF DISPERSION AND MACHINERY

Millbases are dispersed in various equipment depending upon the nature of the pigmented millbase, and the quality and volume required. It is not the intention here to deal with the design, characteristics, and performance of the large variety of dispersing equipment available for the paint-maker, but to give sufficient guidance for the most commonly used machinery.

8.5.1 Ballmills

Ballmilling is a batch process and probably the oldest form of manufacture. It still has a useful role to play in the paint industry. It is eminently suitable for the manufacture of high-quality paints, and particularly with pigments that are difficult to disperse. Owing to their enclosed mode of operation no change in the constants of the millbase can take place. Ballmilling is a reliable system which requires little supervision and maintenance. However, the processing time can be long, 8–24 hours, and sometimes in excess of 36 hours.

There are restrictions regarding the solvents allowed in the mill, because of vapour pressure. Cleaning requires time and a lot of cleaning solvent between colour changes.

A ballmill is a cylindrical unit with roughly equivalent length/diameter ratios, and it is rotated about its horizontal axis. The inner wall of the cylinder is lined with non-porous porcelain, alumina, or hard silica blocks, to prevent contamination of charge due to abrasion. The mill is filled with grinding media (balls) to specified volume

Table 8.1

Ball diameter in mm	Surface area of 1-litre of balls m^2	No. of contacts between spheres
0.5	6.90	
1.0	3.44	
2.0	1.72	576 000
3.0	1.17	
4.0	0.87	
5.0	0.68	37 000
6.0	0.57	
7.0	0.50	
8.0	0.43	
9.0	0.38	
10.0	0.35	4620
12.7	0.27	
25.4	0.13	
30.0	0.12	171
40.0	0.08	
50.0	0.07	37

loading. Some mills are still filled with pebbles. Other than spherical-shaped grinding media (cylinders) may be used for specific purposes.

The balls may be different in diameter and made of porcelain, steatite, alumina, and different grades of steel, all of which have different densities. The process of dispersion is, in fact, the simplest in a correctly loaded ballmill rotating at the correct speed. The grinding medium takes up a concentric and approximately parabolic line of motion, and the individual rotation of the balls results in, through shear and abrasion, the reduction of agglomerates and aggregates to primary particles. Impact of balls is less important in wet grinding. From this it follows that the smaller the size, the greater will be the number of points of contact and higher the active surface area of the balls and greater the efficiency of work by shear and abrasion. Relevant data are given in Table 8.1 and Figs. 8.8a and 8.8b).

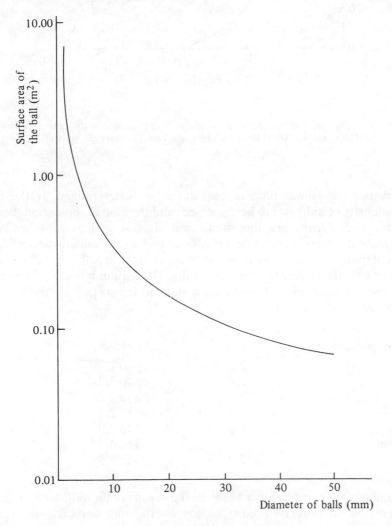

Fig. 8.8a — Characteristic effect of ball diameters on surface area of ball. 1 m³ T_v mill at 60% loading (D_v).

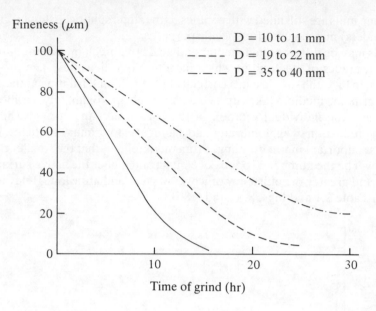

Fig. 8.8b — The effect of ball diameter on the rate and fineness of dispersion. (Example:
Carbon Black/Alkyd Resin millbase. 400 Ltd. Tv. porcelain ballmill).

In practice, the lower limit of ball diameter is determined by the average viscosity/density of millbases to be dispersed, and the ease of separation thereafter. In general, high density grinding media will disperse high viscosity and/or high density millbases faster. In practice it is difficult to allocate ballmills according to the densities of millbases, therefore the general use of steatite balls ($\rho = 2.9$) or alumina balls ($\rho = 3.6$) will give very satisfactory results. The various types of grinding media, listed below, are most efficient when used within the density limits of the corresponding millbase.

Grinding medium	ρ	Millbase ρ
Glass	2.0	up to 1.7
Porcelain	2.4	up to 2.0
Pebbles	2.6	up to 2.2
Steatite	2.9	up to 2.5
Alumina	3.6	2 to 3.0
Zirconium	5.0	3 to 4.0
Steel	7.9	4 to 6.5

A thin millbase or large difference between the density of the millbase and that of the grinding media will cause reduced efficiency, overheating, and excessive wear. This should be kept in mind when using steel ballmills with steel grinding media, despite their detuned mode of operation. The rheological characteristics of a pigmented millbase change during the process of dispersion owing to a number of factors, and they are influenced considerably by the nature of the pigment, its particle size, shape, and surface characteristics. Therefore it is difficult to define an ideal viscosity in absolute units for this stage of the process.

Water-cooling plays an important part in plant-scale milling, as this helps to keep the consistency of the millbase within the right shear range.

Laboratory ballmills are essential tools to carry out meaningful processability tests on millbase formulations, provided that they are operated according to the conditions described previously. It is important to remember that owing to geometrical considerations, the time needed to achieve a specified fineness will be shorter in a larger mill than in a small mill.

Critical speed
N_c = critical speed is the speed at which the balls are just held against the wall of the mill by centrifugal force. It is expressed by the mathematic formula

$$\frac{54.14}{r}$$

where r = the internal radius of the mill measured in feet.

The critical speeds of different size laboratory ballmills are:

Nominal capacity	Volume(ml)	r	Critical speed (rev/min)
1 pint	500	0.385	140.6
2 pints	1000	0.428	126.5
½ gallon	2500	0.460	117.7
1 gallon	5000	0.559	100.5
2 gallons	9300	0.614	88.2

(*Editor's Note*: It is still quite common to find a mixture of Imperial and metric units in use in the paint industry. SI units have not gained general acceptance.)

The optimum ballmill speed of rotation, N rev/min is a fraction F of N, and

$$\frac{N}{N_c} = F_c,$$

where F_c in rev/min is given by the formula

$$F_c = 43.3 \sqrt{\frac{1}{D-d}} \quad \text{where}$$

D = internal diameter of mill in feet

d = diameter of grinding medium in feet.

The optimum speed is about 65% of N_c, but acceptable conditions are obtained between 60–70% of N_c. Above 75% there will be overheating in larger ballmills, therefore this figure should not be exceeded. It is essential to check and adjust, if necessary, the rotation of plant-scale ballmills.

Calculation of millbase volume c and weight of grinding medium w at different charge-to-voids, C/V, ratio and dispersion volume D_v

The dispersion volume is a percentage of the total volume of the mill, and 60% is recommended as the optimum. However, there are circumstances where changes may be made in D_v and/or in C/V, owing to mechanical conditions to ease the load, or in non-critical millbase formulations to increase the yield. In these cases the following will help to obtain the conditions required:

$$D_v = \frac{T_v \% \ D_v \ \text{required}}{100} .$$

The voids (v) is the air space by volume % on the total apparent volume of the grinding medium. For convenience it is usually taken as 40% for close-packed spheres.

The weight of the grinding medium is obtained by

$$\text{mass volume} \times \rho$$

where the mass volume, M_v = the apparent volume − voids, for example $100 - 40 = 60 \ M_v$. The weight of the grinding medium essential to give correct mill conditions is obtained by the following equation:

$$w = \frac{3 \ D_v \ \rho}{3 + 2 \ C/V}$$

C = the volume of millbase charge to be dispersed, and it is obtained by

$$C = \frac{D_v \ (2C/V)}{3 + 2C/V} .$$

A ballmill is shown diagrammatically in Figs. 8.9 and 8.10 and a typical mill in Plate 8.2.

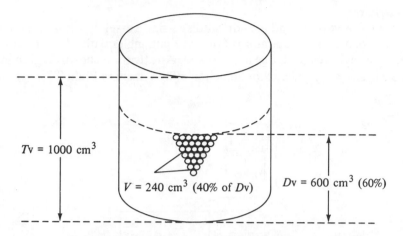

$T_V = 1000 \text{ cm}^3$

$V = 240 \text{ cm}^3$ (40% of D_V)

$D_V = 600 \text{ cm}^3$ (60%)

Fig. 8.9 — A ballmill at 60% loading and C/V ratio of 1/1.

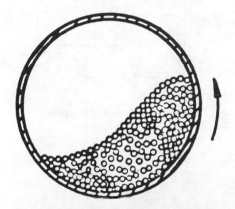

Fig. 8.10 — Operating principle of a ballmill.

Plate 8.2 — Ballmill (Sussmeyer).

8.5.2 Attritors
The attritor is a vertical and cylindrical dispersion vessel which is stationary. The centralized rotating shaft agitator is fitted at right angles with bars (spokes) evenly spread, which agitate the charge in the vessel. An attritor is shown diagrammatically in Fig. 8.11, and a typical attritor and circulating system in Plate 8.3.

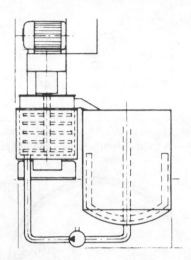

Fig. 8.11 — Diagram of an attritor.

Plate 8.3 — Attritor and circulating system (Torrence & Sons Ltd.).

The units are available in sizes up to about 100 gallons total capacity. They can be used as batch or as a continuous process. The original Szegvari attritor, in fact, was a continuous unit. Loading is by hand or by pipeline of liquid ingredients if used as a batch process. The following empirical conditions apply:

Loading	$= 70\%$ by volume
Charge-to-voids ratio	$= 1 : 1$
Tip speed of agitator spokes	$= 165\text{--}190$ metres per min (an empirical figure only)
Vessel diameter-to-agitator ratio	$= 1 : 0.75$
Size of grinding medium	$= 1/4\text{in}, 3/8\text{in}$ to $1/2\text{in}$ steatite balls.

Sufficient recirculation of millbase during dispersion and adequate water cooling are necessary. The charge-to-voids ratio is rather critical; for example, 0.9:1 may lead to excessive wear on working surfaces and may cause ball breakage. The system requires inspection at frequent and regular intervals, in particular with respect to ball weight and general wear of the agitator. The process is on average about three times faster than a ballmill, but requires constant supervision. It can handle higher viscosity millbases than ballmills. A definite disadvantage of the process is that if no top cover is used, solvent evaporation due to the heat generated during dispersion changes millbase conditions, to the detriment of productivity. This should be checked several times by measuring the viscosity of the millbase during dispersion, especially when longer runs are necessary, the lost solvent being replaced as required. If this is not done the constants of the subsequent finish may be affected. It is not ideal for millbase systems containing low-flash solvents.

8.5.3 Sand/bead mills

Sandmills have been used for the dispersion of pigmented millbases since the early 1950s, using on average 30-mesh Ottawa sand; but during the following decade improved results were obtained with glass beads and other synthetic grinding media, hence the process is now called beadmilling.

Typical mills are shown in Fig. 8.12 and Plate 8.4 and 8.5.

The operating principle is to pump the premixed, preferably predispersed, millbase through a cylinder containing a specified volume of sand or other suitable grinding medium which is agitated by a single or multidisc rotor. The disc may be flat or perforated, and in some units eccentric rings and used.

The millbase passing through the shear zones is then separated from the grinding medium by a suitable screen located at the opposite end of the feedport.

There are several makes, for example Torrance, Sussmeyer, Drais, Vollrath, Netzsch, and Master, with vertical dispersion chambers. These can be divided into two types, open-top and closed-top machines, all being multi-disc agitated in a fluidized bed-type situation. The latest machines use horizontal chambers and are available in various sizes from 0.5 to 500 litres. All beadmills, with the exception of the latest Vollrath twin, have cylindrical dispersion chambers and employ glass, ceramic, alumina, and, in certain cases, steel balls as grinding media. The factors affecting the general dispersion efficiency of these units are known reasonably well. The selection of the right type and diameter of grinding medium and the product range, in particular the right millbase formulation for a given type, is important for

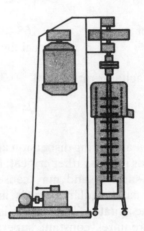

Fig. 8.12 — Original Du Pont de Nemours open unit.

Plate 8.4 — Sandmill (Sussmeyer).

Plate 8.5 — Horizontal beadmill (Netasch LME series).

maximum utilization. In theory, the smaller the grinding medium and higher its specific gravity, the more efficeint it becomes, because of the much increased surface area, as shown in Table 8.1, and because the centrifugal force transmitted to the grinding medium at the tip of the rotating disc increases considerably its weight, and applies greater shear to the millbase. The speed transmitted to the individual members of grinding media at the tip of the disc assumes that speed, and therefore the force, can be calculated. Let tip speed of the disc be 670 metres/min = (2200 ft/min) and the radius of the disc 0.10 metres (3.94 in). Then

$$F = \frac{V^2}{rg} = \frac{(670/60)^2}{0.10 \times 9.81} = \frac{124.69}{0.981} = 122 \times weight\ increase$$

where

F = centrifugal force

r = radius of disc

v = velocity

g = acceleration due to gravity (981 cm/s).

In most paint media and in practice, however, 1 to 1.2 mm diameter balls is about the limit, because of separation problems, and 2 to 3 mm is the best general-purpose grinding medium. In aqueous low-viscosity media, however, 0.7 mm or smaller diameter balls will further increase the efficiency of the mill, subject of course to relatively Newtonian flow characteristics of the millbase. However, the upper viscosity limitations of vertical units, irrespective of whether or not they are pressurized systems, seems to be 5 to 6 poises of millbase viscosity. In practice they cannot handle difficult-to-disperse systems, for example iron oxides, extenders, and Prussian Blue satisfactorily, and in many instances multi-passing of the millbase is necessary. The efficiency of these units could be improved by the selection of the right type of grinding medium for the manufacture of more difficult systems, but these media have probably reached the end of development. The difference between these and horizontal units seems to be that in the horizontal beadmill the kinetic energy of the same bead charge is greater, therefore they are more efficient and can handle higher viscosity and more difficult-to-disperse millbases. These units have not yet reached the end of the development stage, and seemingly have more potential with regard to the range of pigmented systems they can handle, and are easier to clean. All units must be water-cooled to keep operating temperature below 50°C, as even the best-formulated millbase will lose its dispersibility potential above that temperature, owing to a drop in viscosity which can lead to considerable wear of the working surface and bead breakage, and to badly discoloured dispersion. Similar problems arise if the charge to voids ratio is, or drops below 1:1. The normal operating charge to voids ratio is 1:1 or 1:1.2. The accepted shell diameter-to-disc diameter ratio is 1.4:1. The speed of the rotor is about 67 metres/min (220 ft per min). All units need premixes, but more efficient use of the machines can be made if HSD type predispersed millbases are used. Auxiliary equipment consists of pumps, pipes, storage vessels, and mixers. Maintenance can be costly. Cleaning is effected by passing through the additional resin or solution of resin, essential for the final or part make-up of the paint. Final cleaning is effected by solvent, later to be recovered or used to adjust viscosity. A spare shell per unit is advisable to maintain continuity of production and for ease of colour change. Horizontal units seem to be more flexible in operation; they can handle larger sections of paint dispersions, and probably have the potential to replace ballmills in the long run.

8.5.3.1 Correlation between laboratory and plant-scale units
Single disc batch or multi-disc continuous beadmills from 1 pint to $\frac{1}{2}$-gallon vessel volume are ideal for laboratory investigation of the processability of millbases, but at least a 5-gallon machine is required (with a 20 gallons/hour flow rate) to establish scale-up factors with some confidence.

8.5.4 Centrimill
The centrimill is a batch type beadmill with centrifugal discharge. It can be difficult to load the large units. It is a good system if smaller batch size production is required. Owing to the centrifugal discharge facility it gives very good yield, with the exception

of very thixotropic bases. It has been found sensitive to millbase formulations and charge to voids ratio. Wrong conditions lead to excessive wear on glass beads and contamination of colour. A diagram of a centrimill is shown in Fig. 8.13, and a Sussmeyer machine in Plate 8.6.

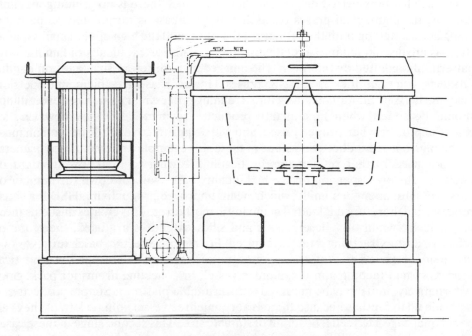

Fig. 8.13 — Diagram of a centrimill.

Plate 8.6 — A centrimill (Sussmeyer).

8.5.5 High-speed kinetic energy types of disperser

There are several makes of high-speed disperser (HSD): Torrance, Cowles, Vollrath, etc., and lately, Mastermix and Silverson. The operating principle is a free rotating disc of 'limited' design (saw type blades) in an open vessel. The exception here is the Silverson type machine where the rotor is enclosed in a stator device which is a somewhat pressurized dispersion chamber. Since there is no grinding medium present, the pigment disperses on itself if the millbase is formulated properly by being accelerated on both the surfaces of the rotor, and the high-speed small-volume streams are discharged into a relatively slow-moving vortex, ideally of laminar flow pattern, also created by the rotor. The units come in different sizes and can handle products with viscosities up to 32–35 poises. However, as a true pigment dispersion machine it has its limitations, therefore a refining process, for example, beadmilling, should be added when high-quality products are made. HSD is, however, an essential process when properly used, and will do an effective predispersion on most of the pigmented products; and some of these, for example undercoats and primers, will not need further refinements by beadmill, owing to their low standard of fineness. The predispersion on the high-quality bases will cut down the number of passes in subsequent beadmills, which means improved productivity. HSD is a batch process, and therefore it is loaded manually in most cases. This creates dust nuisance, in particular when organic pigments and silica extenders are used, therefore an effective dust-extraction system is required. For a more effective batch turnover two dispersion vessels per unit are required unless the unit is arranged in a vertical process system feeding a mixer, storage vessel, or collecting drums for packaging. Alternatively, units may be mounted onto a suitable platform. Simple products can be made up to finishes because there is adequate free vessel volume left, or they can be second-staged to feed a beadmill of suitable size. This second stage in these cases is a formulation modification and ideally should provide the optimum millbase formulation for the subsequent beadmilling process.

For efficient operation, the following specifications are ideal:

Horsepower	15	20	30	45	60
Rotor diameter in inches	12	13	14	16	20
Vessel diameter in inches	30	33	38	40	43
Height of vessel in inches	30	37	40	46	47
Batch size in gallons	30	46	77	100	140
Maximum rotor speed in feet/min.	5000 in all cases				
Horsepower per gallon of millbase	0.5 0.44	0.43		0.45	0.43

The vessel should be water-cooled to give an operating temperature $< 45°C$. The temperature rises rapidly during the dispersion, and contact temperatures at the tip of the rotor can exceed 70°C. This causes rapid loss of solvents, changes the viscosity of the millbase, and reduces dispersion efficiency. The speed of the shaft should be infinitely variable to ease loading and second-staging. The rotor position should also be vertically adjustable. In practice the distance between the bottom of the vessel and the position of the rotor in the vessel in centimetres equals roughly the viscosity of the millbase in poises; for example, if a 25 poises millbase is to be dispersed, then the distance should be 25 cm to obtain a vortex with laminar flow at around 4000–5000 ft/min rotor tip speed. When operating conditions are right the vortex is rolling steadily without surge and splashing. Rise in millbase temperature will, however, lower its viscosity, which is indicated by the onset of splashing, and under these conditions no further dispersion of the pigment will take place. To correct this: reduce rotor speed, lower the position of the rotor, and step up water cooling.

The process needs constant supervision. The average time of dispersion is about 30 mins, and batch turnover about 90 mins. Dispersion efficiency does not significantly increase with the increased rotor speed above 5000 ft/min., but increases only to a limited extent with reduced vessel diameter. This, however, reduces the batch size and considerably increases the power requirements. Scraping the inside wall of the vessel should be carried out several times during the process of dispersion, as the millbase tends to cake at the top edge of the vortex, which is much worse with non-water-cooled vessels owing to the rapid loss of solvent. For this operation, of course, the unit must be switched off.

Practical operation
Cleaning is relatively straightforward by using solvents. This process is rather wasteful, although the solvent is recovered later. It creates, however, the potential hazard of exceeding the *TLV* values in the atmosphere in the immediate vicinity of the unit. Adequate ventilation is therefore required.

In conclusion, high-speed dispersers have an important function in production. They are relatively simple, and they have improved little in efficiency during the last decade. Nevertheless they should be exploited to the full and not used as mixers only. They are too expensive for that purpose, and there is room for improvement.

Correlation between laboratory 1-horsepower machines and up to 60-horsepower production units is quite good, but one has to bear in mind that the amount of power available on the laboratory scale is about twice as much as can be obtained from production units per volume unit of millbase.

Two HSD machines are shown in Plate 8.7 and 8.8.

Plate 8.7 — High-speed disperser (Sussmeyer).

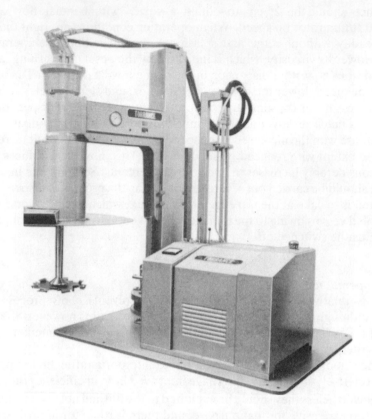

Plate 8.8 — High-speed disperser with hydraulic transmission (Torrance).

8.5.6 Rollmills

The operating principle of rollmills is to feed a premixed millbase between a roller and a bar or between two or three rollers rotating at different speeds, and to apply high pressure onto the thin film in the nip.

The pigmented film which is spread onto the rollers is subjected to very high shear. The dispersed material is removed with a scraper tray (apron), the blade of which lightly touches the front roller.

Roll mills are not suitable for large-volume production; they are labour-intensive, requiring high skill and are therefore costly to operate. Their use is confined to the manufacture of very-high-quality paints, printing inks, and high-viscosity pigmented products where the fineness of dispersion, cleanliness of colour, and colour strength are of great importance. In this sphere of activity it is hard to beat the performance of a well set up triple roll mill.

Efficient extraction is essential over each unit, to eliminate hazards associated with solvent vapours in the atmosphere.

Two rollmills are shown in Plate 8.9.

Plate 8.9 — Two Torrance rollmills.

8.5.7 Heavy-duty pugs

Heavy-duty pugs are used mainly for the manufacture of very-high-viscosity putties, filled rubber solutions, and fillers where the fineness of dispersion is not critical. Pugging is a batch process and may be carried out under thermostatically controlled conditions, circulating water or oil through the jacket. The most commonly used rotors are Z-shaped (sigma blade), and two per unit rotate in an intermeshing mode inwards at different speeds in relation to each other to maximize shear in the dispersion chamber. The bottom of the chamber is curved to form two half-cylinders with a dividing ridge in the middle. The shear forces developed by the rotors can be very high, and they depend on the viscosity of the mix. The discharge of the processed batch is carried out by tilting the dispersion chamber or by a screw transporter which is built into the chamber. The machines are available from 0.5 kW to several hundred kW units.

8.5.8 Heavy-duty mixers

Paste mixers are designed to handle intermediate and smaller batches of high-viscosity pastes, and are used mainly for the dispersion and manufacture of sealers, PVC plastisols, certain printing inks, and pigmented intermediates for paints.

The advantages of the latest hydraulic drive units are that high torque and infinitely variable speeds can be obtained from 0 rev/min., which ensure that optimum mixing conditions can be established for the different systems. The vessel is cylindrical and must be clamped securely in position owing to the high torque

Plate 8.10 — A Torrance paste mixer.

characteristic of the unit. The rotor is of variable design and mounted onto a central shaft which is adjustable for height.

In general, scale-up from laboratory to plant production is not linear, owing to the greater power/unit volume of mix available on laboratory-scale units. However, the correlation curve from the Plastograph Pug mill can quickly be established and matched to plant machinery.

A Torrance paste mixer is shown in Plate 8.10.

8.6 MIXING

The process of paint-making requires extensive use of various types of mixers for the blending and mixing of resins, solvents, additives, and intermediates in order to complete the make-up of paint formulations and obtain a homogeneous product after the dispersion stage. The mixer is a cylindrical vessel which may be fitted with paddle, propellor, turbine, or disc-type agitators to cover a variety of applications and systems. It can be portable or fixed and controlled by automatic timing devices to give periodic mixing if the vessel is used for storage.

Mixing is still often considered as an art, but it has become more sophisticated during the last decade. Many articles have been published on the design and efficiency of mixers; references [5,6,7] are especially useful.

The selection of mixers to do a specific job, and the corresponding data regarding power, tank turnover, and performance, are best known by those engaged in the design and sale of such equipment, with the exception of qualified engineers and scientists working for large industrial organizations. We offer here only general guidance. The main factors affecting the efficiency of the mix are:

(a) viscosity of material(s),
(b) specific gravity of material(s), and
(c) solid content.

These factors determine the geometrical dimensions of the mixer, the shear rate, and the power required. The mixing velocity should be adjustable in order to overcome the settling-out of solid particles. In practice this speed is at least twice that of the settling-out rate, and is predetermined for a particular product to be processed.

The intensity of mixing using the tip speed of the impellor as the unit can be classified as

Slow	up to	500	ft/min
Medium		500–900	ft/min
Fast		1000–1200	ft/min

It is important to use the right speed to avoid the introduction of excessive air into the mix. The viscosity of the finished paint is seldom Newtonian; the mix may possess this property at different stages of the process, but it ultimately becomes pseudo-plastic or thixotropic.

Since mixing is not necessarily a thinning process, the viscosity may increase during certain stages, which means increase in the inertia of the material leading to the flattening of the flow pattern and a reduction of the height up to which the impellor will pump.

The viscosity handling characteristics of the various types of impellers used in the paint industry are such that at a cut-off point of 100 or even 200 poises, the paddle, turbine, and propellers can easily deal with the mixing problems.

However, on the basis of equal vessel volume circulation, with at least two vessel turnovers/minute, the turbine type impeller is more economical as it needs 30–50% less power than the propeller at the same peripheral speed. Furthermore, on an equal power basis the pumping/circulating capacity of the turbine is 30% higher than that of the propeller, again at the same tip speed. One of the least efficient impellers in this respect is the paddle which perhaps requires less power than the turbine for the same size of vessel, but produces liquid movement tangential to the device itself with reduced tank turnover. The paddle is probably the oldest mixing implement and the simplest one. It consists of a shaft with one or more arms either horizontal or pitched to 45°, operating in a vessel which may be baffled to improve efficiency.

The paddle diameter is usually up to 0.9 of the vessel diameter, and it operates in the tip speed range of 250–450 ft/min with a blade width-to-paddle diameter ratio of 1:8 to 1:12.

The anchor-type agitator is a variation on the paddle, giving better sweep of the bottom of the vessel.

Baffles, where required should be placed not higher than the level of the mix, and an oversweep blade on top may be advisable to stop the build-up of dry bits which develop because of evaporation.

Disc-type impellers are not very efficient in an open vessel, owing to their low pumping efficiency. The flow is generated by the surface friction and by the centrifugal force created by the rotation.

Stator/rotor type impellers can subject the mix to very high mechanical and hydraulic shear.

Portable and fixed mixers are shown in Plates 8.11 and 8.12 respectively.

Plate 8.11 — Portable mixer (Sussmeyer).

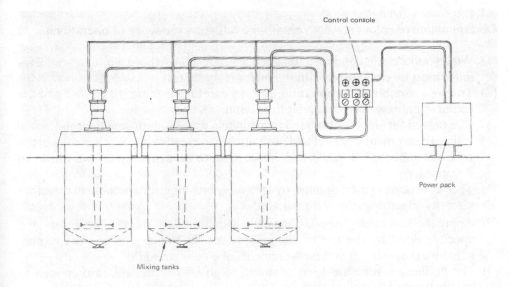

Plate 8.12 — Fixed mixer with hydraulic drive (Torrance).

8.7 CONTROL TECHNIQUES

8.7.1 Quality control of dispersions

8.7.1.1 Fineness

To assess the degree of dispersion of the pigments and extenders in pigmented products the use of a fineness gauge is common practice in the industry. It is no more than a control test, and the figures specified for a particular product must be established with a batch known to be satisfactory in practice. The gauge does not measure particle size distribution but displays only the largest undispersed aggregates. It is not beyond criticism regarding the size of the aggregate (particle), which may not correspond to the depth of the groove, but proper use by trained operators produces good agreement in the results obtained.

The gauge consists of a case-hardened steel block 7in × 2½in × 15/32in, all faces being machined flat. A tapered groove is machined down the middle of one of the broad faces of the block. It starts from zero depth at one end and increases to either 50 or 100 μm at the other end (1 μm = 0.001 mm). One side of the groove is graduated in 10 μm intervals, with a dot for each 5 μm division. A set of two fineness gauges, with groove depths 0–50 μm and 0–100 μm, is advisable, and is available from Sheen Instruments (Sales) Ltd, 9 Sheendale Road, Richmond, Surrey.

A liquid sample is placed at the deep end of the groove, and is drawn to the shallow end with a doctor blade. The distance from the deep end at which continuous pepperiness in the film becomes apparent is a numerical indication of the state of dispersion of the pigment particles in the material.

Variations in ambient illumination can influence the operator's decision; strip lighting, rather than daylight, is therefore advisable. The groove and blade edge must not be unnecessarily handled; and the gauge and knife must be stored in the box which is supplied with them.

8.7.1.2 Colour strength
Determination of colour strength entails the following sequence of operations.

(1) Weigh into a clean container 100 g ± 0.1 g of a standardized white finish†. This finish must be compatible with the millbase under test.
(2) To give a suitable reduction ratio, add, by careful weighing, 10 g ± 0.1 g of the coloured millbase to the standardized white.
(3) The reductions should be stirred immediately to avoid inducing flocculation.
(4) Using a convenient stirrer ‡, and mix until homogeneous. Check that none of the coloured millbase has adhered to the stirrer or to the can sides, and scrape down if necessary.
(5) The reductions can be applied by spraying, brushing, or spreading, therefore viscosity adjustments may be required.
(6) Apply the reductions to suitable panels § and build up the film weight to give opacity, aided by the use of chequered tape ¶. The degree of opacity may influence the instrumental measurements of colour strength.
(7) The finishes must be air-dried or stoved to give films that are hard enough to handle without damaging.
   (a) Most air-drying paints can either be dried at room temperature or 'force dried' e.g. 30 min at 65°C.
   (b) Stoving systems are cured at higher temperatures, e.g. 30 min at 127°C for a typical alkyd/MF finish.
   Before stoving, time is allowed for solvent evaporation from the paint film; for example, 30 min 'flash-off time' is to avoid bubbling etc. while curing.
   Before examination the panels must be completely cold, as hue changes occur during the cooling process; thus, allow for example, 30 min.
(8) The panels can now be assessed for strength either visually or instrumentally‖. For colours checked repeatedly, a panel can be selected as a reference standard. In this case the reduction must be made in a non-yellowing system, e.g. nitrocellulose.

---

†Standardized white finish:
The TiO_2 pigment is dispersed and made up to a standard recipe. The finish is then adjusted with vehicle to give the same tinting strength as the preceding batches.
‡Stirrer type:
Multispeed Stirrer from Anderman & Co. Ltd, Battlebridge House, Tooley Street, London SE1.
§Suitable panels:
6in × 4in tinplate panels.
¶Chequered tape:
¾in black and white chequered tape from Sellotape.
‖Instruments used for analysis:
(a) Absorption or reflectance spectrophotometer.
(b) Spectrum/profile instruments.
(c) Photoelectric filter colorimeter.
The values of colour co-ordinates X, Y, Z of standard and test panels may then be calculated.
(a) by a suitable computer program for hue and strength;
(b) by plotting the values as illustrated, obtaining the same;
(c) by a less satisfactory experimental reduction technique adding small amounts of the standard white to the appropriate proportions, and from the measured weight calculate the difference.

NOTES
When comparing millbases containing different % weights of the same pigment, corrections can either be made by:

(a) altering the weight of millbases added, or
(b) adjusting the strength figures using the simple proportions method.

To test white millbases, make the millbase up to a finish and tint using a standardized coloured millbase.

8.7.1.3 Strength and hue determination
Using the graphs for strength and hue determination (Fig. 8.14), the standard with which the sample is to be compared is plotted on the cross wires of both the Y against X and Y against Z graphs. The differences in the X, Y, and Z values between the sample and the standard are calculated and plotted on two graphs. For the sample to be within specification the plot of Y against X must fall between the parallel lines, and the plot of Y against Z must fall in the solid line box. Dependent upon the 'tightness' of the specification required, the size of the box can be altered.

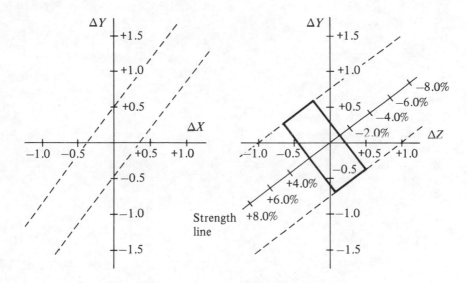

Fig. 8.14 — Strength and hue determination graphs.

ACKNOWLEDGEMENTS

Thanks are due to Torrance Ltd, Bitton, Bristol; Netzsch GMBH and Sussmeyer GMBH for permission to reproduce photographs and diagrams of their respective dispersion machinery. I would also like to thank Ms Eunice Burridge for her help in the compilation of data included in the chapter.

REFERENCES

[1] Daniel, F. K. A system for determining the optimum grinding composition of paints in ball and pebble mills. *Natl. Paint, Varn, Lacquer Assoc. Sci. Sect. Circ.* No. 744, (1950).

[2] Guggenheim, S. Data on pigment dispersion with high speed impeller equipment. *Off. Dig.* **30** No. 402, 729(1958).

[3] Wirsching, F., Haug, R., & Hamann, K. Dispersion of pigments in binders *Deutsche Farben Zeits* Oct. (1957).

[4] Farkas, F. K. Unpublished work.

[5] Parker, N. H. 'Mixing', modern theory and practice *Chem. Eng.* June 1964.

[6] Jeanny, G. H. Blending and agitation in the paint industry *Amer. Paint Journal* Aug. (1971).

[7] Oldshue, J. Y. and Sprague, J. Theory of mixing *Paint and Varnish Prod.* May (1974).

9

Coatings for buildings

J. A. Graystone

J. A.   Graystone

9.1   INTRODUCTION

An important and ubiquitous sector of the coating market is that concerned specifically with buildings. This chapter is addressed to aspects of technology relevant to dwellings ranging from private houses to schools; hospitals; offices; and light-industrial buildings. Coatings for chemical plant, oil rigs and marine environments have more stringent protective requirements (see Chapter 13). A characteristic feature of the coatings discussed here is that they should enrich the surfaces to which they are applied with colour, texture, or other appearance characteristics. Indeed, this market is often known as the 'decorative market', though some favour 'architectural coatings', 'house paints' or 'trade paints'.

Ever since the first cave paintings it has been possible to discern in man a creative urge, which goes beyond the purely utilitarian; this manifests itself in the field of art, but is reflected even in simple room decoration. Building paints would be different were it not for the advent of colour vision!

Emphasizing the importance of aesthetic factors is not to overlook the important protective role that coatings play. Our use of, say, ferrous metals, would be restricted without protective coatings. There is also a more subtle aspect to protection in that coating and substrate must sometimes be protected from each other. Coatings on an inert substrate will normally last longer than they do on many common building substrates. Movement in wood, for example, can cause flaking of paint; conversely, an impermeable paint contributes to the causes of decay in wood if moisture becomes trapped.

Another important aspect of paint, especially on the interior of buildings, is its effect on illumination. White and pastel colours play a major role in increasing the availability of both natural and artificial light, and the actual colours chosen can influence mood and feeling in other ways.

9.1.1 Types of decorative coating

There is no absolute way of categorizing coating types. Hence the classification terms used by marketing, development, or production departments can be quite different, and these may differ again from proprietary and descriptive terms used in the marketplace. Product descriptions can deal with only a limited number of parameters, and will be chosen to reflect the needs of different groups.

Table 9.1 illustrates some of the terms that are commonly used as part of a

Table 9.1 — Typical product descriptions for building paints

Type	Examples
Generic	Paint, stain, varnish
Appearance	Sheen (matt, silk, eggshell, gloss)
	Opacity or transparency
	Colour, build
Market sector	Trade, retail
System function	Primer, undercoat, finish
End use	Ceiling, wall, window, floor
Composition	Acrylic, alkyd, solventborne, waterborne
Property	Permeable, flexible, fungicidal, anti-corrosive
Substrate	Wood, metal, masonry

product description, reflecting some of the attributes that are considered significant. The wide choice of classification presents a difficulty in the logical presentation of building paints which might be considered from several viewpoints.

For the purposes of this chapter, some of the choices and factors which influence the selection or development of coatings for buildings are considered. This will include some general consideration of the influence that binder type and pigment volume concentration (PVC) relationships will have, but specific detailed formulations are not covered, as much information is readily available in the literature [1] and from raw material suppliers. In the latter half of the chapter emphasis is placed on understanding the nature of the substrate, since this is a major factor to be considered in coating design.

9.2 FORMULATING CONSIDERATIONS AND CONSTRAINTS

The role of the formulating chemist is usually to provide a coating or coating system which meets specific market needs, though some may be seeking ways to exploit a particular raw material. Some of the interactions involved are illustrated schematically in Fig. 9.1. Typically, a market opportunity is identified, using appropriate market research, and will relate to an identified area of use. In most cases there will be an accompanying need for certain appearance characteristics. Taken together, these requirements dictate the need for specific properties.

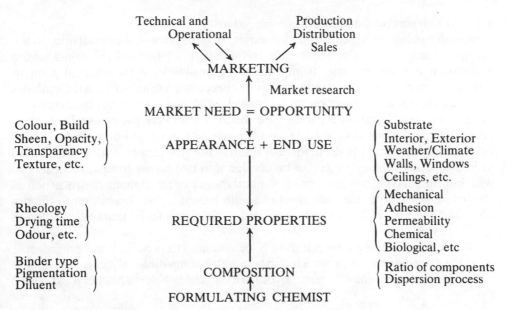

Fig. 9.1 — The role of the formulating chemist.

Having identified necessary properties, including application and subsequent performing characteristics, formulators must choose a suitable starting point and appropriate raw materials. Before considering composition further, attention is drawn to constraints which are of particular importance to the decorative market.

9.2.1 Storage stability
A characteristic feature of the decorative market is that the products must have a long shelf-life. Stock turnover may be slow for specific items, and users generally expect products to remain usable for several years after purchase. Particular attention must be paid to property retention on storage, including application and drying characteristics. Products must resist heavy settlement and irreversible changes such as skinning or coagulation. Where low temperatures are expected, it is desirable to build in freeze–thaw stability. In aqueous paints, hydrolysis of ester linkages is possible and will disqualify the use of some polymers as binder Many coating compositions show rheological changes with the passage of time, with pronounced effects on application. Aqueous, and to a lesser extent, non-aqueous paints will often show a steady increase in viscosity after manufacture. Allowance for this may be made in formulation, but it is vital to confirm that a stable viscosity plateau will be reached. A consequence of these constraints is that confirmation of good storage stability becomes an important part of the formulating process and may even become the rate-determining step for a new product launch. It is possible to speed up the aquisition of some data, for example, by incubator storage (60°C) and 'travelometers' (a vibrating device to simulate the effect of transport), but in general it would be most unwise to launch a new product into the mass market with less than 12 months' storage data. Even so, accurate records must be kept, so that trends can be extrapolated from an early stage.

9.2.2 Environmental and health considerations

Decorative paints are applied by a wide variety of users, from skilled craftsmen to the complete novice. Interior paints are frequently used in confined areas where ventilation is poor, placing an important responsibility on the chemist both to minimize toxic hazards and to ensure that any necessary hazard warning information is stated clearly and simply on the container. DIY users in particular cannot be expected to read complex separate data sheets. Decorative paints usually contain around 50% by volume of volatile 'solvent' whose vapour will be present at high concentration after application, thereby limiting the type of solvent that can be used in house paints. The majority can be divided into two broad groups, according to whether the carrying solvent is predominantly water or an aliphatic hydrocarbon of the 'white spirit' type. Consideration of health hazards must also extend to the dry film; and the uses of toxic substances, such as lead pigments, must be avoided in general use.

While waterborne paints can usually be assumed to provide a more agreeable working environment, there are a number of water-compatible solvents, for example some glycol ethers, where concern over toxicological problems has been expressed [2].

9.2.3 Film formation

Coatings must undergo a process where the liquid film is converted to a dry solid. It is an obvious, but nonetheless significant, factor that the means by which this can be achieved are considerably constrained for building paints in general. Unlike motors and industrial markets, the use of high temperature as a route to conversion is ruled out; also, many chemical reactions of the sort used in 'two-pack' products are unpractical, or too expensive, or have unacceptable toxic hazards. In consequence, the conversion and curing mechanisms used in general decorative paints are almost entirely confined to some combination of evaporation, coalescence, and oxidation, the implications of which are discussed in section 9.6.

9.2.4 Application

Method of application is another constraint operating in the decorative market. Although it is sometimes feasible to design an applicator to suit specific paints, in general, coatings must be designed for application by established tools such as brush, roller, paint pads, and to a lesser extent, by spraying. This situation is demanding; avoiding drips, brushmarks, roller spatter, etc. while maintaining good application characteristics requires subtle rheological properties. Many of these are difficult to measure and derived from highly subjective opinions on, for example, lapping, brushing, and viscosity. It is not unusual for the professional tradesman and the DIY user to have different preferences in this area.

Finally, on the topic of constraints, it is important to stress that the needs of a particular situation will never be met by one single 'correct' formula. Paintmaking is an heuristic process, and different formulating routes will produce alternative solutions, whose acceptability must be tested against market expectation.

9.3   PIGMENT BINDER SOLVENT RELATIONSHIPS

Stripped to the barest essentials, decorative coatings may be considered as consisting of pigment binder and solvent combined in various ratios. After conversion to a 'dry' film the composition is dominated by the pigment-to-binder ratio, but it should also be noted that under some circumstances air will be present in the film, and this aspect is of especial importance to the decorative market. The binder will also have undergone physical and chemical changes, and the possibility of internal stress must not be overlooked.

   If it is assumed that the total liquid formulation represents 100%, it is convenient to represent compositional relationships on a triangular diagram such as Fig. 9.2.

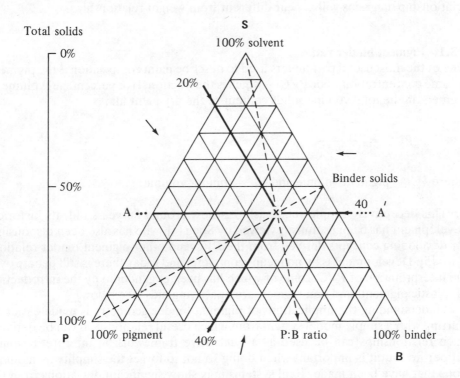

Fig. 9.2 — Pigment/binder/solvent relationships.

These diagrams are useful conceptual devices for illustrating relationships between coatings. For specific compositions they may be used as a framework onto which properties may be superimposed.

   In Fig. 9.2 composition 'X' comprises 20% pigment 40% binder, and 40% solvent. A line from S through X indicates compositions which have a constant pigment/binder ratio (which can be read off the PB axis) and increasing solids content, while the line AA' indicates compositions at constant solids but varying pigment/binder ratio. Because the solvent axis is horizontal, total solids content is indicated by distance from the base line which equals 100%. A line such as P through X is at a constant binder solids which may be read off the SB axis. A useful property of the diagram is that it allows the geometric solution of certain formulating problems and lends itself to simplex lattice design experiments [3].

It has not been stated whether Fig. 9.2 represents weight or volume relationships. Both are important. When actually making up a paint in the laboratory, it is normal to weigh the ingredients, and a chemist's recipe will usually be in % weight terms which must be divided by the density of each ingredient, in order to obtain the required volume. In many cases the properties of both wet and dry coating are more readily visualized in terms of volume relationships, and paint is generally sold and used by volume. Because most pigments have a density greater than unity, volume relationship diagrams will appear different from weight relationships.

9.3.1   Pigment/binder ratios

One of the most useful parameters used to describe paint composition is the pigment volume concentration — or *PVC* — which is the fractional (or percentage) volume of pigment in the total volume solids content of the dry paint film.

$$PVC = \frac{V_p}{V_p + V_b}$$

9.1

where V_p = pigment volume and V_b = binder volume.

This deceptively simple relationship is now around 50 years old; its historical development has been usefully reviewed by Stieg [4], who has also been responsible for developing concepts derived from the *PVC* and other pigment binder relationships [5]. Development is by no means complete, and today there is still debate over the interpretation of *PVC*, some of which has been engendered by the introduction of 'plastic pigment' and non-pigmentary opacifying aids (see below).

Understanding the implications of pigment binder geometry is made easier by starting with a simple model system from which useful relationships can be derived. Such relationships can be used as a qualitative framework in interpreting paint properties, but it is important when doing so not to forget the simplifying assumptions that have been made. Real systems may show significant deviations from the model, especially when interpreted in quantitative terms. For example, calculated *PVC* will differ from actual *PVC* if the binder undergoes shrinkage on drying or curing.

The simple *PVC* definition given above presupposes that the composition of the dried paint film can be expressed only in terms of pigment and binder, whereas in actuality some additives do not fit unequivocally into these categories. The difficulty can be avoided by defining $V_b + V_p$ as total dry film volume (excluding air). V_p (the pigment volume) is usually easier to define, but paint chemists are at liberty to include that part of the total pigmentation in which they are interested. Clearly, there are a number of variants of the basic *PVC* which could be useful for a specific purpose. Unless these are defined, it is assumed that the simplest definition (i.e. percentage or fractional volume of pigment in the dried film) applies.

9.3.2 The critical *PVC*

In conceptual terms it is instructive to consider the changes which take place when binder is progressively added to a bed of dry pigment until the binder is in excess. The process can then be continued by removing pigment until only binder is present. The initial and final conditions represent the extremes of 100% and 0% on the *PVC* scale.

During the process of adding binder to a bed of dry pigment, the point will be reached when added binder just fills the voids between the pigment particles. The *PVC* at this point is known as the critical *PVC* or *CPVC*, and represents an important transition point in paint film properties. Decorative paints contain solvents which are lost during the drying process. It is thus possible to formulate them to have *PVC*s which are above, below, or equal to the *CPVC*. In the dry film, paints formulated above their *CPVC* will contain insufficient binder to fill all the space available, and the film will become porous. In deflocculated, well-dispersed paints below their *CPVC* the pigment particles are not in contact with each other.

The process described above is largely fictitious. Real systems will be subject to many modifying variables including adsorption of binder onto the pigment surface. There are important differences between solution and dispersion binders in their penetration of voids and interaction with particle surfaces.

PVC and CPVC as reference points

It is a common practice to use the *PVC* of a coating as one of the parameters in its qualitative description. Typically, gloss paints have a low *PVC* (15–25%), while primers, mid-sheen wall finishes, undercoats, and flat paints are progressively higher. Knowledge of a coating's *PVC* may indicate an expectation of other properties such as opacity and sheen level, and it is common to use *PVC* as abcissa in many graphical property representations. But in general it would be more useful to relate *PVC* to the potential *CPVC* of a coating even when the composition is formulated below the *CPVC*. The latter acts as an internal reference point automatically compensating for some of the differences between alternative pigmentations. In recognition of this, the term 'reduced *PVC*' Λ has been derived where

$$\Lambda = \frac{PVC}{CPVC}$$

The reduced *PVC* will often prove a more appropriate parameter than *PVC* when comparing formulations or displaying them in graphical form. Λ will be greater than 1 for coatings that are formulated above critical [6].

9.3.2.1 *Pigment packing factors*

The possibility of air being part of the paint film can be expressed as a packing factor which relates the volume of pigment to unit volume of the pigment/binder/air mixture in the dried film.

$$Pf = \frac{V_p}{V_p + V_b + V_a} \tag{9.2}$$

where V_a = volume of interstitial voids.

There will be no voids below the *CPVC*, and in this region *PVC* = *Pf*. Above *CPVC* the pigment packing factor should remain constant and equal to the *CPVC* (for V_b is progressively replaced by V_a). In real systems this may not be exactly so, because pigment packing — e.g. as a result of flocculation — may change. *PVC*, in contrast, will increase steadily to the end point of 1.0 (100%).

9.3.2.2 Particle spacing

Particle spacing is influenced by a number of factors, including particle size distribution. However, it is instructive to consider the consequences of geometry for an ideal simplified system where it is assumed that particles are spherical, equidistant, and equal in size. In this situation for rhombohedral close packing, the void volume is 26%, and if this was exactly filled with binder the *PVC* would be 0.74. It is readily shown that the 'stand-off distance', that is the separation expressed in particle diameters, is given by:

$$ S = \sqrt[3]{\frac{0.74}{PVC}} - 1. \qquad\qquad (9.3) $$

Close approach of particles as *PVC* increases will affect the light-scattering efficiency of pigments (section 9.4.3).

Pigment particles have an enormous surface area. For example, one kilogram of TiO_2 could easily have a surface area of 5000 m^2. Thus considerable amounts of binder can be adsorbed onto the pigment surface. Such binder-coated pigments will act as larger particles with an influence on particle packing.

Considerations of particle packing become more complicated when consideration is given to mixtures of different-sized particles. Suppose, for example, that two beakers were respectively filled with 1 cm and 1 mm diameter beads; the void volumes would be equal at approximately 26%. But if the contents of the beakers were mixed in a ratio of about 3:1 the small beads would fit inside the voids of the larger, and the total voidage would be considerably reduced, in effect raising the *CPVC*. Real pigments and pigment/extender mixtures are much more complex than this, and it becomes difficult to calculate the *CPVC*, which must then be measured. The *CPVC* of two mixed-particle systems with a different particle size will often be higher than that of either of the single components.

9.3.2.3 CPVC and oil absorption

A simple way of establishing the approximate *CPVC* for a given pigment is to add linseed oil to the dry pigment until a coherent mass is formed. The test may be carried out as a spatula rub or by using mechanical mixing against a torque gauge. By tradition, the result (expressed as weight of linseed oil absorbed by 100 g pigment) is known as the 'oil absorption'. If expressed in volume rather than weight terms, then this is equivalent to the *CPVC*. A high *CPVC* implies low oil absorption and vice versa. For the reasons discussed above it is not possible to predict the *CPVC* of mixtures from the known values of single pigments, and the spatula rub must be carried out on the actual mixture. For many combinations of prime pigment and extender there is a condition of maximum packing which is likely to increase with the particle size of the extender.

To use oil absorption as a general predictor of *CPVC* assumes that the pigment will behave similarly in all liquid binders, and this is clearly unlikely. Not surprisingly, variations in oil absorption are found with different liquids. Chemists must make informed judgements as to the value of OA-type measurements and how the test is to be carried out, but simple linseed oil values remain a useful qualitative yardstick for making certain pigment selections, though these will rarely translate directly into the *CPVC* of another system.

The binder associated with a given pigment at its *CPVC* can be considered as comprising two components: that absorbed onto the pigment surface V_{ab} and that necessary to fill voids between the particles V_f. This may be illustrated as [5]:

$$CPVC = \frac{V_p}{V_p + (V_{ab} + V_f)} \tag{9.4}$$

In principle, V_f can be calculated from particle packing models. When this is done, V_{ab} is found to vary more than can be accounted for, by absorption theories. Huisman [7] has emphasized that equation (9.4) can only be used to calculate V_{ab} if the pigment paste at the end point of an oil absorption determination consists of single dispersed particles. In practice, aggregates of unknown surface area are generally present. The OA end point will be controlled by the effective density of particle aggregates, their packing, and the wettability of aggregates by binder. Huisman proposes a model based on effective particle density which allows for the presence of aggregates. Asbeck used the expression 'Ultimate critical pigment volume concentration' to denote the monodispersed state, and proposes an expression which relates *CPVC*, *UPVC*, and the diameters of primary pigment particle, d, and pigment agglomerate, D, respectively [8]:

$$CPVC = UPVC - (UPVC)^2 \left(\frac{D-d}{D}\right)^3 \tag{9.5}$$

If the *PVC* of a coating is progressively increased above the *CPVC*, a second transition-point may be reached where there is only sufficient binder to coat the particles (i.e. the term V_{ab} in equation (9.4)), and all the 'free' binder is replaced by air. Castells *et al.* [9] have published experimental data which corroborate the existence of this point, which may be of value in establishing interrelationships between pigment and binder and hence factors which influence optical efficiency. Cremer has used a rheological method to investigate the thickness of absorption layers on pigment and extender particles [10].

9.3.2.4 *Density methods for determining CPVC*
Coatings formulated to be 'above critical' will contain air, once the carrying solvent has evaporated. Accurate determination of the point at which air first appears in an increasing *PVC* ladder is thus an unequivocal way of determining the *CPVC*. The

presence of air in a dried coating will lower the density, and the volume may therefore be estimated from difference between measured D_a and theoretical density D_t;

$$\% \text{ air} = \frac{D_t - D_a.100}{D_t} \tag{9.6}$$

Density can be measured by a number of techniques including hydrostatic weighing [11] and mercury porosimetry [12].

A plot of density against PVC should show a fairly sharp point of inflection which locates the $CPVC$. Relationships between air, pigment, and binder in a dried film can also be represented on a triangular diagram which has the advantage of giving more information. Stieg has described how such diagrams can be used to estimate the PVC without the need for extensive PVC ladders [13].

In Fig. 9.3 the composition of any *dry* film formulated below critical will lie along

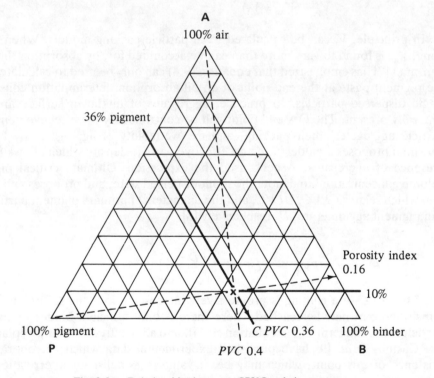

Fig. 9.3 — Relationships between *CPVC* and air content.

the base line P–B. Any composition formulated above critical will contain air (as indicated by the scale A–B) and will lie within the triangle. Any line drawn from the apex A to intercept P–B represents compositions of constant PVC and decreasing air content.

If a coating is made at a given PVC above critical and the air content is determined, then the composition of the dry film is at the intersection of the *PVC* and % air axes. This is illustrated in Fig. 9.3 for a film containing 10% air and a *PVC* of 0.4. If it is assumed that voids between pigment particles do not change in volume as binder is removed, then the % pigment can be extrapolated to the base line of zero air content, indicating a *CPVC* of 0.36. A *PVC* ladder would result in a steadily increasing air content whose intercepts with the appropriate *PVC* line should lie on the same straight line also indicating a *CPVC* of 0.36.

For real, rather than hypothetical, compositions, the *CPVC* line might not be absolutely straight. For example, the action of making paints at very different *PVC*s may introduce new factors which could influence pigment packing. But for many practical purposes a good approximation to the *CPVC* may be estimated from a single measurement.

9.3.2.5 *Porosity*
Inspection of Fig. 9.3 shows that the proportion of air present may be expressed as a total porosity (as indicated by the horizontal scale) or alternatively as a % of the volume, excluding the pigment. The second way of expressing porosity is usually known as the porosity index *PI*, and may be read off the graph by extrapolating the line from point P through the composition to the air–binder axis. In the example shown in Fig. 9.3 the composition has a total porosity of 0.1 and a porosity index of 0.16. If it is intended to change a given formulation, it may be possible to change one or both of these parameters. As they have a different effect on paint properties, this is an important degree of formulating freedom.

Porosity and porosity index are conveniently expressed by numerical expression (see [14]) such as:

$$\text{Overall porosity} = 1 - \frac{CPVC}{PVC} \tag{9.7}$$

$$\text{Porosity index} = 1 - \frac{CPVC\,(1-PVC)}{PVC\,(1-CPVC)} \tag{9.8}$$

9.3.2.6 *Critical PVC for dispersion binders*
Up to this point the discussion of *CPVC* has assumed that binder can readily penetrate and totally fill spaces between pigment particles; by implication the binder is a solution. But for many coatings and especially for the majority of decorative waterborne paints, the binder is, in fact, a dispersion comprising particles that must undergo deformation in forming a film. As there will be situations in which this deformation process is not complete, there will be a need to modify the above equations. Intuitively, it may be anticipated that packing of pigment and dispersion binder will be influenced by particle size of binder (as well as pigment) and the glass transition temperature (T_g) of the dispersion polymer. T_g is a useful, though not absolute, guide to the ability of a dispersion binder to coalesce and form a film. This ability is further modified by fugitive plasticizers — usually known as coalescing agents — which temporarily increase flow and deformation of polymer particles. However, even with the use of coalescing aids the penetration of particles into

intimate contact with pigment is likely to be less than that of a solution, with the result that more dispersion polymer than solution polymer is needed to bind a fixed quantity of pigment, i.e. the *CPVC* of a dispersion paint is less than that of a solution equivalent.

The ratio of solution to dispersion polymer needed to bind a given mixture has been referred to as the 'binder efficiency' or 'binding power index' *e* [15]. That is,

$$e = \frac{\text{volumetric oil absorption}}{\text{volumetric (dispersion) absorption}}.$$

Where the binder index is known then the porosity index achieved with a dispersion binder is obtained for oil absorption data by multiplying equation (9.6) by *e*. Alternatively, the actual volumetric latex absorption is measured and converted to a latex *CPVC (LCPVC)* [17, p. 201].

$$\text{Latex porosity} = 1 - \frac{LCPVC\ (1 - PVC)}{PVC\ (1 - LCPVC)}.$$

Graphical methods for obtaining *e* or *LCPVC* have been described [16].

The introduction of some dispersion polymer (latex) of deliberately low binder index is a way in which porosity may be introduced into coatings (see 9.5.3).

9.4 CONSEQUENCES OF POROSITY

As the pigment volume concentration of a given vehicle system is increased, there will be a progressive change in properties. Clearly the rate at which the property changes will be markedly affected by the onset of porosity, and this will be indicated by a point of inflexion around the *CPVC*. Properties which exhibit a sensitivity to both *PVC* and the onset of porosity can be conveniently grouped under three main headings:

Mechanical
— density
— strength
— modulus
— adhesion

Permeability
— rusting
— blistering
— staining
— enamel hold-out

Optical
— light-scattering
— contrast ratio
— tinting strength
— gloss

Generalized graphical representations of characteristic property/*PVC* relationships have been widely published [17]. In principle, the point of inflexion or break at the *CPVC* presents ways of detecting the *CPVC* itself; but, in practice, the various properties show a different sensitivity to the volume of air voids, and there will be a spread in the apparent *CPVC* value.

9.4.1 Mechanical properties

Film density reaches a peak at the *CPVC*, and, as has already been discussed, this is a sensitive and appropriate method of determining the *CPVC*. It is reported [17] that mechanical properties and adhesion also reach a peak at the *CPVC*, but this information may require some care in its practical utilization.

The effect of filler content on the mechanical properties of polymer has been widely studied. In general, the modulus of elasticity (stress/strain) will show a steady increase; at the same time there is usually a decrease in the extension-to-break. The net result may be an overall rise in the strength at break, but this will depend on a number of features including the efficiency of stress transfer between pigment and binder. Where higher strength is achieved it will often be at a lower elongation. A progressive rise in strength up to the *CPVC* is not invariably found.

Doubts must also be expressed as to whether adhesion will always be maximized at the *CPVC*. It is important to draw a clear distinction between adhesion and adhesive performance (or practical adhesion). The phenomenon described as adhesion is a complex combination of mechanical adsorption, diffusion, and electrostatic factors where consequences will be greatly modified by the way the test is carried out and by the condition of the substrate (cleanliness, porosity, etc). Bikerman has stressed the role of weak boundary layers. Specific examples of using adhesion to detect *CPVC* have been described [18].

9.4.2 Permeability-related properties

The permeability of coatings to liquids, gases, and vapours — and in particular water vapour — is an important property. Moisture vapour permeability is of especial importance in the corrosion of metals and in controlling the moisture content of wood.

Since vapour passes through the binder and not the pigment, it would be predicted that an increase in *PVC* would at first decrease permeability — this is usually found to be the case unless the pigment is not wetted by the binder. Anomalous permeability can be used as a sensitive measure of pigment wetting [19]. With or without good pigment wetting a fairly abrupt increase in permeability will occur at the *CPVC* as definite micropores appear, and this will be reflected by changes in rust prevention and blistering behaviour. Enamel hold-out (i.e. resistance to penetration by glossy finishing coats with consequent gloss loss) will also change.

9.4.3 Optical properties

White paints achieve opacity and whiteness from their light-scattering ability (in coloured paints absorption becomes more important). Functionally, titanium dioxide is the major source of such scattering. It is convenient to break down the scattering potential of particles into three groups of phenomena: surface reflection, refraction, and diffraction.

9.4.3.1 Diffraction

Wave theories of light can account for bending phenomena which give rise to scattering. Light-scattering by this mechanism increases with decreasing particle size until an optimum is reached at approximately half the wavelength of light.

Particles of the optimum size can bend approximately four times as much light as actually hits them, because the scattering cross-section is around four times the geometric cross-section. The crowding of TiO_2 particles will therefore detract from TiO_2 particles achieving their full scattering potential. This makes it difficult to increase the opacity of a given paint in a way which makes the most efficient use of extra titanium. Optical particle size for scattering can be calculated, and also depends on wavelength. Practical results will also be influenced by particle size distribution and shape. In paints which contain extender the titanium will be packed into the voids between the larger particles — thereby acting as though it were at an even higher TiO_2 *PVC*. However, the TiO_2 cannot be packed into the voids of the extender system any more tightly than at its own *CPVC*. For this reason prime pigments designed to be used in high-*PVC* formulations can, with advantage, be high in oil absorption. The adsorbed layer provides spacing to retain efficiency under conditions of close packing. Alternatively, fine particle size particles can be used as 'spacers' to improve TiO_2 efficiency. The two approaches show differences in the final property balance and offer a further degree of formulating freedom. Film porosity is a function of the packing, and oil absorption of the total pigmentation, whereas the spacing term is related only to the oil absorption of the prime pigment.

9.4.3.2 Refraction

Light may be refracted by particles and hence scattered; the change in direction is dependent upon the refractive index. The smaller the radius of the pigment particle, the greater the total area available to interact with light. Scattering by refraction increases down to a critical point where light rays no longer 'see' the particle and go past it as though it were in solution.

9.4.3.3 Reflection

Surface reflection, R, obeys Fresnel's law, which states that for small angles of incidence

$$R = \left(\frac{n_2 - n_1}{n_2 + n_1}\right)^2$$

for light which has travelled through a material of refractive index n_1 to strike the surface of a material with refractive index n_2.

Since air has a lower refractive index than any binder, its introduction into a film will increase light-scattering by the introduction of pigment/air interfaces. Thus paints formulated above their *CPVC* show enhanced opacity known as 'dry hiding'. This can be taken advantage of only if the other consequences of porosity are not a major problem. Dry hiding paints are frequently used on ceilings and walls. Optical properties offer another way of detecting *CPVC* as there is a fairly sharp inflexion in a plot of scattering coefficient (or contrast ratio) against *PVC*. The apparent tinting strength of a fixed amount of tinter will also change fairly sharply around the *CPVC*.

When formulating flat wall paints which may be applied over surfaces of differential absorption, it is sometimes difficult to achieve colour uniformity. For alkyd paints it has been found that a *PVC* just below the *CPVC* gives optimum colour uniformity.

9.5   AIR AND POLYMER EXTENDED PAINTS (MICROVOIDS AND PLASTIC PIGMENT)

Whiteness, when it occurs in nature as in snow, flower petals, or sea foam, does not require the presence of TiO_2! Such whiteness arises from the interaction of light with interfaces and voids between air and a transparent medium as described above. Clearly this is a potential source of cheap opacity for the coatings industry and one which has attracted a great deal of attention in the decorative market sector. The introduction of microvoids gains opacity in a manner analogous to the dry hiding described previously; but in principle, depending on the mechanism used, it is possible to increase opacity without sacrificing other paint properties (such as dirt retention) which are coincident with an increase in porosity.

When a microvoid containing air is totally encapsulated, it is conceivable to have coatings of low TiO_2, high opacity, and high gloss; but the greater part of work is more appropriate to be used in conjunction with conventional pigments and extenders and is especially relevant to matt and silk paints.

The use of plastic pigment, encapsulated air, etc., poses interesting questions in interpreting *CPVC* and similar calculations. It is not always clear whether a plastic particle should be considered as part of the pigment or as part of the binder, and to some extent it is because these materials fall between the two that they offer the chemist new formulating options. To get the best from these options requires an understanding not only of the new pigments but also how best to exploit the old! It will frequently be found that claims made for a new organic pigment (microvoid or solid) are based on a formulation box which is disadvantageous to conventional pigmentation, or vice versa.

Weighing up the pros and cons of two dissimilar formulations each 'optimized' within its own preferred formulation box is by no means straightforward. Formulae may be optimized around a number of properties, each with a different market appeal. However, interest in these new approaches is likely to rise if, as seems likely, the cost of titanium dioxide continues to increase.

Air/particle configurations

The use of microvoids in man-made materials was extensively reviewed by Seiner in 1978 [20]. Seiner has also published an interpretation of theoretical analysis by Kerker *et al.* as applied to coatings [21]. Since then there have been several different practical approaches to the problem resulting in much patent, and some commercial, activity. Chalmers & Woodbridge reviewed the better known commercial routes in 1983 [22]. Four methods of pre-forming polymeric beads have been extensively described, and are discussed in the following sections.

9.5.1   'Spindrift' [23]

Dulux Australia Ltd have described the use of both solid and vesiculated polyester/ styrene beads. The solid beads are comparatively large (up to about 40 μm) and may be clear or pigmented. A major use has been to formulate matt paints with exceptionally good polish resistance. Unpigmented beads act as windows in the film and are used only for highly saturated colours. A requirement of very flat finish combined with good polish resistance has created a market for beads of this type in Australia and South Africa, even though they are more expensive than conventional extenders. However, for cost savings in terms of opacity, the vesiculated bead is used.

Vesiculated beads are prepared by a double emulsion technique in water, starting with an unsaturated polyester/styrene precursor and using a redox free radical system for curing. The manufacturing route requires careful process control. Beads are produced in a size range up to around 25 μm, though 11–14 μm is more typical, and the vesicles are around 0.7 μm and contain TiO_2 within the vesicle. In use, Spindrift beads enhance TiO_2 efficiency while maintaining film integrity, and with a suitable latex very high PVCs can be achieved (i.e. counting beads as pigment). More recent developments have led to both smaller beads and a higher degree of vesiculation, thereby extending the range of application. As with most specific formulations the usefulness of this technique depends on relative costs and local market requirements. Beads which have been successfully exploited commercially in one market may not have the same perceived advantages when translated to another. The difficulty in choosing between an optimized bead formulation and a suitably formulated conventional paint is highlighted by the correspondence reported in [24,25]. Guidelines on the effective use of vesiculated beads have been published by Goldsbrough *et al.* [26].

9.5.2 'Microbloc' (Berger Jenson & Nicholson)

'Microbloc' particles are actually aggregates of finer particles formed by shearing an addition polymerization reaction in aqueous medium; they are used in a similar way to Spindrift. It has been claimed that their irregular shape gives higher film strength than spherical beads. Unlike Spindrift vesiculated beads, it is not claimed that internal pigmentation with TiO_2 is efficient and 'Microbloc' is usually combined with external TiO_2 and film extender to produce high PVC paints, which are very flat.

9.5.3 'Plastic pigment' (Glidden Coatings and Resins Division)

Ramig *et al.* [27, 28] propose fine particle size polystyrene 'beads', i.e. a polystyrene latex in the range 0.1–0.6 μm, to cause microvoids in paint films and therefore enhance opacity. The so-called beads are of the same order of magnitude as the latex in emulsion paints, with which they are typically blended in a 1 : 1 ratio. They are used in silk and matt paints effectively to raise the PVC above critical and gain opacity by dry hiding, but with less loss of film integrity.

The use of polystyrene pigment in this way is a good example of the difficulties encountered in applying conventional PVC calculation to these organic pigments. On the one hand the particles from a high T_g hard latex could be seen as fine particle size extender (i.e. above the line in the PVC equation), and on the other hand as a latex with zero or low binding power. The truth probably lies between these extremes. By appropriate choice of coalescing aids it should be possible to ensure better integration of the hard polymer particles than would be possible with an inorganic extender, but without causing the particles to coalesce totally. Careful matching of coalescing, and other paint constituents, would be necessary to optimize performance, but the route to an alternative formulating box is indicated [29].

9.5.4 'Ropaque'®

Rohm & Haas [31] have developed opacifying aids comprising hollow acrylic styrene beads suspended in water (typically 37% by weight, 52% by volume). As paints containing this 'opaque polymer' dry, water is lost from the cores of the particles to be replaced by air. Resulting air voids act as scattering sources and make a direct contribution to hiding. It is claimed that four parts by volume of opaque polymer is approximately equal in hiding power to one part of titanium dioxide [30]. Commercially available material ('Ropaque'® OP-62) has a uniform fine particle size around 0.4–0.5 μm and may also contribute to hiding power by uniformly spacing the titanium dioxide particles and helping to prevent crowding. Because the particles have less surface area than corresponding volumes of titanium dioxide, it has been argued that there is a reduction in binder demand. In other words, the *CPVC* has been increased, allowing formulation at a higher *PVC*. Clearly this argument defines opaque polymer as pigment rather than quasi-binder.

As with all new opacifying aids, reformulation of existing paints must be carried out sensibly and on a volume rather than a weight basis. Detailed formulating protocols have been published by the manufacturers.

9.6 THE NATURE OF THE PAINT BINDER

9.6.1 General

In the preceding section attention was paid to the consequences of changing the pigment-to-binder ratio in a paint formulation. An equally important stage in considering the formulation steps that are open to the paint chemist is to consider the implications of the physical and chemical nature of both pigment and binder. This is an extremely complex area, much of which is covered in more detail elsewhere in this book. Here we are concerned with some of the more generalized aspects and their consequences in the decorative building paint market with the objective of devising a frame of reference for assessing the value of specific formulating areas.

Within the building market there are a number of specialized sub-sections, e.g. in flooring, concrete protection, swimming pools, etc. which employ a wide variety of paint binders, but the bulk of decorative paints used in the high-volume markets can be broadly divided into two major groups often known as 'waterborne' and 'solventborne' (more accurately, 'aqueous' and 'non-aqueous'). The former may contain a percentage of other polar solvents, including alcohols and glycol derivatives, while the latter term usually denotes an aliphatic solvent, though often containing a percentage of aromatic solvent.

It is pertinent to ask why the balance of waterborne to solventborne varies between market sectors and is different between countries.

In general there is a continued swing from solventborne to waterborne, which proceeds at different rates around the world. It is influenced by practical considerations (e.g. climate), exploitation of specific properties, environmental aspects, economic aspects, and sometimes historical preferences. Currently around 60% of

UK decorative paints are waterborne, while in the USA the figure is higher at 70% [32].

An understanding of how the properties of a paint are related to the nature of the binder is necessary for formulation purposes, and may offer insights into potential routes to product improvement through new materials. In the first instance these can be related to the nature of the solvent and the physical and chemical processes by which the binder is converted from a wet to a dry film.

Non-aqueous decorative paints might be expected to be flammable, have strong primary odours, and be resistant to freezing. Aqueous paints, on the other hand, will be non-flammable, might cause rusting, and be susceptible to freezing. These inferences are important, but they do not go far enough in explaining paint properties. A next stage is to consider the physical nature of the polymeric binder. Of especial relevance are the differences between 'solution' and 'dispersion' polymers.

One of the more obvious differences between solutions and dispersions concerns their rheological behaviour. Owing to particle interactions, dispersions generally show poorer flow properties which, in paint, might manifest itself as poor brushmark levelling. Solutions and dispersions also show differences in the way their viscosity builds up as solvent evaporates. The viscosity rise with dispersions is usually faster than solutions, and the irregularities caused by, say, brushmark disturbance, become trapped as a consequence of the rapidly changing viscosity profile [33]. On the other hand, the generally better flow of solutions can lead to problems such as sagging or runs on vertical surfaces, and a failure to cover sharp edges properly. Extremes of flow behaviour encountered in decorative building paints are illustrated by comparing a matt emulsion paint with a liquid gloss paint. It is generally true to say that in order to minimize flow problems the solids content of an aqueous dispersion paint is usually lower than that of a solution paint.

The next stage in building a model which describes some aspects of paint formulation is to consider the chemical nature of the binder.

To be useful as a paint binder, a polymer must be strong and tough. However, both everyday experience and theoretical considerations show that the high molecular weight polymers likely to meet these criteria are not likely to be very soluble, least of all in the solvents that could be used in the decorative market. It is therefore necessary to use a lower molecular weight material which is soluble and then increase its molecular weight by crosslinking during the drying process. Although there are very many types of crosslinking reactions, only a few are safe or convenient for the purposes of this market. In view of the nature of our atmosphere, oxidation reactions are the most convenient source of crosslinks. Thus, for many years autoxidation of solutions of oil or oil-modified polymers in white spirit have been a mainstay of the paint industry.

The exact mechanisms of oxidation have been the subject of considerable work, but the broad characteristics are explained by the hydroperoxide theory postulated by Farmer in 1942. During the primary stages of oxidation, viscosity increases are small, during which time hydroperoxides are formed on active methylene groups. This is also accompanied by an increase in conjugation due to a free radical mechanism, where detachment of hydrogen from active methylene groups is followed by rearrangement. Secondary oxidation reactions, which are catalysed by metal 'driers' in surface coatings, involve peroxide decomposition followed by recombination forming C—C, C—O—C and C—O—O—C bonds [34].

Unfortunately, the autoxidation of oils leads to unwanted side effects causing odours and yellowing. Furthermore, the autoxidation process is difficult to stop and will eventually lead to embrittlement, with an obviously detrimental effect on durability.

The need for cross-linking can be avoided by using a preformed high molecular weight polymer, but the aforementioned solubility considerations dictate that this should be in dispersion, rather than solution, form, in which case the film formation process is one of coalescence. Not all polymeric dispersions will coalesce, but by a suitable balance of T_g and other properties, film integration is achievable for a useful range of polymers. Such films do not need to crosslink, and remain thermoplastic and tough. According to Dillon *et al.* [35] the driving force for coalescence is surface tension followed by viscous flow of the polymer particles. Brown stresses the role of capillary pressure [36]. Subsequent work has combined features of both theories [37].

Although films cast from dispersion may look completely integrated to the naked eye, microscopic examination will often reveal the original dispersed state, and the films are actually heterogeneous with water-soluble material in pockets or as a network throughout. This influences their properties. Permeability, for example, is usually higher than that of a film cast from solution, and will, in any case, be higher for uncrosslinked films.

The implications of the above discussion are summarised in Table 9.2 which

Table 9.2 — Some formulation 'boxes' available to the decorative coatings market.

	Aqueous (with minor co-solvent)	Non-aqueous (especially white spirit)
Polymer dispersion, film forming through coalescence only	I 'emulsion' or latex paints	II Non-aqueous dispersions 'NAD'
Polymer/oligomer, film forming through autoxidation	IV solubilized alkyds	III 'oil' or alkyd paints

defines four major formulation boxes.

9.6.2 Matching technology to market needs

In practice a high percentage of decorative paints fall into box I (Table 9.2) — emulsion or latex paints — and box III — oil or alkyd paints; and in many ways their properties are diametrically opposed, arising from the three main underlying causes described above. Some of the more general consequences are summarized in Table 9.3. Clearly there is no universally overwhelming advantage to either type,

Table 9.3 — Generalized differences between waterborne and solventborne decorative coatings

Main cause of property difference	Waterborne coatings ('emulsion' paints)		Solventborne coatings (oil or alkyd paints)	
	Positive traits	Negative traits	Positive traits	Negative traits
Nature of thinner: aqueous or non-aqueous?	Non-flammable	Poor early shower resistance.	Films resist water at an early stage.	Flammable
	Easy clean-up	Freeze–thaw stability problems		Needs special thinner
	Quick-drying	Poor dry in cold, damp conditions	No low-temperature storage or can corrosion problems	
	Low primary odour	Rusting of ferrous fittings		High primary odour
	Low toxicity	Can corrosion	Less grain-raising	Relatively low irritancy threshold
	Cheap, readily available	Grain-raising	Longer open time	
	Thinner	Poor wetting of linseed oil putty	Compatible with linseed oil putty	Usually overnight recoat
	Same-day recoat	Prone to biodegradation		
Physical state: dispersion or solution?		Relatively high permeability	Relatively low permeability	
	Easy application	Poor lapping	Good flow	Stickier application
	Good edge cover	Poorer flow	Good lapping	
		Lower solids (build)	Higher solids (build)	
		Diffult to achieve full gloss	High gloss possible	
		Forms crust on container rim	Good penetration	
Chemical state: thermoplastic or crosslinking?	Non-yellowing	Dirt pick-up	Harder film	Yellowing tendency
	Little embrittlement on ageing, can be very flexible	Blocking	Easier to clean	Will embrittle with age
	Retains initial sheen	Transparent to UV	Less prone to blocking	Secondary odours
	No secondary odour			May skin in part-full can

and in many market sectors both aqueous and non-aqueous coatings coexist. Selection of either waterborne or solventborne technology will depend on the intended application and the required appearance, tempered with marketing, economic, and legislative aspects.

Consider, for example, the choice of technology for interior matt, satin, and gloss paints in the UK. Matt paints are used on large areas such as walls and ceilings, where the easy application, quick dry, and lower odour of waterborne paints are major advantages. The consequences of poorer flow and lapping are not so readily visible in matt paints, and so are not seen as serious disadvantages. With silk (satin and eggshell) finishes the consequences of any flow deficiencies are more visible, but the standard of appearance is usually acceptable on wall areas. On trim (e.g. skirting boards) and doors, however, a higher standard is required, and more use is made of solvent borne paints. This is also true in kitchens and bathrooms, where condensation can cause problems. Mid-sheen paints in the trade market also retain a sizeable solventborne element where high resistance to wear and a need for cleaning is required, e.g. in corridors of schools, hospitals, and factories; so the picture for mid-sheen paints in the UK is of a split between waterborne and solventborne, with waterborne continuing to gain ground as the technology improves.

For exterior gloss paints, however, the bulk of the market has remained with solventborne paint. Here the requirement of a very high gloss and the need, therefore, for good flow, has so far outweighed the disadvantages of drying odours and yellowing. For external gloss paints used on windows, solventborne paints are perceived as having other advantages which include greater adverse weather (during application) tolerance, and less blocking or adverse interaction with linseed oil putty. This perception of the needs of an external wood paint is clearly different in countries with different climatic and/or constructional methods. Countries with a high percentage of wood cladding are more likely to have moved to waterborne exterior paints, on the basis of durability and ease of maintenance (see also section 9.9.6).

9.6.3 Future development

From the preceding discussion, it is clear that to interpret the expected properties of a given composition into their market potential is by no means a simple matter. It is, however, always useful, when considering a new formulation, to analyse the likely consequences of change with reference to an established composition in terms of any plus's and minus's. Breaking down properties in terms of physical and chemical causes is often a useful starting point, and although Table 9.2 is a considerable oversimplification, it may still be used to indicate certain broad formulating possibilities.

Turning first to the two remaining boxes, II and IV, it is pertinent to ask why these areas are relatively under-utilized and whether they offer any potential advantages over established technology?

The area represented by box II is usually referred to as a non-aqueous dispersion (NAD) [38]. To a first approximation, the polymeric dispersions in this box may be considered as solventborne counterparts to the waterborne dispersion (latices) of box I, though there are numerous differences of detail in the modes of manufacture and stabilization. NAD technology was later than aqueous dispersion technology in commercial development. In comparing two broadly similar polymers in either aqueous or non-aqueous media, it might seem inevitable that aqueous would be preferred. Why change to a flammable and more expensive solvent? However, there are certain advantages, such as a wider range of evaporation rates, new formulating possibilities in terms of compatibility with other plasticizers and resins, increased weather tolerance, resistance to rusting, etc. Clearly Box II has a balance of properties which is different from I and III, and combines some elements of both. Analysis of the expected property balance might suggest that the main area of application would be outside (i.e. to take advantage of early shower resistance) where a quick-drying flexible coating was required. Possibilities include coatings for masonry and wood, and developments in this area may yet be expected [39].

The remaining area of Table 9.2 — box IV — would include oils and alkyds which have been solubilized in water. This may be achieved by incorporating either acidic or basic functional groups into the alkyd (or other polymer) backbone and then incorporating the appropriate neutralizing agent to produce a water-soluble 'soap'. One method is to produce a maleic adduct which can be neutralized with ammonia or amine. Trimellitic anhydride is also widely used to confer water-solubility. The choice of neutralizing agent is critical and can affect viscosity, drying stability, and film properties [40].

A problem with water-soluble polymers of this type is hydrolytic stability. The vulnerability of ester linkage to hydrolytic attack has reduced their usefulness in a market requiring very long shelf lives. Furthermore, the odour of neutralizing ammonia is often unacceptable. In general, in moving from box I to IV, any advantages gained in, say, flow properties, have been undermined by picking up many of the disadvantages of box III (yellowing, odour, etc.) plus some extra ones, as just mentioned. Consequently, formulation of general-purpose waterborne decorative paints has tended to stay firmly based in the waterborne dispersion box. However, continued development may change this, and increasingly there is interest in the combination of solution polymers with dispersion polymers in both aqueous and non-aqueous systems. Hybrid products of this sort can be interpolated into Table 9.2, and may show a new advantageous property balance, but this is difficult to predict and can be ascertained only from practical measurement. Some water-soluble polymers, for example, show anomalous viscosity/dilution behaviour resulting from the formation of swollen polymer aggregates which can be turned to advantage.

Combination of aqueous dispersions and solutions are under investigation as a way of improving flow in gloss system. Hybrid systems have also found applicability in exterior coatings on wood. Here the role of the alkyd component is to improve adhesion to chalky surfaces, control permeability, and improve penetration to bare wood. The alkyd component may be either a solution or a true alkyd emulsion. (Alkyd emulsion in water would have a property set intermediate between boxes I and IV.)

9.7   IN-STORE TINTING SYSTEMS

9.7.1   General
In most countries a significant proportion of decorative paint is white, followed in order of popularity by pastel colours and the stronger darker colours, in order of rapidly diminishing volume. Naturally, there is a national variation, and there will

also be differences reflecting the intended application. In the UK black and deeper colours are more popular for exterior painting of, for example, doors and garages. All countries face the problem of balancing the limited demand for specific colours against the level of service which is required to sustain the credibility of a major manufacturer. The problem is exacerbated if, as is normal, there are a number of product lines to support.

A solution to this problem is to strike a balance between ready-mixed coloured paint (i.e. a tint, co-grind, or blend made in the factory) and that which is tinted at, or near, the point of sale. Essentially, a tinting system comprises a device for dispensing a tinter or colourant in quantized units to paint bases which are then mixed, normally by shaking.

Tinting systems are widely distributed, but it is interesting to note that their development and exploitation has occurred differently around the world. Currently, for example, the bulk of coloured paint in North America, Australia, and Scandinavian countries is delivered via tinting systems, whereas in the UK the established practice has been to have a limited range of popular colours in ready-mix form supplying the bulk of the volume, which is supplemented by a tinting system which provides less than 10% of the total volume.

There are no simple reasons for the differences found between countries, which reflect geographical, economic, and social differences. For example, the UK has a high population density which enables a daily delivery service for centralized warehouses to be viable in a way which is less practical in a situation where the consumers are widely dispersed. It is also possible that the demand for a very wide colour choice has been slower to develop in the UK. Another point to note is that the fashion for a high-gloss paint in the UK is more at odds with the technical constraints of tinting both waterborne and solventborne paints with common tinters, without some loss of quality.

Whatever the reason for past differences, it is certain that a period of renewed technical change has been entered as the possibilities offered by microprocessors and computers are fully exploited. The perceived benefits in new technology are, to some extent, relative to the starting point and to the marketing operation and outlet size. Tinting systems require a high capital investment, and change is very expensive. Paradoxically, it may be more difficult to justify the benefits of new technology in situations where tinting is the established method of colour delivery, than in a situation where the potential benefits have been under-exploited.

The changing pattern in retailing must also be considered. Basic manual tinting systems require the assistance of a trained operative, and have an operating cycle which will take 5–10 minutes. This will not be sufficient to meet peak demand in a busy 'superstore', though at other times the machine will be lying idle. Operating cycles can be shortened by various degrees of automation, though at an increasing cost penalty, but even so, the degree of automation necessary for user-friendly self-selection has yet to be approached, and may never be economically justified. Outlets operating some sort of tinting scheme will face an increasingly difficult choice of competing options with a need to balance savings in shelf space and reduced stockholding, against capital and running costs, and the intangible 'halo' effect of a visibly modern system.

Decorative paint tinting systems differ considerably from those used in the automotive refinishing market. The latter will normally deal with a much wider colour range which must include very deep colours and metallics, and also cover technologies ranging from nitrocellulose to 2-pack isocyanates. Consequently, refinish systems have developed as intermixing schemes; that is the operative starts with an empty container into which is combined (usually by weighing) pigmented bases and blending clears. By contrast, decorative systems start with a container which already contains the bulk of the paint which is coloured by the volumetric addition of tinters. A problem facing decorative systems is to control the final volume of paint. Tinter additions are usually limited to a maximum of around 7%, and one solution is to fill base containers to 95% of their nominal size, which must be labelled accordingly. Consequently, the paint as purchased may contain between 96% and 102% of the nominal size, presenting some costing problems and an inconsistency in comparison with ready-mixed paint. An alternative solution is to aim for a 'full fill' by adjusting the volume of tinter to, say 5%, with 95% base. Such an approach further constrains the number of colour hues that can be achieved, and if it is not to be too restrictive, usually requires some increase in the number of bases and tinters.

9.7.2 The dispensing machine

With few exceptions the heart of any tinting scheme is the dispensing machine, which must accurately dispense known volumes of tinter. Such machines can be divided into manual, semi-automatic and automatic groupings. Paradoxically the development of automatic machines tended to precede that of semi-automatics, and in any case, the boundaries between the two have become somewhat blurred [41].

All machines must contain a reservoir capable of holding several litres of tinter and usually a mechanism — manual or automatic — to provide periodic stirring. Typically, tinter is dispensed through a piston pump, with volume being controlled by the length of stroke. In manual machines a common arrangement effectively connects the piston rod to a notched scale which is graduated in fractions of a fluid ounce (a measure which varies from country to country). A common system in the UK is graduated on $\frac{1}{96}$ fluid ounce increments. The persistence of archaic or hybrid units reflects the difficulty and cost of changing either machine or formula once a system has become established. This method of dispensing imposes a scale of quantized colour differences which will be modified by tinter and base strength. With automatic machines, electronic control of stepper motors can give a continuous range of colour additions, but there will be practical constraints imposed by the minimum drop volume that can be controlled.

The arrangement of tinter reservoirs in a machine varies between manufacturers. One of the more common is to use a carousel allowing each reservoir to be rotated to a common dispense point. Alternative arrangements in banks may require moving the paint container. Carousel layouts are usually combined with a sequential tinting sequence; with other layouts a simultaneous dispense may be optional. This speeds up part of the overall cycle, but long pipe runs can be troublesome if a blockage occurs.

In a typical manual tinting sequence, the operative must look up the recipe in a file which might be in printed card or microfiche format. Having selected the correct paint base, this must be opened and placed at the tinting station where the various operations to set and activate pumps are carried out. With many colours requiring addition of at least three separate tinters, there is considerable scope for operator error! After dispensing, the container must be resealed and taken to a suitable mixing machine. The co-called 'fully' automatic machines, of which there are several variants on the market, will normally have an electronic recipe bank, and through electric motors the ability to carry out the dispense cycle automatically. Operatives, usually prompted by the machine, must select the correct base type; but many machines have the ability to confirm that the correct size in relation to the selected recipe is in position.

Frequently, tinting is carried out through a hole punched in the lid at the beginning of the tint cycle, which is later sealed by a plastic plug. Using either internal or externally linked computers it is possible to print labels and other product information at the same time. Despite their advantages automatic machines are more than an order of magnitude more expensive than their manual counterparts, and therefore not cost-effective for small or medium-size outlets. Semi-automatic machines have evolved which bridge the gap between the two extremes. Such machines differ considerably in detail but often include features such as an electrically driven pump and tint sequence after the recipe has been entered manually. This is an advantage for small multiple orders.

In terms of future development it is clear that the continued price fall in real terms of microchip technology, will provide opportunities for more sophisticated control systems which, as well as providing unlimited recipes, will provide facilities for other activities. Possibilities include stock control and a databank to provide product information. There are already commercially available packages [42] in which a spectrophotometer is linked with an automatic tinting system, providing a colour-matching service. But many customers are looking for complementary colour schemes rather than an exact match, and here too there are possibilities as the field of colour graphics advances. To fully utilize these advances will require strong tinters and dispensing pumps of very high accuracy.

Incorporating all the features mentioned above will increase capital costs, probably putting them out of reach of smaller outlets.

Advances in dispensing technology will also increase the pressure for a corresponding improvement in paint-mixing which, for automatic and semi-automatic machines, is already the rate-determining step. The difficulties of cleanly mixing a full can in the confines of a shop requires a non-intrusive mode of mixing. Machines based on either a shaker or some biaxial rotation mode are widely used, and to date experiments with a magnetic, ultrasonic, or vibrational mixing have proved impracticable. Improved engineering has, and will, continue to lead to quicker and quieter mixing. Some machines increase throughput by a capability for multiple loading, which may be combined with automatic hydraulic clamping. The energetic nature of the mixing process creates formidable design problems in combining the mixer with the dispenser.

For completeness, it should be noted that although dispensing machines are integral to nearly all tinting schemes, there are some non-machine alternatives. A possibility used in some stores is to use graduated syringes containing tinter which customers use to create their own colours. Clearly this gives only a limited choice and reproducibility. A commercially successful alternative used in the USA is based on small plastic cartons containing a powder tinter which is tipped into paint bases at the point of sale [43]. Less successful commercially was a system introduced in Germany in which pre-weighed quantities of tinter were sealed in water-soluble films protected in transit by an outer blister pack. Although such systems can, in principle, produce a reasonable colour range, there is a practical limit to the number of sizes and colours that can be stocked, and such systems cannot offer the wide colour choice offered by the machine routes [44].

9.7.3 Tinters and bases

9.7.3.1 Colour and compatibility

The demands placed upon tinters used 'in-store' are similar in many ways to those upon tinters in general, but there are some additional constraints. Of these, one of the most difficult concerns the question of compatibility. Factory tinters are often dedicated to a single product line, and even where multipurpose tinters are used they will be used in product groups of broadly similar solvency. In the decorative paint-tinting market, cost and space consideration have dictated common tinters for both water and hydrocarbon thinned products. This almost impossible demand has led to compromises, and product quality will be affected, especially at high levels of use, leading to differences between 'ready-mixed' and 'tinted' versions of the same colour. The balance between solvent and water paints is continually changing in favour of the latter, and it will eventually be possible to service the market with water-based systems alone. Even so, the requirement of different product groups will not be easily met.

The most common formulating basis for broad spectrum tinters is ethylene glycol with various surfactants. Such compositions can retard dry (especially in aklyds) and increase the water-sensitivity of waterborne paints. In countries such as the UK, where a structured paint is preferred, the effect of tinters on the viscosity profile is more marked, a problem further exacerbated by mechanical agitation. Minimizing these effects is a further reason for limiting the maximum tinter addition. Some tinter manufacturers have reduced the problems of adverse tinter reaction by producing a new generation of 'low glycol' tinters [45]. While generally successful, they are more prone to caking and may require dispense modification to prevent the nozzles becoming blocked.

An important consideration when designing a tinting system concerns the number of tinters and bases, and their hue tone and strength. In an ideal world it would be possible to have one colourless base per product line and five very strong tinters in white, black, and the three primary colours. In reality there are practical and economic reasons why this cannot be so. The effect of pigments on hue is a complex combination of subtractive and additive mixing, and there are no pure 'primary' pigments. Even so, a surprising number of colours can be produced from, say, phthalocyanine blue, azo red, and arylamide yellow (plus white and black), but there remain large areas of colour space which are inaccessible. Adding further tinters increases the range, but a point of diminishing returns is reached where the addition of each new tinter introduces so few new colours that the increased cost of maintaining another reservoir is not justified.

Pigments in tinters must also be chosen on the basis of economy, opacity, and durability, and there are likely to be common features between the ranges used by different manufacturers.

A typical core of tinters would include black, white, organic red, organic yellow, inorganic red, inorganic yellow, and an organic blue. The choice of further colours begins to diverge, and will depend on the selection of bases available, fashion trends, and constraints inherited from the recent past (bearing in mind the cost of changing an established range). Additional tinters will be chosen for a group which includes violet, orange, green, brown, and other hues of organic yellow and red. (See Chapter 3 for details of pigment types used in the decorative market.)

Another important consideration is tinter strength (which will be closely related to pigment concentration). For many purposes it would be true to say the higher the pigment concentration, the better. Within the constraints of a maximum addition of 5–10%, it is necessary to have high pigment concentration tinters in order to achieve the maximum number of deep colours. Increasingly, manufacturers are meeting this range with high-strength tinters; even so, it is not possible to reach all the deeper colours. A disadvantage of high- and, in some cases even normal-strength tinters, is that they must be dispensed with a high accuracy if pale shades are to be reproducibly achieved. Such accuracy is outside the reach of most machines and inherently strong tinters — such as phthalocyanine blue — are usually held in reduced as well as full strength form. For a given level of accuracy the addition of stronger tinter might have to be matched with further reduced tinters, thus placing further demands on the limited amount of reservoir space.

To exploit fully a given set of tinters requires a corresponding set of bases, but as with tinters a point of diminishing returns is reached. All systems will include around 3–5 bases which are in effect a TiO_2 *PVC* ladder covering a range from zero to an upper TiO_2 level typical of ready-mixed white paint, thus enabling deep, medium, and light tints to be produced. To widen further the availability of colours requires the addition of coloured bases, with red and yellow being the most common, although others are used, including some mixed pigment bases. Since each new base must be held in a variety of paint sizes and product types, there is a disincentive in introducing too many new ones, and generally there is a trend away from coloured bases. A major benefit of strong tinters is a reduction in the need for coloured bases.

9.7.3.2 Consistency and reproducibility

Once a paint has been tinted, no further adjustment is practical, and the system is required to give a close and reproducible match to colour cards at point of sale displays. This, in turn, requires a higher level of standardization in both tinter and base, which must be accurately strength controlled. Because tinters are usually dispensed by volume, it is important that their specific gravity is known and stable; de-aeration equipment is often used during manufacture.

As well as showing reproducible strength from batch to batch, it is highly desirable that tinters produce the same hue in different product bases, thus enabling the same recipe to be used for different products. Where this is not so, the size of the recipe bank is increased, with additional possibilities for error in manual system. In practice, it is difficult to achieve identical hues in different media, and it is not unusual for waterborne and solventborne paints to require separate recipes. Once the composition of a tinter has become established it becomes very difficult to substitute alternative pigments since alternatives will often show a different pattern of colour development across a product range.

Finally, it is noted that tinters must also have satisfactory rheological properties with sufficient fluidity to pass easily through nozzles and tubes. At the same time they must neither cake nor settle though most dispensing machines include a stirring cycle. The effect of the tinting medium on any materials of construction should also be taken into account. Aluminium, for example, is corroded by glycol.

9.8 MEETING THE NEEDS OF THE SUBSTRATE

Decorative coatings can sometimes be chosen with little regard for the underlying substrate, but often it will be the substrate which has a major influence in deciding their suitability. Choice of a coating may be dictated by the need for a specific protective function, or it may arise from potential interaction between substrate and coating. In some cases these needs will require a totally dedicated system, while in others the needs of the substrate may be met by specific primers which are subsequently overcoated with coatings of a more general-purpose nature.

It is convenient to divide the substrates for building paints into three broad categories, namely wood, metals, and masonry. Organic plastic represents a small but growing sector of the building market which may require more consideration in the future, as plastic components become weathered and discoloured.

In redecoration it is seldom possible to ignore totally the substrate, but it will also be necessary to take the nature and condition of previous coatings into consideration. For example, bitumen coatings and some preservative pretreatments on wood can present problems of bleeding and discoloration when overcoated, while strong solvents may lift previous coatings. A build-up of film thickness will change some properties, including permeability.

In addition to meeting the needs of the substrate, a coating must also resist climatic and environmental influences, and these may interact with the substrate. Thus it is often desirable to formulate interior and exterior paints differently; this partly arises from the direct effect of weather on the coating; also, the need to protect the substrate is different internally and externally. Each of these has, or creates, characteristic problems in service which require specific properties from the coating. For example, wood must contend with biodegradation and moisture movement; concrete is a highly alkaline surface; and metals are subject to corrosion. In each of these processes water plays a major role, and *it would be true to say that one of the prime purposes of the coating is to control some aspect of a process in which water plays an essential part* — hence the importance of permeability. However, the permeability required for each coating type will be significantly different.

In the remainder of this chapter particular attention is paid to the nature of the substrate, the influences which will cause deterioration, and some of the consequences this has for the design and development of surface coatings in the decorative market sector.

9.9 WOOD BASED SUBSTRATES

9.9.1 Introduction
Wood is widely used in buildings for roof trusses, timber frames, and joists, and non-structurally in doors, window frames, cladding, and fencing. This widespread use reflects wood's numerous attractive properties which include ease of working and high strength and stiffness-to-weight ratios. It is also an attractive material, and when properly husbanded is naturally renewable; but wood is vulnerable to a variety of extraneous degrading influences which include light, moisture, and biological attack. Surface coatings have a major role to play in preventing or reducing the consequences of these influences, but must be considered in conjunction with design and preservation.

9.9.2 Characteristics of wood
Wood is a naturally-occurring composite material with a complex structure which occurs over a wide range of magnification. It is necessary to understand something of this complexity in order to select or design the most appropriate surface coating [46]. The properties of wood will differ greatly from one species to another, but it is often useful to divide wood into two broad groupings — hardwood and softwood. This is a botanical distinction; by convention the wood from coniferous trees is known as softwood, and that from broadleafed trees as hardwood. A majority of hardwoods are indeed harder than softwoods, but this is not always so. Balsawood, for example, is defined as hardwood, despite its soft physical character. Hardwoods and softwoods show significant differences in their cell structure.

9.9.2.1 Macroscopic features
A slice taken across the trunk of a tree shows a number of characteristics which are apparent even to the naked eye. Beneath the outer and inner bark is a thin layer of active cells (cambium) which appear wet if exposed during the growing season. It is here that growth takes place. In spring, or when growth begins, the cells are thin-walled and designed for conduction, whereas later in the year they become thicker, with emphasis on support. The combined growth forms and annual growth ring which is usually divided into two distinct areas known as springwood (or earlywood) and summerwood (or latewood). Each type has different properties, and this can lead, for example, to differential movement and is a potential problem to coatings.

When viewing a transverse wood section it is often noticeable that the central area is darker than the outer circumference. This is a characteristic difference between heartwood and sapwood. Sapwood is the part of the wood, the outer growth rings, which is actively carrying or storing nutrient. Young trees have only sapwood, but as the tree matures the cells in the centre die and become a receptacle for waste matter including tannin and gum. Once it starts to form, the growth of heartwood keeps pace with the growth of sapwood, and the ratio of heartwood to sapwood gradually increases. Some heartwoods are distinctly coloured, and they all show important differences from sapwood. Heartwood is less porous, shows greater dimensional stability to moisture movement, and generally shows greater resistance to fungal and insect attack.

The weathering and durability of coatings is greatly influenced by the consequence of the differences mentioned above. When comparing the performance of different coatings, it is not unusual to find that the difference between coatings is masked by the variability of the substrate, and this must be taken into account in the experimental design, in that the wood itself must be treated as a variable.

Because wood is anisotropic, there is the further complication of the way in which the wood is cut. Transverse, tangential, and radial sections each present a different alignment of cells, resulting in the familiar grain patterns. The two main types of cut are known as plain sawn and quarter sawn, but a commercial through and through cut will produce both plain and quarter sawn timber. Quarter sawn boards are less inclined to shrink in their width and give more even wear. The weathering of many paints is better on quarter, rather than plain, sawn timber.

Another macroscopic feature of timber which may interact with the coating, is the presence of knots which arise from junctions between trunk and branch. Knots may be considered a decorative feature, but they can be an unwanted source of resins and stains when coated.

9.9.2.2 Microscopic features
At low magnification the cellular structure of wood is readily discernible. A majority of the cells align vertically with a small percentage horizontal in bands or rays, giving a characteristic 'figure' to quarter sawn wood. The functions of cells include storage, support, and conduction. In softwoods these are carried out by two types of cell (trachieds and parenchyma); hardwoods have a more complex structure, with four main types. A characteristic feature of hardwood is the presence of vessels (e.g. the pores in oak). Vessels can absorb large quantities of paint and can transport water behind the coating. Individual cell walls are made up from cellulose microfibrils which form a series of layers as the cell grows. Fluid moves between cells via tiny pits which occur in matched pairs. Softwood pits may close after seasoning, making it more difficult to force preservatives into the wood.

9.9.2.3 Molecular features
Wood contains 40%–50% cellulose which is in the form of filaments or chains built up in the cell walls from glucose units. Amorphous regions of cellulose are hygroscopic. Cellulosic components are embedded in a matrix of hemicelluloses and lignin, the latter a three-dimensional polymer built up from phenyl propanol blocks. Lignocellulose is largely inert to coatings, but wood may contain up to 10% of extractives, some of which are highly coloured and chemically active. They can cause staining or loss of dry in coatings. Examples include stilbenes, tannin, and lignans.

9.3.3 Causes of wood degradation
Wood is affected by moisture, light, and living organisms, often acting simultaneously, with harmful consequences for both substrate and coating.

9.9.3.1 Moisture
The moisture content of wood is conventionally expressed as a percentage based on the weight of dry wood. When newly felled, timber can contain up to 200% moisture contained in cell cavities and walls. As wood is dried or seasoned, water is lost from the cell cavities, with no corresponding change in volume. Eventually the point is

reached where cavities contain no liquid water, and remaining water is held in the cell walls; this is known as the fibre saturation point which occurs around 30% moisture content; further loss causes dimensional change. Wood will establish an equilibrium moisture content dependant on the temperature and humidity of the surrounding air.

It is customary to describe the changes in dimension which occur in drying timber from the green state as *shrinkage*, while changes which occur in seasoned wood in response to seasonal or daily fluctuations are known as *movement*. Typical shrinkage values are 0.1% longitudinally and up to 10% tangentially; radial movement is around half that of the tangential. By convention, movement is reported in the UK as a percentage change occurring beteeen 90% and 60% RH at 25°C. Tangential movement can be up to 5%, but differs considerably between species.

In service, timber is subjected to a constantly fluctuating environment; this includes changes in humidity internally and externally, and the effects of both rainwater and condensation. This will lead to movement, and when considering the applicability of specific coating types it is useful to divide wooden components into those such as joinery where dimensional stability requires control, and those like fencing, where movement is less critical. The rate at which movement occurs depends on the permeability of the coating, and this is an important factor to be considered when designing a wood-coating system.

9.9.3.2 Effect of water on coating performance
Although coatings can influence the rate of movement, they cannot physically restrain it. Clearly a wood coating must have sufficient extensibility to expand and contract with the wood and adequate adhesion to resist interfacial stress between substrate and coating. Coatings with a low modulus of elasticity will generate less interfacial stress for a given movement than those with a high modulus.

Water acts as a plasticizer when absorbed by a coating, and although this increases extensibility it will also aid removal by peeling. Furthermore, it is possible for low molecular weight soluble material to diffuse to the interface where it concentrates, giving a weak boundary layer. While adhesion is low the coating will be vulnerable to blistering, especially if vapour pressure is high and permeability low. Blisters are caused by a loss of adhesion combined with sufficient flexibility, but there must also be a driving force which might be vapour pressure, swelling, osmotic pressure, or resin exudation. Some polymers, for example those containing ester linkages, are prone to hydrolysis and are best avoided in permanently damp areas.

9.9.3.3 Sunlight
Solar radiation causes wood to undergo physical and chemical reactions which are largely confined to the surface. Many of these reactions also require the presence of water and oxygen. In dry conditions unfinished wood tends to turn brown, but becomes progressively grey on normal wet weathering. Such changes reflect radiation-catalysed reactions which involve oxidation, depolymerization, and general breakdown of lignocellulose. Free radicals have been detected in irradiated wood. Breakdown products are soluble and are leached out by water, leaving a grey denatured surface. Although the high energy of UV radiation is responsible for much

of this damage, recent work by Derbyshire & Miller [47] has underlined a significant contribution from ordinary visible light. The implications of this are serious, for it means that the interface between any transparent coating and wood may be prone to photodegradation, undoubtedly a factor contributing to the tendency of conventional varnishes towards flaking. Problems may also arise with opaque coatings if they are applied to an already degraded surface.

In considering the effect of sunlight on wood-coating interactions the role of infrared, i.e. heat, should not be neglected. Dark coatings will reach temperatures up to 40°C higher than white ones, and this may exacerbate problems such as resin exudation.

9.9.3.4 Biodegradation
Wood is vulnerable to attack by bacteria, fungi, insects, and marine borers.

Bacteria play little obvious part in the decomposition of wood, though their presence is sometimes a condition for more serious attack. Wood which is stored in water, or ponded, can be come very permeable as a result of bacterial attack, and this may affect the uptake of water and preservatives with consequent influence on paint properties. Fungal attack is more serious and can cause both disfigurement or structural damage, according to the species. Softwoods are particularly prone to a blue staining which is caused by several types of fungi, including *Aureobasidium pullulans*. 'Blue stain in service' is caused by further infection arising from colonization of surfaces during or after manufacture, and is not the final stage of growth of blue stain fungi originally present in the timber.

In addition to the disfiguring surface moulds and blue stain fungi, there are a substantial number of fungal species which cause serious structural damage by a process which, in its advanced stages, is known as rot or decay.

Soft rots including fungi from the Ascomycotina and Deutermycotina groups are limited in their growth by the availability of nitrogenous nutrient, but degraded surface layers can have an effect on coating performance. More serious in their effect on structural properties are the fungi which destroy wood. Most are from the Basidiomycetes group. They attack both lignin and cellulose and are less restricted by nitrogen availability. This group includes 'wet' rot (*Coniophora puteana*) and 'dry' rot (*Serpula lacrymans*).

Decay in wood can be prevented by keeping the moisture content below about 22%, and by suitable preservatives. Coatings have an important role to play in controlling moisture uptake, but are counterproductive if moisture is trapped. This has furthered the interest in coatings with a higher vapour permeability than traditional oil-based systems.

9.9.4 Defensive measures
Although this discussion is concerned primarily with coatings, it is important to stress that in isolation coatings cannot ensure maximum life for timber components. This can only be achieved by an alliance between good design, appropriate preservative, and the most suitable protective coating.

9.9.4.1   Design aspects
Good design includes selection of the appropriate timber, attention to detailing, and the adoption of good site practices. It is outside the scope of this book to consider design aspects in detail, but a few pointers to some of the more important aspects are appropriate.

Joinery
Design considerations with serious implications, from the point of view of decay, include the nature of joints, profiles, and the way the wood is handled and fixed on site. Water is most likely to enter at joints and exposed end grain. Moisture uptake through end grain, that is through transverse sections, is very much greater than through other faces, Work carried out at the Building Research Establishment and elsewhere has shown that the moisture content of uncoated wood with sealed end grain is, on average, lower than that of coated wood with unsealed end grain. This arises because joint movement can break the protective coating seal, allowing access of water which cannot readily escape. It underlines the need for joints and any exposed end grain to be sealed using boil-resistant glues and an impermeable end-grain sealer.

Flat surfaces allow water to collect against joints and window profiles, and must be designed with a run-off angle of 10–20°. Sharp corners are a frequent source of coating failure, and should be rounded. The water resistance of windows is greatly improved by weathersealing using, for example, Neoprene or PVC weatherstrips. Where timber is in contact with brick or blockwork, a damp-proof course should be provided [48].

Cladding
Properly designed timber cladding should make provision for both movement and ventilation. Failure to allow for movement leads to warping, while inadequate ventilation can lead to dangerously high moisture levels being reached. Trapped water will support decay, and has a generally disruptive effect on film-forming coatings [49].

9.9.4.2   Glazing
Paint and other coating failures are often localized at the joint between timber and glass, underlining the care that must be taken to ensure that coating and glazing methods are compatible. A wide variety of glazing combinations is available, but for wooden windows the glass normally sits in a rebate, with or without a bead. The simplest and best established method of glazing low-rise buildings which are to be painted, is to use linsed oil putty which, when properly painted, gives remarkably good sevice, but putty glazing is not suitable for frames that are to be stained, varnished, or coated with waterborne paints. The apearance will be marred and the putty will not be protected, leading to early failure. Bead glazing is generally recommended instead, but compatibilty of coating with sealant must still be considered (BS 6262).

9.9.5   Preservation of timber

Timber species vary in their resistance to decay, with nearly all sapwood vulnerable. Heartwood of some species, such as oak and teak is very resistant, but ash and beech have little resistance. A widely accepted classification divides timber into five grades in increments of five years [50]. Preservation should be considered for all sapwoods and non-durable heartwoods, where the equilibrium moisture content is likely to rise above 20%. Such a situation is likely where ventilation is poor, where the timber is in ground contact, and where design features allow contact with water. BS 5268 provides guidance on categories of hazard in relation to preservation. Preservation is essential where insect or fungal attack is endemic.

The three main classes of preservative are tar oils, waterborne preservatives, and organic solvent preservatives.

Tar oils are typified by creosote which may be derived from coal or wood distillation. Timber treated with creosote is not suitable for painting but can be stained, preferably after a period of weathering.

Waterborne preservatives are frequently based on copper, chrome, and arsenic compounds, with sodium dichromate as a fixative. An alternative type employs disodium octaborate which is not fixed and must be protected during and after installation.

Organic solvent preservatives include pentachlorophenol, tributyl, tin oxide, and copper and zinc naphthenates as the active ingredient. Additives include waxes, oil, and resins.

Preservatives are applied by a variety of methods which include brushing, spraying, dipping, and — more effective — double vacuum or vacuum pressure processes (see, for example, BS 5589 and BS 1282).

Although most preservative manufacturers attend to compatibility problems between treated wood and paints, glues, or glazing compounds, it would clearly be unwise to assume compatibility in every case. Formulating chemists should be alert to possible problems of intercoat adhesion, especially of waterborne coatings over water-repellent preservatives. Other compounds, including some of the copper-based ones, can inhibit autoxidation.

9.9.6   Coatings for wood

Coatings for wood are conveniently described by three archetypal terms — paint, varnish, and stain. These terms cover some expectation of appearance and performance, but are not exact. They offer the user a choice of appearance, and while individual products vary, each has certain inherent advantages and disadvantages.

In the UK, until comparatively recent times, the terms paint, varnish, and stain for external use, unless otherwise qualified, would probably be interpreted as implying:

Paint — An oil or alkyd based (solventborne) opaque paint system comprising primer, undercoat and a glossy topcoat.

Varnish — A solventborne, transparent, clear glossy coat.

Stain — A low-solids penetrating composition containing preservative — semi-transparent.

Of these three types the third group was the last to find widespread use in the UK, influenced by growth in North America, Scandinavia, and continental Europe. Clearly, each of the three product types offers a different appearance, and might be chosen for purely aesthetic reasons. However, a major reason for the growth in exterior woodstain was an expectation of easier maintenance. To many users, though, it became increasingly clear that each group had characteristic disadvantages. Thus:

Paint — Two common criticisms of paint were (i) first, that its relatively low moisture permeability meant that water could become trapped, leading to decay beneath a perfectly sound coating; (ii) a second major failing of traditional paint systems was a propensity towards cracking or flaking — the oil or alkyd continued to embrittle. This led to severe maintenance problems.

Varnish — Varnishes were often even more prone to flaking than paint. They suffered from the same embrittling problem exacerbated by the effect of light on the wood surface which could weaken the surface and reduce adhesive performance. Being transparent made it more difficult to restore appearance on redecoration.

Stains — Experience in the UK showed that the low-build/low-solids penetrating woodstain did not adequately control moisture movement in joinery. After a relatively short period of weathering, moisture movement caused splitting and loosening of glazing. Smooth-planed cladding absorbed little, and uncontrolled moisture movement caused warping and splitting problems.

Solving these problems has led to both improved products and the the creation of new product types. This, in turn, created marketing opportunities. In many countries there has been a proliferation of products intended for the exterior wood market, and this has often caused confusion in the marketplace. In the UK this has led to pressure for the drafting of a British Standard giving guidance on the classification of exterior wood coatings [51].

To a first approximation, certain differences between stain, varnish, and paint can be illustrated using triangular relationships as discussed earlier. This is shown schematically in Fig. 9.4 [52].

Taking low-solids stains as a reference point, these will be clustered at the top right of the diagram. To improve build and moisture control, products with higher solids have evolved with characteristics midway between conventional stain and coloured varnish. These are often known as 'high solids' stains, though their solid content will usually be less than that of a conventional paint. Another formulating option is to move along the axis between stain and conventional paint. Again, this gives a product of intermediate character whose final properties will obviously depend on the specific ingredients used. This area, more than any other, has been a source of confusion to users. Products have been described variously as opaque stain, Continental paint, breathing, ventilating, microporous, self-priming, as well as a host of proprietary and other descriptive names. In essence they are best thought of

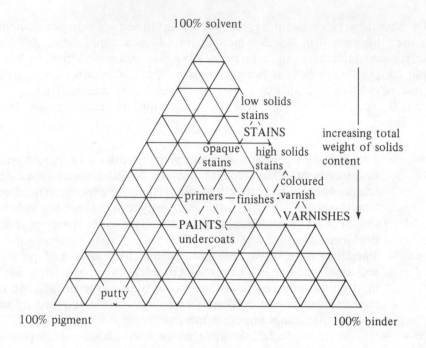

100% solvent

low solids
stains
STAINS   increasing total
                          weight of solids
opaque   high solids   content
stains   stains
          coloured
primers —finishes ·varnish
                    VARNISHES
PAINTS
undercoats

putty

100% pigment   100% binder

Fig. 9.4 — Relationship between wood-coating types according to relative
weight of main components.

as different types of wood paint, just as there will be different types of masonry and
metal coatings.

Designing a wood coating clearly starts with a perceived requirement in terms of
appearance and the projected end use. Decisions must be taken in terms of binder
type, pigmentation, and functional ingredients such as fungicides.

There are no single solutions to the formulating problem raised, but some general
remarks may be addressed to the three main generic classes.

9.9.6.1   Stains
Archetypal stains are designed to penetrate and will generally have a low (<20%)
volume solids. Penetration is usually higher with solventborne compositions,
whereas water causes swelling. Penetration will be greatly retarded by any particulate matter, including pigment. It is important to distinguish between the functions of
a stain and a wood preservative. The latter is designed to prevent decay, and requires
special application methods to ensure deep penetration. Stains should contain
fungicide to inhibit surface colonization, but normal brush application will not give
sufficient penetration to ensure protection against decay. Stains, and indeed all wood
coatings on non-durable wood, should be applied to preservative treated wood.

Penetrating stains can be formulated with little binder, often no more than is
needed to stabilize the pigment. In solventborne products oils and alkyds are often
used, but care must be exercised when formulating at low vehicle solids as there can
be problems of phase separation or oxidative gelling. Waterborne stains can be based
on low volumes of aqueous dispersion polymer. Recent years have seen a prolife-

ration of low-cost waterborne stains, especially for fencing and garden applications. These are often positioned against creosote and have handling benefits with little or no effect on adjacent plant life. Such products are essentially decorative and have little protective or preservative effect.

Low-solids stains may be modified with silicone oils, waxes, etc. to give some, albeit temporary, water repellency, but they exert virtually no control over water vapour. Their use is thus best confined to situations where free movement of the substrate is acceptable. Typical applications might include fencing and some types of cladding, though the latter must be of a design that allows for movement. Good performance on cladding requires a sawn, rather than smooth-planed, surface. As a generalization it would be true to say that virtually all wood coatings, including gloss paint, will perform better on a sawn surface, though naturally the appearance of gloss products will be affected.

In the case of joinery, especially softwood, there *is* a need for moisture control, otherwise constantly fluctuating movement can cause splitting, loosening of glazing, and the jamming of opening lights. On the other hand too low a permeability will allow moisture to become trapped and will increase the chance of decay. Such considerations suggest that the best compromise will be between an upper and lower limit defined by the nature of substrate and end use. Agreement has yet to be reached on what these limits should be. As a guideline, a film permeability between 30 and 60 $g/m^2/24$ hours should be adequate to control movement in joinery, but actual values are very dependent on the method of measurement [53].

Moisture control in woodstain is improved by raising the volume solids, i.e. moving to a high-solids stain, thus effectively giving a thicker film. However, care is needed for a given binder; raising the film thickness could change the failure mode from one of erosion to flaking [52], which defeats the object of using a stain. Binders must be selected with good extensibility and adhesion. Because penetration of higher build stains is reduced it is usually beneficial to pre-treat the wood with a lower-solids product.

Solventborne high-build stains can be successfully formulated using alkyds, bearing in mind the remarks made above. A number of proprietary alkyds have been developed specifically for this purpose [54,55]. In the UK, waterborne *transparent* stains have been less successful. There are three major problems to overcome. Firstly, the much higher permeability creates moisture control problems; secondly, fungal attack is more common; and finally, transparency of acrylic and vinyl polymers to UV radiation makes it more difficult to prevent photodegradation of the surface. It has been found that many of the vinyl and acrylic latices which perform well in paints (i.e. including opaque stains) are not suitable for transparent coatings. Attempts have been made to overcome this problem by incorporating components capable of absorbing UV, and at least one proprietary product is now on the market [56].

Pigmentation of woodstains is relatively straightforward; the demand for 'natural' wood shades makes red and yellow iron oxides particularly useful. The best colour, brightness, and transparency are achieved with synthetic iron oxides which achieve full transparency when ground to colloidal dimensions. However, these are expensive materials, and the standard grades require long dispersion times. Their high surface area is quite active, and some care in selection of dispersant must be exercised in order to achieve long-term stability. Not all alkyds are suitable. When fully dispersed these grades confer exceptionally good UV screening.

The choice of fungicides for woodstains is vast, but their compatibility and long-term storage must always be checked very carefully. For fencing and ground contact applications copper and zinc soaps perform well, and have relatively low mammalian toxicity. Stains and other transparent coatings for wood should include fungicides which are specific against blue stain.

9.9.6.2 Varnish

Exterior varnish usage has declined steadily in favour of woodstains, owing to a high incidence of flaking and discolouration — with consequent high maintenance costs. However, there remains a demand for good-quality varnish to set off the aesthetic qualities of wood. Varnishes also have better wear characteristics in high traffic areas such as doors and door frames.

Problems with conventional varnish arise from the fact that they are high-build films of low permeability, becoming brittle with age. As mentioned earlier the problem is exacerbated by their transparency which allows photodegradation at the wood–varnish interface. Traditionally good-quality varnishes were based on tung and linseed oils modified with phenol formaldehyde resins. Straight alkyds are generally less effective. It is probable that the good performance associated with tung phenolics is associated with their UV absorbing characteristics, and improving this aspect of performance has received considerable attention, notwithstanding the fact that visible light is also detrimental to the substrate. Improvements in performance have been claimed with several types of UV absorber including the benzophenone groups and hindered amine light stabilizers. Benefits have also been noted with inorganic fillers, including pulverized fuel ash [57]. Yellow transparent iron oxide, as used in woodstain, also confers a marked improvement, though clearly the colour will darken. In practice, this may be no worse than the yellowing which occurs with many varnishes on weathering.

In the past it has not been customary to include fungicide in varnishes, but logically they should also contain agents to inhibit blue stain and other fungi. Many fungicides are themselves degraded by UV radiation, so the benefits are seldom realized unless the varnish does contain some protection.

Adhesive performance, and hence resistance to flaking, can be improved by thinning the coat first applied to the wooden surface or by substituting a low-solids woodstain. Such an approach further blurs the distinctions that can be drawn between high-solids stains and coloured varnish.

Both two-pack and one-pack polyurethanes have been investigated as varnishes. Externally the results are disappointing, but urethane alkyds are useful as the basis for quick-drying interior varnishes. Demand for exterior varnishes is usually in the full gloss form, but satin and matt are also popular for interior use. Settlement of matting agents requires careful formulation, for some can destroy the structuring effect of polyamide-based resins.

By far the majority of varnishes remain solvent-based, reflecting the difficulty of good flow and high gloss with aqueous dispersion binders. Externally there is the problem of durability already mentioned in the context of stains. Aqueous latices with in-built UV protection could provide the basis of an exterior varnish, and the type of composition mentioned above has shown considerable promise [56].

9.9.6.3 Paint and paint systems

As noted earlier, the term 'paint' may be used as a generic umbrella term to cover opaque pigmented coatings, regardless of sheen, build, or system. In the UK there has been a long tradition of using high-build full-gloss systems on woodwork especially outside. During the past decade the supremacy of gloss systems has been challenged first by woodstains, and more recently by paints with a sheen level lower than full gloss. Despite these changes, the demand for a full-gloss system remains high, especially for redecoration, and this will remain an important market sector for the immediate future. In other parts of the world, including North America and Australia, the move away from high-gloss exterior systems has gone further. This situation can be explained in a number of ways, but it is likely that different constructional methods and the greater use of wooden siding would show up potential defects of the traditional system (such as flaking) on a larger scale, thus paving the way for greater use of more permeable off-gloss waterborne systems.

Traditionally, high-gloss paints were part of a three-product system comprising primer, undercoat, and gloss. Many of the new paints are two-product, or one-product multicoat systems. Before embarking on the design of a new coating, some consideration must be given to the merits of a system as opposed to a single-product approach.

An obvious advantage of single products is simplicity, both to the user and the stockist. Against this it must be recognized that a coating has many different functions to perform which might include sealing, adhesion promotion, and filling, as well as protection and appearance attributes. Combining these into one product is likely to entail compromises which may or may not be acceptable to the final customer.

To illustrate this point further, consider a traditional liquid-gloss paint. If two or three coats are applied directly to bare wood the durability can be remarkably good, better in many cases than a traditional system. What then has been lost? Answers to this question include build (affecting appearance), speed of re-coat, and flexibility in redecoration. The weighting given to these factors varies between individuals and user groups, but it is a mistake to underestimate them. Many of the newer products that have appeared on the UK market will inevitably be used for redecoration of existing coatings rather than new wood. They may have good durability, but often fail to measure up to the practical problems encountered in both trade and retail use. Whatever the detail of the final chosen system, formulators must consider how the needs of priming, filling, and finishing are to be met.

9.9.6.4 Wood primers

An important function of wood primers is to provide adequate bonding between the substrate and subsequent finishing coats, especially under damp conditions. Primers should be able to seal the grain while resisting hydrolytic breakdown over long periods. Current building practices often exposes building components for long periods without the protection of a full system, hence primers should have good intrinsic weather resistance.

The archetypal wood primer was based on linseed oil and white lead. Lead carbonate has the useful property of forming, with linseed oil, fatty acid soaps which have good wetting properties and yield tough, flexible films. White lead was often admixed with red lead (lead tetroxide), giving the well-known, and still imitated, pink wood primer. Recognition of the cumulative toxicity of lead has led to the withdrawal of BS 2521 (the standard for lead-based primers) and the development of products without lead pigment. These are usually based on oil, oleoresinous, or alkyd binders with conventional pigmentation. To maintain flexibility the *PVC* should be relatively low (typically 30–40%). Early low-lead primers were often overfilled to reduce blocking, and this, combined with fast-drying inflexible alkyds, resulted in poor performance, a factor contributing to dissatisfaction with solvent-borne systems in general. In the UK, standards for low-lead solventborne primers are set by BS 5358 (recently revised), including a six months' exposure period, which uses a white lead primer as control. The corresponding standard for waterborne primers, BS 5082, has also been revised and includes the same durability test. To reduce variability introduced by the wood itself, the test paint is exposed alongside the control on the same piece of wood. Differences between good and poor paints are quickly shown up. Solventborne paints tend to fail by cracking of the primer, while waterborne paints (owing to higher permeability) allow cracking of the wood itself. Such cracks, when they do occur, usually stop at the junction between the test paint and the lead-based control. With careful formulation both types of primer will pass the test which correlates reasonably well with subsequent system performance. For conventional solventborne products a key requirement is to use an extensible binder. Weathering trials have shown advantages in using a combination of free-drying oil, as well as that which is chemically bound to the polyester backbone of an alkyd resin. This may reflect the ability of free oil to penetrate the wood while the alkyd is retained on the surface to maintain flexibility. It used to be said that the oil was 'feeding the wood'; however, over-penetration by the binder will leave the primer underbound.

An important difference between BS 5358 and BS 5082 is that the latter includes a blisterbox test. This need stems from the practice of overcoating a waterborne primer or primer undercoat with solventborne finishing coats. Such hybrid systems can give outstandingly good performance provided that the aqueous dispersion latex has been modified to give good adhesive performance under wet conditions. To use an ordinary interior latex is disastrous, the inferior primers have given the whole group an unfounded bad reputation. Products which do conform to BS 5082 and pass the important blisterbox test have a good record for durability. Improving the adhesive performance of aqueous dispersions has been the subject of much patent activity [58], with the chemistry of the β-ureido group being especially productive. The term 'adhesive performance' rather than 'adhesion' is used here quite delibera-tely. Not only is the concept of adhesion rather elusive, but products which perform well on the blisterbox test do not necessarily give the highest apparent bond strength when measured on peel and other tests. The role of so-called adhesion promoters is more subtle than meets the eye and may involve a dynamic interaction with water. Blisterbox tests correlate less well with the durability of solventborne solution paints, but there is a good correlation with solventborne dispersion binders. Non-aqueous dispersion (NAD), fitting box II in Table 9.2, may have a useful role to play for a quick-drying, flexible exterior primer with good early shower resistance [39]. Exterior wood primers will benefit from the inclusion of fungicide.

Aluminium sealers and wood primers
Primers containing aluminim flake offer good barrier properties both to water and certain types of staining. They are useful for sealing end grain and against resin exudation, creosote, bitumen, and coloured preservatives. Barrier properties are dependent on the amount of flake pigment used. Established vehicles for this type of product include phenolic modified resins; typical formulations are specified in BS 4756.

Aluminium primers are not advisable on large areas where high movement may be expected. A high incidence of flaking has been found on softwoods, reflecting poor intercoat adhesion. Performance is usually better on hardwoods. The darker colour of aluminium primers can be a disadvantage when it comes to overcoating with white or light colours. However, some specifiers have seen this as advantageous as a means of ensuring a full finishing system over a primer that is readily detected!

Aluminium primers suffer from two specific defects on storage — binder exudation and gassing. The former results from the low-viscosity, low-surface-tension binder separating from the pigment and passing through the most minute gap, especially welded seams, rivet holes, and the lid seal. Prevention is usually achieved by a good-quality lacquered container. Gassing is the result of contamination by water or alkali residues from mixing equipment. Contamination of this sort must be rigorously excluded; the consequences of exploding lids can be very damaging! The gassing propensity of aluminium has inhibited commercialization of waterborne aluminium primers.

Preservative primers
This term is sometimes used to describe an unpigmented fungicide containing binder. Such products are designed for maximum penetration at the expense of any contribution to film build. It is thus easier to achieve penetration of fungicidal components, especially into the vulnerable end grain. Inclusion of a suitable low-viscosity binder enables the product to seal end grain to a greater depth than can be achieved with a pigmented product, though more than one application may be necessary. The penetrating nature of these products can help stabilize a denatured surface and improve subsequent coating performance. For completeness, it should be noted that the term 'preservative' has also been used as an adjective to describe any products including stains and primers which contain fungicide.

Dual-purpose stain primers
Because there is a possibility that new joinery may be finished with either stain or paint, it is an advantage to joinery manufacturers to prime with a primer that can be overcoated with either type of product. This has created an apparent niche for dual-purpose primers. Their appearance is inevitably dictated by the needs of the semi-transparent stains, and such products tend to be closer to stain primers than traditional wood primers. In consequence they offer much less temporary protection

than a wood primer, and this has caused problems. There is also a danger that this approach obscures from users the fact that stains and paints are not always fully interchangeable for a given construction. Highly-permeable products may require a better quality of joinery wood if excess movement is to be avoided, and will require alternative glazing and non-ferrous fixings.

9.9.6.5 Undercoats

Undercoats can play an important role in contributing build and opacity to the traditional paint system. They also improve adhesive performance when old gloss is repainted, and help to provide cover on sharp edges. Achieving satisfactory appearance when renovating weathered gloss paint is more difficult without an undercoat to fill damaged areas and provide a contrast between costs. To fulfil these functions, undercoats are normally heavily filled and habitually have the highest *PVC* of any exterior coating. Moreover, to aid sanding, they are usually formulated on a brittle binder. As a consequence they lack extensibility, and from the point of view of durability, are the weakest link in a paint system. Replacing undercoat by an extra coat of gloss significantly improves the durability of many paint systems, though this would be regarded as less practical by many decorators.

The fact that some undercoats are inextensible does not invalidate their useful-ness, but exterior undercoats for wood should be formulated differently from those for interior or general purpose. In particular, they must have greater extensibility which can be achieved by reducing the *PVC* or by using a more flexible binder. This will, of course, increase sheen and makes sanding more difficult. Requirements with respect to adhesion are less stringent if a good primer is used, and there are many potential formulating routes via waterborne or solventborne technology.

9.9.6.6 Finishes

Full-gloss solventborne gloss finishes have remained an important feature of the UK market, especially in maintenance painting. As noted above, when used with primers and undercoats (optional) which have been correctly formulated for use on wood, they can give performance with a typical maintenance period of 5–7 years. There are sufficient differences between interior and exterior conditions to make separate formulations worth while. In particular, exterior gloss finishes for wood will benefit from increased flexibility and the presence of fungicide to inhibit blue stain and other surface moulds. Gaining the optimum balance between longer-term durability and initial drying properties is a difficult and long process. Alkyds, that is, oil or fatty acid modified polyesters, lend themselves to almost limitless modifications, both by varying their chemistry and method of preparation. Such changes will alter mechani-cal properties, permeability, adhesion, etc., and can have a marked effect on durability. Over the 4–5 decades that alkyd resins have been in commercial use there have been many attempts to relate their structure and chemistry to properties such as durability, but owing to the complexity of the situation, clear and unequivocal relationships have not been established. A diligent search of published literature shows many contradictions in respect of oil, polyol, and fatty acid type. Not infrequently important data, such as the molecular weight distribution, is omitted. Alkyds with long oil lengths often show greater initial flexibility but are more prone than shorter oil alkyds to rapid change on weathering, though exceptions to this 'rule' have been reported [59].

Urethane alkyds have specific advantages in terms of quicker surface and through-dry, but this is accompanied by a decrease in extensibility which invariably leads to a greater incidence of cracking and flaking on exterior wood. The same problems beset silicone alkyds which show excellent gloss retention but poor extensibility. Silicone alkyds have outstandingly good chalk resistance, but although this may seem an advantage it can result in very high dirt pick-up and there is no self-cleansing action. This highlights the careful balancing of properties that must be made in developing an exterior coating. The choice of TiO_2 grade can have far-reaching effects, and much information has been published by major TiO_2 manufacturers.

There are now an increasing number of alkyds on the market which have been formulated specifically for wood (e.g. [54]). In selecting one of the many fungicides now offered for wood coatings, it is extremely important to assess the longer-term storage stability; some have shown seeding or discolouration effects, which may take several months to appear.

Off-gloss finishes for exterior wood are usually formulated to be applied direct to bare wood or over an appropriate primer. To some extent they combine the function of both undercoat and finish. They do not form a very clearly defined product group in the UK, and may be described by terms which include 'continental paint' and 'opaque stain', (see section 9.9.6). Ideally, they should be based on a fairly permeable binder and, like gloss finishes, be flexible and fungicidally protected. The choice of matting agent is critical, as it is necessary to raise the *PVC* but without sacrificing mechanical properties. Pigments with a high oil absorption will, as discussed in section 9.3.2.3, lower the *CPVC*, in effect reducing the availability of 'free' binder. To compensate partly for any loss of extensibility in this way, extenders with a reinforcing effect such as talc or mica, may be employed.

Waterborne acrylic gloss finishes suitable for exterior woodwork are well established in many parts of the world, especially where large areas of wooden cladding are found. A major advantage, compared with alkyd-based paints, is the good extensibility which is maintained for long periods of weathering. Advances in the development of new thickeners, such as the associative types, have greatly improved flow properties, though considerable optimization is required to obtain a balance formulation for a specific group of raw materials. Although the highest initial gloss levels and distinctness of image achieved by alkyd systems are not fully matched by waterborne systems, they normally show a much slower rate of gloss loss, and are generally superior after a period of weathering. The thermoplastic nature of acrylic dispersion polymers means that dirt pick-up is high with soft polymers. Dirt becomes engrained into the film itself, and becomes very difficult to remove. This problem was more apparent when gloss finishes were based on dispersions designed for semi-gloss finishes where the higher level of pigment reduces thermoplasticity. More recently, latices based on harder but still flexible resins have become available, effectively eliminating the problem. A related problem is that of blocking, which can be caused by a soft polymer and will be exacerbated by water-soluble material being resolubilized. A major problem area is the space between rebate and opening window lights. This problem is also reduced by the newer generation latices available, but care in the selection of other ingredients must be exercised. Blocking resistance is sometimes improved by the incorporation of a proportion of non-film-forming hard latex which, in formulation terms, is considered as part of the extender *PVC* [60].

Off-gloss waterborne finishes are readily developed from either similar or softer polymer dispersions than those used for gloss finishes. In comparison with solvent-borne mid-sheen finishes it is easier to compensate for the consequences of a higher *PVC* on mechanical properties.

The higher permeability of waterborne dispersion paints is usually seen as an advantage for wood coatings, but for some situations it can prove too high (leading to excess movement), and pigmentation or choice of polymer should be adjusted accordingly. Surface moulds may grow readily on water-soluble components in the coating, and it is essential that they are fungicidally protected.

9.10 MASONRY AND CEMENTITIOUS SUBSTRATES

Among the most widely encountered surfaces in buildings are plaster, concrete, external rendering, and brick. Although these are individual materials with their own characteristics they also have a number of general similarities which influence coating formulation. Of particular significance are alkalinity, the porous friable nature of the surface, and the general consequences of moisture and its interactions with the substrate.

9.10.1 Implications of moisture

Water is often present in large quantities in the materials referred to here, especially in new buildings. This is particularly true with hydraulic cements and plaster, but will also result from the storage of materials in the open during construction. Surfaces created from 'wet' materials of construction may require an initial coating of very high permeability to allow drying out, though subsequent re-decoration with less permeable coatings is possible.

Moisture content is conveniently quantified by quoting the relative humidity in contact with the surface. BS 6150 divides surfaces into four groups according to their equilibrium moisture content;

> Dry   — <75% RH
> Drying —   75%–90%
> Damp —   90%–100%
> Wet   –   100% (with visible surface moisture)

Wet surfaces are very difficult to paint, but damp and drying surfaces can be coated with emulsion paints which are usually formulated above the critical *PVC* to increase permeability.

As with wood, the presence of water which has not been allowed to dry out will reduce adhesive performance, may cause blistering, and creates the possibility of mould growth. Moisture is also a key factor in relation to alkaline attack, efflorescence, and staining.

Alkaline attack
Portland cement is highly alkaline, as are some plasters. Such alkalinity in the presence of water will saponify many oil-based paints and may discolour pigments.

Efflorescence
Unsightly deposits of salt on the surface of plaster or brickwork, etc. are known as efflorescence and appear in bulky and dense form. Bulky efflorescence is usually a form of sodium sulphate, and may disrupt coatings of low permeability, although oil paints have a limited ability to actually hold back efflorescence. Permeable coatings, such as above-critical emulsion paints, allow efflorescence to pass through, where it can be wiped from the surface, but heavy eruptions may cause adhesion failures, and under these circumstances painting should be deferred until efflorescence has largely ceased.

The denser type of efflorescence is usually calcium carbonate (known also as lime bloom), and is more difficult to remove but easier to overpaint after light abrasion.

Staining
Brown stain may appear on emulsion paints on some types of brick, clinker, and hollow clay blocks. The colour derives from soluble salts or organic matter capable of reacting with alkali. Alkali-resistant primers will normally prevent this type of staining. A useful summary of typical defects in walls will be found in [61].

9.10.2 Cement and concrete
Cement, as a component of concrete, finds very widespread application in all types of building, and, indeed, its per capita consumption has been suggested as an indicator of economic development. Such widespread use reflects the attraction of a material which is relatively cheap, can be moulded or cast, and will withstand high compressive stress. Iron or steel reinforcement of concrete also allows the carrying of tensile loads, although cracking of the concrete is not necessarily prevented. Load carrying is much improved by putting the concrete into compression, which is achieved in prestressed concrete through tensioning the steel reinforcement. In reinforced concrete the function of the concrete is to resist compression and buckling, and also to protect the reinforcement from corrosion. Insufficient thickness or inadequately prepared concrete has shown failures in this area, and, in turn, has created markets for both remedial and preventive coatings. A potentially large market, though strictly outside the scope of this chapter, is in reinforced concrete bridge decks. These are especially vulnerable to the corrosion exacerbated by de-icing salts. Epoxy powder coated reinforcements for bridge decks are mandatory in several USA states [62].

An alternative to the massive reinforcement described above is to use a fibre reinforcement. Asbestos cement is a well-established example, though under increasing pressure for health reasons. Cement panels and components can also be made using glass, metal, or plastic reinforcing fibres, and such materials are becoming more common. Although described as a reinforcement, the function of the fibres is to prevent crack propagation rather than to carry a major tensile load. Coating requirements of such products are largely dictated by the nature of the cement matrix, though surface appearance will be modified by the presence of fibre.

9.10.2.1   *Characteristics of cement and concrete* [63]

Concrete is made by binding an aggregate of sand, broken stone, gravel, etc. in a matrix of cement paste and water. The most widely used binder is Portland cement, though high-alumina and other more specialized cements find a number of important specific uses. Hardened cement paste may be regarded as a cement gel matrix which will contain unhydrated cement particles, air, and water; mortar is a combination of sand and cement paste.

Cement is a well-established building material; both the ancient Romans and Greeks used cements based on hydrated lime, and this type of cement still persists. Strength development depends partly on an initial slaking reaction:

$$CaO + H_2O \rightarrow Ca(OH)_2,$$

but long-term durability derives from the conversion of colloidal hydroxide to carbonate by reaction with atmospheric carbon dioxide:

$$Ca(OH)_2 + CO_2 \rightarrow CaCO_3.$$

Portland cement is produced by heating a lime-bearing material such as limestone with a material containing silica, alumina, and some iron oxide — typically clay. After compounding, the material is fused to a clinker at around 1400°C and ground with gypsum and other materials (which act as retarders) to produce cement powder.

The main compounds in anhydrous Portland cement correspond to tricalcium silicate, dicalcium silicate, tricalcium aluminate, and tetra calcium alumino ferrite (C_3A, C_2S, C_3A, and C_4AF in cement notation). When hydrated separately these compounds gain strength at different rates. C_3A is the most rapid hardening, producing considerable heat evolution; its percentage is reduced in cements which are to be used in thick sections. C_3A is attacked by sulphate and must be omitted for sulphate-resistant cements.

9.10.2.2   *Hydration mechanism of Portland cement*

Hydration of cement is a reaction in which a solid of low solubility reacts with water to form products of even lower solubility. There has been much controversy as to the mechanisms of cementing action. Competing theories from the past include Le Chatelier's crystallization hypothesis (1887) and Michaelis' gel hypothesis (1893). Today there has been a tendency for some synthesis of crystallization and gel theories, but there is still no general agreement for all the mechanisms involved.

Initial mixing of cement and water produces a dispersion, the particles of which become quickly coated with hydration products. Hydration products are largely colloidal (10–100 Å), though some large crystals of calcium hydroxide may be formed. The solution becomes saturated with Ca^{2+}, OH^-, SO_4^{2-} and other ionic species. On further reaction the coating of hydration products extends partly at the expense of the grains and partly at that of the liquid. A 1 cm^3 of cement produces 2.32 cm^3 of cement gel; 45% of the gel must fall inside the boundary of the cement grain and 55% in the surrounding capillary space. Rupture of the coating may take place (according to Powers [68]) as a result of high osmotic pressure. When the coatings begin to meet, the cement is at the setting stage. With further reaction the particles become increasingly densely packed; this further hydration and crystallization from supersaturated solution involves a complex diffusion process, in which water from capillary space diffuses 'inwards' through the gel pores, while hydration products move in the opposite direction. No further cement gel forms in the gel pores, and it is assumed that these are too small to allow nucleation of a new solid phase. Crystalline particles such as calcium hydroxide are disseminated through the gel, and may form in the pores by crystallization. Calcium hydroxide crystals are usually thought to be detrimental to the strength of cement, but it has been suggested that it could be a major factor in bonding inert particles and fibres.

The latter stages of hydration are slow, and can take 25 years to complete. During the early stages of hydration free water is in the form of capillary channels; but as these block, once cavities become filled, hydration must cease because there is no room for gel formation (although in certain cases, formation of hydrate by disrupting the existing gel may occur — leading to very weak cements). If all the capillary cements are blocked it should be virtually impermeable.

The space structure of cement gel as first formed is expanded and unstable with a tendency to change, which is accompanied by a diminution in surface area and shrinkage. Properties of hardened cement paste are considerably modified by aggregate, water/cement ratio, and degree of hydration.

Water requirement
Stoichiometrically the requirement of cement powder for complete hydration is met by a water/cement ratio of 0.23. However, if all the cement is to have room physically to hydrate (remembering that the volume nearly doubles), then a water/cement ratio of 0.38 is required. Another constraint is that combined water undergoes 'compression', with specific volume reduced from 1.0 to 0.74; even gel pore water only has a specific volume of 0.9. This reduces the relative vapour pressure within a sealed sample, and if it falls below 80% hydration virtually stops. To avoid this a water/cement ratio of 0.5 is required, though this will not necessarily produce the strongest concretes. Even in the presence of excess water it is common to find unhydrated cores of the original cement particles.

Dried cement gel characteristically has the chemical composition $3CaO \cdot SiO_2 \cdot 2H_2O$, but in saturated gel an extra mole is present as intercrystalline water, and can be removed only at water vapour pressures less than 0.1. Gel pore water is strongly held, and is present at relative vapour pressures of 0.1–0.5. Smaller capillaries are full at vapour pressures of 0.5–0.8, but larger capillaries require a vapour pressure above 0.8 to fill.

The chemistry of the hydration reactions taking place in Portland cement is extremely complex . Of especial relevance to the coatings technologist is the reactive and alkaline nature of the surface, combined with its variable porosity and permeability.

Retarders
Calcium sulphate, usually as gypsum, is universally added to ground cement to control the otherwise rapid 'flash set'. Many other compounds have a retarding effect, and these have been put on a systematic basis by Forsen, according to their effect on the solubility of alumina. Following his categorization, retarders may

be divided into four sets depending on their action as a function of concentration. Typical materials which may be in the groups are:

(1) $CaSO_4 \cdot 2H_2O$ $Ca(ClO_3)_2$ CaI_2
(2) $CaCl_2$ $Ca(NO_3)_2$ $CaBr_2$ $CaSO_4 \cdot \frac{1}{2}H_2O$
(3) Na_2CO_3 Na_2SiO_3
(4) Na_3PO_4 $Na_2B_4O_7$ Na_3AsO_4 $Ca(CH_3COO)_2$

Type (4) retarders may hold up setting and hardening indefinitely if used in sufficient quantity, but they are not all harmful; and some, such as the calcium lignosulphonates, are used as water-reducing agents.

9.10.2.3 Structure and morphology

At the macroscopic level the structure of concrete, etc. is dominated by the dimensions of aggregates and the extent to which they have been compacted. Poor compaction and mixing leads to obvious heterogeneity. Within the cement paste, air entrainment pores are typically around 50 μm. Typically capillary and gel pore 'diameters' are 500 Å and 15 Å respectively. Under the optical microscope hardened cement paste appears amorphous and semitransparent, but the electron microscope reveals a wealth of micro structure [64]. Care is needed in interpreting some micrographs if the hydrates are produced at untypically high water/cement ratios. Various cement hydrates have been identified, and their structure related to that of known minerals such as 'Tobermorite', and they have been observed in crumpled sheets or rolled-up tubes. Midgley [65] described splines and plates as important morphologies. A model described by Double & Hellawell [66] explains the formation of hollow tubes by a similar mechanism to that seen in the silica 'gardens' that form when metallic salt crystals are placed in solutions of sodium silicate.

9.10.2.4 Volume changes in concrete

Volume changes during hydration and subsequent sensitivity of cement gel to moisture content ensure that the overall shrinkage or expansion of cement paste is complex. Re-immersion of dried cement paste causes swelling as water penetrates interstices within the gel, but not all the drying shrinkage is recovered. In the presence of carbon dioxide any calcium hydroxide present may be converted to calcium carbonate with a subsequent irreversible carbonation shrinkage — this can be as much as 50% of the initial drying shrinkage, but is normally limited to external surfaces. The inclusion of aggregate changes the magnitude of volumetric strain largely in proportion to its volume fraction. Concrete is a multiphase material, and some of the shrinkage of the paste may show up as cracks at the aggregate/paste interface. The range of length change in hydrated cement paste specimens subjected to wetting and drying cycles is characteristically spread around 0.5%.

9.10.2.5   *Porosity and permeability of concrete and cement*

Moisture movement in concrete (as in timber) may be usefully considered in terms of both permeability and diffusion. Although these are derived from the same physical processes, the mathematical forms differ. Permeability is associated with a pressure difference and is associated with saturated materials, while diffusion is more useful in considering partly dry materials with the fluid driven by chemical or moisture potential [67].

Permeability is influenced by porosity and hydration of material within pores, and capillaries will greatly reduce flow. Normally, water movement will occur within capillaries, rather than the pores. Powers [68] has published data showing the relationship between permeability and capillary porosity. Drying cracks or flaws at the aggregate interface are likely to increase permeability. Porosity for hardened cement is characteristically 25%, and once capillaries are blocked the permeability falls to about 10^{-12} cm/s, which is less than for many natural rocks.

The most important driving force for diffusion is the gradient between internal moisture and surface or capillary forces. Solutions to the diffusion equation are discussed in [67, Chapter 8].

The permeability of concrete is a major indicator of its potential durability, both in the sense of mechanical strength and resistance to chemical attack. Permeability may be tested by measuring the flow through a saturated specimen subjected to pressure; a penetration test is more appropriate in cases where moisture is drawn in by capillary action. A test known as the Initial Surface Absorption Test (ISAT), which can be applied to concrete *in situ*, is described in BS 1881. It can be used to assess the rate of water absorption — clearly this is relevant to some coating properties.

9.10.2.6   *Resistance of concrete to destructive processes*

In service, concrete may be subject to a physical process such as freezing or heating, which can cause damage. Other potential disruptive reactions include expansive alkali–aggregate reactions which are an increasing cause for concern. However, these are outside the preventative scope of coatings. In principle, concrete is a highly durable material which should not require coating to achieve good weathering, but there are circumstances where the protection offered by a coating becomes necessary. These include protection against acidic environments, and to prevent attack of the reinforcement if the thickness of concrete is insufficient or otherwise unable to provide protection. Coatings may also be necessary to provide protection against specific agents to which the concrete is sensitive.

The resistance of concrete to chemical attack is very dependent on its permeability, the nature of the aggregate and any additives, and the care with which it is made and placed. The life of good-quality concrete can be twenty times that of indifferent material in identical conditions.

Portland cement (but not high-alumina) is markedly attacked by sulphates, with magnesium and ammonium sulphates being particularly severe. Strong aluminium sulphate will also attack high-alumina cement. Attack by seawater stems from the presence of magnesium sulphate which is modified by the presence of chloride and can also attack metal reinforcement. Pure water can dissolve the lime from set cement, but the action is slow unless water is able to pass continuously through the mass. Attack increases with pH as caused by the presence of CO_2 above the amount needed to maintain the equilibrium — $CaCO_3 + H_2CO_3 \rightleftharpoons Ca(HCO_3)_2$. Displacement of this equilibrium means CO_2 is more aggressive in saline solution.

The resistance of all hydraulic cements to inorganic acids is low, but they are also attacked by many organic acids, though there are anomalous concentration effects — acetic (vinegar), lactic (dairies), butyric (silage), and tartaric (fruit) acids all have adverse effects, as do higher molecular weight acids, including oleic, stearic, palmitic, and most aliphatic acids.

Portland cement is resistant to strong alkalis including sodium and potassium hydroxide. High-alumina cement, in contrast, undergoes severe strength deterioration. Even alkaline detergents should be avoided when washing HAC floors. Other agents known to attack concrete include sugar, formaldehyde, and the free fatty acids in vegetable oils and animal fats. Glycerol reacts with lime to form calcium glycerolate [69].

9.10.2.7   *Organic growths on cementitious substrates* [70]

Although not biodegradable in the sense that wood is, the rough surface of most cementition and masonry substrates leads directly to conditions which will support organic growth; that is collection, and holding, of moisture and nutrient. Organic growths found on building surfaces include those requiring light such as algae, lichens, and mosses; and fungi, which do not. Sulphate-reducing bacteria in gypsum plasters can cause staining of lead-containing paints, and have been known to promote corrosion of steel.

It is vital that, before coating, any organic growth should be physically removed by scraping and brushing to prevent rapid re-infection. Surfaces should also be treated with a toxic wash which has been approved for safety in use. In the UK a 'Pesticide Safety Precaution Scheme' is now replaced by *Control of Pesticide Regulations* (1986) [71].

9.10.3   Other cementitious substrates

External renderings [72]

Cement rendering: Many renderings are based on Portland cement, possibly with incorporation of lime; they are therefore highly alkaline. Sand/cement renders become extremely friable with time. Lime/sand renders, as found in older buildings, are usually known by the term 'stucco'. BS 5262 contains recommendations for repairs to rendering, including stucco. Characteristic properties of other renderings including 'roughcast', 'pebble-dash', and 'Tyrolean' are described in [73].

Lightweight concrete blocks

Not normally used outside, these materials present problems in consequence of their open pore structure.

Asbestos cement sheets

When new, these sheets are strongly alkaline and vary in porosity, even within the area of a single sheet. Good resistant sealing will be necessary under many coatings. It is important not to seal one face of a sheet with an impervious coating as there is a risk of warping caused by differential carbonation.

9.10.4   The market structure for cementitious substrates

Market sectors for cementitious substrate coatings provide equal, if not greater, difficulty than those for wood in drawing up a neat, logical framework. This reflects the complexity of the many substrates and products with an inevitable degree of overlap. Jotischky [62] has reviewed aspects of the market structure which can be divided, for example, by substrate (masonry, concrete, etc.), by application (civil engineering, residential, local authority, etc.), by purpose (protective, remedial, etc.), function (e.g. chemical resistance, anti-graffiti).

Probably the most visible market sector in the UK is the masonry sector, essentially decorative products for stucco, rendering, pebble-dash, and brick. The largest product offering in this sector is offered by exerior emulsion paints based on vinyl and acrylic latices. The extension of this market sector to cover architectural concrete has been very slow, though in Germany the percentage is higher. Many architects would maintain that concrete does not require painting, but apart from aesthetic aspects there are situations when a coating can prevent water penetration and reduce attack by carbon and sulphur dioxides [72]. Leading on from this is the repair market for reinforced concrete mentioned in section 9.10.2. This has called for a variety of remedial and preventative products, with liquid epoxy resins finding a useful niche. The market is largely in the hands of specialists, but the value of buildings at risk is staggering [62]. In the UK the market for coating new reinforcement coatings is small, and is more likely to be met by galvanising; but in North America the protection of new reinforcements accounts for an appreciable component of the epoxy powder coatings market.

9.10.5   Plaster and related substrates

Plasters are used extensively for finishing internal walls; characteristic problems include variable porosity and alkalinity. The latter will be especially high with lime and cement plasters, and in practice lime is often added to gypsum-derived plasters to facilitate spreading.

Most commonly, 'plaster' is based on calcium sulphate hemi-hydrate (plaster of Paris) formed by partial dehydration of gypsum (the dihydrate) which will be reformed on addition of water. The reaction of water with plaster of Paris is rapid, and to aid application on large areas it is retarded with alum or borax. In those plasters containing lime the presence of potassium sulphate (alum) and calcium hydroxide (lime) producing potassium hydroxide may affect paint.

BS 1191 broadly distinguishes between the above possibilities with:

> Grade A — plaster of Paris
> Grade B — retarded hemi-hydrate plasters
> Grade C — anhydrous

Grade C are likely to contain added lime. On occasions, plasters fail to hydrate fully; not only will this lead to a powdery surface layer, but subsequent wetting results in expansion with adverse consequences for plaster and coating. All plasters are likely to soften if wetted, and this too leads to coating failures.

Keene's cement (Grade D in BS 1191) is untypical in being slightly acid, though the acidity will often be neutralized by more alkaline backing materials.

9.10.6 Brick and stone

Brick and stone are generally regarded as durable materials which do not require coating for protection (however, see section 9.10.7.2). Coatings are more likely to be required for aesthetic reasons; this is especially the case internally where painting will also facilitate cleaning and improve lighting.

Although most bricks are not alkaline, the mortars used in construction are, and present the problems associated with a cementitious substrate. Efflorescence is always a possibility on brick and stonework. Paint is best avoided in any situation where major moisture penetration is possible, such as below the damp-proof level.

9.10.6.1   Clay bricks (classified by BS 3921 as common, facing and engineering)
Common bricks include 'Fletton' which are notoriously difficult to paint unless sand-faced or rustic [75]. One potential difficulty is coating those areas of bricks which were in contact during baking. 'Kissmarks' appear almost glazed, and have a coarse pore structure. Not only are they more difficult to wet, but their greater porosity is more likely to allow efflorescence.

9.10.6.2   Facing and engineering bricks
Facing bricks are made to give a specific surface texture or colour, and do not present the same adhesion problems as Flettons.

Engineering bricks are much denser and virtually non-porous, and are more likely to give adhesion problems. They may have similarities with glazed surfaces, as does non-porous stone, especially if polished, and may require specific primers to promote adhesion. This area is not well documented, but silane coupling agents are worth consideration in difficult situations.

9.10.6.3   Calcium silicate bricks
Calcium silicate bricks are classified by compressive strength, reflecting porosity (BS 187). Normally more uniform than clay bricks, they are not regarded as problematical in painting.

9.10.6.4   Stone masonry
Many different stones are used in building, with extremes ranging from limestone to granite. The choice of stone is affected, to some extent, by the type or types available from local quarries. BRE Digest 177 discusses some of the factors affecting decay of stone masonry and the conservation measures used in protection [76]. Normally, stone masonry is not painted; but where a need exists, coatings must be formulated to suit the characteristics of the material in question.

9.10.7   Coatings and formulating practice for masonry and cementitious substrates

The difficulty of classifying coatings in general was noted earlier. Inevitably, classification represents a particular point of view which may seem arbitrary when seen in a different context. BRE Digest 197 lists the major coatings established in this area, and divides them into broad categories of water-thinned and solvent-thinned. Within this broad distinction could be found other groupings such as the physical nature of the binder (solution, dispersion, emulsion, etc.) or the chemistry, including a division into one- and two-pack products. A feature of this market sector is the existence of filled and textured finishes which form another recognizable subset. BS 6150 splits coatings into an internal and an external group, with subdivision relating to the moisture content of the substrate [77].

For the remainder of this section a selection has been made which highlights some of the more important and representative groupings, but with no attempt to be comprehensive.

9.10.7.1   Sealers and colourless treatments

The porous and sometimes friable nature of masonry surfaces has created a market for water-repellents and in some cases for sealers to act as primers prior to painting.

9.10.7.2   Water-repellents for masonry

These materials are intended to improve resistance to rain penetration with minimal effect on appearance. They function by inhibiting direct capillary absorption, but do not normally provide a continuous surface film. Properties of interest include resistance to water penetration, water vapour transmission rate (permeability), resistance to efflorescence, and longevity of the effect. Such treatments will not necessarily decrease water uptake through cracks, which may, in fact, increase as the treatment causes more water to run across the surface.

Waxes, oils, and metallic soaps have been used as the basis of water-repellents, but these have tended to be supplanted by silicone resins in various forms. A survey carried out by the Greater London Council [78] showed silicones to be the largest group of proprietary agents available in the UK, but also noted were siliconates, silanes, epoxy resins, and acrylates. User guidance and performance standards for silicone-based repellents only are given in BS 3826. When using proprietary silicone resins as the basis of a formulation, attention should be paid to manufacturers' literature [79], as the different grades may have restrictions in use; some, for example, are not suitable over limestone. Silicone resins are available in both solvent and waterborne form; the latter can be highly alkaline.

It must be emphasized that although water-repellents can be effective in reducing rain penetration, there is a danger that under some circumstances they can cause serious spalling, caused by trapped crystallization salts (normally showing as efflorescence) forcing the surface off. It has also been suggested that differential thermal and moisture movement between bulk and surface is another cause [77]. The vulnerability of repellents to efflorescence pressure reflects their high vapour permeability. Although this has the benefit of allowing trapped moisture to escape, it allows rapid absorption of water when humidity is high. There are parallels with the problems of high permeability in woodstains, discussed earlier. This problem underlines the difficulty in striking a correct balance between permeability and efflorescence resistance. Water-repellents are best avoided if efflorescence is a known problem [80]. In an extensive evaluation of water-repellents carried out by the GLC [78] it was noted that, with a single exception, *all* the products that passed the water penetration test caused spalling of bricks in a simulated efflorescence test.

9.10.7.3 Sealers
Masonry sealers, also known as stabilizers, are intended to consolidate friable surfaces. Typically, they are based on alkyd solutions carried in white spirit; a tung-modified alkyd is advisable in order to improve alkali resistance; tung-phenolic resins are also suitable. Paradoxically, some masonry paints, including the emulsion type described below, do not adhere well to a continuous stabilizer film. It is, therefore, important to ensure (a) that the viscosity is sufficiently low to aid penetration, and (b) that stabilizers are applied only to truly friable surfaces. If the surface is sound, then a stabilizer should not be necessary.

An alternative to alkyd or other resin solution is to use a very fine particle size latex at relatively low concentration; styrene acrylic latices and acrylic latices have proved suitable.

9.10.7.4 Alkali-resisting primers
Another product associated specifically with the masonry market is that of the alkali-resistant primer designed to hold back alkali attack on essentially dry alkaline substrates. Although normally used below oil finishes, they can sometimes be used to improve the adhesive performance of emulsion paints on plaster surfaces. Variants specifically for preparing plaster are also marketed.

Alkali-resistant primers have been successfully formulated for many years on tung-phenolic and tung-coumaroune resins. Resistance may be further upgraded with isomerized rubber. *PVC* and volume solids content are typically around 30% and 50% respectively.

Waterborne primers can be formulated on acrylic resin dispersions. To aid penetration the latex should be of fine particle size and the primer pigmented to a low *PVC* and solids content.

9.10.7.5 Paints based on cement
The market for 'cement paints', which were once widely used, is declining, but is still worth around £14m in the UK [62]. Normally used outside, they have a specific advantage in being applicable to wet surfaces. Cement paints are based on white Portland cement with further additions of titanium dioxide and coloured pigment as appropriate (BS 4764); they also require agents to control flow and structure.

The rough surface of cement paints encourages dirt pick-up and algal growth; they will be eroded rapidly in polluted acidic environments.

Interaction between cement and gypsum precludes their use over this substrate. They are supplied in dry form, to which water is added before application. Stored product must be tightly sealed to prevent hydration in the container.

Although often regarded as old-fashioned, cement paints are unusual in being essentially inorganic, and they clearly have a very different balance of properties to organically-bound paints. In principle, their properties could be modified with additives such as spray-dried polymer particles, to give coatings outside the current formulating box with modified properties.

9.10.7.6 Masonry paints — 'normal emulsion'

Masonry (but not reinforced concrete) surfaces, represented by stucco, rendering, brick, and pebble-dash, are a potentially large market for decorative paints where the dominant reason for painting is one of aesthetics. In the UK, about a quarter of the available surface is said to be painted [62], the volume of 20 million litres representing 6–7% of the total UK decorative paint market. Within the sector, emulsion paints account for the major volume.

A distinction is often drawn between 'general-purpose', 'contract', and 'exterior emulsion paints'. General-purpose paints are good-quality paints, as used indoors. Many of them give adequate performance on exterior masonry without further modification, and some manufacturers do not restrict their use to interior only. 'Contract' grades are, in the main, cheaper products taken well above critical *PVC* with extra extender; they are characterised by high opacity and low scrub resistance, and perform poorly outdoors. Exterior emulsion masonry paints should, in general, be formulated specifically for this purpose. Important formulating parameters include the choice of binder, *PVC*, and coalescing solvent, which must be selected or designed to meet specific needs during, and after, the drying process. It is also advisable to include fungicide and/or algicide to inhibit organic growth on the film, though it should be noted that the prevalence of such growth depends also on other factors, including the availability of water-soluble colloid or other material in the film.

Two major groups of binders used to formulate exterior masonry paints are vinyl acetate and acrylic copolymers, common comonomers being 2-ethylhexylacrylate (2-EHA) and 'vinyl versatate' (VeoVa). Maximum alkali resistance is usually associated with acrylic resin, though much depends on individual formulation practice. Acrylic resin has also shown improved chalk resistance in comparison with other types. The ratio of hard to soft monomer in the copolymer is adjusted to minimize dirt pick-up while maintaining sufficient flexibility to cope with substrate movement which may include the opening of fine cracks. Chalk resistance is also strongly influenced by the type of binder, with best resistance given by small particle size binders of low minimum film forming temperature, *MFT* [81]. However, if the *MFT* is too low, dirt pick-up will result (while high *MFT*s lead to inflexibility, expecially at low temperatures). Some other differences between the durability of binders is reported in [81].

A more recently introduced type of acrylic latex binder [82] is modified with vinyl and vinylidene chloride. These latices are anionic in character and have a pH below 2, influencing their compatibility with other polymer emulsions. High alkali resistance makes these polymers suitable for masonry surfaces, and detailed formulation principles are available from the manufacturers. An outstanding feature of these chlorine-modified copolymers is their low water vapour permeability, which is typically between one and two orders of magnitude below that of other waterborne latex films.

The *PVC* of masonry paints must also be chosen to maintain derived physical properties, and can be used to modify the properties of the binder. Typically, the *PVC* of an exterior masonry paint will lie between that of an interior matt or silk — probably in the range 30–45 — but, as discussed in section 9.3.2, *PVC* in isolation is a fairly meaningless parameter which should be adjusted in relation to the *CPVC* of the composition. *CPVC* will, in turn, be influenced by the binding power of the emulsion and the water demand of pigment and extender. These should be chosen to give a high *CPVC*, i.e. extender, of low water demand. Choice of extender is also important for other reasons; laminar/fibrous extenders such as talc can either reinforce the film or provide a stress-relieving mechanism [83].

Application of emulsion paints over a 'hill and dale' topology such as pebble-dash, accentuates mudcracking (simulated pebble-dash wallpaper is a useful test substrate for simulating mudcracking in general) which tends to be a problem at low temperatures. Mudcracking is caused by three-dimensional shrinkage in a coating which has at least one dimension confined, usually by adherence, to the substrate. Internal stress may exceed the rather weak tensile strength characteristic of a dispersion coating during coalescence, resulting in the familiar mudcrack pattern. The role of coalescing solvent in controlling mudcracking is vital, and with new formulations it is advisable to investigate the effect of coalescent level and type.

Emulsion paints for masonry are readily modified with fine sand-like aggregates, and in the UK two marketing sub-sectors are perceived — 'rough textured' and 'smooth' masonry coatings.

The most immediate effect of adding sand to emulsion paint is a change in appearance which is obvious at short range, but less so at a distance. Whether this is important to the final customer is far from clear! Sandfilled paints tend to achieve a lower spreading rate, and in certain specific systems have led to improved chalk resistance [84]. Against this they may show worse dirt pick-up which is more difficult to remove from the rougher surface. Sand is normally stirred into the composition at the end of the other production process at levels in the range 25–45% by weight. Mean particle size of the sand is from 100 μm to 1000 μm. In effect these paints comprise sand embedded in a matrix of pigmented emulsion paint. *PVC* calculations are usually more easily related to other film properties if the volume of the aggregate is omitted from the calculation. Although sand is the most common additive for the fine-textured paints, other materials including polymer powder or fibre have also been used. Fibres have a dramatic effect on rheology, and are likely to present application problems.

9.10.7.7 High-build textured coatings
Emulsion-based 'high-build' textured coatings are known also as 'organic' renderings.

Some aspects of durability are related to film thickness, and, when properly formulated, thick emulsion coatings can give protection for up to ten years, though dirt pick-up remains a problem. A large range of thick-textured coatings is available in Europe, though the market is less developed in the UK where their use is largely remedial on repair. Significant formulating features are a generally higher *PVC* (50%–60%) than the group described above, the presence of much coarser aggregate (up to 2 mm diameter), and the use of rheological modifier to give a very high low-shear viscosity. The applied film thickness can be up to 3 mm and is achieved by trowel or spray, often followed by roller texturing; colour is normally achieved by over-painting with conventional masonry paint. Variants of these coatings include some coloured chips dispersed in an essentially clear latex binder which can give multicoloured texture effects.

Guidance to the formulation of these products is available from the manufacturers of the specialist emulsions used. The vapour permeability of these coatings lies in the middle range, i.e. between the high conventional cost and the low values of bitumen and solventborne two-pack coatings. Water resistance is good — generally better than that of a sand/cement render.

9.10.7.8 High-performance latex-based systems
Related to high build textured coatings is a group of latex-based products designed to give higher durability. Often described as 'high performance', these are designed as a system comprising primer, topcoat, and, in many cases, a separate undercoat. Practices differ between countries, but when used to give maximum protection, up to five coats will be applied. It is usual for these products to be supplied on a contract 'supply and fix basis' — some companies offer up to 15-year guarantees.

Primers in these systems are based on very fine particle size latices with both acrylic and PVA finding use. Bitumen emulsion can be used to upgrade water and vapour resistance either alone or blended with acrylic. In either case, a sealer which can be acrylic based will be necessary to stop the bitumen bleeding through into the topcoat. Several of the high preform systems employ undercoats, some fibre-reinforced, which are used in a remedial basis. Topcoats are generally structured and textured, often requiring specialized application equipment. Acrylic, styrene acrylic, and vinyl acetate variants are all represented in the marketplace. A feature of this market is the need for ancillary products including fillers and mastics. Epoxy-based mortars are used to repair badly damaged surfaces before renovation.

9.10.7.9 Solventborne masonry paints
The solventborne sector of the masonry market, like the emulsion section, has a few clearly defined general-purpose areas, while the higher performance end is more complex and diversified.

General-purpose one-paint products include alkyd, styrene acrylic, chlor rubber, and polyvinyl chloride 'solution' products.

Alkyd masonry paints
Although their alkali resistance is not good, alkyd coatings perform adequately over masonry, provided that an alkali-resistant primer is used. Alkali resistance is improved by the use of urethane alkyds, though this approach will limit flexibility. In the UK both mid-sheen ('eggshell') and gloss paints are used externally on masonry.

The market is not large, but a surprising number of buildings in London are still coated with alkyd gloss. Alkyd gloss paints will be vulnerable to replacement by latex-based gloss paints, though early shower resistance remains a practical problem.

9.10.7.10 Solventborne thermoplastic coatings

Solventborne thermoplastic paints are not widely used in the UK, though they are well established on the Continent, especially in France [86]. Normally they are used as single-product two-coat systems, though special primers may be needed on poor surfaces. Essentially, they compete for the same market sector as the general-purpose 'emulsion' paints described above. Their performance is generally good in terms of film integrity, but with a greater tendency towards chalking than emulsion types. Being closer to a solution, rather than dispersion-based binder system, leads to more restrictive application properties.

A characteristic of matt or mid-sheen solution paints is to show a patchy mottled sheen when applied over substrates of uneven porosity and on overlaps. This is a familiar problem in respect of wall paints, and has been extensively discussed in the literature [87]. The problem is largely caused by loss of binder into the substrate, causing a variable *PVC* in the coatings; it can be reduced by formulating at the *CPVC*. However, if a paint is at its *CPVC* and still loses binder it will go above critical, thus reintroducing sheen variation and increasing the severity of chalking. Other formulation parameters must be adjusted to prevent this, namely high capillary forces within the film and a high solution viscosity. Both are factors which will resist suction of the binder into the substrate.

For reasons of both economy and sheen control, paints are conventionally formulated on a blend of coarse extender finer prime pigment such as TiO_2. Maximizing the capillary forces will require a high *CPVC* achieved usually through the judicious blending of coarse extender, fine extender, and TiO_2. Changes in the particle size distribution or level of any of these ingredients will require rebalancing of the formulation. It is a point worth stressing in these specific formulations, which is also of general significance in the decorative market, that transferring a formula from one country to another may require adjustments to allow for local raw material differences. Particle size distribution of extender pigments is an example of a parameter likely to differ.

Commercially available thermoplastic resin normally requires exterior plasticization to achieve the best balance of mechanical properties, and also to reduce the cost of the binder component. A substantial amount of information, listing the solubility, parameter, and compatibility of other plasticizers, is available [88]. Plasticizer type and level must be adjusted to suit climatic exposure conditions, and is one of the variables which may be adjusted to compensate for other raw material variations such as extender type.

In common with other exterior masonry coatings, these paints will normally contain a fungicide. Some fungicides contribute to chalking, and this is a point which should be ascertained during screening trials.

As noted earlier, these paints require an aromatic content to maintain solubility; however, excess aromatic solvent can cause lifting on re-coat. More recently available resins can be dissolved in aliphatic hydrocarbon [89]. High styrene/acrylate copolymers are available; they function as solution thickeners.

Chlorinated rubber masonry paints
Solution-carried chlorinated rubber and polyvinyl chloride have good resistance to attack and can be used as the basis of masonry coating, with a typical lifetime of between ten and fifteen years.

9.10.7.11   High-performance solventborne systems

Two-pack polyurethanes
Although relatively expensive, two-pack polyurethanes have found applications around the world which call for very high performance. They have been successfully used on all types of masonry and mineral surfaces, including reinforced cement. Coating life is claimed to be 20–35 years. Formulation guidelines are available from resin manufacturers (e.g. [90]). Typically, systems comprise a penetrating primer, basecoat, and topcoat. The first two coats in particular must be alkali-resistant. Normally, the first coat is unpigmented and is based on polyether resin which may also be blended with vinyl copolymers. The basecoat is formulated to a high *PVC* and both it, and the primer, are crosslinked with an aromatic isocyanate adduct. The topcoat can be formulated to a lower *PVC*, and to avoid yellowing, should be based on aliphatic isocyanate. Increasing use is being made of hydroxy acrylic as the polyol component.

The repainting of old polyurethane coatings may lead to a problem of intercoat adhesion, which may be overcome with adhesion-promoting primer.

An obvious disadvantage of two-pack polyurethanes is the limited pot life. More recently moisture-curing products have been developed, based on blocked isocyanates.

9.11   METALLIC SUBSTRATES

The third major substrate encountered in buildings is metal, with ferrous metals being of greatest significance. Whereas wood and masonry are most conveniently grouped under the general heading of 'building paints', ferrous substrates are encountered in virtually all market sectors, including industrial, motors, marine, and heavy duty sectors. Preparing metal surfaces for coating is a specialized subject beyond the scope of this book. However, information concerning the main characteristics of metals will be found in other places in this book. For completeness, and to align with the type of information given in the previous sections on wood and masonry, the main characteristics of metals are also summarized here.

9.11.1   Characteristics of iron and steel

Pure metals have an underlying crystal structure which is substantially modified by impurities or deliberate inclusions used to form an alloy. Alloys will normally show a mixture of grains comprising different crystal structures or phases which have a marked effect on mechanical properties. Controlling the movement of crystal dislocations is a powerful method of altering strength and toughness. The most common way to achieve this is by inclusion of carbon which is extremely efficient in controlling dislocation movement within the iron crystal. But other elements such as silica and manganese have specific advantages. Manufacturing processes will also alter both physical and chemical properties; established techniques include quenching, tempering, and work hardening. Among the most commonly encountered ferrous substrates in general building are mild steel, cast iron, and wrought iron. Mild steel normally contains between 0.2%–0.8% carbon, whereas cast iron contains up to 4% (i.e. 20% by volume!). Wrought iron has been worked so that the morphology is changed and the carbon is present as glossy inclusions.

Mild steel will normally be covered with millscale which, over a period, loosens and may fall away. It can contribute to corrosion, and it presents an unsound basis for coating. Techniques for removal and preparation are well documented [91].

Cast iron has a more adherent scale with some protective value; it corrodes at a similar rate to mild steel, though the residue is less obviously coloured.

Wrought iron is generally similar to mild steel, though corrosion rates may differ.

Iron and steel will frequently be coated with less corrodible non-ferrous metals such as zinc or aluminium, which have other chracteristics as described below.

9.11.2 Corrosion of ferrous metals

By far the major concern in protecting ferrous metals is the problem of preventing them from returning to their naturally occurring state; that is, to prevent corrosion. Iron is far more vulnerable than wood or masonry to exterior exposure damage without some form of protection. There are several categories of corrosion which include dry (oxidative) corrosion and stress corrosion. Here only atmospheric corrosion is considered.

Free energy considerations show that the reaction of iron with oxygen and water is energetically favoured. Ferrous hydroxide is produced which, in its hydrated oxide form, is known simply as rust. Many theories have been invoked to explain corrosion, but today electrochemical theories are standard [92].

In a neutral saline environment the principal reactions are anodic oxidation of iron and cathodic reduction of oxygen. Iron going into solution as ferrous chloride is randomly distributed, whereas areas rich in oxygen — such as the edge of a drop — become alkaline.

$$Fe \rightarrow Fe^{2+} + 2e^-$$
$$O_2 + 2H_2O + 4e \rightarrow 4OH^-.$$

Typically, then, a flow of electrons, i.e. current, will occur between the centre and edge of a droplet which act as anode and cathode respectively. Hydrogen will form a thin layer on the surface, more rapidly in acid solution than alkaline. As the reaction proceeds, ferrous chloride at the droplet centre reacts with sodium hydroxide to produce ferrous hydroxide, and as hydrogen is removed the hydroxide is hydrated.

$$FeCl_2 + 2NaOH \rightarrow Fe(OH)_2 + 2NaCl$$
$$2Fe(OH)_2 + \tfrac{1}{2}O_2 \rightarrow Fe_2O_3H_2O + H_2O$$

Note the regeneration of chloride and the fact that the reaction products are not precipitated directly at anode and cathode, where they might stifle the reaction. However, this principle is behind the use of some inhibitive pigments. Differences between metals in joints or within alloys have a simple, more direct, and usually faster mode of corrosion! Crevices in metal favour differential aeration, which is a prerequisite of the mechanism described above.

Paint defects related to corrosion are complex and not readily related to the above simplified mechanism. Such defects include blistering, delamination, undercutting, under-rusting, and filiform corrosion. Blistering will involve a lack of adhesion which is normal under moist conditions and may be driven by osmosis or alkaline attack. Alkyds are very prone to swelling under alkaline conditions, and this provides and drives the energy for blistering. Undercutting, which occurs laterally under a film, is aided by the cathodic areas which result from a local oxygen concentration, as oxygen diffuses through a film adjacent to existing rust. Filiform corrosion also involves differential aeration, though the mechanism of growth is still unclear. The relationships between these mechanisms have been reviewed by Funke [93].

9.11.3 Corrosion and coatings
Corrosion requires:

— an aqueous phase
— the presence of ions
— a conductive path between anodic and cathodic areas [94].

Ideally, a coating should prevent one or more of these conditions being met, but once a corrosion cell is established the function of the coating is to contain and retard any further spread.

Factors which influence the extent to which a coating will protect ferrous substrates from corrosion depend largely on:

— permeability to water and oxygen
— ionic migration through the film
— electrical resistance of the coating
— adhesive performance under wet conditions
— alkali resistance
— the presence of substances which act in some way to inhibit corrosion
— the absence of substances which promote corrosion [95].

Many of the mechanisms through which the factors operate are still the subject of debate as to the exact mechanisms involved. Refs. [95] and [99] both include useful bibliographies for further reading.

9.11.3.1 Role of water and oxygen
Water plays a major role in the corrosion processes of painted metal, by creating the electrolyte which completes the corrosion cell and through its influence on adhesive performance. However, consideration of the amount of moisture necessary to allow corrosion shows that an excess will be available through the water permeability of typical coatings [96]. Since painted steel normally corrodes much more slowly than unpainted, water permeability alone is not the rate-determining step. Oxygen is

perhaps a more likely candidate, and it has been reported [98] that permeation rates of oxygen are more comparable to oxygen consumption in corrosion. Oxygen diffusion decreases markedly as the *PVC* increases (up to the *CPVC*), but is strongly affected by temperature. Mayne [98] concluded that oxygen availability was no more the common rate-determining step than water, but the question is still an open one. It has been pointed out [95] that there are examples of films with high permeabilities and good corrosion resistance, also examples of low permeability yet poor corrosion. It is probable that different mechanisms can become rate-determining according to circumstance. In this field, as in many others, it is possible to describe factors which may be necessary but are not sufficient to explain the observed phenomena.

9.11.3.2 Ionic migration through films

Coatings can contain the process of corrosion by resisting migration of ions, though opinion differs as to the rate of permeability. Results have been published which show that chloride permeability can, in some coatings, be comparable to oxygen mobility. High electrical resistance will inhibit the flow of mobile ions. Chloride ions are particularly aggressive in promoting corrosion, and are known to break down protective oxide films and reduce the effectiveness of oxidizing inhibitors by competing for adsorption sites. Ammonium ions accelerate rusting, and British Rail report heavy corrosion in steel wagons which carry ammonium sulphate, but there is some controversy over this point [100]. Coatings with high ion exchange capacity tend to have low corrosion resistance.

9.11.3.3 Alkaline attack

As noted during the discussion on masonry paint, some coating binders including alkyds and epoxy esters, are vulnerable to saponification, and are therefore damaged by the alkali formed during corrosion reactions. This may ultimately lead to a cohesion failure within the coating.

9.11.3.4 Pigmentation for corrosion resistance

Although unpigmented films can prevent corrosion, this will require very thick films or an effectively impermeable binder, high electrical resistance, and good resistance to alkali. Corrosion protection is substantially improved by a suitable pigmentation.

One of the most widely used anticorrosive pigments is zinc dust which provides a measure of cathodic protection. Pigments may also act by reducing permeability or interact chemically with the metal surface to inhibit the corrosion process. The term 'passivator' is sometimes used to denote inhibition which specifically promotes the formation of a protective film. Pigments may also increase the corrosion resistance of a coating by reducing permeability through the film.

Barrier pigments

As has already been noted, most pigments, if properly wetted and dispersed, will decrease permeability until the *CPVC* is approached. Part of this effect is a simple lengthening of the diffusion paths available. In principle, such diffusion paths are greater for lamella pigments. It is usually argued that to be effective lamella pigments should 'leaf'; that is, orient themselves parallel to the coating surface. However, the barrier effect might be even greater if the lamella were randomly oriented and in contact.

Any reduction of water and/or oxygen permeability may check the progress of corrosion and should improve other properties such as adhesion performance which is degraded in the presence of moisture. These mechansims of protection have been reviewed by Funke [93], who has demonstrated a number of situations where undercutting and under-rusting tendencies could be correlated with water permeability.

Currently, the most important barrier pigments are micaceous iron oxide and aluminium. Glass flake has also shown promise and could be of use where pale colours are important.

Micaceous iron oxide has similarities to mica; grades vary in particle size and aspect ratio. It is important that the lamella shape is not appreciably damaged during processing. (For a review of formulating procedures see Carter [102].)

Aluminium will also reflect UV to a high degree, and this is of practical benefit in protecting the binder which gives aluminium a practical advantage over other barrier pigments which may not be apparent in some types of accelerated corrosion testing [103].

Inhibitive pigments
Inhibition is a complex subject, the details of which are well outside the scope of this chapter. A useful review of inhibitive mechanisms is contained in [94].

For many years lead and hexavalent chromium compounds have been the most effective inhibitive pigments, and they still provide the yardstick by which others are judged. However, they are increasingly under environmental pressure (with initial legislation generally directed against building paints).

Lead chromate pigments employ at least two inhibitive mechanisms. Chromate acts as an oxidizing inhibitor, and seems to form a thin oxide layer on the surface of the metal to be protected, which also provides the oxide cation.

The lead cation may inhibit corrosion by forming moities such as soaps which are adsorbed on the substrate surface, but note that Leidheiser [94] records eleven mechanisms through which metallic cations may inhibit corrosion.

Calcium plumbate has found widespread use as an inhibitor pigment over galvanized metal and as a tie-coat over inorganic zinc-rich primers. It is believed that its mode of action is related to the alkalinity of the pigment and the formation of calcium, rather than lead soaps. Calcium plumbate is able to inhibit the formation of zinc soaps which act as a brittle interlayer [96].

Among the pigments which have been introduced, as alternative to lead and chromium are phosphates, phosphate borate, and molybdate, and a number of novel composite inhibitors [104]. Hare and Fernald [105] draw attention to good results with magnesium tetroxide used in combination with calcium borosilicate. They also note that alkaline extenders, including wollastonite, have a synergistic effect with some inhibitors, perhaps by acting as pH buffer. Goldie [106] describes new inhibitive pigments which operate on an ion-exchange principle.

New developments in inhibitor chemistry are of major interest in the heavy industrial, marine, and automotive areas. For many general-purpose building applications zinc phosphate has become a useful 'standby'.

The mechanism of zinc phosphate as an inhibitive pigment is not fully established. Proposed mechanisms include the adsorption of ammonium ions, complex
formation, and passivation through a phosphating process. Turner & Edwards [107]
found little evidence of phosphating but noted anodic and cathodic polarization.

9.11.3.5 Cathodic protection with zinc-rich primers
Paints containing high levels of zinc can protect steel by a sacrificial cathodic
mechanism. Zinc dust still remains the most widely used corrosion-inhibiting
pigment. Protection by zinc-rich primers often lasts longer than can be explained by
the cathodic protection ability, and it is thought that this extra protection arises from
precipitated corrosion products which fill pores in the paint and reduce further attack
[97]. The use of zinc pigment is associated with both inorganic and organic binders;
ethyl silicate and two-pack epoxy respectively, are typical. In designing zinc-rich
primers attention should be paid to Λ, the reduced PVC, as there will be a significant
difference in performance according to whether or not the zinc particles are in
contact and the electrical properties of the binder that separates them. The
geometrics of zinc-rich primers and associated effects on pigment loading have been
reviewed by Hare *et al*. [108].

9.11.4 Design of anticorrosive paints
Requirements of an anticorrosive paint include:

— a binder resistant to alkaline hydrolysis (i.e. resistant to cathodically formed
   alkali)
— an adhesion mechanism which is not destroyed by the presence of alkali (thus
   carboxylate groups are likely to be vulnerable)
— the presence of a mechanism through which the corrosion process is actively
   inhibited
— low moisture and water vapour permeability

In addition, it would be an additional benefit if some means of neutralising or
buffering alkali was present.

9.11.4.1 Paint systems for ferrous metals
When designing coatings to suit a specific application, it is usually advantageous to
design the complete system including primers and topcoats, and this will certainly be
true at the 'heavy' end of the building paints market. At the 'lighter' end of the
market it may be sufficient to combine an anticorrosive primer with general-purpose
undercoats and topcoats. Primers for metal are thus an important sub-sector of the
general buildings paint market. For reasons which have been outlined above, low
permeability is likely to be advantageous in protecting metals without any of the
corresponding problems which beset wood or masonry. Binders for metal coating
parts are thus chosen for impermeability and used at high film thickness — up to
250 μm — in a severe environment.

A need for low permeability has restricted the use of waterborne dispersion
polymers, many of which are very permeable, though the cholorine-modified types
referred to earlier [82] are more suitable. Other problems include flash-rusting, and

poor coalescence on cold metal in a damp environment. These problems have restricted the use of waterborne coatings on metal as opposed to other substrates. Technical progress has reduced or overcome some of these problems, and the use of waterborne coatings for metals is increasing. Nonetheless, coalescence under adverse conditions will remain a problem.

In discussing the needs of different substrates, the importance of preparation has been stressed. This is especially true of ferrous metals and in constructional work; steelwork should be prepared and primed before delivery on site. BS 6150 and BS 5493 review major methods of surface preparation which include blast cleaning, acid pickling, and flame cleaning. The manual methods likely to be used on site are not really satisfactory for best performance, and it will be difficult to remove all rust and scale deposits, underlying the need for anticorrosive as well as barrier film properties to provide maximum protection.

9.11.4.2 Primers for (ferrous) metal

Etch primers
'Etch primers' (wash primers) are more appropriately considered as part of the pretreatment process. They are used to promote adhesion (especially with non-ferrous metals) and give a measure of temporary protection to ferrous metals. They must be overcoated with a fully pigmented primer.

Etch primers are normally based on polyvinyl butyral resins, which show good adhesion to metal. This may be improved with phosphoric acid (as a separate component in a two-pack product) or phenolic resins in one-pack products. Traditionally, UK products are tinted yellow and blue respectively for identification purposes. Two-pack products are often pigmented with zinc tetroxychromate, but at a low (10%) total volume solids. The one-pack products frequently contain zinc phosphate with a total volume solids around 20%.

Drying oil primers
A substantial number of general-purpose metal primers have long been based on raw or processed drying oils.

Improved alkali resistance and, therefore, performance is gained by modification with phenolic resins. Problems of drying and re-coat have led to greater use of alkyd resins which may themselves be further modified with phenolic resins, or with styrene or vinyl toluene, to speed dry further. Alkyds can be modified with chlorinated rubber to improve water resistance. Hardness is gained by forming a urethane alkyd. Other typical binders containing drying oils are the epoxy esters which are also fast-drying but prone to yellowing.

Chlor rubber-based primers
Chlorinated rubber, with its good chemical resistance, is a very suitable base for anticorrosive paints, and has become widely used in all aspects of steel protection, including bridge paints, chemical plant, and pipelines. Chlor rubber paints form a film through solvent evaporation, and dry quickly relative to oil and alkyds, especially in cold weather where application as low as −20°C is possible. They have

inherently low vapour permeability and can be applied at high film thicknesses, and when applied by airless spray will give dry film thicknesses of up to 125 μm per coat.

Primers for metal covered by British Standards include red lead (BS 2523, types A, B, C), calcium plumbate (BS 3698, types A & B), and zinc-rich (BS 4652, types 1, 2, 3).

Water-based primers for metal
For good general maintenance of metal in domestic and light-industrial buildings increasing use of waterborne binders can be expected where good drying conditions can be expected, i.e. warm and dry. Both styrene acrylic and acrylate modified vinyl chloride — vinylidene chloride copolymer binders have become established. One problem that must be overcome is that of flash-rusting, which is a different phenomenon from that of early rusting. Early rusting can be expected from any waterborne coating which is applied thinly and/or under unsuitable conditions where the film does not properly integrate. Flash-rusting happens almost immediately on application while the film is still damp, and is more noticeable on fresh metallic surfaces. (It may also be a problem over nail heads for waterborne wood primers.)

Flash-rusting is reduced by designing a resin with barrier properties and incorporating corrosion-inhibiting pigments. The presence of sodium nitrite (also used to reduce can corrosion) is also beneficial.

Starting formulations are widely available from latex manufacturers (e.g. [109, 110]).

9.11.4.2 *Non-ferrous metals* [111]
Non-ferrous metals likely to be encountered as coating substrates in domestic and light-industrial buildings include galvanized steel, aluminium, and relatively minor amounts of copper, brass, and lead. These metals will often have been left unpainted for many years; thus cleaning and surface preparation will usually be an important part of the overall painting process. With the softer metals it is inadvisable to use wire wools for cleaning, as these may become embedded in the surface and even cause electrochemical reactions; nylon pads are an acceptable alternative. Long exposure of coated steels may allow rusting of the base metal which will require careful preparation. Etch primers are often found to improve adhesion. A phosphoric acid treatment known as British Rail 'T-Wash' is sometimes effective (for formula see BS 6150:1962, 28.3.1.5).

When priming non-ferrous metals, many of the ferrous primers are suitable, but some pigments such as red lead and graphite may accelerate corrosion of aluminium.

Zinc-coated metal (galvanized, sheradized, or electroplated)
Metallic zinc coatings present a specific problem, caused by a reaction between the metal and acidic oil fragments. Resulting soluble salts cause intercoat failures which are often apparent on garage doors and window frames. Primers based on calcium plumbate have a proven track record in overcoming this problem, but there is increasing concern over lead in the environment, and this type of primer is being phased out. Suitable alternatives include acrylic latex-based paints, though these may not prevent adverse reaction when overcoated with oil and alkyd-based paints.

The low permeability of vinyl and vinylidene chloride modified acrylics is a specific advantage here.

Coating performance is almost invariably better over weathered rather than unweathered zinc.

Aluminium
Aluminium and its alloys will not normally require painting, but after a long weather period may show unsightly salt deposits which should be removed. Zinc chromate primers are well established, but zinc phosphate is more suited to domestic situations.

Lead and copper can usually be coated directly with alkyd-based gloss paint.

9.11.4.3 *Plastic as a substrate* [112]
The common plastics used in buildings do not present any special difficulties, but it should be noted that abrasion can initiate cracks, and preparation is often best limited to washing. Expanded polystyrene will be attacked by some solvents, and the overcoating of polystyrene ceiling tiles with gloss paints creates a fire hazard; only 'emulsion' paints should be recommended. BRE information paper 11/79 gives more detailed information.

REFERENCES

[1] Flick, E. W. *Waterborne paint formulations* (1975), *Solvent based paint formulations* (1977), *Industrial water borne trade paint formulations* (1980), *Handbook of paint raw materials* (1982), Noyes Data Corporation, New Jersey, USA.

[2] Dawoodi, Z. Ethoxy propanol: a substitute for ethyl glycol ether. *Polymer Paint Colour Journal* **176**, No. 4159, pp. 41 45, Jan. 22 (1986).

[3] Hesler, K. K. & Lofstrom, J. R., Applications of simplex lattice design experimentation to coatings research. *Journal of Coatings Technology* **53**, No. 674, 33 (1981).

[4] Steig, F. B., The influence of *PVC* on paint properties. *Progr. Organic Coatings,* June 1 (4), 3512 73 (1973).

[5] Steig, F. B., Pigment/binder geometry. *Pigment handbook*, Vol. III (T. Patton, ed.), Wiley-Interscience, New York, pp. 203–217 (1973).

[6] Bierwagen, G. P. & Hay, T. K. The reduced pigment volume concentration as an important parameter in interpreting and predicting the properties of organic coatings. *Prog. Org. Coatings* **3**, 281–303 (1975).

[7] Huisman, H. F. Pigment volume concentration and an interpretation of the oil absorption of pigments. *Journal of Coatings Technology* **56**, No. 712, pp. 65 79 (1984).

[8] Asbeck, W. K. Dispersion and agglomeration effects on coatings performance. *JCT* **49**, No. 635, 59 70 (1977).

[9] Castells, R. *et al.* Particle packing analysis of coatings above the critical pigment volume concentration. *Journal of Coatings Technology* **55**, No. 707, pp. 53–59 (1983).

[10] Cremer, M. The measurement and importance of adsorption layers in paints *European Supplement to Polymers Paint Colours Journal* Oct. 5, pp. 85–93 (1983).

[11] Cole, R. J. Determination of critical *PVC* in dry surface coating films. *Journal of Oil and Colour Chemists' Association,* pp. 776–780 (1962).

[12] Rasenberg, C. J. F. M. & Huisman, H. F. Measurement of the critical pigment volume concentration by means of mercury porosimetry. *Progr. Org. Coatings,* Sept. 13 (3/4) 223-235 (1985).

[13] Steig, F. R. Density method for determining the *CPVC* of flat latex paints. *Journal of Coatings Technology* **55**, No. 696, pp. 111–114 (1983).

[14] Steig, F. R. Numerical expressions of film porosity, *American Paint and Coatings Journal,* Oct. 9, pp. 46–50 (1978).

[15] Bernardi, P. Parameters affecting the *CPVC* of resins in aqueous dispersion, *Paint Technol.* **27**, 24 July (1963).

[16] Rowland, R. H. & Steig, F. B. Graphical solution to *CPVC* problems in latex paints. *Journal of Coatings Technology* **54**, No. 686, pp. 51–56 (1982).

[17] Patton, T. C. *Paint flow and pigment dispersion* 2nd ed., John Wiley Interscience, New York, Ch. 7, p. 172 (1979).

[18] Reddy, J. N. *et al.* Studies on adhesion: role of pigments. *Journal of Paint Technology* **44**, No. 566, pp. 70–75 (1972).

[19] Funke, W. On the relation between the pigment vehicle interaction and liquid water absorption of paint films. *J. Oil Col. Chem. Assoc.* **50**, 942 975 (1967).

[20] Seiner, J. A. Microvoids as pigments: a review, *Ind. Eng. Chem. Prod. Res. Dev.* **17**, No. 4, pp. 302–317 (1978).

[21] Kerker, M. *et al.* Pigmented microvoid coatings. *Journal of Paint Technology* **47**, No. 603, pp. 33–42 (1975).

[22] Chalmers, J. R. & Woodbridge, R. J. Air and polymer–extended paints. *European Supplement to Polymers Paint Colour Journal,* Oct. 5, pp. 94–101 (1983).

[23] Dulux Australia Ltd., PO Box 60, Clayton, Victoria 3168, Australia.

[24] Hislop, R. W. & McGinley, P. L. Microvoid coatings. *Journal of Coatings Technology* **50**, No. 642, pp. 69–77 (1978).

[25] Letters to the Editor. *Journal of Coatings Technology* **50**, No. 645, pp. 112–113 (1978).

[26] Goldsbrough *et al. Prog. in Org. Coat.,* **10**, 35 (1982).

[27] Ramig, A. & Floyd, F. L. Plastic pigment: A novel approach to microvoid hiding. *Journal of Coatings Technology* **51**, No. 658, p. 63 (1979).

[28] Ramig, A. & Ramig, P. F. Plastic Pigment, *J. Oil Col. Chem. Assoc.* **64**, pp. 439–447 (1981).

[29] Letters to the Editor. *Journal of Coatings Technology* **52**, No. 663, April and August (1980).

[30] Hill, W. H. *et al.* A more efficient additive for lower cost hiding. *Resin Review,* **XXXV**, No. 3, pp. 3–10 (1985).

[31] Rohm and Haas Co., Philadelphia, PA 19105, Ropaque OP-62, Trade Sales Data Sheets.

[32] Water holds healthier future for coatings. *Polymers Paint Colour Journal,* **175**, No. 4156, p. 825 (1985).

[33] Smith, N. D. P., Orchard, S. E. & Rhind Tutt, A. J. The physics of brushmarks, *J. Oil Colour Chemists Assoc. Sept.* **44** (9) 618–633 (1961).

[34] Solomon, D. H. The chemistry of organic film formers, 2nd ed. R. E. Krieger Publishing Co., Inc. (1977).

[35] Dillon, R. E. *et al.* Sintering of synthetic latex particles. *J. Colloid Science,* **6**, p. 108 (1951).

[36] Brown, G. L. Formation of films from polymer dispersions, *J. Polymer Science,* **22**, p. 423 (1956).

[37] Sheetz, D. P. Formation of films by drying of latex, *J. Appl. Polymer Sci.,* **9** (11), 3759–73 (1965).

[38] Barrett, K. E. J. (ed.) *Dispersion polymerisation in organic media,* John Wiley & Sons (1975).

[39] Bromley, C. W. A. & Graystone, J. A. Coating compositions, *BP,* **2**, 164,050 A.

[40] *Surface Coatings,* Vol. 1, Ch. 22 — Water reducible resins, OCCAA, Chapman & Hall Ltd. 2nd ed. (1983).

[41] Wide, E. G. Advances in paint dispensing. *Polymers Paint Colour Journal,* **175**, No. 4146, pp. 454–461 (1985).

[42] Atkinson, D. Developments in colour matching and dispensing of paints. *Paint and Resin,* Dec. pp. 29–30 (1985).

[43] *Colony powder colorants,* Color Corporation of America, 200 Sayre Street, Rockford, Illinois 61101.

[44] *Herbol colour pill.* Hoechst A. G., Frankfurt/Main, 80.

[45] *Wintermix colourant pastes,* Winter OY, 33270 Tampere 27 puh. 440300.

[46] Graystone, J. A. *The care and protection of wood,* ICI Paints Division, Slough SL2 5DS (1985).

[47] Derbyshire, H. & Miller, E. R. The photodegradation of wood during solar radiation. *Holz als. Roh-und Werkstoff,* pp. 341–350 (1981).

[48] *Principle of weathersealed timber window design,* The Swedish–Finnish Timber Council. 21/25 Carolgate, Retford, Notts, DN22 6BZ.

[49] *Exterior cladding, redwood and whitewood,* The Swedish–Finnish Timber Council.

[50] *BRE Technical Note* No. 40, Princes Risborough, Aylsebury, Buckinghamshire, HP17 9PX.

[51] Graystone, J. A. Wood coatings and the problem of classification. Paper 1. *PRA/BRE Symposium The Care and Maintenance of Exterior Timber, June 1986.*

[52] Graystone, J. A. Formulating for durability. *PRA Progress Report,* No. 5, pp. 27–33 (1984).

[53] Woodbridge, R. J. The role of aqueous coatings in wood finishing. *PRA Progress Report.* No. 5, pp. 33–38 (1984).

[54] Synolac 6005W. Cray Valley Products Ltd, Farnborough, Kent, England, BR6 7EA.

[55] *Jagalyd Antihydro.* Ernst Jäger, Fabrik chem. Rohstoffe GmbH, 4000 Düsseldorf 13, P.O. Box 130 380.

[56] *Experimental emulsion E-1615.* Rohm and Haas Co., European Operations, Chesterfield House, 15–19 Bloomsbury Way, London, WC1A 2TP.

[57] Boxall, J., Hayes, G. F., Laidlow, R. A. & Miller, E. R. The performance of extender modified clear finishes on exterior timber, *J. Oil Col. Chem. Assoc.* (9), pp. 227–233 (1984).

[58] See also, for example, BP 1,209 108; USP 3,719,646, BP 1,088,105.

[59] Vittal Rao, R. *et al*. Mechanical properties of alky resin varnish films. *J. Oil Col. Chem. Assoc.* **51**, pp. 324–343 (1968).

[60] Rhoplex, R. AC-73. Rohm and Haas Co., Philadelphia PA19105, *Trade Sales Coatings* (1979).

[61] Building Research Digest 198 *Painting walls:* part 2 (1982). Building Research Station, Garston, Watford, WD2 7JR.

[62] Jotischky, H., Coatings for concrete: the market framework. *Polymers Paint Colour Journal* **175**, No. 4153, pp. 740–742 (1985).

[63] Lea, F. M. A. The chemistry of cement and concrete, 3rd ed., E. Arnold (1970).

[64] Williamson, R. B. *Progress in Materials Science* **15**, No. 3 (1972).

[65] Midgely, H. G. The microstructure of hydrated Portland cement, *Proc. Brit. Ceram. Soc.* No. 13, pp. 89–102 (1969).

[66] Double, D. D. *et al*. The hydration of Portland cement. *Proc. R. Soc. Lond.* **A 359**, 435–451 (1978).

[67] Ilston, J. M., Dinwoodie, J. M. & Smith, A. A. *Concrete, timber and metals*, Ch. 8, Van Nostrand Company Ltd (1979).

[68] Powers, T. C. Effect of hydration on permeability. *Proc. American Concrete Institute* **51**, 285 (1954).

[69] Robson, T. D. *High alumina cements & concretes*, Wiley, New York (1962).

[70] Building Research Establishment Digest 139 *Control of lichens, moulds and similar growths* (1982).

[71] Pesticides Precaution Scheme. Health and Safety Executive, 15 Chapel St., London, NW1 5DT. (Control of Pesiticides Regulations SI 1510 (1986)).

[72] Building Research Establishment Digest 197 *Painting walls*. part 1 (1982).

[73] Building Research Establishment Digest 196 *External rendered finishes*.

[74] Building Research Establishment Digest 175 *Choice of glues for wood*.

[75] Building Research Establishment Information Paper IP22/79. *Difficulties in painting Fletton bricks* (1979).

[76] Building Research Establishment Digest 177 *Decay and conservation of stone masonry* (1975).

[77] BS 6150 *Code of practice for painting of buildings* (1982).

[78] GLC Bulletin No. 129 *Brickwork waterproofing solutions* (1980).

[79] *Silicone masonry treatments*. Dow Corning Corporation, Midland, Michigan 48640.

[80] Building Research Station Digest 125. *Colourless treatments for masonry* (1971).

[81] Clark, J., Fuller, H. & Simpson, L. A. *Some aspects of the exterior durability of emulsion paints*. Tioxide Group PLC Tech., Service Report D8672GC.

[82] 'Haloflex' *Vinyl acrylic copolymer latices*. ICI (Mond Div.) P.O. Box 14, The Heath, Runcorn, Cheshire, WA7 4QG.

[83] Aronson, P. D. Some aspects of film formation in emulsion paints, *J. Oil Colour Chemists Assoc.* **57**, No. 2, p. 59, (1974).

[84] Simpson, L. A. Influence of certain latex paint variables on exterior durability, *XIV Fatipec Congress Budapest,* p. 623 (1978).

[85] *Rohm and Haas Reporter* **XLIII**, No. 3 and Resin Review, Vol. XXXVI, No. 1 (1986).

[86] *Pliolite resins*, The Goodyear Tyre and Rubber Co., Akron, Ohio, USA.

[87] Myers, R. R. & Long, J. S. (ed.) *Treatise on coatings,* Vol. 4, Ch. 6, Marcel Dekker, New York (1975).

[88] *Solubility fractional parameters for pliolite resins*, etc., Compagnie Francais Goodyear, Avenue des Tropiques, BP 31-91402 Orsay.

[89] Sandford, R. & Gindre, A. A new way to safer paints, *Polymers Paint Colour Journal* **176**, No. 4166, p. 346–347 (1986).

[90] *PU-Paints for mineral substrates*, Bayer A. G., 5090 Leverkusen, Bayerwerk.

[91] BS 5493:1977. *Code of practice for protective coating of iron and steel structures against corrosion.*

[92] Whitney, W. R. *J. Am. Chem. Soc.* **25**, 395 (1903).

[93] Funke, W. Towards a unified view of the mechanism responsible for paint defects by metallic corrosion. *Ind. Eng. Chem. Prod. Res. Dev.* **24**, pp. 343–347 (1985).

[94] Leidheiser, H. Mechanism of corrosion inhibition. *Journal of Coatings Technology* **53**, No. 678, pp. 29–39 (1981).

[95] New England Society for Coatings Technology. Design of waterborne coatings for the corrosion protection of steel. *Journal of Coatings Technology* **53**, No. 683, pp. 27–32 (1981).

[96] Hare, C. H. *Federation series on coating technology — Unit 27.* Federation of Societies For Coatings Technology, USA (1979).

[97] Dickie, R. A. & Smith, A. G. *How paint arrests rust, Chemtech.* Jan. pp. 31–34 (1980).

[98] Mayne, J. E. O. The mechanism of the protective action of paints. *Corrosion 2*, 2nd ed. L. L. Shreir, ed., Newnes–Butterworths, London, pp. 15:24–37 (1976).

[99] Guruviah, S. Relationship between the permeation of oxygen and water through paint films and corrosion of painted mild steel, *J. Oil Col. Chem. Assoc.* **53**, 669 (1970).

[100] Symposium on the protection and painting of structural steelwork, 11–12 April 1984, *PRA Progress Report* No. 6, p. 12 (1984).

[101] Funke, W. Towards environmentally acceptable corrosion protection, *Journal of Coatings Technology* **705**, pp. 31–38 (1983).

[102] Carter, E. Formulation and manufacture of micaceous iron oxide paints, *Pigment and Resin Technology*, March, pp. 8, 12 (1984).

[103] Hare, C. H. Aluminium pigment rated best in barrier system study. *Modern Polymer Coatings*, Dec. pp. 37–44 (1985).

[104] Kalewicz, Z. Novel corrosion inhibitors. *J. Oil Col. Chem. Assoc.* Dec. pp. 299–305 (1985).

[105] Hare, C. H. & Fernald, M. G. Effect of prime pigment on metal primer performance. *Paint & Resin,* Dec. pp. 22–25 (1985).

[106] Goldie, B. P. F. Novel corrosion inhibitors. *Paint & Resin,* Feb. pp. 16–20 (1985).

[107] Turner, W. H. D. & Edwards, J. B. Zinc phosphate as an inhibitive pigment. *PRA Technical Report* TR/7/78 (1978).

[108] Hare, C. H. *et al.* Geometrics of organic zinc rich primers and their effects on pigment loading. *Modern Paint and Coatings* June, pp. 30–36 (1983).

[109] Burgess, A. J. *et al.* A new approach to the design of latex paints for the protection of steel. *J. Oil Col. Chem. Assoc.* **64**, pp. 175–185 (1981).

[110] *Rhoplex® acrylic emulsion for latex maintenance coatings.* Rohm & Haas Company, Philadelphia PA 19105, (1979).

[111] BRE Digest 71, *Painting in buildings*: 2 *Non-ferrous metals and coatings*, June (1966).

[112] Risberg, M. Plastics as a painting substrate, *J. Oil Col. Chem. Assoc.* pp. 197–200 (1985).

10

Automotive paints

D. A. Ansdell

10.1 INTRODUCTION

The demands and performance required for automotive coatings are considerable. There is a need for 'body' protection, such as anti-corrosion and stone chip resistance, and for a durable and appealing finish. Products also have to be appropriate to mass production conditions, and in this respect, must be robust, flexible, and economic to use.

Vehicle construction has gone through significant changes since its inception at the turn of the century. At the present time the substrate is generally of mild steel but may also contain other alloys and include plastic components; the shape is inevitably complex, and certain parts of the vehicle are very inaccessible and difficult to paint. Production rates are high, e.g. 45 units/hour, and this requires methods and material technology to meet the limitations this imposes.

Passenger car production throughout the world in 1983 (excluding USSR and Comecon countries) was 28.1 million vehicles utilizing of the order of 570 million litres of paint. In relation to these figures, the number of passenger cars produced in the United Kingdom was 1.044 million requiring 20.0 million litres of paint.

In world terms, these figures are broken down and represented diagrammatically below:

The paint products used are principally primers and surfacers (fillers), designated the undercoating system, and the finish or topcoat. In a modern painting system the relative use is broadly in percentage terms, primer:surfacer:topcoat :: 30:20:50. In different parts of the world there is often variation in product technology, particularly in topcoats, and this can have a significant influence on paint performance, specifications, and details of the process.

The basic objectives of the painting process are to protect and decorate. In order to achieve this the process is broken down into a number of different component parts. These parts, or 'layers', are applied in a specific order, and although the

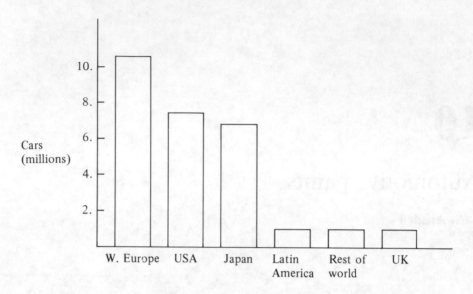

Fig. 10.1 — Passenger car production 1983.

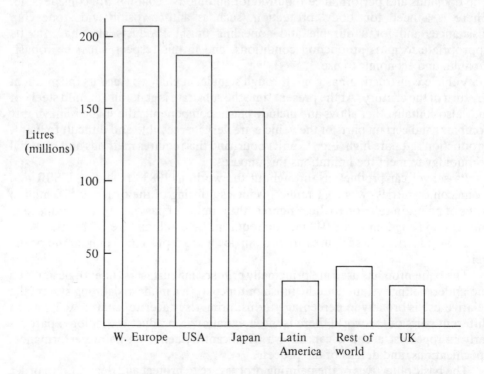

Fig. 10.2 — Paint volumes 1983.

function of each 'layer' is specific it relates very closely to the others to provide the desired balance of properties.

The component parts of the painting system are:

$$\left.\begin{array}{l}\text{Metal pretreatment} \\ \text{Primer}\end{array}\right\} \quad \text{Protection}$$

$$\left.\begin{array}{l}\text{Surfacer (filler)} \\ \text{Finish}\end{array}\right\} \quad \text{Decoration}$$

This fundamental process may be divided into three basic systems which may be classified as follows:

(1) Spray priming system
(2) Dip priming system
(3) Electropriming system (currently most widely used)

The reason for this classification is that the method of priming is the major difference between them; the subsequent process is virtually identical. All these three systems are used over suitable pretreatments.

Spray application is used for surfacers and topcoats. There are a variety of different types of spraying systems, e.g. air atomized (hand and/or automatic) and electrostatic methods. These are described in more detail later.

10.1.1 Spray priming system
This comprises:
Zinc-rich primer : internal sections only
Primer surfacer : 40–50 μm
Finish   : 45–55 μm.

The zinc-rich primer is applied to the internal surfaces of body components before welding of the body is undertaken. This is to give corrosion protection to such surfaces which are otherwise inaccessible for painting in following operations.

This system was used before the advent of dipping primers. It is now used only where low-volume production does not justify the cost, or plant, to dip or electro-prime bodies.

10.1.2 Dip priming system
This comprises:
Anticorrosive dipping primer: 12–18 μm
Surfacer   : 40–50 μm
Finish   : 45–55 μm.

This system was widely used on mass production lines before the introduction of electropriming in the early 1960s. It is still used on a very limited scale where low-volume production or other factors, such as cost, prevail.

10.1.3 Electropriming system
This comprises:
Electroprimer (anodic or cathodic) : 20–25 μm
Surfacer   : 23–35 μm
Finish   : 45–55 μm.

Electropriming is the currently accepted system in mass production plants. It is an efficient, relatively simple operation with a high degree of automation. Introduced in the early 1960s with anodic technology, now being rapidly overhauled by cathodic technology, it has set new standards in processing and corrosion protection.

In the processes outlined above all coatings need stoving; this is required to achieve very high levels of performance and to facilitate processing in a conveyorized production line. Stovings vary from 180°C for cathodic primers to as low as 80°C for repair finishes.

A typical modern paint process/plant is shown in Fig. 10.3 in schematic form.

10.1.4 Performance
Individual manufacturers have their own performance specifications, and while they may vary, for example, dependent on the type of finish and local plant/processing conditions, there are basic aspects which are common to all:

(1) Appearance (gloss and distinction of image)
(2) Durability (colour and gloss retention)
(3) Mechanical properties and stone chip resistance
(4) Adhesion
(5) Corrosion and humidity performance
(6) Petrol and solvent resistance
(7) General chemical resistance, e.g. acid resistance
(8) Hardness and mar resistance
(9) Repair properties.

The products required to achieve these properties are described later. They require extremely sophisticated technology and a lengthy process involving application and stoving. Costs in energy, labour, space, and capital are considerable, and future processes will have to take other factors into account, such as anti-pollution, whilst still continuing to improve performance standards.

10.2 PRETREATMENT

The pretreatment process has three purposes:

(1) To remove the mill and pressing oils ingrained in the steel and any other temporary protective coatings.
(2) To improve paint adhesion by providing an inert surface of metal phosphate which will give a better key for the subsequent primer layer.
(3) To provide a resistance barrier to the spread of corrosion under the paint film.

A normal process for pretreatment may be summarized as follows:

(1) Rust removal
(2) Alkali degrease
(3) Water rinse
(4) Metal phosphating
(5) Demineralized water rinse

Depending on individual requirements and throughputs, processes may be spray, spray-dip, or dip; the latter two are preferred for modern high-volume installations.

(1) *Rust removal*
The best method is by the use of mineral acid particularly where rust deposits are heavy or mill scale is present; phosphoric acid based materials are normally preferred.

Although phosphoric acid is slower acting than hydrochloric or sulphuric, any salt contamination is less, due to the low solubility of most metal phosphates.

(2) *Alkali degrease*
Alkali cleaners are widely utilized in dip and spray installations where the oil and grease are partly saponified and emulsified into the alkali solution. The physical force generated by the jetting of the cleaner significantly assists in the removal of any solid matter present on the surface of the unit.

Formulations vary and are dependent on a number of factors such as the type of oil or grease to be removed, metals to be processed, etc. However, there is a need to restrict the strength of the alkali to prevent deactivation of the metal surface. Typical alkali chemicals are caustic soda, trisodium phosphate, and sodium carbonate: these are used in conjunction with various types of detergents for emulsification of oils and lubricants.

(3) *Metal phosphate (conversion coating)*
There are two basic types of phosphates:

Iron phosphate—coating weight 0.2–0.8 g/m^2
Zinc phosphate—coating weight 0.5–4.5 g/m^2

As a generalization, increasing the phosphate coating weight increases the corrosion resistance and decreases the mechanical strength or adhesion of the subsequent paint coating. Clearly, coating weight is an important parameter, the value being a compromise designed to achieve an acceptable balance of properties.

Iron phosphate produces an amorphous coating of low weight used mainly on components such as refrigerators, washing machines, and metal furniture which do not have to stand up to rigorous corrosion conditions. The prime requirement for such components is mechanical strength/adhesion which makes the low coating weight of iron phosphate ideal for such purposes.

Zinc phosphate is used almost universally in automative paint processes, and is an integral part of the total paint system. Zinc phosphate coatings are unique in that they can be made to crystallize from solution onto the metal surface because of the existence of metal phosphates of different chemical form and the equilibrium that exists in aqueous acid solution.

A chemical equilibrium is set up as follows:

$$3\,Zn(H_2\,PO_4)_2 \quad \rightleftharpoons \quad Zn_3(PO_4)_2 \quad + \quad 4\,H_3PO_4$$

Zinc dihydrogen phosphate	Tertiary zinc phosphate	Phosphoric acid
Soluble	Insoluble	Free acid

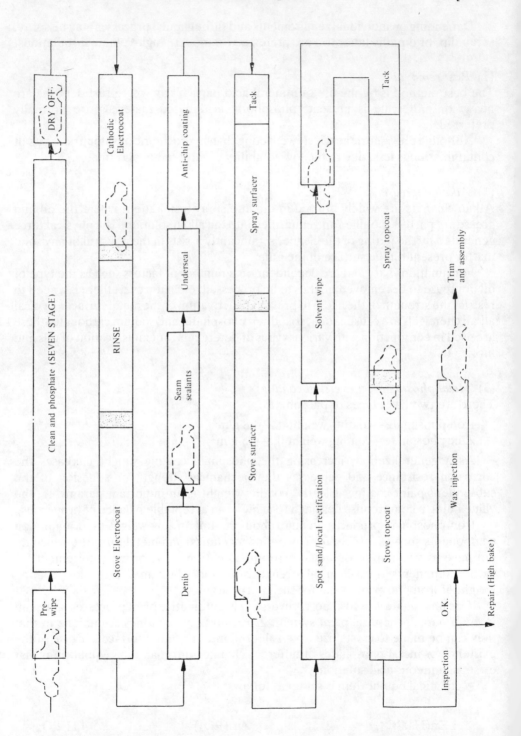

Fig. 10.3 — Paint plant process

This equilibrium will remain unchanged until an influence is brought to bear which will affect the concentration of the chemicals present. The initial reaction, therefore, in the phosphating process is the attack on the metal by the free phosphoric acid.

| Fe | + | $2H_3PO_4$ | $\rightarrow$ | $Fe(H_2PO_4)_2$ | + | H_2 |
| Substrate | | Free acid | | Soluble iron phosphate | | hydrogen |

The equilibrium in the first equation is disturbed, and the reaction moves to the right; the soluble zinc phosphate disproportionates to form insoluble zinc phosphate and free acid. Phosphate crystals start to grow and continue until the surface of the substrate is completely covered, at which time there is no more iron left with which the free acid can react, and the process stops. A build-up of soluble iron phosphate, which would ultimately slow down the reaction and poison the phosphate solution, is controlled by the addition of an oxidizing agent (toner) which precipitates the soluble iron phosphate as ferric phosphate (sludge).

Pretreatment as a corrosion inhibitor:mechanism
The steel surface, or substrate, is made up of a lattice of iron, carbon dissolved in iron, iron carbides, and undissolved carbon particles.

Rusting, or corrosion, is essentially an oxidation process aided by electrolytic action. On a moist steel surface electrolytic cells can be set up by the impurities in the metal (carbides, carbon, etc.) which give rise to cathodic (negative) and anodic (positive) areas. In the presence of water and salts (an electrolyte) these areas become the electrodes in a corrosion cell. Ferric ions form at the 'cathode', where the metal becomes pitted, and combine with oxygen in the air and hydroxyl ions from the electrolyte to form iron oxide and hydroxide. Corrosion consequently starts to take place, and to prevent it the corrosion currents must be prevented from flowing by insulating the metal surface.

Paint alone is a poor insulator since it is not impervious to an aqueous environment. Consequently, any damage to bare metal, by for example stone chipping, results in corrosion taking place which spreads under the film owing to the inherent lack of adhesion of virtually all paint films to untreated metal surfaces.

Phosphate, however, acts as the perfect insulator. The phosphate crystals grow from active sites on the surface of the metal until the coating is completely impregnated and covers the whole surface. This crystalline structure acts as a type of 'ceramic' insulator unaffected by water, particularly when sealed with the paint coating. Even damage due to stone chipping, or other forms of mechanical abuse, is restricted.

10.3 PRIMING

The function of the primer is to provide corrosion protection. This is achieved by formulating a product which incorporates anti-corrosive pigments carried in a resin system providing necessary mechanical, as well as anti-corrosive, properties.

Vehicle construction has changed considerably over the last 50 years to the extent that different techniques in applying the primer layer are necessary in order to ensure that all parts of the vehicle are coated. Table 10.1 identifies the evolution of vehicle priming up to the present day.

Historically there have been two major advances in the priming of motor vehicles. The first was a direct result of the automated techniques developed during the Second World War, i.e. the advent of the spot welder produced the welded unit or monocoque body shell. In this type of construction many areas of the unit are virtually inaccessible, for example, the various strengthening cross-members and the front end near door posts. This meant that spray application of primer was no longer feasible. As a consequence dipping procedures were introduced to make it possible to prime all parts of the vehicle, particularly the underbody.

Initially, dipping primers were solventborne, but in the 1960s waterborne primers were introduced to minimize fire and health hazards. However, although the waterborne systems failed to improve efficiency or economics they did lead on to the second major advance in autobody priming, i.e. applying the primer by electrodeposition.

10.3.1 Spray priming

This system utilises a zinc-rich primer, and sometimes other anti-corrosive coatings, which is applied to the internal surfaces of body components before 'welding-up' of the body. As a result corrosion protection is given to such surfaces which are otherwise inaccessible for painting in subsequent operations.

Before the advent of dipping primers this system was quite widely used. It is now used only where low-volume production does not justify the cost of plant to dip or electrocoat bodies, or where capital is not available to install such plant.

10.3.2 Dip priming
(1) *Products*

These materials are used to provide protection to internal parts of the vehicle body. Since no further coats of paint are applied in such areas, dipping primers are formulated to have good corrosion resistance and good adhesion to various types of phosphating pretreatment, as well as untreated metal.

Because of the size and complexity in shape of autobodies, dip primers must have good flow characteristics. Also, since dip tanks are large, to accommodate car bodies, there is a need for good stability.

There are four basic types of dip primers still in use:

(1) Alkyd
(2) Alkyd/epoxy
(3) Epoxy ester
(4) Full epoxy.

The majority of dip primers are solvent-based but waterborne variants are available. The most widely used in the industry were alkyd-based since they gave an acceptable balance between performance and cost. Epoxy types were introduced later to give improved corrosion resistance, but are more costly and less easy to control in dip tanks.

Table 10.1 — Evolution in car body priming.

Period	Body type	Primer type	Features
Pre-war	Body/chassis	Spray primer	All areas accessible
Post-war-1960	Welded unit construction (Monocoque)	Solvent-based dip primer	Laborious
			Uneconomical
			Poor internal protection
			Fire/health hazard
1960–1966	Welded unit construction (Monocoque)	Water-based dip primer	Reduced hazard
			Otherwise similar to solvent/dip process
1966–1977	Welded unit construction (Monocoque)	Anodic electropaints	Automated
			High production rate
			Reduced hazard
			Efficient
			Uniform coating
1977–	Welded unit construction (Monocoque)	Cathodic electropaints	Superior protection
			Good stability
			Greater penetration

(2) *Pigmentation*

Traditonally, these were red oxide type products, but the more common colour now is either black or grey.

Prime pigments:
    Red oxide and carbon black
Zinc chromate is often included for anticorrosive properties

Extenders:
    Barytes (barium sulphate): hard with good blister resistance
    Blanc fixe (precipitated barium sulphate): helps maintain dip tank paint stability
    Pigment volume concentration (*PVC*): Normally 20–30%

(3) *Process*

This process is now used only on a very limited scale where low-volume production or other factors such as economic considerations are prevalent.

Dipping of the car body can be shallow ('slipper' dipping) up to windows, or full immersion. In the 'slipper-dip' process the short dipping distance gives comparative freedom from runs and sags, and the next coating of surfacer can be applied 'wet-on-wet' after appropriate draining and air-drying. The two materials are then stoved as a 'composite'.

In the case of deeper dips the complex shape of the car body gives more runs and sags which have to be removed by solvent wiping etc., before surfacer application. Sometimes the primer is stoved and rectification carried out before surfacer application.

An interesting form of dip priming used in the mid-1960s was roto dipping. In this process the body was mounted transversely on a spit and, by means of a suitable conveyor, carried through the pretreatment (dip zinc phosphate), primer, and then stoving operation.

The advantage of this method is the thorough pretreatment and even distribution of primer to all surfaces. However, the plant required was expensive and high in maintenance costs, and could be operated at only relatively low line speeds. These and other factors led to it being superseded by electropriming in common with most other dipping processes.

In summary, dipping methods are dirty, laborious, and wasteful. The advent of electropriming showed great advantages (refer 10.3.3), giving greater internal protection and eliminating solvent 'reflux' in box sections, a feature inherent in both solvent and water-based dip primers.

10.3.3   Electropainting

Electropainting of motor vehicles began in 1963, some time after the principle had been established in other areas, although the technology was conceived in the 1930s. The Electrocoat system consists of the deposition of a waterborne resin electrolytically, the resins employed being polyelectrolytes.

At the present time virtually all mass-produced vehicles are primed in this manner; the body shell is immersed in a suitable waterborne paint which is formulated as an anodic or cathodic resin system at very low viscosity and solids.

10.3.3.1   *Anodic Electrocoat*

Until 1977 all electropaints used in the automotive industry were of the anodic type, primarily because the resin chemistry was relatively simple, readily available, and adaptable to the needs of the motor industry.

(1) *Resin systems*
There are four main resin systems:

(a) Maleinised oil
(b) Phenolic alkyd
(c) Esters of polyhydric alcohols (epoxy esters)
(d) Maleinized poly butadiene.

Earlier anodic electropaints were based on (a) and (b), but these were superseded by (c) and (d) because of their superior corrosion performance.

(2) *Pigmentation*
The pigmentation used in anodic Electrocoat (and cathodic) should

(a) be simple;
(b) be non-reactive, i.e. stable (e.g. to stoving) and pure, containing no water-soluble contaminants (e.g. chloride and sulphate);

(c) in one form, or another, reinforce or improve corrosion performance.

Pigment volume concentration — normally in the region of 6%.

Prime pigments:
High-grade rutile titanium dioxide, together with a minimal amount of carbon black, in greys, and a very good quality red (iron) oxide for reds.

Extender pigments:
High grade, free from soluble ionic species, such as chloride and sulphate.

Anticorrosive pigments:
A number are available; a typical example is lead silico chromate.

(3) *Mechanism of deposition*
This may be summarized as follows:
During electrodeposition three basic reactions occur: electrolysis, electrophoresis, and electroendosmosis. Note: Solubilization of the resin:

$$R\,COOH \quad + \quad KOH \quad \rightarrow \quad R\,COOK \quad + \quad H_2O$$
$$\text{(insoluble)} \qquad\qquad\qquad\qquad\qquad \text{(soluble)}$$

e.g. R=epoxy/ester
(The base used here is potassium hydroxide, but others such as amines are also utilised)
(i) Electrolysis
(a) Electrolysis:

$$R\,COOK \rightleftharpoons R\,COO^- K^+$$

(b) Electrolysis of water:

$$4\,H_2O \rightleftharpoons 3\,H^+ + 3\,OH^- + H_2O$$

Reduction at cathode:

$$H_2O + 3H^+ + 4e \rightarrow 2H_2 + OH^-$$

Alkaline reaction from OH^- present: can be removed by electrodialysis.

Oxidation at anode:

$$3\,OH^- - 4e \rightarrow H_2O + O_2 + H^+$$

— acidic reaction from H^+ present

i.e. $R\,COO^- + H^+ \rightarrow R\,COOH$
insoluble (electrocoagulation)
deposited coating

Metal dissolution — minor reaction: $M - ne \rightarrow M^{n+1}$

(favoured by presence of contaminating ions).

(ii) Electrophoresis
When a voltage is applied between two electrodes immersed in an aqueous dispersion the charged particles tend to move towards their respective electrodes. This effect is called electrophoresis.

In the Electrocoat process, when a current is applied negatively charged paint ions move towards the anode (positive). Electrophoresis is the mechanism of movement, and does not contribute to the actual deposition of the paint on to the anode.

(iii) Electroendosmosis
The layer of electrodeposited paint on the anode will contain occluded water. This is driven out through the porous membrane formed by the paint by the process called electroendosmosis.

(4) *Practical considerations*
In practice the process simply involves the passing of a direct current between the workpiece (anode) and the counter-electrode (cathode). The resultant coating deposited on the workpiece is insoluble in water and will not redissolve.

The film is compact, almost dry, and has very high solids which adhere strongly to the pretreated (zinc phosphate) metal substrate. It is covered by a very thin dip layer which is easily removed by rinsing. After stoving (165°C) the resin cures to form a tough, durable polymeric film.

The amount of paint deposited is largely dependent on the quantity of electricity passed. As the film is deposited its electrical resistance increases and the film becomes polarized when the resistance is so great that it stops the flow of current and halts the deposition. Thus once an area has been coated the local increase in resistance forces the electrical current to pass to more remote areas. This effect ensures progressive deposition, or 'throw', into box sections and internal or recessed areas.

The result is exceptional uniformity of film thickness free from runs and sags and other conventional film defects. The process is very suitable for mass production conditions, facilitating a primer coating over the whole surface of a welded monocoque carbody. It can be fully automated and is highly efficient in terms of paint utilisation.

(5) *Basic plant requirements*
(a) A powerful paint circulating system to preclude pigment sedimentation at the low solids (~12%) of operation and maximize paint mixing.
(b) Sophisticated refrigeration and filtration to maintain stability and remove contamination.
(c) A smooth and stable direct current supply with suitable 'pick-up' gear to the carbody.

(d) A paint rinsing system to remove the adhering dip layer. Incorporation of ultrafiltration (see 10.3.3.2) maximizes water and paint utilization as well as serving as a method of decontamination.

(6) *Control methods*

The alkali bases generated at the cathode (see *Mechanism of deposition* above) by electrolysis during electrodeposition will increase the pH of the paint and must be removed to maintain optimum properties. This can be done in two ways:

(a) Electro-dialysis
This uses a selective membrane (ion exchange) around the cathode. Base ions are allowed to pass through the membrane, but the anions (colloidal paint particles) are prevented from doing so.

(b) Acid feed/base deficient feed
This is self-explanatory. The feed, or top-up material, is partly neutralized to compensate for the base generated during the process of deposition.

The difference in method of pH control gives rise to differences in other aspects such as resin system, means of neutralizing, and type of plant. Electrodialysis has the advantage of being fully automated with a consistent feed material to the tank; the ion exchange membrane having a considerable life unless subjected to physical damage.

(7) *Deficiencies of anodic electrocoat primers*

(a) Phosphate Disruption
During anodic electrodeposition very high electrical field forces occur which rupture many of the metal–phosphate bonds (phosphate disruption). This leads to a weakening of the adhesion of the phosphate coating to the steel substrate.

When subsequent paint coats are applied, stresses are set up in the total paint film which, when damaged to bare metal, tend to cause the total paint film to curl away from the damage point. In early anodic systems, as soon as corrosion started at any point of damage the weakened phosphate layer allowed the paint to peel back, exposing more metal already chemically clean and very prone to corrosion. This failure was designated 'scab corrosion'.

This property was improved by reducing the incidence of rupture of bonds by simply increasing their number, i.e. phosphate coatings of densely packed fine crystals.

Such phosphate coatings are of low coating weight ($1.8 \, g/m^2$), and their introduction led to a marked reduction in the incidence of scab corrosion.

(b) Poor saponification resistance
The chemistry of anodic electropaints is such that they are formulated on acid resin systems. As a consequence the deposited stoved film, when exposed to an alkaline environment, will tend to form metal soaps soluble in water.

When damage occurs to bare metal, salt (as caustic) will simultaneously attack the steel substrate and the Electrocoat film, producing rust and the sodium salt (soap) of the anodic resin. This dissolves the primer coating, leading to a loss of adhesion of the remaining paint film and general corrosion problems.

10.3.3.2 *Cathodic Electrocoat*

Cathodic electroprimers had always been acknowledged to be theoretically desirable because of:

(1) Their anticipated freedom from substrate disruption.
(2) The fact that cationic resins being 'alkaline' in nature would tend to be inherent corrosion inhibitors free from saponification.

However, the complexity of the required resin systems precluded their early introduction.

In the early 1970s successive outbreaks of serious motor vehicle corrosion in North America, combined with legislation for minimum corrosion standards, promoted the US Motor Industry to develop a process of guaranteed corrosion protection. This situation gave the impetus for the development of cathodic Electrocoat in the USA, and it was introduced into the North American motor industry in 1978.

(1) *Resin system*

All current cathodic products are based on epoxy/amine resin systems which are stabilized in water by neutralization with various acids. These products require a crosslinking agent to be present with optimum film properties being developed at high stoving temperatures (165–180°C).

(2) *Pigmentation*

The basic properties are similar to those required for anodic Electrocoat — simplicity, high purity, and the ability, in some form, to supplement corrosion performance and catalyse the curing process.
Pigment volume concentration — normally in the region of 10%.
Colour:
The most popular colours vary from medium grey through to black, and the prime pigments and extenders align very closely to those used in anodic technology.

(3) *Mechanism of deposition*

The same principles of electrochemistry apply as in anodic deposition. Key differences are as follows:

Solubilization of the resin:

$$
\begin{array}{ccc}
R & & R \\
| & & | \\
R{-}N{:} + R'COOH \rightarrow & R{-}N{-}H^+ + R'COO^- \\
| & & | \\
R & & R
\end{array}
$$

(R = epoxy) (solubilizing acid)

(a) Electrolysis of water

$$4H_2O \rightleftharpoons 3H^+ + 3\,OH^- + H_2O$$

(b) Reduction at the cathode

$$H_2O + 3H^+ + 4e \rightarrow 2H_2 + OH^-$$

$$\begin{array}{ccc} R & & R \\ | & & | \\ R-N-H^+ + OH^- \rightarrow & R-N: & +H_2O \\ | & & | \\ R & & R \end{array}$$

Insoluble deposited coating

(c) Oxidation at the Anode

$$3\,OH^- - 4e \rightarrow H_2O + O_2 + H^+$$

acidic reaction
H^+ *removed by electrodialysis*

$$R'COO^- + H^+ \rightarrow R'COOH$$

(4) *Performance characteristics*
Cathodic technology rapidly replaced anodic products in North America since results from test track evaluations, and the field, confirmed the superior corrosion protection of cathodic systems. Where the cathodic system is particularly good is in its throwing power, being better than most anodics, and in its thin film ($\sim$12 μm) performance particularly on unpretreated steel. This is found to be extremely important in box sections etc., where the durability of the product is most vulnerable.

The Japanese automotive industry was also quick to change, with Western Europe following, even though European anodic technology was adjudged as being quite satisfactory at the time. Cathodic plants are now to be found in all parts of the world, e.g. Taiwan, Malaysia, Indonesia, Brazil, and the Philippines, with the overall objective to maximize standards.

Fig. 10.4 clearly demonstrates the improvements made in corrosion protection, as measured by the ASTM salt spray test.

(5) *Plant requirements*
Cathodic installations (Fig. 10.5) are basically similar to anodic plants in terms of paint circulation, refrigeration, filtration, and the requirement for a smooth and stable direct current power supply. However, there is a fundamental need to introduce either materials which are fully acid resistant or to use suitable acid-resistant coatings on mild steel. Stainless steel is the most common material, and pumps, valves, and piping are often made of this. The high cost of stainless steel can often preclude its use, and it is replaced with PVC and other plastics where feasible.

The Electrocoat tank and rinse tunnels also need to be lined with acid-resistant

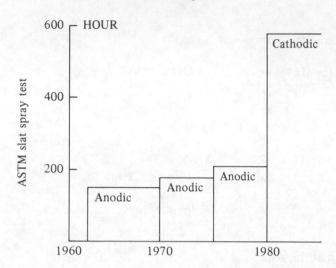

Fig. 10.4 — Improvements in corrosion protection.
Note: Anodic and cathodic electropaints compared at equivalent film thickness (18–22 μm)
over a suitable zinc phosphate pretreatment.

materials, although there are instances where stainless steel is used for rinse tunnel construction.

(6) *Dip rinsing*
Cathodic electropaints operate at relatively high solids (20–22%), and the most effective method of rinsing has been found to be dipping. In this way 'drag-out' materials can be removed more effectively than the normal spray rinse procedure. Clearly, the more effective the rinsing, the better the economics.

The principal feature in design is to ensure turbulence in the dip rinse tank to effectively remove the dip layer (or 'cream coat').

(7) *Ultrafiltration*
General
Ultrafiltration is pressure-induced filtration through a membrane material of very fine porosity. In the case of electropaints it can be regarded as a process which separates the continuous phase (water, solvent, and water-soluble salts) from the dispersed phase (resin and pigments). The velocity of the paint required to drive the process is about 4–4.5 m/s (12–15 ft/s) or 150 litres/min. To avoid pigment precipitation in the tubes, and subsequent flux decay, this flow must be maintained.

Purpose of ultrafiltration

(a) Saving of paint
This is the most important consideration. By utilizing the ultrafiltrate to rinse or wash the 'drag out' paint (dip layer or cream coat) back into the tank the efficiency of paint usage can be improved by 15–20%.

(b) Decontamination

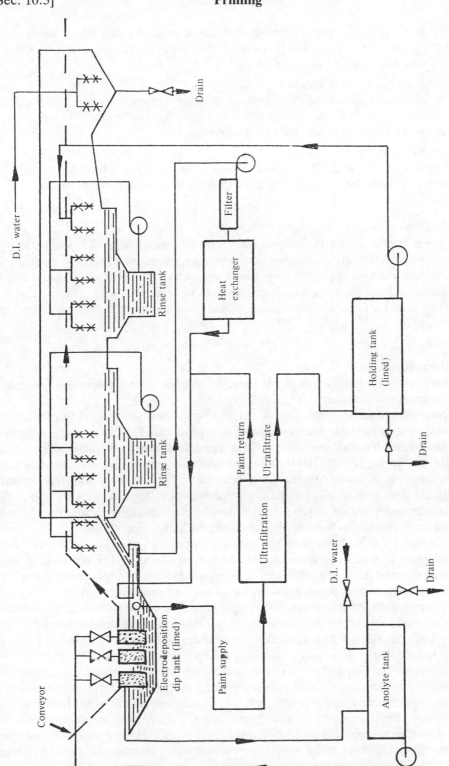

Fig. 10.5 — Cathodic electrocoat plant.

Soluble salts, which are derived from carry-over materials and which can significantly degrade performance and processing, can be purged or removed from the dip
tank by simply discarding ultrafiltrate.

(c) Control of paint performance
In general terms, with cathodic Electrocoat, purging is essential to maintain the paint
within certain specified limits (normally 800–1200 μS). This is needed to avoid any
reduction in performance and for ease of processing.

(d) De-watering
When the tank solids are too low and tank value is high, solids may simply be
increased by discarding ultrafiltrate and replacing with fresh paint.

(8) *Control method*
The pH is controlled by the system of electrodialysis. A selective, ion exchange
membrane is used around the anode, and acid ions, generated during the electrolytic
process, are allowed to pass through the membrane and are simply flushed away.
Control is automated, using a recycling anolyte system, and the conductivity of the
recycling liquid is maintained by integrating a conductivity meter with a de-ionized
water supply.

(9) *Pretreatment*
In relation to cathodic Electropaints there have been quite significant advances in
pretreatment development.

Spray phosphate systems suitable for anodic systems were found to give loss of
adhesion under 'wet' conditions with cathodic products. The cause of this adhesion
loss has been identified as cleavage of a secondary phosphate layer (chemically
known as hopeite, $Zn_3(PO_4)4H_2O$). This secondary phosphate has been termed
'grass' because of its appearance and because it 'grows' on top of the primary
phosphate layer (chemically known as phosphophyllite, $Zn_2 Fe (PO_4) 4H_2O$). In the
anodic electrodeposition process this 'grass' is largely detached because of substrate
disruption, but remains basically intact during cathodic deposition.

The high cohesive strength of the cathodic epoxy film exerts a pull on the weak
secondary layer when swollen with the absorption of water. The net result is that
cleavage takes place at this interface, which manifests itself in loss of adhesion.

As a result of this problem, with its associated mechanism, iron-rich homogeneous zinc phosphate coatings of fine tight crystal structure were developed which
had a minimal secondary phosphate layer. Added insurance can be provided by a
post-phosphate rinse with a weak chrome solution. This acts as a 'knitting process'
which simply 'knits' the crystals together, aiding homogeneity.

European and North American manufacturers saw this post-phosphate route as
an essential feature to maintain consistency of quality and performance. However,
because of environmental restrictions it was unacceptable in Japan. This led to dip
phosphating being adopted in Japan, since it had already been established that, by
utilizing the more controlled diffusion conditions afforded by dip processes, a higher
degree of homogeneity in the phosphate coating could be achieved, i.e. the process
favours the formation of the primary phase.

(10) *General appraisal and current developments*
Cathodic electroprimers have led to major advances in corrosion protection. The
levels of improvement have been significant; and, in association with wax injection
which is often applied in box sections, six years of anticorrosion and freedom from
perforation can be confidently expected. The product/process has shown itself to be
stable and consistent in performance, a most important feature in mass production
terms.

Newer developments are coming on to the market. These include a higher build
(35 μm) product which minimizes the use of primer surfacer, and lower stoving
versions. The latter products, apart from offering modest savings in energy, also
facilitate the use of certain plastic components in the body shell.

10.4   PRIMER-SURFACER/SEALER

When an assembled bare steel carbody arrives for painting, the paint process, apart
from protection and decoration, is expected to provide a degree of 'filling'. This is to
hide any minor imperfections and/or disc marks arising from the pressing and
assembly operations.

The phosphating process has no filling potential, and conventional dip primers
and electroprimers, because of their low solids, have minimal filling properties. In
addition, with electroprimers, the inherent nature of the deposition process, tends to
emphasize any metal defects.

As a consequence the primary function of the surfacer, or 'filler' as it was known,
is simply to 'fill', together with a number of other properties. The key properties are
mechanical (stone chip resistance), flexibility, resistance to moisture, and a good
even surface to maximize the appearance and performance of the appropriate finish.

The development of suitable materials has, to some extent, been influenced by
the improving standard of pressings and the evolution of new priming techniques
such as Electrocoat. Early surfacers were often formulated to give very high build
(40–50 μm), were porous in nature, and required heavy sanding to obtain a smooth
surface. Often, because of the porous nature of the surfacer, a further coat or sealer
(lowly pigmented) was applied prior to finish application. Also they were, and still
are, required to perform over dip/spray primers and pretreated metal, and they need
good adhesion to either substrate.

Early surfacers were formulated on alkyds, superseded by epoxy esters and more
recently by polyester products. Solvent based surfacers tend to predominate,
although waterborne materials, because of their pollution advantages and good
levelling, are attracting more attention.

Epoxy-based products have been the most commonly used over anodic electropaints because of their excellent adhesion and water resistance. They were used in
the early days of cathodic electroprimers, and minimal or non-sanding versions were
developed to reduce the labour-intensive sanding operation.

However, during the past few years the advent of basecoat clear metallic topcoats
has had a marked effect on the use of epoxy-based surfacers. Their poor resistance to
erosion from UV radiation (chalking at the interface between topcoat and surfacer)
when used under basecoat/clear metallics led to finish delamination. As a result of
this, polyester surfacers have been developed which have far superior UV resistance.

Polyester surfacers are now universally recognized as the best for all-round properties over cathodic electroprimers. The product, together with accepted standards for chemical and physical properties, is designed for minimal or non-sanding to maximize its suitability for production purposes. There is now a prerequisite for the unsanded adhesion between surfacer and topcoat to be of an acceptable level.

10.4.1 Resin composition

Solvent-based: Alkyd ⎫ Reacted with a suitable

           Epoxy ester ⎬ melamine and/or urea

           Polyester ⎭ formaldehyde resin

Stoved 20 minutes at 150–165°C (effective) to form a crosslinked insoluble film.

In alkyd and/or epoxy ester products the mechanism of cure is often a combination of autoxidation and thermosetting. Esters are usually based on linseed, tall, or dehydrated castor oils. These are used to give an acceptable degree of film flexibility, hardness, and insolubility within the specified stoving limits.

Oil-free -polyester products are 'pure' thermosetting, and the necessary flexibility is built into the polymer structure, for example by using long-chain synthetic fatty acids.

10.4.2 Pigmentation

Primer surfacers are normally highly pigmented with pigment volume concentrations (*PVC*) varying from 30–55%, depending on the physical and chemical requirements and the topcoat to be applied. The basic pigmentation is normally a blend of prime pigment (for opacity and colour) and extender pigments (for economic reasons and where performance considerations allow).

Application
Traditionally, primer surfacers were spray applied in two coats wet on wet over dip primers or phosphated metal (film thicknesses were ~50 μm). The first coat was a red (iron) oxide colour followed by the second, normally grey, which was a guide coat to control sanding. The advent of Electrocoat has meant that the red oxide coating has been dropped, and now only grey or off-white coloured primer surfacers are used, normally at film thicknesses of 30–35 μm.

Prime pigments
Red (iron) oxide or a good grade rutile titanium dioxide (Anatase has poor UV stability, which precludes its use).

Extender pigments
Extender pigments are much cheaper than prime pigments, and, apart from reducing cost, can by careful selection improve certain properties of the finished product or dry coating. Proper choice can improve properties such as consistency, levelling, and pigment settling of the paint. Other extenders can reinforce the dry coating's

mechanical properties, and others the resistance to moisture and blistering. The more common extenders are:

Barytes (barium sulphate): Prime extender, best for blister resistance. Hard, and has poor rubbing properties.

China clay (aluminium silicate): Good rubbing properties. Has a high oil absorption which tends to reduce gloss.

Winnofil (stearate coated calcium carbonate): Provides structure and anti-settling properties.

Talc (magnesium silicate): Early talcs had high water solubles. Now considerably improved. Good for improving sanding and cheapening.

Pigment volume concentration (PVC)

Pigment volume concentrations vary from 30–55% depending on the type of topcoat (e.g. acrylic lacquers require surfacers of *PVC* 55%) and specific properties such as mechanicals, appearance, and sanding properties.

10.5 INVERTED OR REVERSE PROCESS

The established undercoating system, following the introduction of anodic electro-painting, was pretreatment/electroprimer/surfacer or sealer. Later, the concept of inverting, or reversing part of this process, was developed; the objective being to overcome some of the fundamental weaknesses of anodic Electrocoat and to maximize the performance of the paint components of the system. A typical inverted process may be designated as follows:

Pretreatment (spray zinc phosphate)

↓

Spray surfacer to outer skin panels
(stove or partly stove)

↓

Electroprime (normal procedure)

↓

Final stove

The main advantages are seen as:

(1) Higher film build and thus better corrosion protection on internal surfaces (due to insulating effect of coating on external and carbody interior).

(2) Improved filiform, scab corrosion, and blister resistance (no phosphate disruption on external surfaces).

(3) Better adhesion and stone chip resistance of full finishing system — inverted surfacers can be formulated with suitable properties and greater tolerance to phosphate variations than electroprimers.

(4) Easier painting of composite metal bodies, e.g. steel, zinc, and aluminium: the different electrical properties of metals can give rise to problems during electrodeposition.

452

A number of manufacturers did experiment with this process, using solvent-based surfacers or powder coatings. Its adoption was inhibited by a number of factors:
(1) Poor appearance/performance at the surfacer/electroprimer interface. But this problem may be resolved by the use of a waterborne surfacer having suitable electrical and compatibility properties.
(2) Capital cost required to modify existing plants.
(3) The advent of cathodic Electrocoat and the simultaneous development of improved pretreatments and process control.

Electro powder coating (EPC)
This is another example of a reverse process, but incorporating cathodic electro-primer. Originating in Japan, EPC is essentially a process in which a powder surfacer carried as a slurry in water is electrodeposited in a matter of seconds on to the phosphated carbody. Since the EPC has little or no throwing power it is confined to the outer skin panels. After rinsing and a set-up bake the carbody is then cathodic electrocoated.

The cathodic electrocoat does not deposit on the electrically insulated EPC surfacer but rather is used to maximum effect in protecting the floor area, box sections, etc., and it is formulated accordingly.

The process has much to commend it from the standpoints of automation, quality consistency, elimination of a spray booth, and the virtual absence of organic solvents. On the debit side, however, the present generation of EPC surfacers deposit films of variable thickness and exhibit pronounced mottle or surface texture which requires considerable sanding before topcoat application.

10.6 ANTI-CHIP COATINGS

Anti-chip coatings are recently introduced materials with the prime function of upgrading stone-chip resistance on vulnerable areas of a carbody. Particular areas are the door-sills, front and rear ends below the bumpers, and the underbody; and (depending on the design of the car) these coatings can be used up to 'waistline' level. They are applied as high-build products (50–100 μm) in combination with the primer surfacer in a wet on wet process.

Resin type
They are a one-pack product composed of a blocked aromatic isocyanate blended with a cycloaliphatic diamine. The film, stoved as a composite with the primer surfacer (e.g. 20 min at 150°C), has a rubber-like character and high mechanical properties.

Pigmentation
In addition to prime pigments the formulation can also include extender pigments,

although the amount of extender can affect mechanical properties. Thixotropic aids such as 'Aerosil' are beneficial in reducing sagging tendencies at the high film thicknesses used.

10.7 AUTOMOTIVE TOPCOATS

The differing and stringent demands made by the user, and the ultimate customer, have led to a number of different topcoat technologies. All have a different balance of properties closely aligned to particular test specifications and process requirements.

The key necessity of excellent durability, i.e. outstanding colour stability and resistance to UV degradation, combined with use in mass production, obviously calls for sophisticated materials.

There are two basic forms of finishes: Solid (or straight) colours, and metallics. These can be further divided into the following types:

(1) Alkyd or polyester finishes: Solid colours
(2) Thermosetting acrylic finishes (including non-aqueous dispersion — NAD — Technology): Solid colours and metallics
(3) Thermoplastic acrylic finishes: Solid colours and metallics
(4) Basecoat/clear metallic systems

Alkyd finishes remain the most widely used throughout the world for solid colours, mainly because of their low cost and ease of processing. Metallic finishes are normally based on thermosetting acrylic systems (or NAD), since alkyds have poor durability in metallics. Thermoplastic acrylic lacquers are still used to a considerable extent by General Motors for both solid colours and metallics.

More recently, basecoat/clear metallic technology has gained increasing prominence because of enhanced gloss, stylistic appeal, and outstanding durability: the clear being based on a thermosetting acrylic resin 'reinforced' by a suitable UV absorber.

10.7.1   Alkyd or polyester finishes
(1) *Basic chemistry*
Alkyd finishes are based on a class of resins produced by the reaction of alcohols (glycerol, glycol, etc.) and dibasic acids (phthalic anhydride), and modifying with a natural or synthetic oil to give the designed balance of durability, flexibility, hardness, etc. The 'oil' or fatty acid is selected for its good colour and non-yellowing (non-oxidizing) characteristics, e.g. coconut oil fatty acid or 3, 5, 5 trimethyl hexanoic acid, and simply acts as a reactive plasticizer. Typical alkyds are short/medium oil length ($\sim$35%).

The crosslinking reaction, needed to form an insoluble film, takes place between the alkyd (or polyester) and a melamine formaldehyde condensate involving N methylol groups or their ethers.

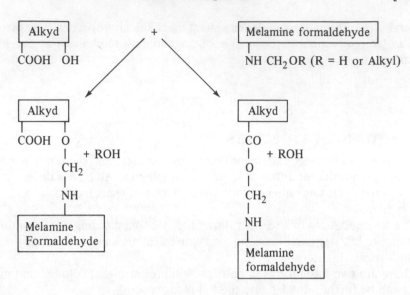

A stoving temperature of 20 minutes at 130°C is suitable for effecting suitable cure.

(2) *General properties*
These may be summarized as follows:

> Widely used/solid colours
> Relatively high solids (~50% w/w)
> Low cost
> Ease of processing
> Good durability
> Tolerant to a wide range of undercoats
> Proneness to dirt pick-up (film wetness)
> Relatively poor polishing properties.

Their high solids, low cost, and ease of processing (40–50 μm in two coats wet on wet application) has helped to maintain their popularity. These properties enable good build (and filling) properties and very good appearance after stoving. Also they have good durability, being equivalent to thermosetting acrylics after two-year tests in Florida (5° South).

However, on the debit side they do have relatively poor solvent release, compared to acrylics, and the wet film is prone to pick up dirt. Clean operating conditions are particularly important as a consequence. Also their somewhat limited polishability makes minor rectification on a production line difficult.

10.7.2   Thermosetting acrylic non-aqueous dispersion finishes
(1) *Basic chemistry*
Thermosetting Acrylics are based on complex acrylic copolymer resins produced by the reaction of a number of acrylic monomers selected to give the desired balance of properties. The required crosslinking reaction, with a suitable melamine formaldehyde resin, is facilitated by the presence of hydroxyl groups in the polymer

backbone. The hydroxy-containing polymers are readily prepared by the use of hydroxy-acrylic monomers.

A stoving of 20 minutes at 130°C produces an insoluble crosslinked film.

$$\boxed{\text{Melamine Formaldehyde}} - \text{NHCH} - \text{OR} + \text{HO} - \boxed{\text{Acrylic}}$$

$$\downarrow$$

$$\boxed{\text{Melamine}} - \text{NHCH} - \text{O} - \boxed{\text{Acrylic}}$$

$$+ \text{ROH}$$

Early thermosetting acrylic finishes were made from these resin systems, and they allowed for the formulation of solid colours and metallic finishes. Durable metallic paints were made by incorporating non-leafing aluminium flake (via an aluminium paste) into the film. The non-leafing flake (15–45 μm in length) distributes randomly through the coating and produces specular reflection from almost any angle of view. Suitably pigmented with coloured pigments, extremely attractive colour ranges were produced.

These types of finish, introduced in the late 1950s and early 1960s, found wide usage throughout the world, and in modified form still find use today. Their deficiencies are the requirement for three coats, because of solids content, and the difficulty in metallic application. The introduction of non-aqueous dispersion (NAD) versions of these finishes in the late 1960s enabled these problems to be overcome.

Whereas with conventional thermosetting acrylics the resin is carried in solution, NAD finishes are based on the dispersion of similar polymers in a solvent mixture such that a significant proportion of the polymer is insoluble, even at application viscosity.

These dispersions are stabilized by the presence of aliphatic soluble chains (i.e. by the use of an amphipathic graft copolymer) which are chemically linked to the polymer particles during the polymerization process. Such dispersions are referred to as 'super-stabilized' to distinguish them from systems where stabilizer is held on the surface of the particle by physical or polar forces.

So long as a substantial proportion of aliphatic hydrocarbon is present in the continuous phase of the dispersion these chains are extended and provide a stabilizing barrier around each particle. To promote eventual coalescence of these particles into a continuous film the volatile portion of an NAD finish also contains some solvents for the polymer itself. The amount of such solvents is controlled such that the stabilizing chains are not collapsed while the finish is in the package or the circulating system at the automotive plant. These solvents can be shown to partition themselves between the polymer (leading to softening and swelling of the particles) and the continuous phase. On application of the finish, evaporation of the aliphatic non-solvents leads to fusion and coalescence of the solvent-swollen polymer particles to give a continuous film, the process being completed by the stoving process. The structure of the final film is almost identical to that laid down from a conventional solution polymer.

Although the basic technology allows the preparation of dispersions where practically all the polymer is in the disperse phase, the rheology of 'all disperse' systems is not necessarily the optimum for achieving the latitude necessary under production line conditions. However, definite benefits can be gained by having some polymer in solution, and an important part of the technology consists of establishing the optimum ratio of disperse to solution resin for each polymer system. The rheological characteristics of such thermosetting acrylic formulations are unique and lead to greatly improved control of metallic finishes, better utilization of paint solids, and greater resistance to sagging and running.

(2) General properties
These may be summarized as follows:

> Durable in solid colours and metallics (single coat)
> Solids lower than alkyds (~30%)
> Applied in two or three coats
> Good polishing properties
> Good solvent release
> High-performance undercoat required, e.g. epoxy–ester or polyester

Thermosetting/NAD acrylic finishes can be formulated in both solid colours and metallics, and it is possible to produce a wide and attractive colour range. Their rapid solvent release minimizes dirt pick-up, and their good polishing properties make them more amenable to local rectification on production lines. They require high-grade undercoats, such as epoxy–ester or polyester, to maximize their performance. One other important feature is that thermosetting acrylics can be used directly over electroprimers because of their inherent adhesion properties. Alkyd finishes cannot, and poor adhesion can result. It is for this reason that acrylic finishes are always used in the two-coat electroprimer/finish system adopted for commercial vehicle production.

(3) Metallic appearance
A good metallic finish is designed to achieve a pronounced 'flip' tone, i.e. the polychromatic effect seen when viewed from different angles. It is simply an optical effect, and depends on the orientation of the metallic flake parallel to the surface, so that the amount of light reflected varies with the angle of viewing. Thus at glancing angles the surface appears deep in colour. In fact, the aluminium flake can be regarded as a small plane mirror. To achieve this optimum orientation it is necessary to ensure:

(a) Uniformity of spray application.
(b) Maximum shrinkage of the film after application so that the aluminium flake is physically 'pulled down' parallel to the surface.
(c) Minimum tendency for the flake to reorient randomly after application.

Typical solution thermosetting acrylic finishes go some way to meeting these requirements by changes in solvent composition, resin flow and design, and differing types of aluminium. High levels of operator skill during application are also necessary since uneven application leads to areas of different colour, and overwetness allows the aluminium flake to move around producing lighter and darker patches (sheariness or mottle). Dark lines (black edging) can also form around the edges of holes in the body shell or along styling lines. To overcome these problems solution enamels were, and are, sprayed as dry as possible consistent with gloss and flow with often low film thicknesses resulting.

Introduction of dispersion systems, often in combination with solution technology, eased the situation considerably by introducing improved rheology to the paint film. As mentioned earlier, a dispersion does not exert such a strong viscosity influence in a film as a solution version of the same polymer. However, in passing from the dispersed state to the solution state as the non-solvent part of the liquid evaporates, a sharp increase in viscosity is introduced. As this change occurs in the wet paint film after spraying there is a faster increase in viscosity at this stage than would be produced purely by evaporation of the liquids. This effectively reduces the amount of movement available to aluminium flakes, restricts any tendency to reorientation, and allows an even appearance to be produced more easily.

(4) 'Sagging'
The same rheological control described above, exerted by the change from disperse to solution phase, has resulted in a marked reduction in 'sags' and 'runs' with dispersion coatings. A higher build can also be applied to fill minor defects. These improvements are particularly marked in highly pigmented colours such as whites and oranges.

(5) 'Solvent popping' resistance
'Solvent-popping' (sometimes designated as 'boil') is caused by the retention of excessive solvent/occluded air in the film which, on stoving, escapes by erupting through the surface. It invariably occurs on areas where there is above-normal wet paint thickness. Solution acrylic thermosetting finishes are very prone to this because of their fast 'set-up' rate immediately after application.

It would be imagined that dispersion systems would suffer the same problem since the polymer compositions are similar. However, dispersion systems do show advantages. The reasons are probably two-fold:

(a) Better atomization during spraying leading to a finer droplet size.
(b) The utilization of non-solvents, not associated with the polymer; this leads to a faster and more effective 'solvent' release.

10.7.3 Thermoplastic acrylic lacquers
(1) Basic chemistry
Acrylic lacquers, in common with all lacquers, dry simply by the evaporation of solvent, and are based on a hard poly methyl methacrylate polymer that is suitably plasticized. Plasticizers are normally external and include butyl benzyl phthalate and linear polymeric phthalates derived from coconut oil fatty acid. The external plasticizer ensures a good balance of properties, i.e. improved crack resistance, adhesion to undercoats, solvent release properties, and flexibility.

Most common acrylic polymers for this type of finish have average molecular weights of approximately 90 000: these give outstanding gloss retention on external exposure. Polymers with average molecular weights greater than 105 000 tend to cobweb or form long filaments when applied by spray at commercially acceptable solids contents. Low molecular weight polymers result in poor film properties and poor durability. Furthermore, the improvement in gloss retention when the molecular weight is increased above 105 000 is proportionally small, and is more than offset by the reduced solids at application viscosity.

Solvent blends used for lacquers are balanced compositions, albeit expensive, chosen to give acceptable viscosity, evaporation, and flow characteristics. To avoid excessive solvent retention in the film it is necessary to use solvents free from high boiling 'tail' fractions, and to balance carefully the evaporation rates of the remainder. The external plasticizer assists in solvent release by maintaining a fluid film for as long as possible; this allows shrinkage stresses (considerable in acrylic lacquers) caused by the drying/stoving process, to be relieved.

Although acrylic lacquers will ultimately air-dry, in practice drying is accelerated either by a short stoving, 30 minutes at 90°C, where polishing is required to achieve acceptable gloss, or by the bake-sand-bake process. In this latter process surface imperfections of the film are removed by sanding after a short set-up bake (15 minutes at 82°C), and the film is reflowed (20 minutes at 154°C) to give a glossy flaw-free film.

(2) *General properties*
These may be summarized as follows:

> Very good durability in solid colours and metallics
> Robustness and adaptability in production
> Good polishing and self-repair properties
> Reflow (bake-sand-bake) process
>> Set-up 15 minutes at 82°C
>> Reflow 20 minutes at 154°C (effective metal temperatures)
> Excellent metallic appearance
> Low application solids (15–20%)/multi-coat process
> Requirement for large quantities of expensive thinners
> High raw material costs
> Special undercoats needed

Acrylic lacquer finishes are still widely used throughout the world by General Motors and prestige car producers like Jaguar. Their enhanced appearance, particularly metallics, proven durability, and in-process flexibility (self-repair and polishability) makes for an attractive technology. Metallics are particularly outstanding for appearance since as a low-solids product, carried in a high viscosity/high molecular weight acrylic polymer they meet the necessary criteria for optimum 'flip' tone, i.e. higher shrinkage, rapid solvent release and thin coats restricting by shear geometry reorientation and movement of aluminium flake.

Their disadvantages stem from fundamental characteristics of lacquers. These are:

(a) Low application solids means up to four coats are required to achieve the film thickness (55–60 μm) necessary for reflow. Multi-coats mean long spray booths with their inherent cost.

(b) Appearance on force drying at low temperatures (80–90°C) is poor, requiring excessive polishing to achieve acceptable gloss.

(c) High raw material costs, especially for solvents.

(d) Poor intrinsic adhesion entailing the use of special undercoats (high *PVC*, 55%, epoxy–ester). In fact the choice of undercoat has a greater influence on general performance than other types of finish. Special adhesion-promoting sealers are also used, which add even more to the cost of the process.

10.7.4 Basecoat/clear metallics

As has been described earlier, metallic car finishes have been used by the majority of large car producers for a considerable time. Metallics, because of their stylistic appeal, have been considered a very desirable feature of finishing by both colour stylists and designers alike. In the form of what are designated 'single coat' metallics they have been supplied either in thermosetting or thermoplastic acrylic technologies.

However, they do have a number of inherent disadvantages compared with solid colours:

(1) Lower gloss levels — particularly in light metallics.
(2) Limitation in certain pigment areas such as organic pigments.
(3) Poorer resistance to acidic environments.
(4) Application difficulties: eased by the introduction of NAD thermosetting acrylics.

The concept of putting the aluminium flake in a separate foundation or basecoat and then overcoating with a clear resin was first thought of and applied several decades ago. 'Flamboyant' enamel technology used on bicycle frames is a case in point.

Certain European car manufacturers saw this type of technology as overcoming the weaknesses described above, and introduced basecoat/clear technology into production in the late 1960s. The basecoat provided the opacity and metallic appearance, whilst the clear imparted gloss, clarity, and overall durability. The use of this technology has subsequently grown considerably in use, and is now widely used in Western Europe, Japan, and to a growing extent in the USA.

(1) *Basic chemistry*

The function of the two components has been described above, and this is achieved in the following manner:

(a) Basecoat
Properties:
(a) High opacity (~10 μm) to facilitate application in thin films.
(b) Rapid solvent release — short drying time (2–3 minutes) before application of clearcoat.
(c) Low solids (<20%) to achieve maximum metallic effect.
(d) 'Compatibility' with the clearcoat, i.e. good adhesion with no sinkage of clear.
Resin Composition:
Basecoats are thermosetting products modified with resins such as cellulose acetate butyrate to promote 'lacquer dry' and to accelerate solvent release. The basic resin component is either an oil-free polyester or a thermosetting acrylic polymer suitably

reacted with a nitrogen (melamine) resin. The features of these two types may be summarized as follows:

Features

Polyester type:　A very low-solids product (10–12% w/w) recognized for outstanding metallic appearance and ease of application. (Main basecoat used in Western Europe).

Acrylic type:　Higher solids than polyester type (15–20%), less pronounced metallic effect, but better filling properties and shorter processing time. (Main basecoat used in Japan and USA).

(b) Clearcoats

Properties:

(a) Good clarity of image and 'compatibility' with basecoat

(b) Offers a high level of protection to ultraviolet i.e. >3 years Florida exposure.

Resin Composition:

Clearcoats are formulated on crosslinking thermosetting acrylics reacted with a melamine resin and modified with UV absorbers and light stabilizers. These materials are applied at a film thickness of 35–50 μm to achieve maximum gloss and UV protection.

There are two types of clearcoat:

Features

Solution acrylic type:　A high-gloss product giving outstanding clarity. Normally applied in two coats.

NAD acrylic type:　A single-coat product with lower clarity than solution type.

(2) *Application/process*

The basecoat is applied as a 2-coat wet on wet process with a short air-drying time between coats. This is necessary to give acceptable opacity and evenness of appearance. After application of the second coat of basecoat has been completed, a short air-drying time (2–3 minutes) is allowed, sometimes supplemented by a warm-air blow, before the clear is applied in one or two coats.

Typical film thickness and stoving schedule for the system is as follows:

Film thickness

Basecoat: 15 μm

Clearcoat: 35–50 μm

Stove 30 minutes at 130°C to effect crosslinking.

(3) *Colour/pigmentation*

The pigments used in metallics in general are chosen for their potential transparency, realized when they are correctly dispersed and stabilized. Full transparent coloured pigments leave the metallic flakes free to contribute the maximum of brightness, sparkle, and flip tone.

Provided that satisfactory transparency exists, metallic appearance will depend upon the orientation of the flakes. As described earlier, if each flake is parallel to the substrate then this will give the optimum metallic effect or 'flip tone'. The light-reflecting quality of metallic coatings can be measured by a goniophotometer. This instrument is particularly useful for measuring the reflectance of unpigmented silvers, since the performance of different silver paints can be compared without any additional reflection and absorption by coloured pigments.

In Fig. 10.6, instrumental comparisons are made of various metallic technologies demonstrating the excellence of low-solids polyester basecoats. At a standard angle of incidence (45°) reflectance is measured at various viewing angles, and a curve is plotted; the higher the peak the better the reflectance, indicating the extent of parallel metallic orientation.

(4) *Aluminium flake orientation*

It has already been stated that the principal factor regulating aluminium flake orientation is film shrinkage during the drying process. Loss of solvent from the applied film during the flash-off and baking periods presents the flakes (typically of length from 10–25 μm) with an ever-decreasing freedom of movement. Surface tension, together with the large size of the flakes, ensures that the flakes will align more or less parallel to the substrate.

However, the state of 'almost parallel' alignment can still cover great differences in visual appearance, and metallic systems where all the flakes are aligned nearly perfectly parallel to the substrate will exhibit a far brighter appearance than systems containing many flakes at angles of, say, up to 20° to the surface. This is particularly true of very bright low-solids basecoats. In addition, dry film thickness is of the same order as, or even frequently less than, the flake length. This constrains flake mobility very effectively, particularly as the dry film thickness in the final stage of the drying process involves a very high degree of shrinkage.

All these factors combine to give low-solids basecoats such an attractive, stylistic appeal.

(5) *Undercoats*

In general terms, basecoat/clear technology requires polyester surfacers (oil-free) to maximize performance, particularly in the resistance to delamination. As has been described earlier, epoxy products tend to 'chalk' at the interface between undercoat and basecoat owing to UV radiation. In the past this has led to breakdown in the field of early basecoat/clear systems.

Nowadays the use of polyester surfacers, combined with UV absorbers and light stabilizers in the clearcoat, has resolved this problem. In fact the performance of basecoat systems both on test at Florida and in service is exceptional (see below).

(6) *Performance/durability*

Durability testing at Florida (5° South) is a universally accepted measure of exterior durability in the automotive industry. Florida is very suitable for such testing because it is high in ultraviolet and humidity.

Early clearcoats were based on alkyds, but failed owing to cracking (UV degradation) within twelve months. Modern thermosetting acrylic clearcoats have quite outstanding durability — a minimum of three years to five years can be

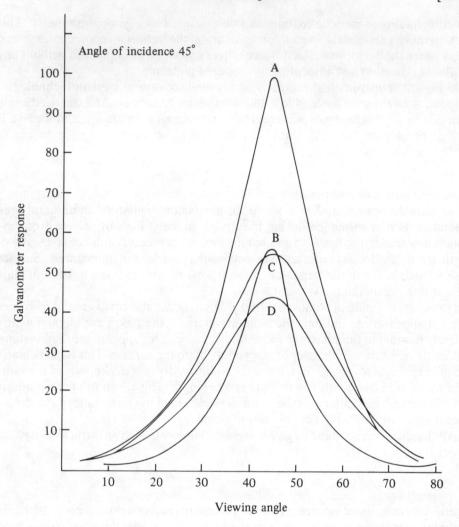

Fig. 10.6 — Goniophotometric curves.

Colour silver metallic
Key: Application solids
Curve A: Polyester basecoat   12%
Curve B: Thermoplastic acrylic lacquer   15%
Curve C: Acrylic basecoat   18%
Curve D: Thermosetting acrylic   25%

confidently expected. Such high levels of durability are unique to basecoat/clear technology, since the normal accepted standard is two years at Florida free from defects.

It is also possible to use a wider range of pigments in basecoat technology than thermosetting or thermoplastic acrylics; not only because of the pronounced face/flip contrast but also because it is feasible to use a much wider range of organic pigments than hitherto without sacrificing colour stability on exposure.

Organic pigments have high transparency but poor coverage and often poor durability. However, since there are no gloss constraints with basecoats, high levels of such pigments can be used at low film thicknesses (15 μm). The clearcoat provides gloss and offers the necessary ultraviolet protection.

10.7.5 Pigmentation of automotive topcoats

The choice of pigmentation for any particular colour must be considered in the context of the requirements of the market in which the product is to be used. In this respect the motor assembly market is probably the most demanding of all, requiring class 'A' matches at all times, excellent durability under severe conditions, and good opacity at minimum cost. Thus, for most types of topcoat many of the lower-cost pigments with inferior lightfastness cannot be considered, and the range of colours which can be used is limited in comparison to other markets, e.g. decorative paints.

10.7.5.1 *Solid colours*

Pigmentation practice for all types of automotive technology is very similar. As a general rule (apart from some relatively minor differences in performance) alkyds/ polyesters and thermosetting acrylics can be treated as identical. However, in acrylic lacquers whites are more prone to chalking, reds and violets tend to fade more in pale shades, and colour retention of some phthalocyanine blues is not as good as in the other technologies. The following general constraints apply:

Durability
Many organic pigments are completely excluded, or can be used only at certain concentrations or when combined with other pigments of excellent durability.

Opacity/gloss
Very clean, bright, pure colours are often non-feasible because the pigment loading required to achieve opacity reduces gloss to an unacceptably low figure. Inorganic pigments should always be used where possible, because of their opaque nature; but stronger organics such as blues, greens, and violets are also helpful.

Cost
Most organic pigments are very expensive and in many cases of relatively poor tinting strength. The most notable exceptions are phthalocyanine blues and greens which are less costly than most and, because of good tinting strength, can be used in lower concentrations.

Bleed
Many organic reds and yellows are excluded because of a tendency to bleed.

Metamerism
Metamerism is the phenomenon observed when two samples are similar in colour under one set of lighting conditions but different under another (commonly daylight and a tungsten filament lamp). This usually occurs where pigments of a different type

from those in the original 'master pattern' have been used. Pigments especially prone to this are iron oxides and phthalocyanines.

Use of lead chromate pigments

Automotive topcoats are heavy users of lead chromate/sulphate/molybdate pigments because of their brightness of colour, good tinting strength, and low cost. Their limitations are, however, considerable.

(1) On exposure to sunlight, lead chromate pigments darken. This is particularly obvious in the bright yellow shade, and it varies with seasonal changes at the exposure site and the degree of pollution in industrial atmospheres.

(2) Lead chromates are also susceptible to attack by dilute acids, the colour being bleached out. This can show as white spots or as an overall effect.

(3) Acid-catalysed repair films will also darken more readily because of the influence of the acid on the surface coating of the pigment.

Environmental pollution problems are leading to more and more demands for automotive paints to be lead/chromate free. Unfortunately, many of the alternatives available for production of clean, bright colours are very costly, it is difficult to achieve opacity/gloss to meet specifications, and, particularly with yellows, durability is poor. However, a reasonable range of lead-free reds can now be produced, although there are still many problems to be overcome before the totally lead-free situation is reached to the satisfaction of all concerned, i.e. colour stylist, paint manufacturer, end-user, and environmentalist.

10.7.5.2   *'Single-coat' metallics*

General pigmentation practice is the same for both thermoplastic and thermosetting acrylics, although some organic pigments, particularly certain yellows and phthalocyanines, perform slightly better in the former technology. The following constraints apply:

Durability

There are even more constraints on organic pigments than in solid colours, particularly with pale shades. These are often borderline for feasibility unless performance is boosted by the use of a UV absorber or the inclusion of transparent iron oxide pigment. This is normally the only inorganic pigment used in metallics. Both red and yellow shades are available, can be prepared in highly transparent form, are of excellent durability (apparently having some UV-absorbing property), and cheap. Their only disadvantage is weakness/dullness of colour.

Opacity/gloss

In any metallic finish aluminium flake contributes most of the opacity, since transparency of tinter is essential for tone metallic appearance (flip tone). Deep, clean metallic colours (low aluminium level) are therefore usually more difficult to formulate than pale shades unless some black can be included. The most difficult area of all is the clean, bright red shade where the majority of durable pigments available have poor tinting strength and the aluminium level must be kept low to avoid producing a greyish pink tone.

Cost
The cheapest durable pigments available for metallics are the transparent iron oxides which accounts for the popularity of metallic gold, beige, brown, and bronze colours. Phthalocyanine blues/greens also feature prominently because of their high tinting strength and relatively low cost.

Colour matching
Metallics must match the master pattern at all angles of viewing. The colour achieved from a metallic is very considerably influenced by conditions during film formation. The paint must be formulated and applied such that during drying the aluminium flakes align parallel to the substrate surface to ensure maximum brightness and degree of flip. Obviously any variation from one sprayout to another will result in a different colour, and for this reason automatic application is used wherever possible, particularly when colour matching. Spraybooth conditions, i.e. temperature, humidity, and air-movement, also need to be closely controlled.

Choice of aluminium flake
Various grades are available, differing principally in particle size. As a general rule, as size increases sparkle increases, colour becomes brighter (less grey), and flip tone increases. On the debit side, gloss, image clarity, opacity, and tinting strength diminish. Therefore the usual compromise is between adequate opacity/gloss and degree of brightness. The medium/fine grades are most popular and least likely to give problems. Very coarse flakes should be avoided whenever possible.

10.7.5.3 *Basecoat/clear metallics*
General pigmentation practice is the same for both types. Constraints are as follows:

Opacity
This is normally expected to be in the range 10–20 μm. As with single coat, most of the opacity is obtained from aluminium, therefore the same basic constraints apply, particularly with the low-solids polyester type where a very high degree of flip is expected. Again, bright reds are the most difficult. The big advantage over single coat is that no consideration need be given to gloss, therefore much lighter colours can be achieved because of the higher aluminium levels permitted.

Cost
The same constraints as for single coat apply.

Colour matching/durability
A much wider range of pigments is available for use in basecoat systems because of the protection afforded by the use of UV absorbers and light stabilizers in the clearcoat (which serve a dual purpose in preventing breakdown of clearcoat due to UV and also protect pigments in the basecoat). Many of the organic pigments prone to fade on exposure can therefore be used in base/clear systems, and the range of colours available is much greater. This is particularly true of the bright pastel shades where a combination of wider pigment range and better aluminium laydown produces some very attractive colours.

The same constraints on application parameters apply to basecoats as to single-coat.

Choice of aluminium flake
As with single-coat, medium-fine flake is preferred. Coarse flakes cannot be used. In addition to poor opacity, with the very low (10–20 μm) film thickness of basecoat films applied, the flake size (up to 30 μm) is such that if application is not perfect, flakes may protrude through the surface of the basecoat, giving a seedy appearance and the danger of film breakdown.

In addition to this, coarse flakes can cause safety problems on electrostatic application (discharge to earth through the pipework of the circulating system). Table 10.2 provides a comparison of the basic processing properties associated with automotive topcoats.

10.8   IN-FACTORY REPAIRS

After the initial paint process the carbody is trimmed and final assembly completed. During this latter part of the process minor paint damage can often occur and require repairing; the quality of such repairs must of need align to the quality of the original paint finish.

Final repairs can be effected in two ways:
 (i) Panel repairs, e.g. doors, bonnets, etc.
(ii) Spot repairs

The type of repair possible is directly related to the type of technology used for the original finish.

10.8.1   Thermosetting finishes (panel repairs)
Technologies include alkyds, thermosetting acrylics, and basecoat/clear metallics. The original enamel, or the clearcoat in the case of basecoat clear, is catalysed by a small addition of an acid catalyst and stoved 30 minutes at 90°C/10 minutes at 100°C. Higher temperatures are not feasible because of possible damage to trim, plastics, etc.

At the present time there is some growth in the use of two-pack (polyurethane and acrylic) materials for in-factory panel repairs. Such products have good durability and gloss, and need only minimal temperatures, 15 minutes at 80°C, to cure. However, they do have two disadvantages: (1) toxicity problems due to the isocyanate catalysts which require particular respiratory precautions, and (2) the need for a high standard of colour matching to the parent enamel.

10.8.2   Thermoplastic acrylic lacquers (spot repairs)
The very nature of these products facilitates self-repair. The technique of spot repair is mainly employed, although if the damage is extensive a panel repair can be done just as easily. Their excellent polishability also allows minor imperfections, such as dirt or dry spray, to be conveniently removed by polishing.

This use of the same product, unmodified, makes for considerable process flexibility and minimizes any colour matching problems. A minimum stoving of only 15 minutes at 80°C is required, followed by polishing to maximise gloss.

Table 10.2.— Automotive topcoats: comparison of basic processing properties

Type Property	Alkyd/melamine	Thermosetting acrylic/NAD	Thermoplastic acrylic	Basecoat/clear
Solid colours/metallics	Solid colours	Solid colours and metallics	Solid colours and metallics	Metallics
Appearance	High gloss 85% at 20°	Slightly less gloss. Microtexture 80% at 20°	High gloss 85% at 20°	High gloss 90% at 20°
Solids at spray (weight)	40–45%	30–35%	12–18%	Basecoat 12% (polyester) 18% (acrylic); Clearcoat 35%
No. of coats (air spray)	Two	Two/three	Three/four	Two basecoat, one/two clearcoat
Sensitivity to undercoat	Tolerant (poor adhesion to Electrocoat)	Reasonably tolerant (good adhesion to Electrocoat)	Specific-high *PVC* (~55%) Undercoat required. Sealer optional (poor adhesion to Electrocoat)	Polyester type recommended
Stoving temperature	20 mins at 130°C	20 mins at 130°C	30 mins at 155°C (reflow)	20 mins at 130°C
Polishability	Poor	Good	Excellent	Good
Repair	Panel repair (acid catalysed)	Panel repair (acid catalysed)	Spot (self) repair	Limited spot repair. Clearcoat— acid catalysed

10.9 PAINTING OF PLASTIC BODY COMPONENTS

In recent years there has been considerable growth in the utilization of plastic materials in motor vehicle construction. These products are now finding more and more use as exterior components such as bumpers, 'spoilers', 'wrap-arounds', and ventilation grills. Apart from advantage in weight-saving there are benefits in styling, resistance to minor damage, and corrosion resistance. The use of plastics for major body parts, such as door panels or bonnets, is seen as the next logical step forward.

The use of plastic does, however, bring particular painting problems such as adhesion, stoving limitations, solvent sensitivity, and matching for appearance. Currently, the autobody painting process is designed to coat a mild steel monocoque construction, and such processes involve high temperatures such as 165°C for primers and 130°C for finishes. As more plastics are used, either they will have to align with current practice or new systems will have to be developed.

There are many different types of plastics with a variety of different properties which affect their painting. The principal ones are as follows.

10.9.1 Sheet moulded compound (SMC) and DMC
The main advantages over other plastics are: (1) its high flexural modulus (stiff enough for horizontals), (2) its high distortion temperature which will withstand the 180°C electrocoating schedule, and (3) its high solvent resistance.

Disadvantages are (1) a surface profile prone to waviness, and (2) 'outgassing' on stoving, causing topcoats to bubble. Normally, it is usual to seal the surface either by an in-mould coating or with a spray-applied polyurethane.

10.9.2 Polyurethane: PU RIM and PU RRIM
Reaction injection moulded polyurethane and reinforced reaction injection moulded polyurethane have the following main advantages: a very wide range of moduli (from rigid to rubber-like), and toughness. Disadvantages are: (1) variable porosity which is sometimes at, or very near the surface, (2) dimensional instability (the material 'grows' when overstoved particularly when glass reinforced), and (3) solvent sensitivity, which is aggravated by the presence of glass fibre.

10.9.3 Injection moulded plastics
These materials include polycarbonate, ABS, polyamide (mainly glass reinforced), and polypropylene (modified with EPDM, ethylene propylene diene methylene). Their general advantages are: (1) toughness and strength, (2) good surface quality, and (3) a wide range of flexibility from rigid to ductile. Their general disadvantages are: (1) some are excessively brittle at lower temperatures, (2) Heat distortion temperature (HDT) is fairly low in many cases, and (3) the amorphous types are notch-sensitive.

10.9.4 Painting problems
(1) *Adhesion*
It is difficult to get consistently adequate adhesion to some modified polypropylenes, and special adhesion-promoting primers are required. With some other plastics adhesion is no problem.

(2) *Heat distortion*
Plastic parts will warp or sag at elevated temperatures. The temperature to which a part can be taken without distorting depends not only on the polymer but also on fillers, reinforcement, shape, size, and degree of mechanical support. Thus for a given polymer there is an absolute upper limit at which it begins to melt or decompose, but well below this temperature there will be a practical heat distortion temperature for a particular moulding and mounting.

Depending on the heat distortion temperature the plastic may have to be painted entirely off-line (e.g. most PU RIM) or fitted after the electropainting oven (e.g.

some PBT, glass-reinforced polypropylene, most polyamides) or can be fitted in the 'body in white' (e.g. some polyamides, SMC, and related materials).

Where painting is off-line, colour matching to the body is difficult.

(3) *Surface texture*
For true appearance matching of different materials meeting in the same plane, it is virtually essential to use a common undercoat.

(4) *Solvent sensitivity*
Some plastics are affected excessively by common paint solvents, causing the surface to craze and degrading the mechanical properties of the component. On the credit side a mild degree of solvent attack can be beneficial to adhesion.

(5) *Degradation of mechanical properties*
If the paint film fails by cracking when the painted part is impacted or flexed, the effect in some cases is to induce failure by cracking of the plastic substrate. Thus an unsuitable paint system will weaken the part.

10.9.5 Paint processes and products
(1) *On-line*
The conventional painting process, using a monocoque construction, has already been well described. Of the possible plastics only SMC related materials, and some grades of polyamide can withstand electropaint stoving schedules (165–180°C) without distorting, and also accept standard body-finishing systems.

A typical process for an SMC component is to use either an in-mould coating or a sprayed polyurethane coating to 'bridge' or seal the surface. Subsequently the component is fitted to the 'body in white', passed through cleaning, phosphating, and electropainting (where it remains relatively unaffected) before receiving the standard spray surfacer and topcoats.

(2) *Off-line*
In this process the plastic part is painted off-line, maybe at a moulder's or painting sub-contractor, and fixed to the carbody after the final paint oven. This practice enables the paint system and process to be designed to suit the plastic, but if the part is to be in body colour it makes colour matching, particularly in metallics, difficult.

PU RIM is always painted off-line. If the parts are fully jigged they can be painted with a one-pack flexible polyurethane/melamine formaldehyde finish and stoved at 120°C. This is standard practice in the USA, but in Europe two-pack flexible clears and non-metallics are used for RIM and ductile thermoplastics. The two-pack method gives lower stoving temperatures (100°C), more formulation scope, and a shorter painting process. Problems of toxicity are diminishing as automatic processes are introduced.

Colour matching in off-line painting may be more or less critical according to the shape and position of the component. A styling break or a change of plane where the off-line and on-line painted parts meet will conceal minor differences. With basecoat/clear metallics use of the same basecoat on plastics and metal is a major help; alternative clears (normal and two-component) complete the finishing process. In the longer term a common clear is an important objective.

10.9.6 'Part-way' down paint line
A final method is to fit the particular plastic component between the electropaint oven and the colour spraying station, provided that the paint system is suitable for the part. This procedure is commonly used for small parts such as ventilation grills in polyamide. Some other reinforced plastics will withstand normal topcoat schedules and also retain satisfactory impact performance when painted in a standard system.

10.10 SPRAY APPLICATION

The introduction of the automobile mass production line, in combination with the development of new synthetic paint technology, was the major factor in the adoption of spray application. The principle of spray painting is to atomize the liquid paint into a fine spray and subsequently to direct this spray on to the carbody. Originally, compressed air was the atomizing medium, but other techniques such as electrostatic spray are being more and more employed because of improved transfer efficiency.

This method of application, however, generates overspray, and provision must be made to carry off this overspray and exhaust it to the atmosphere or, in some instances, to retain it through a recovery system. Nowadays, water-washed spray-booths are used, being efficient and free from fire hazard. In this type of booth the exhaust is drawn through water which 'carries' the overspray into a tank for effluent disposal.

The main basic properties required for spray applications are:

(1) To offer a fast, flexible, reliable, and robust method of application which will achieve the necessary film thickness in the limited time available.
(2) To provide a smooth film with good flow free from defects such as mottle (orange peel), 'popping' (air entrapment), sags, and craters. Surface-active agents are often used to good effect to ease problems such as cratering.
(3) In the spraying process, produce a wet film capable of absorbing its own overspray.
(4) With metallics, to produce shear-free films.
(5) To maximize material utilization/transfer efficiency without impairing film appearance and processing.

10.10.1 Air spray
The principal method used in mass production is the 'pressure feed' system, although suction feed techniques are used for small-scale operations such as minor repairs and laboratory work.

In pressure feed, thinned paint (suitably stored and circulated) is fed by pressure through paint lines to the spray gun. It leaves the gun through a needle valve — the amount of paint being controlled by a trigger and the pressure applied. The fine stream of paint leaving the gun is atomized by jets of compressed air flowing out of openings in a removable air-cap at the head of the gun. The jets can be directed to produce an even spray pattern. This is diagrammatically represented in Fig. 10.7.

Because of its versatility and speed the spray gun has dominated since its inception. In spite of significant paint losses caused by the fact that not all the atomized paint is deposited, it still remains widely used for surfacer, sealer, and topcoat, including metallic application.

Spray losses
As has already been described, the spray gun is essentially a device for atomizing liquid with compressed air and projecting this cloud of fine droplets so formed onto the surface of the carbody. When the fan of paint droplets sprays from the gun, part of the fan passes to the side of the unit as it hits the surface being coated. This loss is acceptable, as edges need to be well covered; but a portion of the atomized paint is pulled away by the compressed air current deflected by the surface, and the paint appears to rebound or 'bounce back'.

At that particular moment the paint droplets are subjected to two actions. One is caused by their forward velocity which tends to push them along their course towards the object; the other is the air current which will deflect them from their trajectory and drag them in a different direction. Obviously this latter action will be stronger if the droplets are very fine and the air current very fast. A coarse spray needs less air pressure and the rebound can be reduced, but the quality of the finish will be poor as a result.

A compromise has to be reached between the air pressure necessary to produce the desired quality of the coating and an acceptable degree of bounce back and overspray. Other factors which affect the air pressure used are the surface tension and the viscosity of the paint. The higher these are, the greater the energy required to ensure atomization and the more powerful the air jet.

Typical viscosities of application are:

Coating type	*Viscosity at 25°C*	*No. of coats (wet on wet)*
Primer surfacer	25 secs. BSB4	: Two
Alkyd finish	25 secs. BSB4	: Two
	(25 secs. BSB4 (solid colours)	
Thermosetting acrylic/NAD	(33 secs. BSB3 (metallics)	: Two/three
Thermoplastic acrylic lacquer	35 secs. BSB3	: Three/four
Basecoat (polyester)	25 secs. BSB3	: Two
Clearcoat	35 secs. BSB4	: Two

(BSB=British Standard 'B' type viscosity cup)

The actual efficiency of an air spray gun under normal production conditions is low. Only 40% of the paint reaches and remains on the car body. Of the 60% lost, 20% is due to 'bounce back' and 40% to overspray.

The need to improve efficiency is obvious. Automatic machines are beneficial, but there have been other developments in recent years which have offered very significant paint savings. The means of reducing losses and associated developments are:

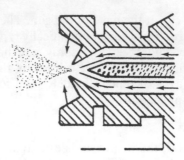

Fig. 10.7 — Spray air-cap.

Air spray gun $\begin{cases} \text{Lower pressures (already described)} \\ \text{Low-pressure hot spray} \\ \text{Air-assisted electrostatic} \end{cases}$

Electrostatic spray $\begin{cases} \text{Air-assisted} \\ \text{Bells} \\ \text{Discs} \end{cases}$

The newer techniques, which are described later, are often fully automated. They offer a range of benefits such as manpower savings and consistency of performance, apart from marked improvements in transfer efficiency.

10.10.2   Low-pressure hot spray

This technique has found some use in the application of primer surfacers, but it is rarely used in modern installations. Heating of the paint to 60–80°C reduces viscosity and surface tension, making atomization easier. Consequently, lower pressures can be used, and higher solids at application reduce bounce or rebound losses. However, overspray losses can increase because of the wastage of high solids, and can limit solvent selection. Evaporation losses can also be a problem.

10.10.3   Electrostatic spray

The principle of electrostatic spray is simple. If paint particles are atomized in an electric field they will become charged and drawn towards the article to be painted, which is usually at earth potential. There are many different types of electrostatic spray systems, but the most widely used in the motor industry are rotating bells or discs, covering the whole range of undercoats and topcoats.

Figure 10.8 represents a typical system. The paint is pumped to the atomizers from pressure-feed containers via a hollow drive shaft. The rotation of the atomizers spins the paint to the periphery where it is partly atomized by centrifugal force but mainly by the electrostatic field.

As the paint particles leave the atomizer, under the attraction of the electrostatic field, they are drawn to the unit being painted. Any particles which pass are attracted to all sides, giving electrostatic its very high efficiency of paint utilization. Generally adopted parameters are rotation speeds of 30 000–40 000 rev/min and high voltages in the region of 90–110 kV for maximum efficiency. The reduced overspray also

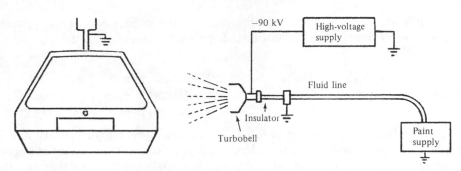

Fig. 10.8 — Electrostatic spray installation.

means less need for efficient and expensive spraybooths, and reduced effluent. This system also lends itself to full automation, with the attendant benefits.

However, certain recessed and other areas are difficult to paint with this method since paint is attracted to the nearest part of the complex-shaped carbody, and does not have sufficient velocity to penetrate further. Air-assisted electrostatic guns can overcome this weakness. These use compressed air and electrostatic techniques for atomization.

Air-assisted equipment has an insulating barrel having an earthed handle with a high-voltage electrode in the air-cap which is usually charged from 30–60 kV. The discharge current flowing from the high-voltage electrode creates a region, adjacent to the atomization zone, rich in uni-polar ions that attach themselves to, and charge, the paint spray particles.

In the vicinity of the spray gun the inertia of the aerodynamic forces usually dominates, but as the paint particles get nearer to the earthed carbody the forces on the charged particles (due to the electrostatic field between the charging electrode of the spraygun and the unit) become more significant. These forces drive the spray particles towards the workpiece.

Maximum paint savings are generally obtained by maintaining the charging voltage as high as possible, giving a high depositing field strength between the gun and the workpiece, and by keeping the spray velocity in the vicinity of the article being sprayed as low as possible to give good atomization and flow of the 'deposited' film.

Air-assisted guns can be used either manually or automatically. They offer a measurable compromise between the less efficient air spray and the highly efficient electrostatic system, and compared to the latter process give better film uniformity and smoothness.

Electrostatic spray — metallic appearance
The controlling factors in metallic appearance have been well described in previous sections. In practice, electrostatic spray diminishes the high flip (tonal contrast) of metallic finishes quite significantly. There has been much investigation into the cause of this phenomenon, and the considered view of the mechanism is as follows.

In comparing the velocities at which paint is applied, and the degrees of atomization, there is a considerable difference in the various application systems employed. For example, typical flow rates for a low solids basecoat are:

		Atomization
Full electrostatic (high speed bells)	300–500 cm³/min	Fair
Air-assisted electrostatic	450–500 cm³/min	Good
Air spray	800 cm³/min	Very good

In a hand-spray system the basecoat is applied at very high velocity and is extremely well atomized. There is evidence to suggest that at the moment of impact of a spray droplet with the substrate, the aluminium flakes are thrown flat against the surface by a considerable shearing force. Subsequently a certain amount of reorientation takes place as the wet film builds, but shrinkage and surface tension effects quickly take over and 'pull' the flake into ideal alignment (parallel to the substrate) for optimum effect.

The lower velocities and relatively poorer atomization of electrostatic systems means that, on impact, shearing forces are lower and the flakes are not forced totally into parallel alignment, and they tend to be more randomly oriented. The shrinkage and surface tension effects improve the orientation, but, because of the initial starting point, the final film does not align in the perfect manner of hand spray. It is for this reason that metallic basecoats are applied by a 'mix' of application methods to optimize appearance and efficiency and to minimize repair colour matching.

10.10.4 Application efficiency: practical considerations

In terms of absolute efficiency for any particular type of equipment it is difficult to be specific. There is considerable variation from plant to plant, depending on whether the system is manual or automatic, the available spraybooth space, and the differing shapes of the unit being sprayed. Practical figures can differ significantly from the ideal.

However, some mean figures are quoted below, based on data available from various installations. They are intended only to give the reader a realistic indication of the transfer efficiencies of systems currently in use.

	Variance ±5%
High-voltage electrostatic (high-speed bell)	75%
Air-assisted electrostatic	60%
Air spray	40%

In practice the principle of 'mixing' or 'blending' the various methods is widely used in mass production. A fairly typical process is shown in Table 10.3. Such a system compromises the various strengths and weaknesses of the individual spray procedure, realizing the following benefits:

(1) Maximizes appearance standards at the required film thickness (45–55 μm).
(2) Maintains an optimum level of transfer efficiency.
(3) Copes with the complex nature of a monocoque unit which comprises both internal and exterior surfaces.
(4) Effects (1), (2), and (3) above within the time and space available.
(5) Requires minimal labour.

Table 10.3 — Typical spray process (simplified).

	1st station	2nd/3rd station	4th station	Air blow	5th station	6th station	7th station
	Electrostatic (high-speed bells)	Hand spray (air)	Hand spray (air)		Electrostatic (high-speed bells)	Air-assisted electrostatic.	Air-assisted electrostatic
	Automatic	Hand	Hand		Automatic	Hand	Hand
Basecoat/clear metallic	Basecoat	Basecoat	Basecoat		Clearcoat	Clearcoat	Clearcoat
	Exterior	Interior	Exterior		Exterior	Interior	Exterior
	1 coat	2 coats	1 coat		1 coat	1 coat	1 coat
Alkyd solid colour	Alkyd	Alkyd	Alkyd				
	Exterior	Exterior	Interior				
	1 coat	Hand reinforce	1 coat				

Note: (i) Hand spray is used to maximize metallic appearance.
         (ii) Full electrostatic is used for 1st coat only for best efficiency.
         (iii) Air-assisted electrostatic is used for appearance/efficiency compromise.

10.11 STOVING PROCEDURES

The various types of material used in autobody painting have already been described. They vary from water-based epoxy technology for electropainting to sophisticated solventborne finishes such as alkyds/polyesters and acrylics. The organic solvents used are complex blends of aliphatic and aromatic hydrocarbons, alcohols, esters, and ketones. These paint materials are either applied by some form of dipping process or spray-applied by one technique or another. Subsequent to application all of these products require stoving, for the following reasons:

(1) To achieve the high level of performance demanded by the motor industry, i.e. to form durable and protective films.
(2) To facilitate processing in the limited time available on a conveyorized production line.
(3) To control or accelerate solvent release and minimize dirt pick-up.

The stoving operation simply cures the film by effecting certain chemical or crosslinking reactions which form a protective and durable coating. During such film formation, solvent and by-products are released which can, and do, provide pollution problems.

Stoving temperatures vary. Modern cathodic electropaints require 20 minutes at

180°C to cure, while at the other end of the scale repair requirements can be as low as 15 minutes at 80°C. In the initial stages of stoving the temperature must be increased at a controlled rate to prevent possible film defects such as 'solvent popping' or 'boil' (due to entrapped solvent or occluded air). Once the required temperature is reached it is maintained for a period sufficient for effective cure to occur. These times are typically 8–10 minutes heat-up of the paint film, and 20 minutes held at a specified temperature.

10.11.1   Oven technology
There are two types of oven that can be used to stove automotive coatings, the first being by far the more commonly used:

(1) Convection heating ovens.
(2) Radiant-heating ovens, e.g. infrared.

Convection ovens rely upon air movement and its even distribution over the unit — ideal for the complex shape of a carbody. Radiant heating is better suited to regular-shaped items, but it is often used in conjunction with convection ovens, for example to offer preheating in the initial zone to minimize dirt pick-up or to supplement the heating of heavier sill areas of the carbody.

10.11.2   Design considerations of convection ovens
Optimum design requires that a number of parameters be considered to effect the best compromise between process requirements, plant layout, energy consumption, anti-pollution requirements, maintenance, and capital cost.

(1) *Oven configuration*
There are two stages in the stoving of the painted body: the controlled heat-up followed by a 'hold' period during which the resin system either reacts or, in the instance of acrylic lacquer, reflows. This requires specific oven zones, each with their own heating and control systems. The convected air is introduced at high velocity into the oven enclosure, normally via distribution ducting mounted at roof level. Individual heat-up and 'hold' zones may be further divided into two or more zones.

(2) *Oven ventilation*
The stoving of automotive paints releases into the oven atmosphere combustible compounds which can cause an explosion hazard. For this reason fresh air is introduced into the enclosure to ensure a safe installation. This requires that a 'balancing' volume of air be exhausted from the enclosure; obviously this contains the diluted combustible compounds released from the painted surface, which are also a source of fume and odour and generally undesirable.

(3) *Oven heating*
Each zone of the oven is fitted with a recirculation/supply fan which extracts a volume of air from the oven, mixes this with any fresh air required for ventilation, supplies heat to that mixture to satisfy the zone heat load, and then returns it via either distribution nozzles or slots to the oven zone.

(4) *Fresh air requirements*
The fresh air quantity supplied to an oven is calculated from the anticipated solvent quantity entering on the body; it is essential to ensure that a maximum of 25% of the 'lower explosive limit' of that solvent is not exceeded. This air quantity may be introduced into the oven as filtered fresh air into the zone heating system, as air infiltrating at the oven airseal, or as combustion air to direct-heating zone burners.

(5) *Fuel available/heating method*
In the UK, and most of Europe, it is normal to use natural gas as the heating fuel. As this is a sulphur-free fuel it has resulted in the adoption of direct-fired heating systems on installations for all parts of the painting process, since testing and experience have shown this to have no detrimental effect on the quality of the coating. Where natural gas is not available, butane or propane are used, both having negligible sulphur contents.

Other systems involve burning a distillate of oil; this results in the use of a less efficient indirect-fired system (an indirect-fired heater being about 70% efficient).

10.11.3 Fume and odour emission
As has been mentioned, the stoving of automotive coatings releases certain materials during the curing process. These are mixtures of organic solvents, products of the chemical reaction, and some decomposition products. This can lead to visible fume and odour problems from the exhaust stack and to condensation in the stack.

Whilst the problem of visible fume, odour, and exhaust stack condensation are not over-serious, the high level of solvent emission can be a major concern. Legislation exists in the USA to control such emissions, and there are significant safeguards in Europe.

The control of the exhaust emissions from industrial plant, and in this instance stoving ovens, originated in the USA in the 1960s. Since the majority of coatings are stoved and based on organic solvents, such solvents became recognized as potential sources of pollution because of the toxic nature of the products formed by their photochemical reaction in the atmosphere. In the late 1960s the Los Angeles County and San Francisco Bay area in the USA introduced regulations which limited the amount of certain organic solvents to be used in organic compositions. These regulations were called Rule 66 or Regulation 3 by their respective bodies; and where pollution control is required under this code it is stipulated that 90% or more of the hydrocarbons from the process be oxidized to carbon dioxide before exhausting to the atmosphere.

More recently in the USA, Federal regulations have provided guidelines on the permitted hydrocarbon emission. The actual limits depend upon the location of the plant. These can vary from 1.4 kg–2.3 kg per hour and 100–300 tonnes per annum. This level of emission can be achieved by adopting pollution control of the oven exhaust, or by a process modification such as a change of paint formulation/technology, or by a combination of both.

In France and the UK the Rule 66 legislation is still widely adopted, but in other parts of Europe the code of the Federal German Republic is becoming more widely used. In this latter instance precise limits are set for permissible emissions for all solvent types and for various flow quantities.

Besides new paint technology a number of methods of achieving the pollution control required are now available. These include thermal incineration, catalytic combustion, carbon absorption, liquid scrubbing, and odour masking. Of these, thermal incineration is the most widely accepted and reliable method of achieving control; but, more recently, catalytic combustion has proved successful in a number of installations.

(1) *Thermal incineration*
Thermal incineration consists simply of passing the fume-laden exhaust air through a highly efficient combustion system where a primary fuel is burned in order to raise the temperature of the effluent to a critical reaction point, and holding this for a specific period. In this way the complex hydrocarbon compounds are oxidized to carbon dioxide and water vapour. The primary fuel may be gas or distillate oil.

(2) *Catalytic combustion*
The application of catalytic combustion techniques has been gaining acceptance for the control of exhaust gases emitted from various processes. They show a substantial reduction in fuel consumption when compared to thermal incineration systems, with limited or no heat recovery equipment included.

The exhaust gases supplied to the catalyst cell contain organic compounds which, when passed over the catalyst surface, react with the oxygen present in the airstream.

The application of catalyst combustion to stoving in the automotive industry was introduced in the mid-1970s. Catalyst cells do not have indefinite life, but if contamination or poisoning can be avoided then a service life of five years can be expected.

10.11.4   Future stoving developments
The development of alternative curing systems will obviously be linked to developments in paint technology. The two overriding considerations from the point of view of the curing system will be the most economical use of energy and the elimination of the effluent problem.

Any changes in the curing process will entail the adoption of new radiant heating technology, as the present convection type oven is very close to its optimum efficiency. There are a number of radiant curing techniques at present used in the curing of flat stock such as boards, sheet metal, fabrics, paper, and plastics. Perhaps these technologies will be adopted for use in a more general way in the metal-finishing industry. Among these are electron beam and ultraviolet curing and induction heating.

It is not yet certain which of these processes will be adopted. Electron beam and UV curing have the advantage of being suitable for plastic parts which are being used increasingly to reduce body weight. Infrared stoving has the advantage of short curing time, reduced length of line, minimal dirt, and flexibility (such as its use in conjunction with convection ovens). Table 10.4 lists typical stoving schedules.

Table 10.4 — Typical stoving schedules

Dipping primers: 30 minutes at 150°C/20 minutes at 165°C	
Electroprimers : Anodic type 20 minutes at 165°C/20 minutes at 175°C	
: Cathodic type 20 minutes at 165°C/20 minutes at 180°C	
Primer surfacers: 30 minutes at 150°C/20 minutes at 165°C	
Topcoats : Alkyds, Thermosetting acrylics/NAD : 20 minutes at 130°C	
Thermoplastic acrylic lacquers:reflow 30 minutes at 155°C	
Repair schedules: Thermoset products :	
Catalysed enamel : 30 minutes at 90°C/10 minutes at 100°C	
Two-pack enamels : 15 minutes at 80°C	
Thermoplastic products:	
Self-repair : Minimum stoving 15 minutes at 80°C	

Note: All temperatures quoted are metal (effective) temperatures, not air temperatures

10.12 PERFORMANCE/TESTING

The performance required of automotive coatings has been outlined. The testing of the paint system, and the component 'layers', is designed to simulate conditions likely to occur in practice in order to give some measure of performance in the field. Obviously the testing of undercoats has a different emphasis from that of topcoats; but the interaction of these products in a total system is of equal, or more, importance.

Performance standards have improved dramatically in recent years. For example, the ten days (240 hours) salt-spray tests used as an acceptable standard for anodic electropaints has been extended to greater than 20 days (480 hours) and beyond for cathodic electropaints. The acceptable standard of exterior durability was to be satisfactory after 12 months' Florida testing; for modern basecoat/clear systems it is 3–5 years. There is now a great emphasis on chemical (environmental) resistance properties such as resistance to acid.

The subject of testing and durability is a complex and detailed subject, and is dealt with in Chapter 19 of this book. However, key properties, performance, and testing of automotive paints/systems are summarized below, and comparisons are made. Also included is a brief description of important and suitable test methods.

A comparison of various topcoat processing properties was made in section 10 7

Testing may be divided into two categories: appearance and performance. The qualities to be assessed are given in note form in the next two sections.

10.12.1 Appearance

Colour; opacity; smoothness, i.e. freedom from defects such as mottle or orange peel; gloss and distinction of image, (important for consumer appeal).
Note: Apart from the formulation and quality of the finish, the undercoating system has a significant effect on the final appearance.

10.12.2 Performance

(1) *Physical properties*
Hardness; flexibility; impact resistance; adhesion; stone-chip resistance; cold-crack

resistance, i.e. stability to extremes of temperature and humidity; curing efficiency.

(2) *Chemical resistance*

To petrol, acid, alkali, water, humidity, corrosion, scab corrosion.

Exterior durability, i.e. resistance to ultraviolet irradiation and humidity.

10.12.3 Test procedures

Typical test procedures and performance procedures are listed in Table 10.5.

Table 10.5 — Typical performance properties: various finishing systems

Performance Test	Type of topcoat Alkyd melamine	Thermosetting acrylic/NAD	Thermoplastic acrylic lacquer	Basecoat/ clear metallic
Gloss (20°)	85%	80%	85%	90%
Hardness (Tukon–Knoop)	5–7	8–14	14–18	8–12
Adhesion (cross hatch)	<5% removal	<5% removal	<5% removal	<5% removal
Petrol resistance (some Super Shell—slow drip)	Excellent	Very good	Fair	Very good
Acid resistance (Non-staining 1N $H_2SO_4 \times 48$ hr)	Pass	Pass	Pass	Pass
Alkali resistance (Non-staining 1N NAOH×48 hr)	Pass	Pass	Pass	Pass
Surface distortion	—	—	55–65°C	—
Impact resistance	Pass	Pass	(Depending on colour). Pass	Pass
Stone-chip resistance	Excellent	Very good	Good	Excellent
Water-soak (40°C)	No blistering >240 hr	No blistering >240 hr	No blistering >240 hr	No blistering >240 hr
Humidity resistance (100% RH at 40°C)	No blistering >96 hr	No blistering >96 hr	No blistering >96 hr	No blistering >96 hr
Corrosion (salt spray) resistance	>480 hr	>480 hr	>480 hr	>500 hr
Scab corrosion resistance	Pass	Pass	Pass	Pass
Florida exposure/ 20° gloss (washed) after 2 years	70%	65%	75%	80%
Primer	Cathodic electrocoat	Cathodic electrocoat	Cathodic electrocoat	Cathodic electrocoat
Surfacer	Epoxy ester/ polyester	Polyester	High *PVC* (~55%) epoxy ester	Polyester
(Pretreatment)	Zinc phosphate	Zinc phosphate	Zinc phosphate	Zinc phosphate

The following details are in summary form and are fairly typical. Different car manufacturers have particular variants and different emphases. These procedures are used for full painting systems and for primer only, as appropriate.

(1) Cure (test for crosslinking products)
20 double rubs with a clean white cotton cloth soaked in MIBK (methyl isobutyl ketone).
Pass: no removal or marking of paint film.

(2) Adhesion: Cross hatch test (1.5 mm or 2.0 mm template)
Test panels are evaluated by cross hatch before, and after, 120, 240, and 480 hours' water immersion (refer (9) below also).
Pass: <5% removal.

(3) Hardness: (Tukon, 'Indentation': Knoop Hardness)
The Tukon hardness is the generally accepted test in the automotive industry, particularly for topcoats. Other tests include Pencil Hardness, Sward Rocker, Pendulum Hardness (König, Persoz, etc.).

(4) Stone-chip resistance
There are a number of pieces of apparatus in the market place, but a simple and efficient one is as follows:
$100 \times \frac{1}{4}$ in Whitworth nuts are dropped down a 15 feet (457 cm) pipe (2.5 in or 6.0 cm in diameter) onto painted panels held at 45 degrees. Panels are graded by degree of removal of paint film.
Pass: <5% removal of paint.

(5) Impact resistance
2 lb weight (1 kg) dropped from 10 in (25 cm) and 20 in (50 cm) onto a painted test panel (impacted from reverse side).
Pass: No cracking of paint film.

(6) Acid resistance
Test panels are immersed for 96 hr at room temperature in 0.1N sulphuric acid.
Pass: Film is not affected.

(7) Alkali resistance
Test panels are immersed for 240 hr at room temperature in 0.1N Sodium hydroxide.
Pass: Film is not affected.
Note: (6) and (7) above are alternative tests to those shown in Table 10.5.

(8) *Humidity resistance (continuous)*
Panels placed in a cabinet (100% RH at 40°C).
Pass: No loss of adhesion, no blistering, and no colour change after a minimum of 96 hr.

(9) *Water immersion (continuous)*
Panels immersed in a stirred/agitated demineralized water bath at 40°C.
Pass: No loss of adhesion, no blistering, and no colour change after a minimum of 240 hr. In excess of 500 hr expected.

(10) *ASTM B117 salt spray (5% salt solution at 35°C).*
Test panels in a cabinet at the conditions stated above. They are 'X' scribed before test.
   Results are evaluated at 240, 480, and 840 hr by inspection and tape strip (over 240 hr normally only applicable to cathodic electroprimers).
Pass: Not more than 2 mm corrosion at scribe.
This is a key test for primers, and is carried out in primer only and a full system situation.

(11) *Scab corrosion test*
This test is designed to simulate scab corrosion of a carbody after water contact and stone chipping.
   Initially, the painted panel (full system) is immersed in a demineralized stirred water bath at 40°C. It is then subjected to a stone-chip test and subsequently to ASTM 117 salt spray. In addition, it can be further exterior exposed.
Pass: No scab corrosion.

(12) *Florida exposure (5° South)*
Florida is the site preferred by the majority of motor manufacturers for the exterior durability testing of automotive finishes. It has a high humidity combined with a high level of ultraviolet; this provides an extremely rigorous environment for topcoat technology. Ideal for testing colour stability of pigments, film (polymer).
   Although this can be considered a good absolute test of a coating, certain factors should always be taken into account. Performance can often be affected by the time of year when exposure begins and the particular weather conditions in that year. Another hazard which can often be misconstrued as microblistering is fungal growth — characterized by small 'pits' in the film and threads or filaments leading from this defect. Consequently, testing is normally 'relative', with known standards included in any test programme, to avoid misinterpretation of results.
Pass: Typical results are shown in Table 10.5.
The accepted standard is good gloss and colour retention, free from defects after up to two years (e.g. alkyds, thermosetting/NAD acrylics, and acrylic lacquer). Base-coat clear systems have an expected Florida performance up to 5 years.

(13) *Accelerated weathering*
The development of a new paint formulation requires an early appreciation of the exterior durability characteristics, and it is not possible to wait for a 2–3 year exposure at Florida to obtain this information. Recourse is therefore made to an accelerated weathering device which can indicate likely durability performance.

A variety of machines are available which subject the paint film to UV radiation in combination with humidity and temperature. No machine can accurately predict the Florida performance, because of the difficulty of producing a UV spectrum identical to natural sunlight by artificial means. Nevertheless, a composition which has withstood 2000 hours on one of the more severe accelerated cycles can confidently be predicted to show acceptable Florida performance.

Typical machines are marketed by Atlas (Carbon Arc or Xenon lamp UV source), Q-Panel Co. (QUV apparatus using Xenon lamp), Xenotest (Xenon lamp). Natural sunlight concentrated by mirrors is the basis of an accelerated weathering process devised by Desert Sunshine Exposure Tests of Arizona (EMMA and EMMAQUA cycles).

10.13 FUTURE DEVELOPMENTS

The future development of automotive coatings technology and/or processes will depend on a number of factors, namely economics, energy saving, environmental considerations, and consumerism (the demand for improvements in durability and appearance). There are a number of options open, each having their respective advantages and disadvantages; and a compromise, to satisfy the various conflicting demands, will undoubtedly evolve.

From an environmental viewpoint certain progress has been made with the stoving effluent released to the atmosphere. As has been described, mechanical, thermal, and chemical methods are already available for this purpose, including 'after burners', scrubbers, and carbon absorption units.

However, there still remains the vast amount of solvent-laden air to be dealt with from spraybooths, and the high amounts of energy consumed during the painting process (>50% of the energy is, in fact, caused by the spraybooths simply moving and heating the air and water). High curing temperatures are also an area requiring consideration, particularly if the concept of plastic components is to be fully realized.

In economic terms the cost of organic solvents will continue to rise in line with oil prices. If it is considered that some spray coatings are used at ~12% solids, then, although there are other desirable features, this makes them unacceptable from an environmental and economic viewpoint.

Product development
Three basic routes are being followed as means of reducing or, ultimately, eliminating organic solvents in spray coatings: high-solids technology, powder coatings, and waterborne products. These will now be discussed.

10.13.1 High-solids technology
The benefits and problems associated with this type of technology may be summarized as follows, in relation to existing products.

Benefits	*Problems/disadvantages*
Use current manufacturing techniques.	More difficult to achieve equivalent film properties.
Use current application equipment.	More difficult to achieve equivalent appearance in metallic finishes.
Use similar process parameters.	Rheological control difficult.
	Automatic spray equipment with high transfer efficiency essential for good economics.

Solids range
Pigmented products — up to 70%, metallic (basecoats) 20–40%, clearcoats up to 55%.

Resin types
Alkyd, polyester, acrylic

Types of product involved
Primers, surfacers, sealers, anti-chip coatings, solid colour and metallic finishes.

The obvious advantages are that high-solids products normally have similar manufacturing, application, and process methods to present low(er)-solids technology. Problems and disadvantages arise from the fact that lower molecular weight/viscosity resins must be used. These, however, tend to give excessive flow on application, leading to lower run/sag resistance; therefore some form of rheological control is essential to facilitate suitable application properties. Furthermore, lower molecular weight resins require higher degrees of reactivity to form coatings with acceptable chemical and physical properties.

Rheological control can be effected by the use of microgel technology. Microgels are essentially organic extenders and are used in small amounts; they also have crosslinking potential and some form an integral part of the film after curing.

Basecoat/clear metallic finishes provide an example of how microgels can be used in higher-solids compositions to excellent effect. In Europe, current basecoats are formulated at very low solids (~12%) to provide the required level of shrinkage which is the main factor controlling orientation of the metallic flake parallel to the surface. This makes such finishes very appealing stylistically.

In raising the solids, less shrinkage results, which reduces the glamour effect. However, the inclusion of a small amount of a suitable microgel restricts any reorientation of the aluminium flake after application, and there is less dependence on film shrinkage for appearance. Thus the application solids can be raised to 20–40% without the expected loss in aesthetic appeal.

10.13.2 Ultra-high-solids coatings
This relates to products with solids contents in excess of 70%. There are, however, severe constraints on their development. For example, metallic finishes cannot be considered in this context, since the loss of cosmetic appearance would be so dramatic as to make them unacceptable.

Pigmented (solid colour) products would have to be formulated as two-pack products using a reactive monomer or high-solids activator because of their limited pot life. In any mass production plant such products would undoubtedly call for dual-feed application equipment with a complex metering system. While equipment manufacturers have made certain progress, the associated cost and complications have severely limited the exploitation of such products, and they are, at present, of limited interest.

10.13.3 Waterborne products

Benefits	*Problems/disadvantages*
Environmentally acceptable; cheap and available diluent.	Water release on stoving.
Use normal manufacturing methods.	Need for tighter control of temperature/humidity for application.
Use normal application equipment.	Excessive flow: need for rheological control.
Use similar process parameters.	Resin limitations (stability).
	Prone to surface defects through contamination.

Solids range
No environmental restriction.

Resin types
Alkyd, polyester, epoxy, acrylic.

Product types
Primers, surfacers, sealers, anti-chip coatings, solid colour and metallic finishes.

The use of water as a 'solvent' for organic coatings has always been seen as an admirable objective. However, the difficulties associated with it have prevented wide exploitation, i.e. 'boil' (popping) on stoving, excessive flow, and sensitivity to oil-based contaminants.

Water-based undercoats have been developed, since the problems mentioned above lend themselves to a greater degree of formulation control in such compositions: finishes have proved extremely difficult. It is only by the use of air-conditioned spraybooths that waterborne finishes could be made to work relatively successfully, and the high equipment and running costs restricted their adoption.

Previous developments have been largely based on water-soluble resins and aqueous slurries. However, recently there have been rapid advances in the stabilization of polymeric particles in water. Dispersion of numerous polymers in water can be readily achieved, and some of these have unusual rheological properties. The latter can be used to impart flow control, 'structure', to aqueous resin systems.

Aqueous polymer dispersions also offer much better water release on stoving, a reduction in limitation on resin types, and smoother films than slurries, because of finer particle size. Also, their flow control aspects makes aqueous dispersion coatings less prone to surface defects, e.g. cratering through contamination.

While the aqueous dispersion route has given greater scope for formulation, it is still not possible to produce aqueous finishes which can be used in the 50 μm range (e.g. automotive solid colour topcoats), without some form of temperature/humidity control. However, such polymer dispersions can be used to good effect in the formulation of aqueous basecoats where basecoat film thicknesses of only 10–15 μm need be applied. In these, the bulk of the film thickness is contributed by the clearcoat (35–50 μm).

The aqueous dispersion controlled rheology approach permits good atomization of the basecoat through a conventional spray gun, while providing suitable flow-control to the applied film. Aqueous metallic basecoats can be formulated to give similar brightness and varied hue to the best low-solids (12%) solventborne type, but at an 'equivalent' solids when related to organic solvent level of 55–60%. Such basecoats are much less dependent on temperature/humidity levels than would be ordinarily expected, and they can be used in similar facilities to the solventborne type, with the assistance of an airblow before application of the solventborne clear.

10.13.4   Powder coatings and aqueous slurries

Powder coatings

Benefits	Problems/disadvantages
100% solids (solvent-free).	Need for specific manufacturing equipment.
High application efficiency.	Need for specific application equipment.
Good film properties (at high film thickness).	Limited colour range for good economics.
	Colour ranges difficult.
	Metallics non-viable.
	Difficult to repair.
	Need for high film thickness to achieve acceptable appearance.
	Surfacers require heavy sanding to achieve acceptable topcoat appearance.
	Difficult to coat complex shapes economically.
	Higher temperature needed for flow/cure.
	Dust explosion risk.

Resin types
Epoxy, polyester, acrylic

There are numerous basic problems inherent in powder coatings, so they can be considered of only limited potential. Their single major advantage is the absence of solvent pollution; but it is difficult to see them being exploited to any great extent beyond their current main areas of usage — primers and one-coat finishes in industrial coatings. Their lower standard of appearance makes them poor contenders in a motors context, except perhaps as a clearcoat in basecoat/clear systems. Even here, however, the higher curing temperature called for would prove an obstacle to their wide adoption.

Aqueous powder slurries
Such materials would overcome some of the inherent problems associated with

powder coatings. However, there would still remain the question of high energy and cost, repair difficulty, and lower standards of appearance because of the coarse powder particles.

10.13.5 Solid colour basecoats
Such basecoats could be formulated either as a high-solids solventborne type or as an aqueous dispersion type. They would have a high pigment coverage in a thin basecoat film, and would be independent of gloss level, the main criteria being smoothness and film properties. However, some limitation on colour range may be necessary through adverse effects on film properties, e.g. 'mud cracking' with certain pigments at high pigment loadings.

The main benefits are the alignment with metallic analogues and more assured durability and scope to use pigments which are difficult to employ in normal solid colours (the durability is potentially dependent on the clearcoat). This latter point would particularly facilitate the use of lead-free pigments.

10.13.6 Clearcoats
Current clearcoats are around 40% (w/w) application solids. To reduce the organic solvent level further would necessitate the use of lower molecular weight resins in conjunction with a suitable microgel to give flow control. The microgel resin would of necessity need to be optically designed to give the desired gloss and clarity for optimum appearance.

The aqueous dispersion approach would also lend itself to adoption in clearcoats. However, the possibility of having to use some form of temperature/humidity control with waterborne clearcoats at higher film builds cannot be ruled out.

Powder and aqueous slurry clears are also seen as contenders, although higher curing temperatures likely with these would be a disadvantage. Another alternative is two-pack clearcoats; the advantages and disadvantages of this type of technology have been described earlier.

10.13.7 Pigmentation
There is a growing trend for cleaner, brighter tones for solid and metallic colours. There is also, for reasons of toxicity, a strong desire to stop using lead chrome pigments. This means that there is a definite requirement for organic pigments that possess the cleanliness and brightness of colour of chrome pigments with high durability; and whilst opacity is important for solid colours, transparency is required for metallics.

In addition, new highly sophisticated technologies place greater demands on the function of the pigment in the total system. Pigments will have to perform in very different media compared with those for which they were designed, and the new systems will be subjected to conditions and forces not envisaged a few years ago.

10.13.8 Painting of plastics
The problems of painting a composite carbody built of different metals and plastic has been described in section 10.9. The particular requirements of such an integrated unit will have to be taken into account in the design of new paint systems as the growth in the use of plastics continues.

A major factor is the temperature constraint, and the development of novel low-curing resin systems is the principal solution to the painting of the wide range of plastics available.

10.13.9 Process/equipment and electrodeposition
In the field of spray application, which applies to the application of surfacers and finishes, new equipment such as robotics and turbo (high-speed) electrostatic spray is growing in use. Such equipment provides higher transfer efficiency and less solvent emission. To realize the implications of this paint, utilization for manual hand spray is of the order of 40–50%, while automatic electrostatic is 60–70%.

Electrodeposited (EDP) coatings represent a high percentage of the total coatings volume in terms of the primer. The EDP process is an extremely elegant method of coating complex units, being virtually solvent-free, and is the most efficient and economic of all application methods. Its virtues are such that increased adoption is foreseen.

The recent introduction of high-build EDP primers (35 μm) into the automotive industry offers an opportunity for lower thickness of subsequent surfacers/sealers to be spray applied, thereby reducing the volume of organic solvent. It can be envisaged where the total undercoating system is done by an EDP process, so eliminating organic solvents in that part of the car painting process.

10.13.10 Summary
The high-solids approach has given rise to microgel technique for rheological control, and novel technology such as higher-solids basecoat/clear systems has resulted. Solid colour basecoat/clear systems using a similar technique are likely to become prominent in the future. Such products are seen as offering significant benefits before the wider use of waterborne materials.

The waterborne approach offers the widest scope for the future in all types of coating. High-build electroprimers will provide an opportunity to reduce, and possibly eliminate, the need for subsequent spray undercoats which themselves would be water-based.

In the topcoat area, controlled rheology water-based polymeric dispersions have already given rise to certain novel products, i.e. automotive waterborne metallic basecoats. Solid colour basecoats also lend themselves to similar technology as do clearcoats for both types of basecoat.

The combination of low-polluting coatings with new equipment, i.e. electrodeposition and the automatic high-speed electrostatic spray, is clearly the way ahead. Such a package should satisfy the growing demands for energy constraints, ecological considerations, and the need for economical and shorter processes.

BIBLIOGRAPHY

Payne, H. F., *Organic coating technology*, Volume I (Oils, resins, varnishes, and polymers), New York, John Wiley, (1954).

Payne, H. F., *Organic coating technology*, Volume II (Pigments and pigmented coatings), New York, John Wiley, (1961).

Solomon, D. H., *The chemistry of organic film formers*, Robert E. Krieger Publishing, (1977).

Waring, D. J., *Automobile paint systems* (paper for Institute of Production Engineers), February 1984.

Baylis, R. L., *Trans. Inst. Metal Finishing* **50** 80–86 (1972).

Waghorn, M. J., Non-aqueous dispersion finishes, *IMF/OCCA Symposium Warwick University* pp. 228–233 (1973).

Ansdell, D. A., Painting of plastic body components for cars, *Industrial and Production Engineering* **3** 30–35 (1980).

Strong, E. R., The reduction of lacquer waste in spraying, *FIRA Technical Report 21.* (November 1965).

Eaton, M. J., *Product finishing,* July 1983.

Inshaw, J. L., *Jour. Oil and Colour Chemists Association,* July 1983, 183–185.

ACKNOWLEDGEMENTS

The author would like to acknowledge the assistance given by many colleagues in the Automotive Group at ICI (Paints Division) either through discussion or by providing suitable reference literature. Particular thanks are extended to Michael Waghorn, my associate for many years, for his advice and consultation on various key aspects, and to Barrie Wood for his contribution on the pigmentation of motors finishes. Finally, to my wife Kathleen and son David, much gratitude is due for producing some excellent illustrations to augment the text.

11

Automotive refinish paints

A. H. Mawby

11.1 INTRODUCTION

11.1.1 Definition

The term 'automotive refinish' refers to paint products applied to a motor car at any time subsequent to the initial manufacturing process. In the factory, a car body is finished as a piece of metal only, permitting the use of products which are cured at any temperature, commonly up to 150–160 °C.

Once the vehicle has been fitted with plastics, rubber tyres, and fabrics, and indeed may be on the road with petrol in the tank, it is no longer feasible to cure finishes at these temperatures. Hence different types of products must be used for *re*-finishing cars, and the refinish paint manufacturer must strive to achieve finishes of equivalent performance to the original finish within this constraint.

The size of the market for refinish paints, in volume terms, is comparable with that for original automotive products. Approximately 250 million litres of topcoat colour is sold worldwide (1984), with a roughly equal additional volume of primers, solvents, and ancillary products.

Commercial vehicles — trucks and buses — are usually painted in materials which are more closely related to automotive refinish products than to either original automotive or general industrial products. The volume market for paint products on commercial vehicles is approximately 40% of automotive refinish. The performance requirements for commercial vehicle paints are generally less critical than for car refinishing, although in some applications resistance to corrosive chemicals or environments is needed.

Automotive refinish products are applied to motor cars for three different purposes:

Repair or respray following accidental damage in use.
Rectification of damage or errors during manufacture or during transit before sale.

Voluntary respray to improve the appearance or change the colour of a vehicle.

In developed countries the bulk of the market is covered by the first two. Voluntary respray is largely limited to the secondhand car market, where a new coat of paint can make the difference between selling or not selling the vehicle. A change of vehicle colour is very rare. A large proportion of car refinish work is paid for by accident insurance.

In developing countries cars tend to be older, the accident insurance business is less well developed, and voluntary resprays with colour change are much more common.

Even within developed countries cultural differences in attitudes to motor cars are quite marked. These show up in the average standard of appearances of vehicles, and can be traced through to effects within the refinish market. The difference in the average standard of bodywork appearance between, say, Lisbon and Zurich is very marked, and it is reflected in the level of sophistication of bodyshop equipment and techniques.

The purpose of most car refinish work is to return a motor vehicle to a standard of appearance equal to that of a new car direct from the production line.

In most Western countries, a damaged motor car represents a severe blow to the ego, morale, and lifestyle of the owner, and its repair to 'original' conditions is correspondingly significant. When the repaired car is returned, the only standard by which the owner can assess the quality of work done is the exterior finish. Every owner will inspect the paint job, and above all will judge the work by the standard of the *colour match*. Colour capability and accuracy is the most important feature of car refinish systems.

Naturally the colour to be matched stems from the car manufacturer, and eventually from the supplier of automotive paints. However, car colours are chosen for their customer appeal, not for their repairability, and the car refinish suppliers, although often the same as the automotive paint suppliers, have historically had to react to changes initiated by them.

11.1.2 Air-dry and low-stoving

Automotive refinish products are applied under a very wide variety of conditions. In many developing countries it is common to find cars being resprayed in the open air, even at the side of a street. At the other end of the scale many large body repair shops are highly sophisticated industrial installations. This diversity of conditions, together with the frequently low level of operative skill and training, represents a very different challenge to the paint formulator in comparison with the automotive original market, where a production line has highly trained operatives and a strictly controlled environment. The major diversity in refinish workshops as it affects automotive refinish products, lies in the division betwen air-drying (drying at ambient temperatures) and stoving. In the automotive refinish context, stoving temperatures range up to 80° metal temperature, although modern refinish materials normally have optimum stoving schedules which reach 60 °C.

The absence of stoving facility in a refinish workshop usually also implies a relatively poor environment for painting, often dusty and variable in temperature and humidity. It is essential for an air-drying product used in such an environment to be 'dust free' in an extremely short time, normally a few minutes. This implies 'lacquer' products drying by physical processes. Products curing through reaction with atmospheric oxygen or 2-pack materials cure relatively slowly at ambient temperatures (since otherwise they would inevitably be unstable in storage or have short pot-lives), but may be used in refinish if a dust-free environment is available. In proper conditions such products have great advantages over lacquers, in both application and performance.

Hence, increasingly in all markets, and almost exclusively in the most highly developed, cars are refinished in full spray booths which are either linked to a 'low-stoving' oven or can themselves be converted into an oven (a combination spray booth/oven).

11.2 TOPCOAT SYSTEMS

1.2.1 Introduction

The restriction on stoving temperatures eliminates all the common automotive original topcoat technologies from use in automotive refinish, i.e.,

> Thermosetting acrylic
> Reflow thermoplastic acrylic
> High-bake alkyd melamine

In addition, it is a consequence of the structure of the refinish industry, with a very large number of small body repair establishments, that the pace of technology change is slow, so that all the main basic technologies used in the industry throughout its history are still in use today.

The driving force for changes in technology has been two-fold:

> The need to improve productivity in body repair shops.
> The requirement by many car makers to use materials capable of sustaining extended anti-corrosion warranties.

The result has been a long trend to higher-solids, higher-build materials needing fewer coats and giving 'gloss from the gun', i.e., requiring little or no polishing, and towards more durable resin systems, particularly acrylics.

11.2.2 Nitrocellulose

In the early days of the motor vehicle, essentially all automotive original and refinish paints were nitrocellulose based. On the production line nitrocellulose was early replaced by acrylic and alkyd systems. In refinish, nitrocellulose has persisted until the present day, though by now it has been superseded in most developed markets. The major advantage of physically drying products like nitrocellulose (and thermoplastic acrylic) in automotive refinish is the very rapid air-drying which, combined with easy polishing, permits good results to be obtained in ill-equipped and dusty workshops. The major disadvantage is poor durability. In summary, the strengths and weaknesses of nitrocellulose refinish products are as follows:

Strengths: Very fast drying
            Easy application
            Polishability
            Recoatable at any time

Weaknesses: Poor durability, poor UV resistance
            Brittleness
            Low gloss — polishing required
            Poor gloss retention
            Low solids/low build
            Poor solvent (petrol) resistance

Nitrocellulose lacquers are usually plasticized with a combination of solvent plasticizers, polymeric plasticizers, and non-drying alkyds.

11.2.3 Thermoplastic acrylic (TPA)

TPA automotive refinishes have many of the same characteristics as nitrocellulose products, and, except where durability and resistance to UV is essential for climatic reasons, are seen as largely interchangeable with nitrocellulose by the refinisher.

Like nitrocellulose, TPA lacquers harden by solvent evaporation only. Unlike the products used for original finishes, refinish TPA products are not subjected to stoving to achieve 'reflow'. Polymers used in TPA lacquers always contain a high proportion of methyl methacrylate, copolymerized with several other acrylic monomers to achieve the appropriate blend of hardness, flexibility, and adhesion, and blended with solvent plasticizers.

The strengths and weaknesses of TPA for refinish are as follows:

Strengths: Very fast air-drying
            Good durability (UV resistance)
            Easy rectification, polishable
            Excellent metallic effects

Weaknesses: Poor solvent (petrol resistance)
            Low solids/low build
            Brittleness
            Low gloss — polishing required

11.2.4 Alkyd

Alkyd finishes used in refinishing are typically short oil, fast-drying by nature. They may be air-dried with conventional cobalt and lead driers or applied as two component products by the addition of a melamine or polyisocyanate resin. When used with melamine the alkyd must be stoved at 70–80 °C metal temperature:

addition of isocyanate accelerates hardening at any temperature, but is commonly associated with 'force drying' at 40–60 °C.

Alkyds were the first 'enamels' (as opposed to physically drying 'lacquers') to be used for automotive refinish, bringing with them the great advantages of higher solids on application and hence higher build and 'gloss from the gun', eliminating the necessity for labour-intensive polishing.

Alkyds are, however, slower to become 'dust-free' than lacquers, whether nitrocellulose or TPA, and hence must be applied in a dust-free environment. Their introduction was accordingly accompanied by the development of suitable spray booths for the car repair industry.

A major disadvantage of alkyds is the difficulty in rectification, arising from the tendency of the partly cured film to soften if overcoated. As a result, any error in painting or damage caused subsequently, cannot be rectified until the film is fully cured. For alkyds dried conventionally by air oxidation this process can take several days, but films crosslinked by melamine or polyisocyanate addition can be recoated much more quickly.

In addition, the characteristics of the alkyd resin system are not well suited to the inclusion of aluminium flake for metallic finishes.

Metallic pigmented alkyd finishes give a poor 'flip' effect because the orientation of the aluminium particles is too random.

11.2.5 Acrylic enamel

Acrylic enamels are alkyd/TPA copolymers which represent a hybrid product type, developed to improve the build and gloss of conventional TPA without losing the very rapid dust-free performance of the lacquer. The properties of these products fall accordingly between the two types. Like conventional alkyds they may be used as 2-pack materials with a polyisocyanate second component. This confers benefits of through-drying speed, durability, and hardness, but loses all the benefits of the 'lacquer' drying, thus requiring dust-free application conditions.

11.2.6 Acrylic urethane

The most recently introduced major product type in the refinish industry is based on a hydroxy functional acrylic resin, used exclusively in association with a polyisocyanate second component. Although slow to be dust-free compared with lacquers, in all other respects acrylic urethanes have the best range of those properties which are most relevant to automotive refinish, and are hence rendering other products obsolescent in most developed markets for other than specialized applications.

The acrylic urethane combination is particularly well suited to the refinish market. Acrylic resins confer benefits of:

Durability, gloss retention
Hardness
Flexibility
Easy pigmentation
High gloss.

By using a relatively low molecular weight acrylic resin, solids can be high by refinish standards, while the crosslinking reaction with polyisocyanate takes place at rates which allow acrylic urethanes to be used across the range of temperatures required by the market, curing at an acceptable rate even below 5 °C.

Application of heat accelerates through-drying, and for optimum bodyshop throughput, acrylic urethanes are typically cured for 30–40 minutes at 80–100 °C air temperature, leading to a metal temperature of about 60 °C maximum.

11.2.7 Basecoat/clear systems

The development of basecoat/clear metallics for automotive finishing (section 10.7.4) had necessarily to be matched by refinish systems capable of repairing and respraying them to achieve an identical appearance.

The main requirements for basecoat formulation are high opacity and good intercoat adhesion to primer and to clearcoat. Gloss and durability are conferred by the clearcoat, and are not needed in the basecoat. Conventional refinish systems are designed to have gloss and durability and cannot satisfactorily be used to repair basecoat/clear original finishes. On the other hand, the main resin constituent of automotive basecoat technologies, CAB, hardens by solvent evaporation only, and is therefore equally suited to refinish use.

The introduction of basecoat/clear metallics therefore brought about the development of refinish basecoat repair technologies based on CAB, modified with polyesters, acrylics, or nitrocellulose. Many refinish basecoat metallics also contain polyethylene wax or other aids to aluminium flake control.

Indeed, automotive basecoat colour is commonly supplied for refinish use with little if any modification. Although some automotive basecoats contain melamine, which almost certainly contributes little to the curing process in a refinish environment, the hardness, durability, and intercoat adhesion performance of automotive basecoats under acrylic urethane refinish clearcoats is more than adequate.

Where refinish basecoats are not identical to the original, the technology is nonetheless similar, based on CAB blended with polyester or acrylic resins and with similar additives to automotive products.

11.3 COLOUR

11.3.1 Introduction

Colour is the central feature of refinish paint technology. The number and diversity of car colours on the roads rose very rapidly from about 1970, when colour became a significant part of the marketing package used by the car makers in an increasingly competitive market. This was particularly true of European markets where at the same time the expectations of colour-matching quality in bodywork repairs also increased.

It is estimated that car manufacturers' production colours on the roads in the main European markets in 1985 exceeded 10 000, a five-fold increase from 1965.

The colour range required to be matched in the process of repair is further extended by the variation in colours coming from the car factories. These variations may be caused by:

Drift over time in the standard or technology of paint supplied, conditions, or equipment on the production line,
Differences between the paint, equipment, or techniques used at two or more units producing cars in nominally the same colour.

Refinish paint makers monitor the colours of cars actually coming from the factories on to the roads, rather than simply matching the original 'styling pattern' which is the car maker's master for the colour.

11.3.2   Factory-matched colour and mixing schemes

The supply of colour matched at the factory to specific car colours is the most obvious and simplest approach to the refinish market. This approach is viable so long as a manageable number of colours represents the major part of the total market demand. Once it became unpractical to supply the whole of the colour demand in this way, paint manufacturers introduced the concept of the refinish mixing scheme to permit colour mixing initially at distributor level, subsequently at end user level. At first the mixing scheme supplemented the factory-matched range by providing a means to supply the less common, less frequently needed, colours. However, in many markets the mixing scheme has almost wholly supplemented factory-matched colour, because the enormous diversity of colours on the road renders the potential sale of an individual factory-matched product too small to be worth manufacturing or stocking.

Factory-matched colour is still a significant feature of markets which are either less well developed, with a majority of small repair shops, and/or a low expectation of colour accuracy, or else are dominated by a relatively small number of car makers, so that each individual colour can still command a reasonable sale. The biggest market in the world, the United States, remains in this position, though the rapid increase in imports of cars has initiated the trend towards mixing schemes which is by now nearly complete in Europe. The sheer size of the USA market naturally allows the paint maker to achieve an economic batch size for an individual colour in the range of one of the giant car makers: the eventual limitation lies in the size of stock inventory which the distribution outlet is prepared to hold. Experience in developed markets suggests that once a range of 700–1000 colours is able to service less than 60–75% of the demand, then the trend to mixing schemes becomes steady and irreversible.

11.3.3   Refinish mixing schemes

A typical refinish mixing scheme consists of a series of single-pigment paints, very carefully controlled for batch-to-batch consistency of both hue and tinting strength. These are mixed together, according to formulae supplied by the paint manufacturer, to match car colours. Formulations are normally supplied on microfiche in the form of parts by weight or by volume required to achieve a specific volume of accurately matched colour.

The hardware of a refinish mixing scheme consists of a stirrer bank, capable of simultaneously stirring all the tinters to maintain homogeneity within the tins, a microfiche reader for formula presentation, and a weigh scale or volumetric measuring device.

The number of pigments which can be included in a mixing scheme is naturally finite, generally amounting to 30–40 pigments, including a selection of grades of aluminium flake for metallic colours. Often a number of pigments are duplicated at 'full' and 'reduced' strength, the latter allowing small colour adjustments, for example in tinter white car colours. A typical range of pigments included in a refinish mixing scheme, and designed to give maximum coverage of typical car colours, is shown in Table 11.1.

Naturally, the range of pigments available to automotive paint formulators for new colour styling is much wider than that available within a refinish mixing scheme. As a result, refinish colour matches are frequently to some extent metameric to the original car colour. A satisfactory repair may then require a large area of the car to be painted; in extreme cases the entire car may have to be resprayed, because no acceptably close colour match can be achieved from the available pigments.

Refinish paint suppliers have to invest heavily in providing formulae to match all the car colours which might require to be mixed, in disseminating this information, and in giving the refinisher the means to identify the car colour. Production of mixing formulae is greatly assisted by the use of computer-linked reflectance spectro-photometers. By comparing the reflectance spectrum of the car colour with the spectra of the available mixing scheme pigments, the computer can suggest a range of possible mixing formulae from which the colour matcher selects according to the criteria he chooses to impose, e.g. level of acceptable metamerism, pigment cost, or total number of pigments used. Computer-matching is not yet capable of handling metallic colours fully satisfactorily, because of the angular dependence of the reflectance in such colours; but it is nevertheless useful in shortening the process of mixing formula production even for metallics.

11.4 FUTURE DEVELOPMENTS

11.4.1 Relationship to automotive technology

The basic difference between automotive and refinish technology which arises from the constraint on stoving temperature in refinishing a complete vehicle, is likely to continue to divide the two technologies, in spite of pressure to reduce stoving temperatures in car manufacture. This pressure comes from, first, a need to reduce energy costs, and second, the trend to the use of plastics in vehicle construction. Few plastics are capable of withstanding conventional automotive stoving schedules without distortion. Hence, unless the plastics are to be painted separately in a different, lower stoving process (with obvious dangers of poor colour matching), the new low-stoving processes are required for automotive finishing. Nevertheless, stoving temperatures of 100–130 °C can be used without distortion of most structural plastics, and these temperatures are still well in excess of those available to refinish.

However, we have seen that the introduction of basecoat/clear processes has reduced the difference between automotive and refinish technologies in these finishes to the clearcoat alone. It is apparent that in the next few years a much larger

Table 11.1 — Typical mixing scheme pigmentation

Colour area	Pigment type	Example
White	Titanium dioxide	SCM Tiona RH472
Blue	Copper phthalocyanine	BASF Heliogen Blue L6975F
	Copper phthalocyanine	Hoechst Hostaperm Blue BFL
	Indanthrone	ICI Monolite Blue 3R
	Potassium ferro/ferricyanide	Manox Blue 10SF
Green	Chloro-brominated copper phthalocyanine	ICI Monastral Green 6Y
	Chlorinated copper phthalocyanine	ICI Monastral Green GN
Yellow	Lead chromate	
	Lead chromate/sulphate	
	Flavanthrone	
	Tetrachloroisoindoline	Ciba Geigy Irgazin Yellow 2 GTLN
	Azomethine copper complex	
	Hydrated iron (III) Oxide	Bayer Bayferrox 3910
Red/orange	Iron (III) oxide	Bayer Bayferrox 110–160
	Molybdated lead chromate	BASF Sicomin L3135S
	Monazo naphtol AS	Hoechst Novoperm Red F3 RK-70
	Quinacridone	Ciba Geigy Cinquasia Red Y
	Dibromoanthrone	ICI Monolite Red 2Y
	Naphtol AS type	Hoechst Hostaperm Brown HFR
	Perylene	Bayer R6418/R6436
	Dimethylquinacridone	Hoechst Hostaperm Pink E
	Thioindigo derivative	Sandoz Sandorin Bordeaux 2RL
	Linear, trans, quinacridone	Ciba Geigy Cinquasia Violet RT–887–D
	Dioxazine	Hoechst Hostaperm Violet RL Special
Black Blue	Carbon Black/Furnace Black	Degussa Special Black 10
Jet	Carbon Black/Channel Black	Degussa Carbon Black FW200
Aluminium	Fine	Eckert 601
	Medium	Toyo 8160
	Coarse	Silberline 3166

proportion of new car colours, whether conventional solid colours, metallic, or pearlescent (mica) in pigmentation, will be formulated on basecoat/clear technology. Possibly only pastel colours and whites will be conventional one-coat technology, so that the colour technology in automotive and refinish will be equivalent, if not identical.

The trend towards more variety in automotive pigmentation, rendered possible by the separation of the decorative and protection functions of the paint finish implied by basecoat/clear, could start to reverse the trend towards on-site mixing of colour, and refinishers may find themselves increasingly using colour identical to that applied to the car on the production line. There is a limit to the number of pigments which may be added to a mixing system to permit an acceptable match to be achieved, and the addition of coated mica pigments, in particular, may cause that limit to be exceeded. However, the convenience of the mixing scheme principle to the bodyshop should not be underestimated, and much research effort is likely to be put into further extending the usefulness of refinish mixing schemes.

11.4.2   New resin systems

The development of acrylic urethanes has resulted in a resin system which meets most of the requirement of refinish. But for one major drawback it would be difficult to foresee a fundamentally new technology superseding these products in a generation, especially since substantial further improvement in already excellent performance is almost certainly possible.

The major drawback is the inherent toxicity of the isocyanate. There is a constant danger to operatives of becoming sensitized through inhalation of spray mist (aerosol), arising from bad practice or equipment malfunction. Subsequent exposure to isocyanate even at extremely low concentrations can cause asthmatic symptoms, and sensitization is therefore likely to oblige an operative to leave the refinish industry.

Research into refinish resin systems is therefore concentrated on achieving the performance of acrylic urethanes without the need for isocyanates. The great flexibility and established durability of acrylic systems make it likely that any new generation of refinish products will still be acrylic based.

The patent literature contains a number of alternative approaches to the chemistry of crosslinking in hydroxy-acrylics, but the urethane reaction is singularly difficult to match in its flexibility and capacity to proceed even at very low temperatures. Such patented crosslinking resins include those based upon:—

Melamine
Polyepoxides
Polyanhydrides.

All of these should be suitably externally or internally plasticized.

It remains to be seen which, if any, of these technologies has the capacity to make obsolete the present generation of acrylic urethane automotive refinish products.

BIBLIOGRAPHY

The refinishers handbook: published by Imperial Chemical Industries in association with the Vehicle Builders and Repairers Association (1983).

12

General industrial paints

G. P. A. Turner

12.1 INTRODUCTION

To the paint industry 'industrial paints' are those coatings used by industry at large, as opposed to painters and decorators, painting contractors, and do-it-yourselfers. Some parts of the industry prefer, as in this book, to treat automotive, automotive refinish, marine, and heavy-duty coatings separately, leaving the remainder as 'general industrial paints', the subject of this chapter.

General industrial paints are therefore all paints, except those excluded above, which are used by industry in factory finishing processes. They include coatings and end uses as diverse as wire enamels, clear and pigmented furniture finishes, can lacquers, tractor finishes, paints for toys, paper coatings, aircraft finishes, domestic appliance finishes, protection for under-body automotive parts, coatings for plastics, and so on. Industrial articles may be as large as road-grading machines or as small as dice. They are often made of metal, but may frequently be made of wood, wood composites, paper, card, cement products, glass, or plastic. Metals may be steel in any of its forms, with or without protective surfacing, such as galvanizing or tin, or they may be aluminium, zinc, copper, or any of the numerous alloys. Each substrate and end use is a different painting problem, which must be solved within the commercial and other constraints of factory processes.

There is therefore no such thing as a typical general industrial paint or painting system. Most sub-classifications of general industrial paints are based upon the industries served with those paints, e.g. drum paints for the steel and plastic drum industry. These classifications are often used in statistical data on paint usage. Some typical data are shown in Tables 12.1–12.5 illustrating the volumes of paint used and the industrial markets served in the UK and elsewhere.

Let us now consider the factors which shape an industrial painting process, leading to such a diversity of paints and systems.

Table 12.1 — The UK market distribution of paint (from Government statistics), 1980

Product type	Volume litres ($\times 10^6$)		%
Decorative/building applications			
Oil and/or synthetic-based, non-aqueous	152.7		
Oil and/or synthetic-based, aqueous	17.4		
Emulsion paints	173.4		
	——	343.5	53.4
Industrial applications			
Oil and/or synthetic-based, non-aqueous	102.9		
Composite thinners	29.7		
Catalysed materials	17.0		
Oil and/or synthetic-based, aqueous	6.3		
Cellulose-based systems	40.2		
	——	196.1	30.5
Specialized uses (including maintenance)			
Marine	25.8		
Spirit-based applications	35.0		
(knotting, stains, etc.)			
Bituminous applications	28.6		
(pipelines, roofing, etc.)			
Others	14.3		
	——	103.7	16.1
Total		643.3	100.00

12.2 FACTORS GOVERNING THE SELECTION OF INDUSTRIAL PAINTING PROCESSES

12.2.1 The substrate

Industries make products for sale, and these articles need finishing. Often the finishing process does not involve painting, e.g. the metal-plating of plastic car trim or costume jewellery. More often it does, or it involves a number of finishing processes, including painting or lacquering. Thus the specification of an industrial painting process begins with an article to be finished. The surface of this article is the substrate to be painted. This substrate has a major influence on the finishing system that is finally chosen for the article.

12.2.1.1 *Physical and chemical characteristics*

Some questions need to be asked about the substrate:

(A) Is it susceptible to attack by features of the normal everday environment (heat, light, oxygen, water, microorganisms)?

(A.1) If so, will it need protecting from these things by the paint system?

Table 12.2 – The UK market distribution for industrial finishes (1979)

		%
Transport		
Refinishing	21.1	
Automotive	14.3	
Other vehicles	4.0	
		39.4
Furniture		
Wood	7.0	
Metal	5.0	
		12.0
Prefabricated items		
Wood	5.0	
Metal	3.0	
		8.0
Office equipment		7.0
Packaging		6.0
Decorative goods, toys, sports equipment, etc.		5.0
Machinery		4.0
Household appliances		3.5
All others (including paper, board, special plastics, and powder coatings)		15.1
Total:		100.0

For example, the sensitivity of iron and steel to water and oxygen may lead not only to the design of a quality protective coating system, but also to the use of a passivating chemical pretreatment prior to coating. Sensitivity to light may compel the use of a pigmented coating for UV protection, though a clear coating would otherwise give sufficient protection and an attractive appearance.

(A.2) Will the substrate have been altered by environmental conditions immediately before painting? As a result, will it need conditioning, preparing, cleaning, or treating first?

For example, wood absorbs moisture from, or surrenders it to, its environment. If, after painting, it is to be stored or used in an environment of markedly different humidity, it can warp or distort. It should be conditioned to the right moisture content first by storage in air of the appropriate relative humidity.

(B) Is the substrate susceptible to attack by the paint or is the paint likely to be adversely affected by the substrate?

For example, a plastic substrate may be susceptible to attack by the solvents in

Table 12.3 — European market distribution for industrial finishes (1980)

		%
Transport		
Refinishing	15.0	
Automobile production	13.5	
Other vehicles	3.5	
	——	32.0
Wood finishes		
Furniture	7.5	
Prefabricated items	5.0	
	——	12.5
Metal goods		
Furniture	5.5	
Prefabricated items	3.0	
	——	8.5
Office equipment		8.0
Packaging (drums, can coatings, and linings)		7.5
Machinery		4.5
Decorative goods, toys, and sports goods		4.0
Household appliances (white goods)		3.5
Coil and strip coatings		3.0
All others (including engineering, paper and board, special plastics, and powder coatings)		16.5
Total:		100.0

some types of paint, or a paper substrate may be susceptible to attack by water in waterborne coatings. This may lead to avoiding certain coatings, or to the use of sealer coats to insulate the substrate.

Alternatively, the substrate may have such a low critical surface tension (e.g. untreated polyethylene or polypropylene) that most — possibly all — coatings will not wet it or adhere to it. In this case, the chemical nature of the surface must be changed, e.g. by oxidation by corona discharge, so that paint will wet and adhere.

(C) Will the substrate be adversely affected by the paint application or curing conditions, or will it have an adverse affect on paint application or cure?

For example, plastic or wood substrates may be distorted by a hot oven, but accept short bursts of radiant energy directed at the coating. The melting point of tin (232°C) determines the maximum metal temperature acceptable in the stoving of coated tinplate in can manufacture. Conversely, the high heat capacity of heavy metal castings can make the stoving of coatings on them impracticable or very inefficient, and some woods contain addition polymerization inhibitors which can be

Table 12.4 — USA estimated end-market distribution of industrial paints (1979) (in m gallons)

	Volume (million gallons)	%
Metal finishes		
Cans and containers	45	
Coil coatings	30	
Furniture and fixtures	20	
General metal	15	
	110	33.1
Wood finishes		
Furniture and fixtures	50	
Prefinished	30	
	80	24.1
Transportation		
Cars	40	
Trucks and buses	10	
Aircraft, railway, marine	25	
	75	22.6
Machinery and equipment	25	7.6
Appliances	20	6.0
Packaging (paper and plastics)	10	3.0
Electrical, electronic	5	1.5
Miscellaneous	7	2.1
Total:	332	100.0

extracted by the solvents into the coating, where they inhibit its crosslinking drying mechanism.

(D) Is the physical nature of the surface as supplied suitable for wet painting?

If it is too porous the pores must be sealed first. If it is too rough it may require abrasion or sanding to make it smoother. If it is too dirty it will require cleaning or degreasing. For example, chipboard is too porous and must first be sealed, either with a layer of paper impregnated with melamine resin, or with a putty-like chipboard filler applied by reverse roller coater.

12.2.1.2 Shape and size

Finally, there are the shape and size of the article to be painted to consider. These have a major influence on the methods of application and cure to be used.

For example, metal screws present problems of efficient paint application and subsequent handling, and the answer may be to use the old-fashioned technique of tumbling the screws in a drum with just sufficient paint to coat them. A jumbo jet, on

Table 12.5 — Japan end-market distribution of paint (1980–81)

End use	Volume ('000 tonnes)	%
Building industry	330	21.3
Structural steel	90	5.8
Shipbuilding	150	9.7
Railways	24	1.5
Automobile industry	326	21.0
Electrical industry	97	6.3
Appliance industry	81	5.2
Metal products	126	8.1
Wood products	94	6.1
House paints	47	3.0
Export	21	1.4
Others	164	10.6
Total:	1550	100.0

the other hand, cannot be housed in any practical oven, and will require a paint that dries at room temperature. Flat articles, like sheets of paper, hardboard, or coils of strip steel are very amenable to a wide variety of painting and curing techniques, which are readily automated.

12.2.2 The finished result

The industrial paint user has an article to be painted, and the nature of this article, as we have seen, has a major impact on the paint system and painting process chosen. The paint user also has in mind an end result that he wishes to obtain from the finishing process. This finished result is the next major factor determining the finishing system and process finally chosen.

12.2.2.1 Appearance

Undoubtedly the manufacturer of the article is after a certain finished appearance. Colour will be important, as will uniformity of colour and consistency of colour from article to article. Gloss also has a major effect upon appearance, and the paint user may require anything from the high gloss and crisp image reflection that gives showroom appeal to a motor car, to the satin-smooth look and feel of a furniture finish, to the dead-matt appearance used to hide minor imperfections in the substrate.

12.2.2.2 Cost

The process of selection of a new industrial finishing system is a matter of give and take between paint user, paint suppliers, and suppliers of application and curing equipment. Initially, the user will have a general concept of the desired appearance, which he will examine further in discussion with paint suppliers. Overhanging this and all subsequent discussions will be the factor of cost. The finished appearance must be achieved within a given cost, which may be tightly or loosely defined. If, for example, the finisher has envisaged a high-gloss, high-quality appearance and presents the paint supplier with a porous surface of rough texture, the latter will be obliged to suggest a through surface preparation followed by a multistage finishing system; possibly thorough sanding, a filler or sealer, followed by two further coats. The total film weight may also have to be relatively high. The finisher may be forced to conclude that he cannot afford this, and will have to think again.

12.2.2.3 Protection
This re-think will also have to take account of his other main requirement in the finished result: the protection afforded by the finishing system to the final product. This may range from protection from damage in the shop or showroom, to a complex specification including flexibility: impact and scratch resistance; wear resistance; water, humidity, and corrosion resistance; resistance to specified chemicals; and exterior weathering resistance. To combine, for example, good corrosion resistance with good weathering properties may require the use of a system, e.g. an anti-corrosive primer followed by a durable topcoat, as in aircraft finishing. Extremes of performance, such as 15 years or more outdoors without repainting, are also likely to narrow dramatically the choice of film formers available to the paint formulator, as in the coating of strip metal for cladding of buildings. The options will, as before, be subjected to the constraint of cost.

12.2.3 The required output rate

The manufacturer has an article which he wishes to paint to get a certain appearance and a certain amount of protection. We have seen that these factors alone almost define the options open to him. However, there is another major factor: he wishes to achieve a certain output rate of painted articles from his factory, and again he must do it within a certain cost.

Given unlimited capital resources, this output could probably be achieved in a number of different ways. In practice, though, unless manufacture is commencing on a 'green field' site, there will be the constraints of an existing factory and site, as well as limitations on finance. Perhaps the existing factory already has a paint line with existing equipment installed. The question arises as to how much money the manufacturer is prepared to spend to improve his paint line to finish the new article, and what constraints of space and services there will be. Looking ahead to running costs, manpower will be an important factor, and the manufacturer is likely to define at an early stage a constraint on the number of men per shift acceptable for the new painting process.

Once the required production rate has been coupled with the acceptable capital and labour costs for the painting plant, and the nature of the article has been considered as discussed above, a limited number of options for applying and curing the paint will have been identified.

For example, the problem may be that there is an existing paint line for flat boards doing an excellent job, but it has reached the limit of its capacity. The new article can be painted in the same way, but an upsurge in sales is expected. There is insufficient capital to move the painting line to a new and larger site. More production must be obtained from the same space, but there is some money for new equipment. This creates an opportunity for a complete review of the painting process since, as the line cannot be duplicated, it must run faster. This is almost certain to throw strains upon the existing paint system, which it may not be able to endure. A likely outcome is a modernized line with updated application and curing equipment, e.g. infrared lamps in place of hot air, and new paint products or even a new system.

In another example, the complexity of shape of a metal article means that the only suitable methods of application are some sort of spraying or electrodeposition. The manufacturer wants to keep paint and labour costs as low as possible, and is prepared to spend capital to achieve this. He is therefore attracted to electrodeposition, but he wants a high-gloss finish. No electrodeposition coating he has seen meets this requirement. He therefore reconsiders spraying, and may well end up with some variant of electrostatic spraying. the amount of automation he can build in will depend upon the complexity of the shape and whether other articles are to be finished on the same line. Curing will almost certainly have to be by hot air, but ovens may be shortened by high-velocity hot air.

At this stage of the planning *safety, toxicology, and pollution* will also have to be considered. The shape and cost of the paint line are emerging. The paint options are, at this stage, limited. Cost criteria seem to be attainable. But is the finishing process emerging safe to use? Will it produce unacceptable gaseous effluent or require toxic liquid or solid waste disposal? Sometimes these questions are enough to rule out certain paint options. In other cases, they may be resolved, for example, by installing superior fume extraction facilities to protect the working environment, or after-burners to reduce pollution and recover energy from the paint solvents.

12.2.4. Summary

The interactions of the above factors in the selection of an industrial painting process are summarized in the two complementary Tables 12.6 and 12.7. In subsequent sections more will be said about the application and curing processes used in industrial finishing, and examples will be given of how these factors have operated in specific cases, when selected industrial finishing processes are discussed.

12.3 INDUSTRIAL APPLICATION AND CURING METHODS

In Chapters 9 11 the techniques of applying and curing architectural, automotive, and refinish paints have been described. Application in those markets is by brush, roller, dipping, electrodeposition, and a variety of spraying techniques, manual and automated. Cure is under ambient atmospheric conditions or in convected hot-air ovens, or sometimes by radiant heating (infrared). In general industrial painting, because of the wide variety of articles to be painted and other requirements (as discussed above), a number of additional methods have been introduced. As we have seen, these have a significant effect on painting systems and processes.

Table 12.6 — Constraints imposed by or on the paint user, and the paint and process options they affect

	Constraint	Paint and process options affected
A	Substrate	Preparation, pretreatment
		Need for a paint system
		Film former type
		Solventborne or waterborne (solvent types)
		Application method
		Curing method
B	Required dry film properties (finished result)	Preparation, pretreatment
		Need for a paint system
		Film thickness
		Film former type
		Curing method
C	Required production rate	Nature of paint system
		Film former type
		Application method
		Curing method
D	Capital expenditure and existing plant	Preparation, pretreatment
		Nature of paint system
		Application method
		Curing method
E	Running costs	Preparation, pretreatment
		Nature of paint system
		Film thickness
		Film former type
		Paint type: High, medium or low solids
		solvent types
		one-pack or two (film-former)
		Application method
		Curing method
F	Health, safety and the environment	Film former type
		Solventborne or waterborne
		Paint solids

12.3.1 Application methods

In any method of application, either an excess of paint is applied and the surplus is removed, or the desired thickness of paint is put on directly during application. Of the methods described earlier, dipping is in the first category, and electrodeposition, spraying, brushing, and roller application are in the second. The additional methods used in industrial painting can be similarly subdivided.

Table 12.7 — Summary of the paint and process options and the constraints which affect them

Paint and process option	Constraints affecting choice (see Table 12.6)
Need for a paint system	A, B
Preparation, pretreatment	A, B, D, E
Nature of paint system	C, D, E
Film thickness	B, E
Film former type	A, B, C, E, F
Paint solids	E, F
Solvent type	A, E, F
Application method	A, C, D, E
Curing method	A, B, C, D, E

12.3.1.1   *Application of excess paint and removal of surplus*
(a)   *Flow-coating*
Paint is applied under pressure through a number of carefully positioned jets, so that all parts of the article are coated. Excess paint drains from the article and coating chamber walls back into the sump. Articles are usually hung from a monorail; accurate jigging is important. After the coating chamber comes a draining zone, in which the solvent content of the atmosphere can be controlled to delay evaporation and to permit paint flow and the escape of aeration. A sloping floor takes drips back to the sump. Next comes a drying zone, which in most cases includes a stoving oven.

Flow-coating can be used for large articles with relatively complex shapes. Penetration into recesses is better than in dipping, and the paint sump is much smaller (⩽10% of dip tank size). However, solvent losses due to evaporation are much higher, thanks to the large paint/air interface continually being recreated in the application chamber. This increases painting costs. However, with the increased use of waterborne primers in the USA, the cost of replenishing the tank after evaporation losses has become lower, and this method of application has created new interest for painting agricultural machinery.

(b) *Centrifuging*
Small articles, such as jewellery or furniture fittings, are placed in a basket, which is immersed in and then withdrawn from a paint tank. The basket is then centrifuged at several hundred rev/min to remove excess paint. The articles may be practically dry when tipped out onto net drying trays. The film thickness is dependent on centrifuge speed and paint viscosity.

(c) *Vacuum impregnation*
Vacuum impregnation is used for the application of relatively high-solids, viscous coatings to complex articles, such as electrical windings. Penetration into the extremely small spaces would be difficult without the aid of vacuum. The winding is preheated, and vacuum is applied while it is still hot. A varnish cock is then opened,

causing varnish to rise under atmospheric pressure from a lower tank into the coating chamber. When the windings are submerged, the varnish cock is closed, the vacuum is released, and additional air pressure is applied to force varnish into the windings. The varnish cock is then opened, and the surplus varnish drains back into the tank. Solvent may be partly removed from the windings by further application of vacuum.

(d) *Knife-coating*
Knife-coating can be used for the coating of flat and usually continuous sheet material, e.g. paper, plastic film. Excess coating is applied by any suitable technique, e.g. roller coating, and the coating thickness is then reduced by passing the web under a doctor knife (an angled metal blade) or an air knife (a curtain of high-velocity air directed onto the web). This technique is particularly suitable where very thin coatings indeed are required. Viscosity can be relatively high, since the pressure of the knife determines film thickness, and the appearance requirement is not exacting.

(e) *Extrusion*
Extrusion is a form of knife coating suitable for rods, tubes, and wire. The article to be coated is passed through a reservoir of coating and is then extruded through a die or gasket. The die acts like a doctor blade, permitting only a controlled thickness of coating to pass through with the article. Wire enamels are applied in this way.

12.3.1.2 Direct application of the required paint thickness
(a) *Tumbling or barrelling*
Tumbling or barrelling is a method suitable for coating small articles like screws, buttons, knobs, or golf tees. The articles are loaded into a barrel which can be rotated about its axis, the axis being horizontal or inclined at about 45°. By experiment the amount of paint required to coat the interior of the barrel and the surfaces of the articles to the required thickness is determined. This is then added, so that the barrel is one-third to one-half full. It is then rotated at 20–40 rev/min [1] for a predetermined time. Film thickness and appearance are controlled by paint viscosity, solids, and quantity, and barrelling time, speed, and temperature. The coated articles are tipped out onto screens or trays, where the coating dries by solvent evaporation or by thermal cure in ovens. Tumbling and centrifuging are variations on the same theme.

(b) *Forward roller-coating*
Forward roller-coating is a widely used method of applying low to medium thicknesses of coating to flat surfaces, such as board, sheet, or a continuous web of metal coil, paper, or plastic, one or both sides being coated in a single coating station. Coating is transferred from a reservoir via two or more rollers (see Fig. 12.1) to the substrate. Coating thickness is controlled via a polished metal doctoring roller, and application is from a rubber or gelatine-coated roller rotating in a 'forward' direction, i.e. with, rather than against, the forward movement of the substrate. Other factors controlling film thickness are relative roller speeds, roller pressures, and web or sheet speed. Coating solids and viscosity also affect thickness, and viscosity is critical to flow. The roller process imparts a wavy surface to the coating so that, in the worst cases, 'tramlines' of paint are laid down on the article. Paint viscosity must be low and Newtonian, so that these striations may flow out and level

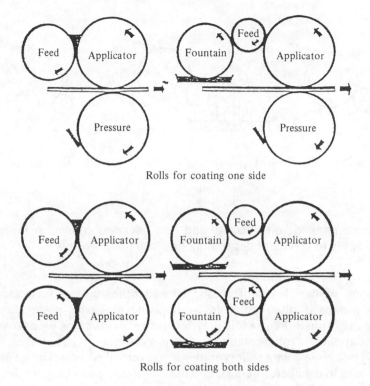

Rolls for coating one side

Rolls for coating both sides

Figure 12.1

to a uniform film. Except when the machinery is emptied and cleaned, no paint is wasted, since what is not applied stays on the roller. Coating speeds in excess of 100 m/min are possible.

A variant on this process, borrowed from printing, is the use of a gravure rather than a plane or grooved roller to transfer paint to the applicator roller. The thickness of coating is then directly related to the depth of the gravure cells etched in the roller surface; excess is returned to the reservoir by a doctor blade. Coaters of this type are used to apply very accurate coats of a few micrometers, and are sometimes called 'precision coaters'. The coating is laid down as a series of closely spaced 'dots', which must flow out to produce a continuous film.

(c) *Reverse roller-coating (coil)*
Reverse roller-coating is essentially the same as forward roller coating, but with the web of strip metal running in the opposite direction, i.e. counter to the direction of the application roller (see Fig. 12.2). Thus the paint is almost scraped off the application roller, and a high shear is applied, leading to better flowout of the coating. Since web speeds in coil coating are extremely fast (30–200m/min) and the coating enters the oven only seconds after application, every effort is made to avoid tramlines and to encourage flow. Consequently, reverse roller coating is the preferred method for topcoats, and is also used for some primers.

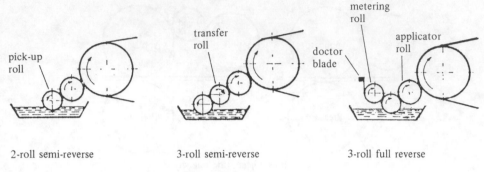

Figure 12.2

The comments made on viscosity and film thickness control in the previous section also apply to reverse roller coating on a coil line.

(d) *Reverse roller-coating (wood)*
In the coating of flat sheets of chipboard and other porous surfaces, a level impervious surface is often obtained by applying a viscous paste filler to the substrate, curing it and then sanding. A reverse roller coater is used to apply the filler, but the applicator roller rotates in a *forward* direction. It is, however, followed immediately by a large doctor roller rotating in a reverse direction. The reverse roller applies pressure to the filler, forcing it into the board and smoothing out the surface.

Coat weights of filler are high (50–120 g/m^2), and the material must be high in solids content and relatively viscous. The usual controls on film thickness are supplemented by varying the pressure and speed of the doctor roller.

(e) *Curtain-coating*
Curtain-coating is an ideal method of applying thicker coatings (60 μm and upwards) to flat boards or sheets. A curtain of paint is allowed to fall from a head (which may be pressurized) through a V-shaped slot into a trough below (see Fig. 12.3). From the trough it is returned to a reservoir tank, from which it is pumped again to the head. The curtain width depends upon the size of machine, and widths up to two metres are common. The curtain thickness is controlled by adjusting the width of the slot between 0.1 and 5 mm. Excessively thin curtains are unstable.

Running at right angles to the curtain on either side are two conveyor belts. The leading belt picks up the panel from the main conveyor line and accelerates it at up to 100 m/min so that it passes through the curtain and is coated. It is then picked up by the conveyor on the other side and carried away.

Some machines have twin parallel heads, and these can be used to apply similar or different coatings.

Coating film thickness is controlled by the width of the slot, the pressure in the head, the viscosity of the coating, and the speed of the conveyor belt.

12.3.2   Curing methods
There are two stages in the drying of paint films:

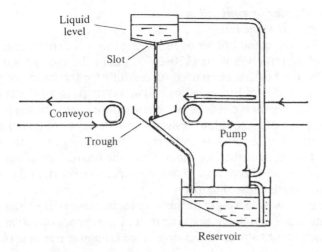

Figure 12.3

(1)   Removal of solvent or diluent
(2)   (a)   fusion of particulate polymer into a film,
     or   (b)   crosslinking of polymer molecules (curing),
     or   (c)   a combination of (a) and (b)

In lacquers, (1) is the only stage in the drying process. Fast air movement is even more important than heat for this stage. In 100% polymerizable coatings, only stage (2) is required, and air movement is only important in so far as it aids heat transfer.

Of the methods of transferring energy to the coating, conduction from a hot substrate can be used to supplement other forms of heating, and is the mode of heat transfer in induction heating. Forced convection is widely used in conventional stoving ovens. Radiation and oscillating electric fields are used in various special curing techniques.

Drying methods specific to general industrial painting will now be discussed.

12.3.2.1   Energy transfer by conduction : induction heating
If a metal object of simple shape, e.g. a rod or sheet, can be placed in close proximity to an induction heating coil, then the eddy currents induced in the object will heat it. Heating can be localized within the skin or surface of the object. Heating is rapid and readily controlled. If the object is a painted object, then heat is rapidly induced in the substrate and transferred to the coating by conduction from the substrate.

In spite of the attractive features inherent in the method, induction heating has had very limited use in the drying of industrial paints, the main reason being, probably, the need to match very closely the contours of the painted article and the induction heater. The method is suitable, therefore, only for processing similar articles of simple shape day-in day-out on a given painting line.

12.3.2.2   *Energy transfer by convection*

(1) *High-velocity hot-air jet heating*

In this variant of the convected hot air oven, drying times are considerably reduced by directing jets of extremely hot air (180–550°C) onto the moving surface of the painted article. This method has been used successfully for the curing of painted flat boards (e.g. for 'knock-down' furniture) or coil metal strip (in air-flotation ovens).

In the former use, solvent is partly removed in a more conventional hot-air zone, so that the coating is viscous enough to resist the impingement of jets of air at about 180°C, moving at speeds of 25–36 m/s and striking the coating at right angles to the film plane, without being disturbed by them. Since the boards are moving at speeds of up to 30 m/min, the time of contact with the hot air can be short, and boards do not reach temperatures which cause distortion.

In the latter use, the whole weight of the strip metal is supported along its length by hot air from jets located in the base of the oven. Line speeds of 150 m/min can be used with oven stay times of about 15 seconds, so air temperatures of up to 550°C are necessary for cure.

(b) *Flame drying*

In this method, hot air impinging on the film is provided in the form of an air curtain surrounding a smooth high-velocity flame. This air barrier prevents solvent ignition. Contact times are as short as 0.02–0.04 seconds. This method has been used successfully to dry inks and coatings on plastic and metal containers. Equipment is also available for sheet-fed and moving-web processes.

12.3.2.3   *Energy transfer via radiation or electric fields*

Most types of electromagnetic radiation have been employed to cure or dry industrial coatings. The high-frequency oscillations of an electric field are employed in radio-frequency drying, and radiations varying in wavelength from long wave infrared to the ultrashort wavelength β-radiation (better known as electron beams) have been used.

(a) *Radio-frequency drying*

In radio-frequency drying, otherwise known as 'dielectric heating' [2], two electrodes (rods or platens) of opposite polarity are arranged parallel to one another and to the conveyor, but at right angles to the flow of work pieces. The work pieces should be flat boards or sheets, e.g. hardboard sheets, since the coating should pass within about 5 cm of the electrodes. The electrodes are connected to a high-frequency generator so that the polarity of each electrode oscillates at about 20 MHz.

In this arrangement the fringe field of the stray field between adjacent electrodes selectively passes through and heats the material or dielectric in the vicinity which is of highest 'loss factor' (the product of dielectric constant and power factor of the material). Water-based coatings have higher loss factors than hardboard, thus the field passes through the coating and does not heat the substrate. The coating is heated because the water molecules are small dipoles, and, every time that the polarity of the electrodes changes, they attempt to realign to the new field. With millions of field changes per second, this causes great friction between the water molecules, and the temperature rises rapidly, causing water to evaporate. Once the water has gone, heating ceases, so the method is not very suitable for curing thermosetting coatings. However, water can be removed from 50 μm films in about 20 seconds (faster removal may cause blistering).

The method is used for board and paper coatings based on water.

(b) *Infrared heating*

In this method IR heat is emitted by the radiation source and directed at the coating, which uses the energy efficiently if it absorbs most of it.

IR sources are hot bodies which emit radiation over a broad spectrum of wavelengths, the peak wavelengths varying with the temperature of the hot body. White-hot sources at 1200°–2200°C emit with peak wavelengths in the short wavelength region (1.0–2.0 μm), red-hot sources at 500°–1200°C in the medium wavelength region (2.0–3.6 μm), and dull emitters (90°–500°C) in the long wavelength region (3.6–8 μm). The peak wavelength is the highest energy point radiated from the source, with 25% of the waves of a shorter length than the peak and 75% of a longer wavelength [3]. Emitters may be heated by gas (flames impinging on the backs of curved or flat panels) or electricity (heated wires or filaments in ceramic sheathing, lamps with built-in reflectors, or quartz tubes with external reflectors).

Since the high-temperature sources emit more energy, they may seem the natural choice for all uses. However, absorption by the coating is an equally important factor. Most polymers contain groups which absorb in the IR. These groups absorb strongly around 2.9–3.7 μm and above 5.5 μm [3, 4]. Thus medium- and long-wave IR is better for heating clear coatings; short-wave heats the substrate. If the coating contains pigment, pigment colour and scattering are important. Black is highly absorptive, other colours less so. Scatter due to reflection is most pronounced at the shortest wavelengths and declines to zero between 2 and 7.5 μm [5]. Thus a good general compromise for efficiency of absorption lies in the region above 5.5 μm, but faster (though not necessarily more cost-effective) heating can be achieved with short–medium wavelength emitters because more energy is emitted.

Radiant heating has to be directed, and heaters and workpieces have to be arranged to avoid shadowing and cool spots. For the reasons given above, for a given arrangement of emitters and a given exposure time, paint film temperature can vary from paint to paint and colour to colour. It will be higher over thin-gauge metal rather than thick. These problems lead to difficulties of control with more complex articles or mixed products on one line. On the other hand, IR is good for substrates such as chipboard, hardboard, and plastic, where substrate heating is to be avoided and where articles may be in sheet form or of simple shape.

Ovens combining zones of IR and convected hot air avoid many of the difficulties and capitalize on the ability of IR to bring paint to temperature very rapidly.

(c) *Ultraviolet curing*

UV curing involves only the crosslinking of molecules. Ideally, no solvent is present to require evaporation, the coating being kept fluid by the use of low molecular weight polymer (oligomer) and monomers (see Chapter 2, section 2.19). If solvent must be used, it is removed in a flash-off section, possibly with the aid of IR. Curing is triggered off by the absorption of UV by photo-initiator molecules, which decompose, normally to free radicals. These then initiate polymerization.

UV sources are tubular quartz lamps in a suitable housing which contains a reflector. The lamps contain mercury which is vapourized and ionized between electrodes or via excitation by microwaves. The light radiation emitted has the line spectrum of mercury, with principal UV lines at 365 and 313 nm, and several smaller lines at or below 302 nm, merging into a continuum between 200 and 250 nm. Strong lines in the visible occur at 405, 436, 546, and 578 nm. The most widely used lamps are of the medium pressure type with a power rating of 80 W per cm of tube length.

Lamps are set over the conveyor at right angles to the direction of travel and usually at a height such that the radiation is focused in the plane of the coating. The number of lamps is sufficient to give cure at the conveyor speed required, and the curing rate is expressed in m/min of conveyor speed per lamp, e.g. 15 m/min/lamp. Coatings are exposed to radiation for times well below one second; but, since most lamps emit about 15% of their energy as IR, coating temperatures rise at slower line speeds.

UV curing is widely used for clear coatings on flat surfaces such as wood-based boards, paper, card, and floor tiles. Pigments absorb UV strongly, preventing decomposition of photo-initiator and hence cure. However, with inks and with thinner, low-opacity paint layers, cure is possible, and UV has found its widest use for the cure of printing inks. Lamps can be introduced in the very limited space available on printing lines, and can be used to produce cure between printing stations of different colours. UV energy costs are lower than for thermal methods of cure.

(d) *Electron beam curing (EBC)*
In an electron accelerator, electrons are generated at a heated wire or rod within a vacuum chamber, and are directed through a thin titanium window as a narrow beam which scans rapidly backwards and forwards across a conveyor, or as a 'curtain' beam covering the full width of the conveyor. The electrons are highly energetic (voltages of 150–600 kV are used), penetrating the coating and creating free radicals on impact with molecules therein. Photo-initiators are not needed, otherwise coating compositions are similar to those used for UV curing, and free radical addition polymerization ensues. Line speeds of several hundred m/min are theoretically possible.

Since pigments have only a limited effect on beam penetration, thick coatings of any colour can be cured, though the depth of penetration is directly related to voltage. At 150 keV penetration to a depth of 120 μm is possible in a coating of density 1.0. Electron accelerators emit X-rays, so they must be screened to protect the operators. At 150 keV, 5 mm lead screening is sufficient, but accelerators of 300 keV and above need to be housed in concrete bunkers. This increases the cost of an already expensive machine: the cheapest type (150 keV) costs about ten times as much as a UV installation. Running costs are lower than for UV in terms of energy, but are boosted by the need for a continuous supply of inert gas over the coating. As with UV, surfaces must be flat or nearly so. Thermoplastic substrates can be coated, since temperature rises are low.

The very high capital cost of this process limits its use to high-volume production outlets. Electron beam curing has been used on wood panels and doors, car fascia panels, car wheels, and in certain reel-to-reel processes, e.g. silicone release papers and magnetic recording tapes.

Table 12.8 — Characteristics of industrial methods of application

Feature Application method	Applies excess (E) or required amount (R)	Suitable for flat (F), simple (S), or complex (C) surfaces	Suitable (S), Fairly suitable (F), or unsuitable (U) for rapid colour change	Paint economy: E= excellent G=good F=fair	Speed of application F=fast S=slower	Continuous (C) or batch (B) process
Dipping	E	SIC	U	G-E	S	C
Flow-coating	E	S/C	U	G	F	C
Centrifuging	E	S	U	E	S	B
Vacuum impregnation	E	C	U	E	F	B
Knife-coating	E	F	S	E	F	C
Extrusion	R	S	F	G	S	B
Electrodeposition	R.	C	U	E	S	C
Forward roller coater	R	F	S	E	F	C
Reverse roller coater (coil)	R	F	S	E	F	C
Reverse roller coater (wood)	R	F	S	E	F	C
Curtain-coating	R	F	F	E	F	C
Spraying, various	R	C	S	F-G	F-S	C

Table 12.9 — Characteristics of industrial methods of drying and curing

Feature	Energy transfer: conduction (Cd) convection (Cv) Radiation (R) electric field (F)	Suitable for flat (F) simple (S) or complex (C) surfaces	Speed of dry/cure: VF=very fast F=fast M=moderate S=slow	On/off control	Start-up time L=long M=moderate S=short	Capital costs: H=high M=medium L=low VL=very low	Energy costs:	Notes
Induction heating	Cd	S	F	Yes	S	M	M	For metal substrate
Room-temperature cure	Cv	C	S	No	S	VL	VL†	
Convected hot air	Cv	C	M	No	L	M	M/H	
Jet-drying	Cv	F/S	F	Yes	M	M/H	H	
Flame-drying	Cv	F	VF	Yes	S	M	M	
Radio-frequency	F	F	F	Yes	S	L/M	M	For removal water
Infrared	R	S	M/F	Yes	S	L/M	M/H	
Ultraviolet§	R	F	VF	Yes	S	L/M	M	For curing by addition polymerization
Electron beam	R	F	VF	Yes	S	H	L‡	

† air movement needed   § not for thicker pigmented films   ‡ but additional running costs for nitrogen

12.3.3 Summing up

A wide variety of application and curing methods is available to the industrial finisher. In tables 12.8 and 12.9 the main features of these alternatives are summarized.

12.4 FINISHING MATERIALS AND PROCESSES IN SELECTED INDUSTRIAL PAINTING OPERATIONS

12.4.1 Mechanical farm implements

Mechanical farm implements are large and complex machines with many component parts, some very large, some small, some sheet metal, and some cast metal, but mainly steel. Equipment is made by large international companies and painted in their distinctive house colours.

These manufacturers require a uniform and attractive colour on all parts, with a high gloss, giving an excellent showroom appearance. Less emphasis is put on the quality of protection, though the coating must be durable outdoors and must withstand wear and tear, oils, and fuel.

The complexity and variety of size and shape rules out many application and curing options at once. The uniformity of colour makes dipping and flow-coating acceptable and economic options for primers and for one-coat finishing. Airless spraying is used for other topcoats. Room temperature and convected hot air are the only practical drying options, but drying temperatures vary from 15°C to 150°C according to the size and heat capacity of the painted part.

Paint costs are extremely important in this market and must be minimized. Production is on mechanized lines, but output rates need be only moderate.

What is required, therefore, is an inexpensive coating capable of room-temperature drying, force-drying, or stoving to give exterior durability and moderate protection on steel. Solids must be relatively high in the paint, and the overall appearance full and glossy. It would be difficult to think of polymers better suited to meet all these requirements than the short–medium oil length alkyds. With driers, these give the necessary room temperature or force-dried curing. With melamine resin as crosslinker, they provide an even higher standard of protection on parts that can be fully stoved. The coatings remain stable in dip-tanks and flow-coaters, provided that these are monitored and adjusted at regular intervals. The same materials can be thinned to spraying viscosity for spray application to completed machines, assemblies, or parts.

Farm machinery parts are normally degreased, but not pretreated. Trouble is taken to prepare a relatively smooth metal surface for painting, but the filling and natural levelling properties of the medium–low molecular weight, higher-solids alkyd resins are important for giving the final high-quality appearance.

Recent trends are to increase safety, reduce pollution, and avoid loss of expensive solvents by moving to waterborne alkyds, for primers at least, and, in some cases, to electrodeposition. However, waterborne alkyds must be made from more expensive ingredients than their solventborne counterparts (for hydrolytic stability), and the water-miscible solvents still necessary in minor amounts cost more than hydrocarbons. Paint costs do not fall, therefore, but savings can be made on thinners. There is also a move to higher standards of exterior durability, and this could lead to better alkyds, methacrylated alkyds, or even to two-pack sprayed topcoats.

12.4.2 Panel radiators

Panel radiators are the heating panels used in the circulating hot-water heating systems of European houses. In their simplest forms, they consist of moulded, hollow steel panels. More recently double radiators, comprising parallel pairs of the simple panels, have been introduced, and corrugated sheet metal has been spot-welded to the backs of the panels to produce a larger hot 'extended' surface from which convection may take place. Even with these complications, panel radiators remain essentially laminar and suitable for dip application.

Appearance standards are good, but not necessarily of the highest quality. in some countries, house purchasers paint the radiators even before using them, to match their preferred decorating schemes. Thus the coating must have the appearance of a topcoat, but act as a primer and accept overcoating with decorative house paints. Coatings must resist knocks and scratches, but other protective requirements are modest. Good coverage of all parts of the sometimes complex shape is, however, essential. Economy is important in the paint process. Production rates are moderate.

In view of the panel shape, radiators are ideal for dipping. For some years the most popular method was dipping in trichloroethylene paints. This solvent has a high vapour density and, although evaporation from the paint still readily occurs (boiling point 87°C, evaporation rate six times as fast as butyl acetate), the vapour does not escape from a properly constructed tank, but hangs about above the liquid layer. As the radiator is withdrawn from the paint, it passes through this zone saturated with trichloroethylene vapour, and evaporation of solvent from the paint is delayed and flowout of paint on the radiator takes place readily. As the radiator leaves the vapour layer, evaporation proceeds rapidly and flow ceases. The overall result of this is that a much more uniform film thickness is obtained than with conventional dipping. Resin systems are alkyd–amino resin or styrene- or vinyl toluene-modified alkyds.

In recent times manufacturers have been concerned about both the toxicology of chlorinated solvents and also the ability of the coating to cover all recesses of the more complex modern designs of radiators. As a result, electrodeposition has become the most frequently used method of coating, giving complete coverage of all surfaces and good control of film thickness, coupled with full automation. Anodic acrylic or polyester coatings are generally preferred, since pale colours (near white) are usually applied, and cathodic resin systems are prone to discolour on baking. These coatings require high bake temperatures (170–180°C) and stoving times of about 20 minutes for 20–25 μm films. Curing mechanisms usually entail crosslinking of hydroxyl-functional acrylic polymers with amino resins.

Good-quality metal is degreased and then pretreated with iron phosphate to give enhanced adhesion and a sufficient degree of corrosion protection at minimum cost.

There is an increasing trend to fully-finished radiators, on which an Electrocoat primer is topcoated with a spray polyester or a powder coating.

12.4.3 Refrigerators

The refrigerator market has seen fierce international price competition in recent years. There has been heavy pressure on painting costs, with a requirement to achieve more with less paint and to apply and cure it most economically. Coating colours in Europe are mainly white, with a larger colour range in the USA. Colours must be clean and bright, and the finished appearance good. The coatings must be hard and scratch-resistant and must withstand household hazards, such as fats, fruit juices, ketchups, and various polishes. They must perform well in a humid environment. A steady production rate is required.

Good-quality metal is therefore cleaned and pretreated with iron phosphate, and 25–35 μm of finish are applied directly to the treated metal. Good paint utilization is sought by electrostatic spraying from disc or bell. The hardness requirement, resistance properties, and clean white colours call for thermosetting acrylic or polyester/melamine resin finishes. Acrylics are usually of the self-condensing type and contain crosslinking monomer, such as N-butoxymethyl acrylamide. Sometimes epoxy resin may also be included in either type of finish to improve resistance properties in corrosion tests. Although the shape is essentially box-like, it is complex enough to require convected hot-air curing. Stoving schedules for these resin systems (with the aid of acid catalyst such as p-toluene sulphonic acid) are 20 minutes at 150–175°C.

Under pressure to upgrade quality while reducing solvent emissions, some manufacturers have turned to powder coatings. Acrylic, epoxy or polyester–epoxy types are used. Very tough, high-quality finishes are produced, but there are difficulties controlling coat weights to the economic levels required by the industry. Cure times are around 15 minutes at metal temperatures of 170–190°C. Electrostatic powder spray guns are used, and economy is improved by recovering unused powder via cyclones. All this calls for special plant and capital investment, but coatings are virtually non-polluting, and further economy comes from low reject rates, due to the tough finish resisting damage during assembly.

An alternative approach to the needs of the industry has been to remove the paint shop altogether and fabricate the refrigerator from prepainted electrozinc-plated steel coil finished by a coil coater. This has called for a redesign of the doors and casing to deal with the problem of exposed unpainted metal edges where the sheet is cut. Special flexible, but resistant, polyester/melamine coil coatings have had to be developed to withstand the forming operations in refrigerator construction. Some sacrifice in hardness has been inevitable, but General Electric (USA) have managed to maintain hardness by forming the metal at temperatures of about 60°C using coatings less soft and less flexible at room temperature.

12.4.4 Coil-coated steel for exterior cladding of buildings

Many large industrial buildings are constructed with a steel framework, which supports an outer skin of cladding and a roof. The cladding and roof are frequently made from prepainted hot-dipped galvanized steel, which has been formed and corrugated after painting to give extra strength and stiffness. The product from the coil-coater is a coil of painted metal, which may vary in width and uncoiled length.

Coils may be bought in a fairly wide range of colours. The coil-coater is expected to give a prediction (and in some cases a guarantee) of the life expectancy of the painted metal.

A coil line is a large-scale reel-to-reel operation, with steel at one end and pretreated, painted steel at the other (see Fig. 12.4). The capital investment is very

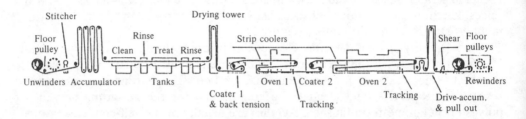

Figure 12.4

high. The process is economic if (a) the final article is a high-quality product, (b) paint thicknesses can be minimized and carefully controlled without waste, and (c) painting can be done quickly (minimizing floor space needed and hence capital costs). Coil lines run, therefore, at speeds between 30 and 200 m/min.

Application by roller-coater (mainly reverse roller-coater) is ideal for these speeds (see section 12.3.1.2(c)) and permits a rapid change of colour, since a 205 litre drum is the paint reservoir, and only the limited amount of paint on rollers and in the pump, troughs, and piping needs cleaning out. In view of the flat nature of the painted surface, a wider variety of drying processes are theoretically suitable; but in practice coil coaters have found UV and electron beam too limiting on the nature of the coating used, and have preferred high-velocity hot air to IR.

The coatings must therefore be suitable for rapid application and must cure in 15–60 seconds, though a lot of heat is available. The metal temperature peaks at 180–250°C just before the paint leaves the oven, and is 'quenched' by water spray. When it has cured, it must be capable of forming around sharp bends, must have excellent exterior durability, and must withstand corrosion in marine or industrial environments. Specific chemical resistance may be needed, and resistance to building-site damage is obviously required.

Although the zinc-layer will protect the steel cathodically in the event of damage that exposes the metal, a pretreatment is required to provide a suitable surface for the coating to adhere to, and to reduce 'white rusting' of the zinc. This is usually a spray or immersion coating of a complex metal oxide pretreatment, followed by a passivating chromate rinse. These are applied in the first stages of the coil line.

A paint system is normally required in order to provide the very high standard of durability and resistance properties. The first coat is 5 μm of primer containing chromate anti-corrosive pigments. The topcoat varies with the durability required. The most expensive coatings are applied at 20 μm, and vary in life expectation from thermosetting acrylic (7 years), polyester/melamine (10 years), silicone polyester/melamine (12–15 years) to polyvinylidene fluoride (20 years). Alternatively, the cheaper PVC plastisol can be applied at 100–250 μm to give a very damage-resistant coating with 10–15 years durability. The primer for plastisol is usually acrylic based, while thinner finishes are applied over epoxy–urea or polyester–melamine primers.

The back of the coil·is coated in the same sequence of operations as the front (see Fig. 12.4) with 10 μm of polyester–melamine backer or 3–5 μm of primer and 8–10 μm of backer.

A considerable amount of work has been done on waterborne coatings for coil, and thermosetting acrylic emulsion-based materials have met with limited success on cladding. However, the use of after-burners to incinerate solvents from the coatings, with recycling of hot gases into the oven, has provided an economical alternative to water-based coatings with fewer problems in use.

2.4.5   Interiors of two-piece beer and beverage cans

The beer and beverage can is a form of food packaging, and must not add excessively to the cost of its contents. Can-makers are constantly seeking ways of making the package cheaper. Once the can was made in three pieces: the body (from a flat sheet) and two ends. Now most beer and beverage cans are two-piece cans. The body is produced from one piece of metal by a process known as drawing and wall ironing. This method of construction allows much thinner metal to be used and the can has maximum strength only when filled with a carbonated beverage and sealed. In the USA, where aluminium is cheaper, most beer and beverage cans are made from that metal. In Europe, tinplate is often cheaper, and many cans are made of this. Modern beer and beverage tinplate has a low tin content at the surface, the main functions of the tin being cosmetic and lubricating (in the drawing process). So a lacquer with excellent protective properties is required, to be used at minimum coat weight (6–12 μm, dependent on metal type).

Can-making is economical only if the cans can be made very quickly. 800–1000 cans a minute will be produced from one coating line, with bodies and ends coated separately. Bodies for beer and beverage cans are lacquered after being made and degreased. The rapid application is achieved by short bursts of airless spray from a lance positioned opposite the centre of the open end of the horizontal can. The lance may be static or may be inserted into the can and then removed. The can is held in a chuck and rotated rapidly during spraying to obtain the most uniform coating possible. Coating viscosities must be very low, and solids about 25–30%. The shape is relatively simple, but interiors are cured by convected hot air, in schedules around 3 min at 200°C.

Carbonated soft drinks are acidic. Resistance to corrosion by such products is provided by coatings such as epoxy–amino resin or epoxy–phenolic resin systems. Beer is a less aggressive filling for the can, but its flavour may be spoilt so easily by iron pick-up from the can or by trace materials extracted from the lacquer, that it also requires similar high-quality interior lacquers.

The majority of these coatings have been successfully converted to waterborne colloidally dispersed or emulsion polymer systems, especially on the easier substrate to protect, aluminium. Water-based coatings have reduced overall costs and lowered the amount of solvent that has to be disposed of by after-burners to avoid pollution.

Most successful systems are based on epoxy–acrylic copolymers with amino or phenolic crosslinkers.

Recently, there has been revived commercial interest in the electrodeposition of water-based lacquers in beer and beverage cans. Such a procedure avoids the need to apply in two coats, and is potentially capable of giving defect-free coatings resistant to the contents of the can at lower dry film weights.

12.4.6   Aircraft

Jet aircraft are made principally from a range of aluminium alloys. Most parts are coated before assembly, but eventually the completed aircraft must be finished, providing a huge surface and bulk, quite unsuitable for any of the stoving or curing techniques described above. Room temperature drying is the only option possible. Nevertheless, a high quality of appearance and protection is required. Resistance to corrosion is essential, and also to the operating fluids of the aircraft, including aggressive phosphate ester hydraulic fluids. Exterior durability is, of course, necessary, with the coating being exposed to shorter-wavelength high-energy UV at high altitudes and to cycles of temperature varying from $-50°C$ in flight to over $70°C$ on a tropical airstrip.

Thus a high-quality room-temperature curing system is essential. Most widely used primers are of the epoxy–polyamide (or polyamine) two-pack type, with leachable chromate anti-corrosive pigmentation. The preferred topcoat formulations are two-pack polyurethanes, incorporating hydroxyl functional polyesters and aliphatic polyisocyanates, also two-pack.

Because of the size and complexity of shape of the aircraft, these coatings are applied manually by airless or conventional spray, with operators protected by wearing air-fed hoods.

Cleaning and pretreatment of the metal is important, with chromate conversion coatings and chromic acid anodizing being widely used.

12.4.7   Flush doors for interior frames

The modern flush door consists of a rectangular wooden frame between two 'skins' of hardboard sheet. These are kept apart by a lattice pattern cardboard spacer, the whole structure being given rigidity by glueing. The skins may be plain hardboard or plywood, or the hardboard may have a wood veneer coating. A thin wooden lipping protects the long edges of the door.

The two skins of the door are large flat surfaces and as such are ideal for coating and curing by the modern techniques described above. Equipment is situated in a conveyorized coating line. The final coating is required to look good, to be hard and abrasion-resistant, and to have resistance to humidity changes and a limited number of domestic liquids. These properties can be provided by materials as long established as acid-catalysed alkyd–amino resin or nitrocellulose–amino resin types or as modern as 100% solids polyester–styrene or acrylic oligomer-plus-monomer types.

In a 'green field' situation the choice between these types will depend on the required production rate, the available floor space and capital, acceptable running costs, etc. The more traditional finishes will involve sealer coat application by forward roller and finish application by curtain–coater. Cure will be by convected hot air, jet-drying, or a combination of convection and infrared. A considerable amount of solvent will require removal. The 100% solids finishes will be applied by roller (possibly even by precision coater) and cured by UV (clears) or electron beam (clears and pigmented).

Like the modern furniture industry, this industry has long accepted two-pack products for use in sophisticated application machinery, provided that they can be formulated to give at least eight hours pot-life. For acid-catalysed products this is easily exceeded (1–4 days). Radiation-cured coatings are one-pack materials.

Some of the earliest commercial uses of electron beam curing were for the finishing of doors by major European door manufacturers.

12.4.8 Record sleeves

For many years sleeves for long-playing records were finished after printing by laminating polyester film onto the card. This gave a glossy finish which withstood continual use.

The same effect can now be produced by coating the print with 4 g/m^2 of UV-curing acrylic overprint varnish. Although the varnish is expensive by conventional paint standards, the low coat weight is economical against plastic film and laminating adhesive, and it can be roller coated and cured with two UV lamps at speeds of 60 m/min. This high output with low capital cost leads to overall finishing economies without loss of quality.

The UV coating must crease without cracking and peeling, must be scuff-resistant, have good adhesion over inks, be solvent-resistant, and withstand over-wrapping at 160°C. The high-gloss finish should have good 'slip' (low coefficient of friction), yet it must accept glueing and foil blocking with metal foils.

Very high standards of worker safety are required in the printing industry, and modern UV coatings are based on materials of such low toxicity that skin irritation indices (Draize ratings) of the monomers used must be no higher than those of long-established coatings solvents.

Other end uses for clear film laminates are also amenable to UV-curing finishing processes. An end use closely related to record sleeves is the finishing of covers for paperback books.

12.5 DEVELOPMENTS AND TRENDS IN GENERAL INDUSTRIAL FINISHING

The last decade has been one of considerable change in industrial finishing, change that has been brought about as the result of the following pressures:

Escalating petroleum prices.
Legislation and pressure groups against atmospheric pollution.
Legislation and pressure for increased safety in the workplace.
Rising manpower costs.
Economic recession.

These pressures have taken industrial finishing in the following directions:

12.5.1 Automation and economic application

For those who are able to afford the capital investment, one response to increasing paint and labour costs was the introduction of automation wherever possible, with application techniques that controlled coat weights closely and minimized paint losses or waste. Electrodeposition, for example, did all these things, but many other options were available.

12.5.2 Lower coating thicknesses

Rapidly rising petroleum prices directly forced up the costs of paint resins, solvents, and organic pigments. The response of most finishers was to seek lower overall film thicknesses. This led, for example, to the halving of the heavy coatings of unsaturated polyester wood finishes used in Western Europe for so long. In other cases, two-coat systems replaced those of three coats, and one coat replaced two. Few, if any, manufacturers were prepared to accept a lowering of finished appearance or performance. This often led to a change to new materials, which might have been more expensive in themselves, but gave superior results at lower coat weights.

12.5.3 Reduction in use of organic solvents

The loss of these expensive materials by evaporation had to be reduced. They also could cause pollution and increase fire risk (and sometimes other health risks) in the working environment. Solvent usage was reduced by changing where possible to waterborne, high-solids, 100% polymerizable or powder coatings. Although waterborne materials contained some solvents, the coatings were generally non-flammable, an especially attractive feature in dip tanks.

12.5.4 Reduction in energy costs

Rising petroleum prices led to rises in the costs of all fuels and to increased energy bills. In some areas local shortages (e.g. of natural gas) led to reduced factory output. All manufacturers looked for savings. These were achieved by reduction in stoving time or stoving temperature (with attendant paint modification), changes to EBC or UV, installation of after-burners to consume waste solvents with recycling of energy, and introduction of novel room-temperature curing processes, such as vapour-curing. The last-named is the acceleration of paint cure by placing the coated article in a cabinet and exposing it to catalyst vapour. Major energy savings were to be made in metal pretreatment, leading to the introduction of 'cold' pretreatments.

12.5.5 Elimination of more toxic paint ingredients

Much more toxicological information has accumulated on a variety of paint ingredients in the last decade. Laws in the USA and in Europe require the notification and registration of all chemicals used in industrial products. Ingredients previously considered harmless came under suspicion. Often, risks could be eliminated by improved handling procedures and safety equipment. In other cases, paint-makers replaced the suspect materials with less-toxic alternatives. Examples are the introduction of air-fed hoods or robots for handling isocyanates, the reduced use of lead and chrome pigments, and reduced *TLV*s for many solvents.

In the next decade the above trends can be expected to continue, with the continuing decline of conventional low-solids solventborne paints and the growth of the 'new technologies': waterborne, high-solids, and 100% polymerizable finishes, powder coatings, and vapour curing. Some new curing chemistry for low-temperature processes is likely. New processing equipment and automation will become more widespread.

Manufacturers will also experiment with new construction materials, which will present painting problems, or remove or present painting opportunities. The use of plastics will increase, also new woods and wood composites, and alternatives to cement asbestos.

Towards the end of the decade attention may return to petroleum as a finite resource and to the manufacture of coatings from renewable resources.

REFERENCES

[1] Tatton, W. H. & Drew, E. W. *Industrial Paint Application*, p. 47, Newnes (1971).
[2] Intertherm Ltd, *A first guide to dielectric heating* (pamphlet, London) (June 1973).
[3] Pray, R. W. *Radiation Curing* 5(3), August 1978, 19–25.
[4] White, R. G. *Handbook of industrial infra-red analysis,* p. 178, Plenum press (1964).
[5] *Kronos guide*, Kronos Titanium Companies (1968). Patton, J. C. *Pigment handbook, Vol. III*, John Wiley (1973).

13

The painting of ships

R. Lambourne

13.1 INTRODUCTION

The painting of ships and structural steel work, such as oil and gas rigs, poses some of the most formidable problems in respect of corrosion protection that the paint technologist has to face. Salt-water immersion or partial immersion, spray followed by drying winds, are extemely aggressive environments. They call for the development of paints which protect the steel from which the ship's hull is constructed, by means of a number of different mechanisms which we shall discuss briefly. In addition to problems of corrosion, the fouling of ships' bottoms with marine organisms leads to increased drag on the vessel and hence to increased fuel consumption. Gitlitz [1] cites an example of the economic effects of fouling, in relation to the operation of a VLCC (very large cargo carrier) with an operational speed of 15 knots. Fuel is consumed at the rate of about 170 tons per day. Assuming that the ship is at sea for 300 days in the year, the fuel cost (at the 1981 price of $80 per ton) exceeds $4M. He suggests that even moderate fouling can increase the fuel required by as much as 30% to maintain the optimum operating speed, i.e. by over $1M in a year. With the continuing increase in the cost of fuel oil it is obvious that the availability of effective anti-fouling compositions is of prime interest to the ship owner. Furthermore, it is not uncommon for 10 lb per square foot of fouling to accumulate in some marine environments within six months in the absence of an effective anti-fouling. Fouling also contributes to the breakdown of protective coatings, and hence to the earlier onset of corrosion. It is the aim of the ship owner to avoid frequent and expensive dry-docking. He has therefore a vested interest in improved anti-corrosive and anti-fouling paint systems.

Of course only the part of the hull that is immersed is subject to fouling, and it is important to identify and classify other types of surface which call for painting, each requiring a different type of paint. We have already mentioned the anti-corrosive/anti-fouling paints that are used for ships' bottoms. The area between the light load line (the water line when the ship is in ballast) and the deep load line (the water line when the ship is carrying a full cargo) is known as the boottopping. This is sometimes below the water level and sometimes exposed to the marine atmosphere, alternating between the two in service. Also, this area is subject to damage, e.g. abrasion by contact with jetties or wharves and with other vessels, such as tugs, in spite of the use of fenders. Above the boottopping are the topside and superstructure, terms that are self-explanatory. In addition to these areas there are speciality paints for decks, interiors, engines, etc. In naval vessels intumescent paints, designed to prevent the spread of fire, have been developed, but doubt exists as to the efficacy of such systems after ageing or subsequent repainting. In the light of the experience gained during the Falklands war, it is doubtful if these paints would have had more than a marginal effect on the spread of fires following a direct hit from a missile. However, they may be useful in reducing the spread of minor fires.

The painting of ships can be considered in two contexes: the painting during construction and the repainting during service. The difficulties of painting ships in service arise from the limited opportunities available to apply the paint under anything like ideal conditions. Paints as supplied must therefore be sufficiently 'robust' to enable them to be applied under adverse conditions, for example, with the most rudimentary surface preparation. Certain areas such as the topside and superstructure may be repainted at quite frequent intervals, the latter often while the ship is at sea.

In this chapter we shall consider briefly the requirements for the main types of marine paints. However, before doing so we shall examine in simple terms the mechanism of corrosion of steel and the methods by which it can be eliminated or at least retarded.

13.2 CORROSION

Corrosion is an electrolytic process involving the reaction between iron metal, oxygen, and water resulting in the formation of hydrated ferric oxide or rust. The overall reaction is:

$$4Fe + 3O_2 + 2H_2O \rightarrow 2Fe_2O_3 . H_2O \tag{13.1}$$

The generally accepted reaction mechanism is one which involves anodic and cathodic processes. The surface of the iron or steel in contact with water develops localized anodes and cathodes at which these processes take place. Electron flow (constituting a 'corrosion current') occurs through the metal, and this is complemented by an equivalent transport of charge through the water or electrolyte by hydroxyl ions. The process is shown diagrammatically in Fig. 13.1. The individual electrode processes are as follows:

At the anode: the formation of ferrous ions by the loss of electrons,

$$4Fe \rightarrow 4Fe^{++} + 8e \tag{13.2}$$

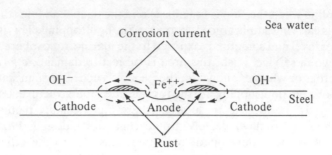

Fig. 13.1 — The corrosion process.

and at the cathode: the formation of hydroxyl ions,

$$4H_2O + 2O_2 + 8e \rightarrow 8OH^-$$ (13.3)

The initial product of oxidation is thus ferrous hydroxide,

$$4Fe^{++} + 8OH^- \rightarrow 4Fe(OH)_2$$ (13.4)

In the presence of excess oxygen the ferrous hydroxide is oxidized to hydrated ferric oxide, the all too familiar red oxide which is rust.

The anodic and cathode regions at the surface of the metal arise from compositional heterogeneity of the surface. This may be due to a number of factors including grain boundaries, stresses, and microscopic faults that cause local concentration gradients of electrolyte or oxygen in solution. Any of these giving rise to a potential difference between adjacent areas in the surface is sufficient to cause galvanic action.

In dealing with ships under construction one is faced with painting steel that has at its surface a layer of oxides up to 60 μm in thickness that is formed when the steel is manufactured. The steel is subjected to hot rolling to the required thickness, and this is carried out in the temperature range 800°–900°C. Oxidation of the steel takes place during cooling. This oxide layer is known as 'millscale'. Millscale, unless removed, can cause corrosion because there is a significant potential difference between the millscale and the bare steel (about 300 mV) when the steel is immersed in electrolytes such as sea-water. The millscale is cathodic, and the bare steel anodic. Thus where there is a crack or fissure extending through the millscale layer to the metal a galvanic cell is formed and the corrosion process begins. The rust is formed in the vicinity of the anode which is in contact with the salt water, but not directly on the anodic site. Thus the anode is slowly dissolved, and pitting occurs. The cathode site undergoes no such dissolution. All millscale is now removed by blast cleaning, and the steel is 'shop primed' before fabrication. However, this does not mean that corrosion is entirely eliminated.

How, then, can corrosion be reduced or eliminated? Two approaches are possible:

(1) The elimination or inhibition of the electrolytic processes.

(2) The exclusion of water or oxygen from the potential corrosion site.

The cathodic process requires the presence of water and oxygen at the metal surface, so that it can be seen that the two factors indicated above are interrelated. In practice it is very difficult to prevent oxygen and water reaching the metal surface because most paint binders have quite high permeability to these agents. Materials that are low in permeability to water and oxygen are, for example, waxes, because of their crystalline nature. Similarly, crystalline polymers exhibit low permeability, but these are difficult to formulate into air-drying compositions. In general, conventional marine paints do not confer cathodic protection. At the anodic surface, for corrosion to take place ferrous ions are required to pass into the electrolyte. This may be suppressed by the use of paint systems of two kinds. The first uses metallic zinc as pigment at a very high pigment concentration, such that contact between the zinc particles and the iron surface takes place. In this case, because zinc is more electronegative than iron, a galvanic couple exists between the two metals, the iron being cathodic and the zinc anodic. The net result is the suppression of the dissolution of ferrous ions and the consequent corrosion of the zinc. The second kind of cathodic suppression can be achieved by using paints containing inhibiting pigments such as red lead, white lead, or zinc oxide. These pigments are used in conjunction with binders based upon drying oils that undergo autoxidation. In addition to the formation of crosslinked films some scission of the drying oil triglycerides occurs with the formation of oxidation products such as azelaic acid. These acids react with the pigment to form lead or zinc soaps as the case might be. The soaps react with the surface oxide film on the metal and prevent the ingress of water which is necessary for the formation of rust. A second group of pigments that act in a similar way are those that exhibit limited water solubility. These include the chromates of zinc, barium, strontium, and lead. In these cases the oxide film is protected by the formation of an iron/chromium oxide complex.

Recently much concern has been expressed about the toxicity of these pigments, and work is in progress to eliminate lead and chromate pigments from all paints.

Two other ways of suppressing the electrolytic processes that are responsible for corrosion have been used on ships:

(1) The use of 'sacrificial anodes', in which anodes are built onto the hull, using alloys that are electropositive with respect to the steel of the hull. These anodes are unpainted and their purpose is to corrode away, being replaced when necessary.
(2) The application of a controlled external current to the hull to suppress the dissolution of ferrous ions. This is called the 'impressed current method'.

Neither method eliminates the need for painting the hull for other obvious reasons.

13.3 SURFACE PREPARATION

To obtain satisfactory performance from a paint system it is essential for millscale to be removed. Not only does it contribute directly to corrosion, but it will eventually become detached from the steel surface and is therefore an unsatisfactory substrate for painting. Surface preparation is vitally important when one is seeking to obtain the best performance from a paint system.

It is only since the early 'sixties that cost-effective surface treatments for millscale-coated steel have been developed. Before World War 2, when ships were built from a keel and a frame and were rivetted, it was common for millscale to be removed by weathering over periods often in excess of a year. The steel plate was simply exposed to the elements and allowed to rust. The rusting enabled the millscale to be detached by scraping and wire brushing. After World War 2, with the development of welded ships, the shorter time to build did not allow for millscale to be weathered off. More recently a number of alternative methods have been used, including acid pickling, flame cleaning, and grit blasting. The last-named is the most satisfactory and commonly used method of surface preparation today.

Two main methods are in general use, mechanical blast cleaning and 'open' or 'nozzle' air blasting. In the former, steel shot or grit are commonly used with the abrasive being recovered after being separated from the millscale particles. The abrasive is normally coarser than the dust that is produced in the process, and it is capable of being separated by a cyclone separator. Some attrition of the abrasive occurs, so that in spite of recycling there is a gradual consumption of abrasive. The machines using this method employ a rotary impeller which projects the abrasive particles at high velocity at the surface. The open or nozzle blasting method uses compressed air to project the abrasive at the surface. This method is more versatile, since it can be used on site on preformed structures, over welds etc., but suffers from the disadvantage of dust generation. A modified method allows the abrasive and the dust to be removed by applying a vacuum to an annulus around the cleaning nozzle, e.g. the Vacublast method. However, this method is much slower and is used only on small areas where dust cannot be controlled or permitted. After blast cleaning it is important to apply a protective coating as soon as possible to prevent rusting of the clean surface. Blast or prefabrication primers are used for this purpose.

13.4 BLAST PRIMERS

Blast primers are applied by spray immediately after blast cleaning. They must dry quickly (within a few minutes of application) to enable the steel to be handled without significant damage to the coating. The primer must not adversely affect the subsequent fabrication process. For example, it should not prevent successful welding. It must also provide a suitable surface for subsequent painting.

Four types of blast primer are in common use:

(1) Phenolic/polyvinyl butyral (PVB), acid cured, one- and two-pack compositions;
(2) Two pack cold curing epoxy compositions, pigmented with red iron oxide (and optionally an inhibitive pigment such as chromate);
(3) Two-pack cold curing epoxy compositions pigmented with zinc dust; and
(4) Zinc silicate one- and two-pack compositions.

The pot life of the paint has to be about 8 hours to be of practical use. The epoxies are usually crosslinked with a polyamide such as Versamid 140 or 115. In the two-pack zinc silicate, pigmentary zinc is stirred into the paint at the time of use.

Typical formulations for these types of primer, as described by Banfield [2], are shown in Table 13.1.

The phenolic/PVB types are the most widely used, and are capable of being overcoated with most systems except in situations where they are likely to encounter strong solvents. The two-pack epoxy is the second most widely used. Both the red oxide and zinc-rich types can be used under practically any paint. They do, however, suffer the disadvantage of being two-pack, and can be more expensive than the phenolic/PVB types. The zinc silicate primers are mainly used when they arc to be overcoated with zinc silicate systems.

Zinc-metal-containing paints give the best overall protection against corrosion, but they do have several disadvantages in use that will sometimes dictate the use of one of the other types of primer. The corrosion products of zinc are water-sensitive, and they must be removed before being overcoated with another paint. This can be done, for example, by high-pressure hosing with fresh water. Another problem associated with zinc-containing blast primers is the formation of zinc oxide fume when coated plate is welded. Welders often object to this, although it is not established that it presents a long-term health hazard. The inhalation of zinc vapours may give rise to a condition known as 'zinc fume fever'. This has been known in the galvanizing industry for over a century.

13.5 PAINT SYSTEMS FOR SHIPS

The protection of steel plate after manufacture has been described briefly above. The subsequent painting of ships during and after construction is a highly complex technology. As has been indicated, different surfaces require different treatment. Each surface requires a *system*, since it is impossible to combine all the requirements for any one surface within one paint. Thus multi-coat systems are essential, and for any one type of paint it has become the convention to apply several coats to build up film thickness to minimize corrosion occurring, for example, through pinholing. Typical systems for topside and superstructure paints, boottopping, and bottoms are summarized for what are regarded as traditional or conventional marine paints in Table 13.2.

In recent years major changes have taken place in painting practices and paint formulation to meet the needs imposed by the development of the supertanker. The development of the supertanker came about with the closing of the Suez Canal in 1967 as a result of the Middle East war. Supertankers were required to make the transport of oil from the Middle East an economic proposition. This could not be achieved with the typical tanker of the early 1950s, having a tonnage of about 30 000 gross. The average supertanker of the 1980s has a gross tonnage of 300 000, and it operates with a smaller crew than its predecessor. This means that fewer crew are available to do maintenance painting on a much larger ship. The overall effect of these changes has required the development of paints that require the ship to undergo less frequent and shorter periods in dry dock. To achieve this, paints giving higher build with fewer coats yet providing improved performance have had to be produced.

Thus a new generation of paints has been developed that enable the periods between dry docking to be increased from about 9–12 months to 24–30 months. Even

Table 13.1 — Formulations for blast primers

One-pack phenolic/PVB blast primer

Paint base	% by weight
PVB (e.g. Mowital B6OH)	4.7
Phenodur PR263 (phenolic resin)	4.7
Red oxide (synthetic)	8.6
Asbestine (magnesium silicate)	1.4
Phosphoric acid, S.G. 1.7	1.0
Methyl ethyl ketone	30.0
n Butanol	19.6
iso Propanol	30.0
	100.0

Two-pack epoxy blast primer

Paint base	% by weight
Epikote 1004	12.0
Zinc phosphate (to BS 5193:1975)	8.0
Red oxide (synthetic)	10.0
Asbestine	1.5
Talc (5 μm)	15.3
Bentone 27 (anti-settling agent)	0.4
Toluene	23.0
iso Propanol	11.5
Crosslinking agent	
Versamid 140 (polyamide)	1.8
Toluene	11.0
iso Propanol	5.5
	100.0

Two-pack zinc-rich epoxy blast primer

Paint base	% by weight
Zinc dust (Zincoli 620 or Durham Chemicals 'Ultrafine')	80.0
Bentone 27	1.0
Calcium oxide	0.3
Epikote 1001	4.0
Toluene	6.6
iso Propanol	3.3
Crosslinking agent	
Versamid 115	1.8
Toluene	2.0
iso Propanol	1.0
	100.0

Note: The calcium oxide acts as a water scavenger, reacting with moisture in the solvents. The moisture would otherwise react with the zinc to form hydrogen in the can.

Zinc/ethyl silicate blast primer

Paint base	% by weight
Ethyl silicate (e.g. Dynasil H500)	20.0
Bentone 38, anti-settling agent	1.4
Talc (5 μm)	4.0
Toluene	5.3
iso Propanol	5.3
Cellosolve	4.0
Zinc dust (e.g. Zincoli 615)	60.0
	100.0

The zinc dust is stirred into the base at the time of use. The product then has a useful life of about 8 hours.

this extended period is not considered to be the limit of modern technology, and some paint manufacturers are aiming to extend the period between dry docking up to five years.

Table 13.2 — Summary of 'conventional' marine paint types

Surface	Paint system	Binder type	Coats	Dry film thickness (μm)	Total average film thickness (μm)
Bottom	Primer	Oil-modified phenolic; bitumen	3–4	40–60	~265
	Barrier coat No.1 anti-corrosive	Bitumen/limed rosin	1	50–60	
	Anti-fouling	'Soluble matrix' based on boiled oil/limed rosin; Cu or TBTO toxicant[†]	1	50–75	
Boottopping	Primer 655	Tung/linseed-modified phenolic	2–3	40–60	~200
	Finish	Tung-modified phenolic	2	40–60	
Topside and superstructure	Primer (quick-drying)	Medium oil length linseed alkyd (55–60% O.L.)	2	50	~220
	Undercoat	Medium oil length linseed alkyd (50–55% O.L.)	1	50	
	Enamel	Long oil length soya alkyd (65% O.L.)	1–2	35–50	

[†]Three grades are available, 'Atlantic', 'Tropical', and 'Supertropical' in which the level and choice of toxicant vary.

These requirements have been met by developing systems which are capable of being applied as thick coats, each dry film being at least 100 μm thick. These modern products are most commonly applied by airless spray. One airless spray gun is capable of spraying between 50 and 80 litres of paint per hour, i.e. covering 150–400 square metres per hour at the required film thickness. To avoid sagging on vertical surfaces the paints must exhibit non-Newtonian rheology, that is, they must be shear-thinning to facilitate flow through the gun and atomization, but must rapidly develop structure in the liquid film. Thus they may have a viscosity of about 5–10 poise at 10 000 sec^{-1}, and will probably reach a viscosity of ~1000 poise at the low shear rate applicable to flow under gravity on a vertical surface, e.g. 10^{-2} sec^{-1}. This type of behaviour is achieved by using thixotropic or gelling agents such as the montmorillo-nite clays or polyamides. Many of the high-performance systems are two-pack. The anti-corrosion performance obtained with these systems relies more on the suppression of electrolytic action by acting as thick barrier coats than the specific action of inhibitive pigments, although the primers used do often contain such pigments. The toxicity of lead and chromates has called for a concerted effort to find replacements for them. In this context, zinc phosphate has increased in use.

13.5.1 Topside and superstructure paints

Conventional paints meeting the requirements of topsides and superstructures come closest to those used in the trade paint market, requiring primers, undercoats, and gloss coats that dry at ambient temperature. These have for many years been based upon oleoresinous and alkyd binders. Thus, a primer would use an oleoresinous vehicle (e.g. an oil-modified phenolic) with an inhibitive pigment, usually red lead. An undercoat would be formulated on a medium oil length linseed alkyd. The topcoat might be based on a long oil alkyd or a vinyl-toluenated alkyd. Pigmentation in the latter case would be to achieve opacity and other aesthetic purposes. The vinylated alkyd confers an improved drying rate, but suffers from the disadvantage that special care has to be exercised in recoating, because the initial dry occurs as a result of solvent evaporation (lacquer-drying), and lifting can arise if the first coat is recoated after 12 hours but before 16 hours have elapsed, particularly if the ambient temperature is low. This effect is due to the relatively slow rate of crosslinking of this type of binder.

Modern high-performance systems are based on a wide range of binders including epoxies, polyurethanes, and chlorinated rubber. The last-named is often used in blends with acrylic or alkyd resins.

A typical epoxy system for topsides would, for example, involve applying two coats of a high-build primer/undercoat followed by one coat of epoxy enamel, giving a total dry film thickness of about 300 μm. An alternative for the superstructure, where gloss retention is more important, would be to use one coat of primer, followed by an epoxy primer/undercoat and a polyurethane topcoat. The dry film thickness would be approximately 200 μm in this case. Examples of a red oxide epoxy thick coating and a white epoxy enamel are given in Table 13.3.

Epoxies can be crosslinked with amines, diethylene triamine and triethylene tetramine commonly being used. Amines are often pre-reacted with epoxides to form adducts which have certain advantages over amines *per se*. The adduct will uaually be formed with part of the epoxide that will be used in the paint base. By using this approach, the odour of the free amine can be reduced or eliminated, and the mixing ratio is more convenient, e.g. 2 or 3:1 instead of about 10:1, Paint:activator.

Two-pack polyurethanes are also used in topcoat enamels. The paint base contains a hydroxyl-containing saturated polyester resin (e.g. Desmophen 650); the curing agent will be an isocyanate-containing moiety (such as Desmodur N) as a 75% solution in a 1 : 1 mixture of xylene and cellosolve acetate. Polyurethanes are used when good gloss and gloss retention are required.

Table 13.3 — Formulation of a red oxide thick coating and a white epoxy enamel

Paint base	*% by weight*
Synthetic red oxide	5.0
Asbestine (magnesium silicate)	22.0
β Crystoballite (silica)	
to pass through a 325 mesh sieve	18.0
Epikote 1001	16.0
Bentone 27	2.0
Xylene	14.0
n Butanol	5.0
Curing agent	
Versamid 115 (polyamide)	8.0
Cellosolve	10.0
	100.0

A typical white epoxy enamel formulation:

Paint base	*% by weight*
Rutile titanium dioxide (e.g. Tioxide R-CR2)	30.0
Aerosil 380 (anti-settling agent)	0.5
Epikote 1001	23.0
Solvesso 100 (naphtha)	16.5
Methyl isobutyl ketone	8.0
n Butanol	3.0
Curing agent	
Versamid (polyamide)	11.0
Accelerator (e.g. Anchor K54)	1.0
Xylene	7.0
	100.0

High-build, high-performance paints can also be formulated in one-pack compositions based, notably, on chlorinated rubber and vinyl resins. Chlorinated rubber paints have gained wider acceptance than vinyls although the latter have been developed for use by the US and Canadian navies. Typical formulae for a primer, undercoat, and topcoat all based upon chlorinated rubber are given in Tables 13.4–13.6.

The composition given in Table 13.4 calls for careful formulation and manufacture. The plasticizer is a preformed blend of a solid (waxy) chlorinated paraffin with a liquid chlorinated paraffin. The thixotropic agent is based on hydrogenated castor oil. This develops structure only when it is incorporated during the milling stage, when the temperature rises, depending on the scale of manufacture, to over 40°C. The temperature should not exceed 55°C, otherwise the thixotropic effect is lost. The aluminium paste is not milled in with the other pigments, but is stirred in when the

Table 13.4 — Chlorinated rubber primer

	% by weight
Aluminium paste, non-leafing (65% in naphtha)	20.0
Zinc oxide	1.0
Blanc fixe	10.0
Thixatrol ST (thixotropic agent)	0.5
Alloprene R10 (chlorinated rubber)	15.0
Cerechlor 70 (chlorinated wax plasticizer)	7.5
Cerechlor 42 (chlorinated wax plasticizer)	7.5
Propylene oxide (stabilizer)	0.1
Xylene	28.4
Solvesso 100 (Naphtha)	10.0
	100.0

batch has cooled. The propylene oxide is an 'in-can' stabilizer which 'mops up' Cl^- ions that are generated slowly from the chlorinated rubber on storage.

In the formulation given in Table 13.6, the reduced amounts of filler (cf. undercoat), the use of Bentone 38 instead of Thixatrol ST, and the higher proportion of Cerechlor 70, are all measures designed to improve the gloss of the topcoat. Even so, the gloss is low compared to epoxide or polyurethane paints. These paints are all applied by airless spray. A typical system would be:

	Dry film thickness (μm)
Primer, one coat	50
Undercoat, two coats	200
Topcoat, one coat	50
Total	300

13.5.2   Boottoppings

As distinct from the traditional boottoppings mentioned previously (Table 13.1), high-performance compositions for this area are based essentially on the same binders as the topside paints. In this category the chlorinated rubbers perform very well because of their excellent intercoat adhesion. This is particularly important because of the damage to the paint system from abrasion and impact with fenders and quays, and the consequent need for frequent repainting. With VLCCs, fouling can also be a major problem in the boottopping area, and it is common practice to use an anti-fouling composition as the finishing paint. Thus it is usual to employ the same anti-fouling as is used on the bottom.

Table 13.5 — Chlorinated rubber undercoat — white

	% by weight
Titanium dioxide (e.g. Tioxode R-CR2)	16.0
Blanc fixe	14.0
Zinc oxide	1.0
Thixatrol ST (thixotropic agent)	1.0
Alloprene R10 (chlorinated rubber)	14.0
Cerechlor 70 (chlorinated plasticizer)	7.0
Cerechlor 42 (chlorinated plasticizer)	7.0
Propylene oxide (stabilizer)	0.1
Xylene	29.9
Solvesso 100	10.0
	100.0

Table 13.6 — Chlorinated rubber topcoat — white

	% by weight
Titanium dioxide (e.g. Tioxode R-CR2)	18.0
Blanc fixe	2.0
Zinc oxide	1.0
Bentone 38 (thixotropic agent)	1.0
n Butanol (swelling agent)	0.5
Alloprene R10	16.0
Cerechlor 70	12.0
Cerechlor 42	4.0
Propylene oxide	0.1
Xylene	30.4
Solvesso 100	15.0
	100.0

13.5.3 Bottom paints

The high-performance/high-build systems described for topsides and boottoppings are also used for bottoms. However, since asethetic reasons are not predominant it is possible to formulate such paints at lower cost, without sacrificing performance, by the incorporation of coal tar. The paints are thus usually chocolate brown or black, but this is of no real significance. The proportions of coal tar in an epoxy coal tar thick coating can be varied over a wide range. The higher the coal tar content the poorer the oil or chemical resistance, but this is less important in underwater compositions. The coal tar pitch in a formula complying with the UK Ministry of Defence (Ship Department) Specification DGS 5051 would represent about 60–65% of the total binder. It would be selected according to its compatibility with the resin; and, although regarded as essentially unreactive, it can react with the epoxy groups because it contains some phenolic hydroxyl groups. For this reason it is often incorporated into the curing agent part of the two-pack composition, otherwise the storage life of the base would be inadequate. A typical formulation, as given by Banfield [2], is given in Table 13.7.

Table 13.7 — Typical formulation for two-pack bottom paint

Base	% by weight
Barytes	20.0
Asbestine	20.0
Epikote 1001	11.0
Bentone 27	1.0
Xylene	10.0
Solvesso 100	5.0
n Butanol	3.0
Curing agent	
BSC Norsip 5 (75% aromatic coal tar in xylene)	25.0
Versamid 140	2.5
Synolide 968	0.3
Xylene	2.2
	100.0

After mixing, the paint is applied by airless spray to give a dry coat thickness of 125 μm per coat. The pot life is about 4 hours at 15°C. The reaction rate doubles, approximately, with a 10° rise in temperature, or is halved with a 10° drop in temperature. Thus the paint can have a pot life varying from about 8 hours at 5°C to only 2 hours at 25°C. The rate of cure also affects recoatability. To achieve satisfactory intercoat adhesion and to avoid lifting, maximum and minimum recoating times are specified, according to the temperature of application.

This type of composition may be used both on ships under construction and those undergoing 'in service' repainting. The former poses a problem in that the application of an anti-fouling will not take place until shortly before the ship goes into service, and this may well be long after the optimum time for overcoating the coal tar epoxy paint. In this case a 'tie coat' is used. Tie coats are also specified by the UK Ministry of Defence (Specification DGS 5954). A composition that meets this requirement is given in Table 13.8.

Table 13.8 — Aluminium tie coat

	% by weight
Aluminium powder, non-leafing	14.0
Basic lead sulphate	14.0
Gilsonite (bitumin)	22.4
Linseed stand oil, 50 poise	5.6
Coal tar naphtha, 90/160	44.0
	100.0

The tie coat is applied by airless spray to give a dry film thickness of about 50 μm. The anti-fouling coating can then be applied shortly before launching.

Alternative bottom paints may also be based on chlorinated rubber and vinyls. A fuller treatment of the formulations commonly used is beyond the scope of this chapter. The reader seeking more information is referred to Banfield's excellent monographs on this subject [2].

13.5.4   Anti-fouling coatings

13.5.4.1   *The problem of fouling*

Marine organisms, both plant and animal, are able to attach themselves to the hulls of ships. Plant growth requires some daylight to sustain it, so attachment occurs to the sides of the hull in the upper regions of the underwater area. The most common of these are of the genus *Enteromorpha*, seaweeds that have long green tubular filaments that grow very rapidly under favourable conditions. The brown and red seaweeds require less light, and these can grow at lower levels or even on the bottom of the ship. Firm attachment of these plants can take place in a few hours unless they are prevented from doing so.

Aquatic animals that are capable of attaching themselves to ships include barnacles, mussels, polyzoa, anthozoa, hydroids, ascidians, and sponges. The best known (and most troublesome) are the barnacles. These animals do not require light to sustain growth, so they are to be found on all immersed parts of the hull. The larvae require about 48 hours to become firmly attached. The consequences of fouling have already been mentioned, but it is worth repeating that the most important consequence of fouling with large modern ships is the increased cost of propelling the vessel through the water as a result of the increased drag that fouling imposes.

The first 'anti-fouling' paints based upon the use of toxicants were introduced in the middle of the 19th century. These were oleoresinous compositions containing mercuric and arsenious oxides. Subsequently, because of the high human toxicity and the increasing cost of mercuric oxide, alternative paints using cuprous oxide were developed. Cuprous oxide remains the most commonly used toxicant in the 'conventional' types of composition today.

13.5.4.2 Conventional anti-foulings
There are two types of conventional anti-fouling paints, classified according to their mode of action. These are generally known as 'soluble matrix' and 'contact' types.

Soluble matrix type
In the soluble matrix type the toxicant is dispersed in a binder that is slightly soluble in sea-water. The toxicant is slowly released as the binder dissolves. The rate of dissolution has to be controlled very carefully since an inadequate concentration of the poison at the surface will not prevent the attachment of the marine organisms, whereas too rapid dissolution will mean that the effectiveness of the anti-fouling will be too short-lived. The binder used in this type of anti-fouling might be a 3 : 1 limed rosin/boiled linseed oil mixture.

A better 'supertropical' grade might have the following composition given in Table 13.9. 'Atlantic' or 'tropical' grades would not necessarily call for the use of tributyltin oxide.

Table 13.9 — Formulation for supertropical anti-fouling composition

	% by weight
Cuprous oxide	24.0
Zinc oxide	5.0
Tributyltin oxide	2.0
Red iron oxide	8.0
Paris white	9.0
Talc (10 μm)	6.0
Bentone 38	0.3
n Butanol	0.1
Limed rosin/boiled linseed oil, 3 : 1, 60% in naphtha	41.0
White spirit	4.6
	100.0

Contact type
The contact type of anti-fouling composition is so named because it is formulated at a very high pigment volume, such that the pigment (toxicant) particles in the disperse phase are in contact with each other in the dry film. The binder is largely insoluble in sea-water, so that when dissolution of the toxicant takes place a porous film of the paint binder remains on the surface. The binder is not entirely insoluble, and the balance between binder solubility/insolubility plays a part in controlling the rate of leaching of the toxicant. Although these paints are formulated to give a close-packed pigmentary structure in the dry film, it is doubtful if this is achieved or is desirable in practice. Any significant flocculation of the particles of toxicant will lead to a more open structure, and the paint may in fact be underbound. This does not appear to be

a major problem in practice. A typical formulation for a contact type anti-fouling is given in Table 13.10.

Table 13.10 — Typical formulation for contact type anti-fouling

	% by weight
Cuprous oxide	57.4
Asbestine	2.4
Rosin	16.1
Modified phenolic/linseed stand oil, 1 : 2, 60% in 90/190 solvent naphtha.	9.0
Chlorinated plasticizer	5.4
Solvent naphtha, 90/190	9.7
	100.0

One of the most important features of anti-fouling compositions is the rate at which they release the toxicant. Various organisms exhibit different degrees of sensitivity to the poison. In the case of barnacles about 10 μg of copper per cm^2 per day is sufficient to prevent their attachment. There is some evidence of synergism between poisons. This is one of the reasons that copper and tributyltin oxide are used together to meet more demanding situations, for example, in 'supertropical' grade anti-foulants.

13.5.4.3 *Modern anti-foulings*
Just as the development of larger vessels (e.g. supertankers) has required paint systems that call for less frequent maintenance occasioned by paint breakdown and corrosion, the prevention of fouling over a longer period is similarly required.

In 1954, Van der Kerk & Luijten [3] reported on an investigation into the biocidal properties of tributyltin compounds. They concluded that these materials had a broad spectrum effect; that is, they were toxic to a wide range of marine organisms. Subsequently, tributyltin oxide was used in combination with copper oxide. However, with the expectation that the effective life of an anti-fouling coating should exceed 18 months, it became clear that this could not be achieved with cuprous oxide at practical film thicknesses. Thus, the use of organo-tin derivatives has increased, and a 'family' of organo-tin derivatives is available to the paint formulator. In laboratory experiments tributyltin derivatives were shown to be one hundred times as effective as copper oxide, weight for weight [4]. Under more practical test conditions later work showed that $\frac{1}{10}$ to $\frac{1}{20}$ the amount of organo-tin compound, compared to copper, would provide equivalent algae and barnacle control [5]. It is important that the organo-tin compound should be *available*, otherwise it will not be any more effective than copper. Thus the paint requires very careful formulation in order to make the maximum effective use of the toxicant.

Several organo-tin derivatives have now become established as toxicants in the marine paint market. The principal ones are tributyltin oxide, tributyltin fluoride, and triphenyltin fluoride.

Tributyltin oxide is a liquid, miscible with common paint solvents. It has a relatively high salt-water solubility, 25 ppm, that enables it to be used when a high leaching rate is required. It has a plasticizing action on the film in which it is incorporated, and this limits the amount that can be used to about 13% by weight in a typical vinyl system. In current commercial use it is usually a cotoxicant with cuprous oxide at about 2–5% by weight of the dry film.

Tributyltin fluoride is a white, high-melting waxy solid that is insoluble in common paint solvents. It is therefore incorporated as a pigment up to about 30% by weight of the dry film. It is less soluble in sea-water than the oxide, 10 ppm, and is often used as the sole toxicant in vinyl/rosin or chlorinated rubber/rosin-based paints. Tributyltin fluoride is available either in powder form or as a paste. The paste offers more convenient handling and can be readily incorporated by high-speed mixing.

Triphenyltin derivatives are also useful toxicants. Triphenyltin fluoride has been used as an agricultural fungicide for over ten years. It is a white powder with a sea-water solubility of less than 1 ppm. Paints containing it can give effective protection against marine growth for up to two years. The hydrolysis products of triphenyltin fluoride in sea-water are triphenyltin chloride and triphenyltin hydroxide, both of which are solids that will not diffuse through the film. TPTF is therefore most effective when used in compositions that dissolve or erode.

Conventional anti-fouling systems, relying on the leaching of a toxicant, exhibit a logarithmic decay in the concentration of available toxicant at the paint/water interface. Thus to achieve a long service life for the coating it must deliver a much higher concentration of toxicant initially than is required for the effective control of fouling. When the concentration of available toxicant falls below that required to prevent fouling there is still some residual toxicant in the film. The process is thus inefficient in two ways, in the use of what is an expensive component of the paint. Looked at another way, doubling the film thickness increases the effective life of the coating by only about 13%, if the only release mechanism is that of diffusion. Clearly, if it is possible to achieve a constant and optimum rate of release of the toxicant throughout the service life of the coating, this would be a great step forward. A breakthrough of this kind was made in the early 1970s through the pioneering work in the UK by International Paint in the field of organo-tin polymer compositions.

Polymers of trialkyltin acrylates were first described by Montermoso, Andrews, & Marinelli [6] in 1958. These polymers were prepared in a search for thermally stable polymers, and they were not tested for their biocidal activity. Leebrick [7] in 1965 was the first to claim the biocidal use of organotin polymers, but it was not until 1978 that they received approval by the US Environmental Protection Agency. By this time they were well established in Europe and Japan.

Polymers that incorporate organo-tin moieties are simple random co(or ter)polymers having, typically, the following structure derived from tributyltin acrylate and methyl methacrylate:

$$
\left[
\begin{array}{c}
Bu_3Sn-O \\
| \\
C=O \\
| \\
-CH_2-CH-
\end{array}
\right]_x
\left[
\begin{array}{c}
CH_3 \\
| \\
-C-CH_2- \\
| \\
C=O \\
| \\
O \\
| \\
CH_3
\end{array}
\right]_y
$$

Binding the toxicant into a polymer offers many advantages over the use of non-polymeric derivatives that generally have to be incorporated into the paint in a pigmentary form. The toxicant content of the polymer can be varied at will between wide limits, and the T_g of the polymer can be varied to meet the requirements of use. Perhaps the biggest single advantage is the opportunity to provide controlled release of toxicant through the chemical process of hydrolysis rather than by physical means alone.

The advantages claimed for the new organo-tin polymer anti-foulings include:

(1) Constant rate of release of toxicant.
(2) Controllable rate of erosion and toxicant release.
(3) Self-cleaning (and at high erosion rates, self-polishing).
(4) High utilization of toxicant is achieved, and as a result of the erosion there are no significant problems due to residues when repainting is necessary.

The formulation of organo-tin polymer-based anti-fouling paints does, however, present very special problems. The use of organo-tin biocides in rapidly eroding systems increases the initial cost to the ship owner, but this may be more than offset by the reduced requirement for dry docking. The main protagonists of organo-tin polymer anti-fouling paints, International Paint (in the UK) and the Nippon Oils and Fats Company in Japan, have developed their products in different directions. International Paint claim the combined benefits of anti-fouling coupled with self-polishing, and they offer products that undergo more rapid erosion than those offered by the Japanese company. Thus, for the polishing/anti-fouling type the rate of erosion is typically 10–12 μm per month. Up to four 100 μm coats are required to give 2–3 years of effective service life. The Japanese system erodes at about 3–6 μm per month, so that two 100 μm coats can give adequate protection. However, as we have observed earlier, surface roughness can affect the energy requirements for propelling large vessels at their service speed. There may be additional savings with the polishing type of composition that are not realizable with the slower-eroding type. Harpur & Milne [8] point out that conventionally-painted ships increase in hull roughness by about 25 μm per year from an initial mean amplitude of 110–125 μm, if well painted. Hydrodynamicists have shown a $\frac{1}{3}$ power law relationship between propulsive power and hull roughness, such that 10 μm of increased roughness can contribute up to an additional 1% in increased fuel consumption. However, the benefits of self-polishing are less likely to be realizable on old vessels which have already undergone considerable corrosion and already exhibit gross pitting.

REFERENCES

[1] Gitlitz, M. H. *J. Coatings Tech.* **53** 46 (1981).
[2] Banfield, T. A. (1) 'Protective Painting of Ships and of Structural Steel', SITA Ltd, Manchester (1978). (2) *J.O.C.C.A.* **63** 53 (1980); *J.O.C.C.A.* **63** 93 (1980).
[3] Van der Kerk, G. J. M. & Luijten, J. G. A. *J. Appl. Chem.* **4** 314 (1954).
[4] Evans, C. J. & Smith, P. J. *J.O.C.C.A.* **58** 160 (1975).
[5] Chromy, L. & Uhacz, K. *J.O.C.C.A.* **61** 39 (1978).
[6] Montermoso, J. C., Andrews, T. M. & Marinelli, L. P. *J. Polym. Sci.* **32** 523 (1958).
[7] Leebrick, J. US Patent 3,167,473 (1965).
[8] Harpur, W. & Milne, A. *Proc. 5th PRA Int. Conf.*, May 1983, 45–50.

14

An Introduction to Rheology

T. A. Strivens

14.1 INTRODUCTION

The science of rheology is concerned with the deformation and flow of matter, and with the response of materials to the application of mechanical force (stress) or to deformation. Such responses include irreversible (viscous) flow, reversible (elastic) deformation, or a combination of both. In the former process, energy is dissipated (mainly in the form of thermal energy (heat)); in the latter, energy is stored and released, when the mechanical force is removed. The balance of such responses is dependent on the speed (time scale) at which the mechanical force is applied as well as the material temperature. Thus, at normal room temperature substances like glass and pitch will shatter when hit with a hammer, but will slowly stretch and deform irreversibly when weights are hung on sheets or rods of the material. Familiar materials, such as 'bouncing putty', can be kneaded and stretched between the fingers (long time scale), but, when formed into a ball and dropped onto a surface they bounce like a rubber ball (short time scale for impact). Plastic sheet like Perspex is reasonably pliable at room temperature, but becomes hard and brittle like glass when immersed in liquid nitrogen and soft and permanently deformable when immersed in boiling water. Even simple liquids like water or substituted paraffin hydrocarbons, such as 6,6-11,11 tetramethyl hexadecane, can show elastic responses if the time scale is short enough: 10^{-9} to 10^{-12} seconds [1]. Many commercially important materials, such as paint, are dispersions of one or more liquid or solid phases in a liquid or solid matrix, e.g. emulsions, dispersions, composites, etc. Such materials often exhibit very complex responses to the application of quite small mechanical forces. For example, thixotropic paints look like solids or very viscous (thick) liquids, when at rest in the can; but, when stirred gently or stressed by the insertion of the paint brush, they become thin, mobile liquids. When left at rest, they recover their original appearance. The science of rheology covers all the complex and varied responses of this whole range of different materials.

Whilst, academically, rheology is often seen as a branch of applied mathematics or physics, proper understanding of the results of rheological measurements (as well as, sometimes, a sensible choice of rheological measurement technique) must involve other scientific disciplines, in particular, physical chemistry. Thus a knowledge of colloid science is essential to a proper understanding of the rheology of emulsions, dispersions, and suspensions, as much as polymer science is to the rheology of polymer melts and solutions. It is (or should be) a truly multi-disciplinary science involving the skills of mathematicians, physicists, chemists, and engineers, as well as others, such as biologists, on occasion. Rheology can involve considerable mathematics, but, in this account, it will be reduced to the minimum necessary to clarify relationships and concepts.

The control of rheology is essential to the manufacture and usage of large numbers of products in a modern industrial society, e.g. food, plastics, cosmetics, petroleum derivatives, and paints. Few manufacturing industries are devoid of material forming or coating processes, mixing operations, transport of liquids or slurries through pipelines, liquid–solid separation processes, such as sedimentation or filtration, etc. All of these processes require, to a greater or lesser extent, control of material rheology. The use of such products often involves rheology in their application; for example, the smoothing of cosmetic creams on the face, the taste and texture of foodstuffs such as sauces and mayonnaise in the mouth, and the application of paints to a surface by spraying, brushing, etc. as well as the flow-out after application to give a smooth, uniform film.

14.2 HISTORY OF VISCOSITY MEASUREMENTS

E. C. Bingham, one of the founders of the modern science of rheology, in 1929 defined rheology as the science of deformation and flow of matter. However, historically, the measurement and definition of viscosity were established well before the foundation of the science of rheology. Some centuries before the Greek philosopher Heraclitus of Samos (*c.* 540–475 BC) categorically stated 'παντα ρει', i.e. everything flows; the Jewish prophetess Deborah talked about 'The mountains flowed before the Lord' (Judges 5,5; the King James authorized version is wrong to say 'melted'; the Hebrew word is quite definitely 'flowed'†). Some understanding of the flow of simple liquid through capillaries was shown by the designers of water-clocks in ancient Egypt (*c.* 1540 BC). However, it was not until our own era that Newton, in his *Principia* (2nd ed. 1731) gave the modern definition of viscosity, and Hooke, in 1678, discovered experimentally the proportionality between stress and strain for elastic materials (*ut tensio, sic vis*). During the eighteenth century came the development of the branch of physics known as hydrodynamics by, amongst others, the Swiss brothers Bernouilli, as well as the measurement of gas liquid viscosity by Coulomb in 1798, using the decay of oscillation of a disk suspended in the test sample. This method is still used by a torque measurement on a rotating disk in the Brookfield viscometer, much favoured by paint technologists. The reference method for measuring the viscosity of Newtonian liquids, namely capillary viscometry, arose

---

† The Revised Version corrects this error

from the work of Hagen (1839) and Poiseuille (1846), the latter of whom was trying to understand the flow of blood in the human body. The use of concentric cylinders as an experimental method of measuring viscosity was established by Couette in 1890, and Bingham's own studies on the viscosity of paint were publshed in 1919 [2].

14.3 DEFINITIONS

Before describing rheological measurements and their interpretation, a number of definitions must be given, which form the conceptual basis of the science of rheology.

14.3.1 Modes of deformation

The commonest mode of deformation for liquid or deformable solid materials is *shear* deformation. This is achieved by confining the material between the walls of the measuring instrument and by setting up a velocity gradient across the thickness of the material, i.e. causing it to flow. This may be by forcing the fluid to flow through a pipe or capillary tube or by moving one wall relative to another. This latter can be achieved in a number of flow *geometries* which will be described under measurement techniques (section 14.4, below).

In simple shear, the change in shape brought about by the application of mechanical force is not accompanied by a change in volume. In other modes of deformation, a change in volume rather than shape may occur (*bulk compression or dilatation*), or both may change together as in *extensional deformation*. This latter mode can be of practical importance in the paint industry; for example, it has been established that it affects spatter and filament formation in roller application of paint [3].

If the macroscopic deformation of the material sample is the same as that of any of its constituent elements, then the deformation is referred to as being *homogeneous*. However, in many experimental deformation geometries, such as torsion between concentric cylinders, flow through a tube etc., the deformation and deformation rates vary from point to point. If the deformations are small, the mathematical relationships between the rheological parameters and the external forces and displacements can be simplified, and the relationships are not dependent on the deformation geometry used. However, if the deformations are sufficiently large, this is no longer the case, and the relationship becomes geometry-dependent, and this can cause difficulties in both the design and interpretation of experiments. This is an important point when considering experimental methods for materials with complex rheological properties, such as the thixotropic paints mentioned above.

14.3.2 Shear stress

Consider the experimental set-up in Fig. 14.1 (simple shear deformation mode). Two parallel plates, each of area A, are separated by a thickness, h, of the material under test. Suppose a force, F, is applied to the upper plate at its left-hand edge, and the bottom plate is held rigid; then the stress applied to the material is given by F/A (cgs units : dyne cm^{-2}; SI units : Pascal, or Newton m^{-2}).

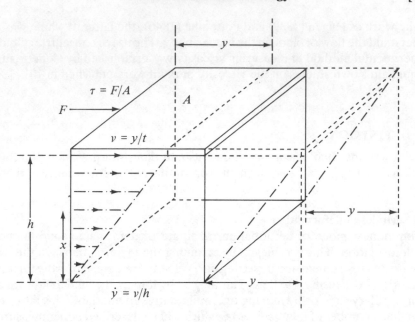

Fig. 14.1 — Simple shear experiment — definitions of stress, strain, strain rate.

14.3.3   Shear strain
When the force is applied as shown in Fig. 14.1, the upper plate responds immediately by moving a distance, y, from left to right. The shear strain (y) is then defined as y/h and is dimensionless.

14.3.4   Shear strain rate (shear rate)
If the upper plate moves through the distance y in time t, then the upper plate moves with a velocity v, given by y/t.

   If the material filling the gap is an ideal liquid, then the plate will continue to move at a steady velocity v for as long as the force F is applied. Furthermore, the velocity, v, will be proportional to the force F applied. As soon as the force is removed, movement will stop (assuming both plate and fluid have negligible inertia). Because the liquid velocity at the surface of the bottom plate is zero, and, assuming that the liquid sticks to both plates, there will be a velocity gradient through the thickness of the liquid. If it is further assumed that this velocity gradient is linear as shown in Fig. 14.1 (i.e. the liquid velocity at any point distance x from the fixed plate is given by xv/h) then a shear strain rate or shear rate ($\dot{y}$) can be defined by the ratio v/h (units $\sec^{-1}$ in both cgs and SI systems).

14.3.5   Shear viscosity
This quantity is simply defined by the shear stress divided by the shear rate. In Fig. 14.1, the shear viscosity (η) is given by

$$\eta = \frac{\tau}{\dot{y}} = \frac{Fh}{Av} \qquad\qquad (14.1)$$

where τ is shear stress and $\dot{y}$ shear rate. Note that $\dot{y}$ is equivalent to dy/dt.

The units are poise or dyne sec cm^{-2} (cgs units) and Pascal seconds or N sec m^{-2} (SI units).

14.3.6 Shear elasticity modulus

Suppose the material between the plates is a perfectly elastic body. Then when the force, F, is removed, the top plate will return immediately to its initial position, assuming again perfect adhesion between material and plate. Moreover, once the material has moved the distance y under the influence of the force, movement will cease; in fact, y will be proportional to the magnitude of F.

The modulus of elasticity (G) is then defined as the shear stress divided by the shear strain, i.e.

$$G = \frac{\tau}{y} = \frac{Fh}{Ay} \tag{14.2}$$

14.3.7 Normal force

This is a difficult process to describe without using rather sophisticated mathematics, such as tensor analysis. Basically, in the experiment shown in Fig. 14.1, assume the gap between the plates is filled with a liquid, possessing elasticity as well as viscosity. Suppose the upper plate is moved at a constant velocity, v, (or strain rate $\dot{y}$). As a result of the resistance to continuous deformation exerted by the material, due to the presence of elasticity in the material, the total force exerted on the moving plate is at an angle to the direction of motion. This total force can then be resolved into its components (Fig. 14.2), which include a force parallel to and in the same plane as the plate, as well as a component in a plane vertical to the plane of the motion and at right angles to the direction of motion. This latter is the normal force. In practical terms, this force will tend to try to push the plates apart while there is motion. In practical instruments, the moving plate must either be held rigidly in the vertical plane or it can be allowed to move and kept in position by applying an equal and opposite restoring force to counteract the normal force. Clearly, this last approach provides a means of measuring normal force. Familiar examples of this normal force effect are the rod climbing effect (Fig. 14.3) and the tendency for some flour mixtures or doughs to climb up the stirrer rod when kitchen mixers are used.

14.3.8 Types of rheological behaviour

14.3.8.1 *Newtonian viscous liquid and Hookean elastic bodies*

If the viscosity, as defined in equation (14.1), is independent of the shear rate (or shear stress) applied to cause the liquid to flow, then the liquid is referred to as a Newtonian liquid, after Isaac Newton, who first enunciated the law of liquid flow in his Principia (1718).

Equally, if the elasticity is independent of the value of shear strain (or shear stress) applied, the material is referred to as a Hookean elastic body (Hooke's Law, 1678).

As has been indicated above, both these materials represent physical ideals to

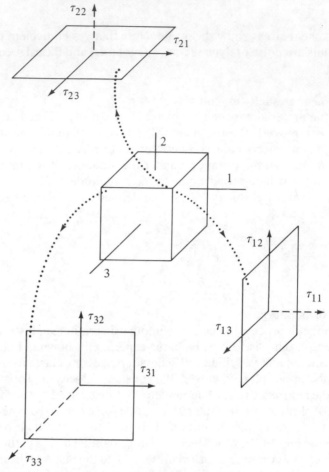

Fig. 14.2 — Components of stress in a viscoelastic liquid under shear. Normal stresses, i.e. stresses at right angles to plane concerned, are shown dashed. For simple shear, $\tau_{13} = \tau_{23} = \tau_{31} = \tau_{32} = 0$.

which most real materials will approximate over limited ranges of stresses, strains, or strain rates. However, if the materials do show variation in their viscosity or elasticity moduli, then it is still often useful to define so-called *apparent* viscosity or elasticity moduli (more usual with the former) using the definitions of equations (14.1) and (14.2).

For the purposes of considering paint rheology, we will examine a number of types of behaviour in which the apparent viscosity is a function of shear rate and time.

14.3.8.2 Pseudoplastic materials

.The important features of pseudoplastic materials are that (1) the apparent viscosity decreases with increasing shear rate (or shear stress) values, and (2) the apparent viscosity value at a given shear rate value is independent of the shear history of the sample.

Fig. 14.3 — Rod climbing (Weissenberg) effect.

   The last statement implies two things: (1) during a given measurement, the apparent viscosity value is *independent of the time* for which the shear rate has been applied to the sample, and (2) the apparent viscosity value is *not dependent on the previous measurements made* (whether these are at higher or lower shear rates). These statements need some qualification in practice. Firstly, there is nearly always some rapid change in shear stresses with time when a measurement is started, due to *inertia* in the moving parts of the viscometer and perhaps also a less rapid change due to elasticity in the sample (the so-called stress overshoot or undershoot effects).

Secondly, either high sample viscosity or high shear rate applied or both can lead to substantial energy dissipation in the form of heat. Unless sample temperature control is very good, this leads to a rise in sample temperature and an apparent drift down in viscosity value with time.

Also implicit in the definition is that the shear stress increases less than proportionally with shear rate. The chief features of pseudoplastic behaviour are illustrated in Fig. 14.4 for an experiment in which shear rate is increased uniformly with time until a maximum value is reached and then reduced at the same rate (shear rate cycle experiment).

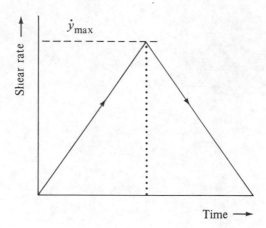

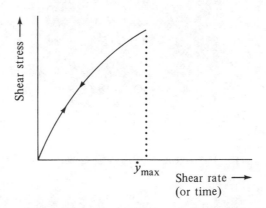

Fig. 14.4 — An example of pseudoplastic behaviour.

14.3.8.3 *Dilatant materials*
The important features of dilatant materials are that (1) the apparent viscosity increases with increasing shear rate, and (2) the apparent viscosity value at a given shear rate is independent of the shear history of the sample. Implicit in the first

statement is that the shear stress increases more steeply with shear rate than proportionality (see Fig. 14.5).

In addition to the qualifications discussed under pseudoplastic materials, there is an important restriction to be added to the second half of the definition. Many dilatant materials, such as concentrated suspensions, behave like hard solids and break or fracture in the viscometer, if the shear rate is too high or it is applied too quickly [4,5].

To this extent, they become history- or time-dependent, but these effects arise from different mechanisms and are more limited in extent than the time effects characteristic of thixotropic or rheopectic materials (see below). Also true dilatancy must be carefully distinguished from the artefacts introduced by the viscometric measurement technique applied to these materials [5].

14.3.8.4 Power law materials

The three types of viscous materials just defined (Newtonian, pseudoplastic, and dilatant) may all be classified as *power law fluids*, because the shear stress — shear rate relationships can all be fitted more or less accurately and usually over limited ranges of shear rate by a power law relationship.

$$\tau = K\dot{y}^n \tag{14.3}$$

or

$$\eta_a = L\dot{y}^{n-1} \tag{14.4}$$

If n equals 1, then the shear stress is proportional to the shear rate, the material is a Newtonian liquid, whose viscosity is equal to the values of the constants K and L in equations (14.3, 14.4). If n is less than 1, then the material is pseudoplastic, and it is dilatant if n is greater than 1.

Such a relationship is often used by engineers for interpolation and (dangerously!) for extrapolation of viscosity — shear rate data.

14.3.8.5 Thixotropy and rheopexy

If the definitions given above for pseudoplastic and dilatant materials are changed in the second half, so that the apparent viscosity value at a given shear rate is now *dependent* on sample shear history, then the materials may be described as *thixotropic and rheopectic*, respectively.

The resulting effect of shear history can be seen for the shear rate cycle experiment by comparing the behaviour shown in Fig. 14.6 for a thixotropic material with that shown in Fig. 14.4, for a pseudoplastic material. It will be seen that a plot of shear stress against shear rate (or, equivalently, time) is a loop, where the shear stress value for a given shear rate value is higher if the shear rate is increasing than if it is decreasing. Moreover, the area of the loop, A, is dependent not only on the rate of shear rate change $(d\dot{y}/dt)$, but also decreases with the cycle number if successive shear rate cycles are applied to the material.

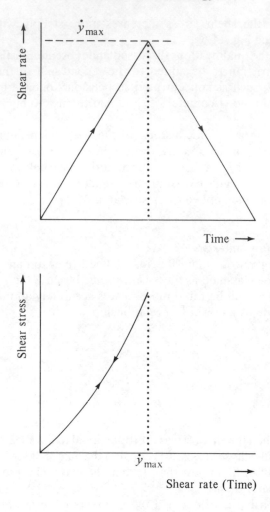

Fig. 14.5 — Dilatant behaviour.

Rheopectic materials are rather rarer in occurrence than thixotropic materials [6,7].

14.3.8.6 Yield stress, Bingham bodies, etc.

If on a shear stress — shear rate plot for a material, the data are extrapolated to zero shear rate and the plot appears to cut the shear stress axis of the graph at a positive stress value, then the material is said to possess a *yield stress*.

In practical terms, this means that it appears that a certain minimum force must be applied to the material before it will flow.

If, above the shear stress value, the shear stress is proportional to the shear rate, then the material is referred to as a *Bingham body*. If not, then it is referred to as a *pseudoplastic, thixotropic, etc. material with yield stress*, whichever description is appropriate.

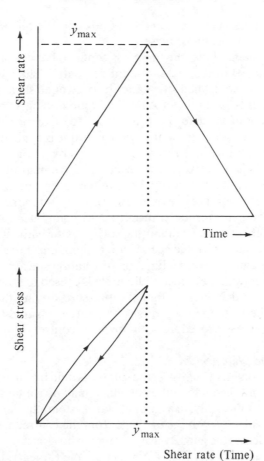

Fig. 14.6 — An example of thixotropic behaviour.

14.4 METHODS OF MEASUREMENT

14.4.1 Geometries
In practice, the simple shear geometry illustrated in Fig. 14.1 is difficult to realize. Most shear deformation experiments are more easily achieved by causing shear by rotation around a central axis (torsional shear mode). Viscometers using this mode are referred to as *rotary viscometers*. There are three main geometries of this kind:

    (1) cone and plate,
    (2) concentric cylinder and,
    (3) parallel plate.

In addition, the capillary flow geometry is frequently used in steady-state measurements (see below) as a reference standard for Newtonian liquids, as well as in other applications. Other geometries have been, or are being, developed, usually for rather specialized applications, and the books by Walters [8] and Whorlow [9] should be consulted for details of these.

In the torsional shear mode, the sample is confined between the two components (components A and B) of the measurement geometry. The measurements can be made in two ways: (a) controlled stress and (b) controlled shear rate. In either case, one of the components (component A) will be mounted on some form of spring or torsion bar, by means of which torque can be applied to, or measured on, component A. In the controlled stress experiment, a known torque is applied to component A by means of the spring, or in more recent instruments by an induction motor system, whilst component B is held fixed. The resulting movement of A is measured as a function of the time and the torque applied. In the controlled shear rate experiment two methods of operation are possible. In the first, component B is rotated at a controlled speed and the torque transmitted to component A by the sample is measured by the deflection of its mounting spring. Alternatively, component A and its spring are rotated at controlled speed with component B fixed and the torque on component A again measured as a function of rotation speed.

In the capillary flow mode, the liquid sample is caused to flow through a capillary of known length and diameter. Either the pressure driving the fluid through the capillary is controlled and the consequent flow rate measured (usually, volumetric flow rate), or the reverse procedure with flow rate control, is applied.

14.4.1.1   Cone and plate geometry
In cone and plate geometry (Fig. 14.7) the sample is held by surface tension in the gap between the cone and the plate, either of which may be rotated in controlled shear rate measurements. Obviously, in a parallel plate system, the velocity at any point on the rotating plate will vary with the distance from the centre, and, consequently, so will the shear rate across the gap. By using a cone, whose surfaces are at a small angle to the horizontal (usually, less than 4°), it is possible (at least, in theory) to ensure

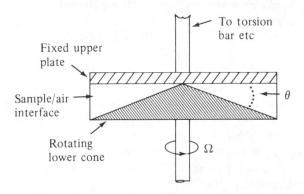

Fig. 14.7 — Cone and plate geometry.

uniform shear rate across the gap and across the whole diameter of the cone and plate system.

With this system, the shear stress (τ) on the sample is given by:

$$\tau = \frac{3T}{2\pi R^3} \tag{14.5}$$

and the shear rate ($\dot{y}$) by

$$\dot{y} = \frac{\Omega}{\theta} \tag{14.6}$$

where T is the measured torque
      R the radius of the cone and plate system
      θ the (internal) angle of the cone (radians)
and Ω the rotational speed of the system (radian sec^{-1}).

In practice, the cone is often slightly truncated to avoid frictional contact between the two components, and then gap setting becomes critical.

Apart from the uniform shear rate, the major advantages of the cone and plate system is the small volume of sample required and the ease of cleaning the system.

Major disadvantages of the geometry are the free liquid edge and difficulties associated with measuring concentrated dispersion systems. Because of the large air–liquid boundary area, the formation of a film or thick viscous layer, due to evaporation of volatile solvent, will exert an extra torque, in addition to the torque exerted by the bulk of the sample, leading to serious errors. Also, at high rotary speeds (shear rates), sample may be thrown out of the gap by centrifugal force.

In concentrated dispersions, apart from the danger of mechanical abrasion or particles jamming in the gap if they are large, there is a subtler effect, due to hydrodynamic forces, and a purely geometric factor. Because the particle centres cannot approach closer to the wall than a distance equal to the particle radius, there is a thin 'particle-depleted' layer separating the component wall from the bulk of the sample. This acts like a layer of lubricant. In addition, it is possible that hydrodynamic forces may prevent the close approach of particles to the walls. These effects are difficult to estimate, but it appears that they can be of the order of the particle diameter for the depletion layer thickness [10]. Unless the gap width average is very much greater than the particle diameter, this will lead to measurement errors. Obviously, the influence of such effects should be assessed by measuring a dispersion sample with several cone and plate geometries of different cone angle and diameter; or, better still, using the parallel plates geometry, in which the gap width may be freely varied [5].

14.4.1.2 Concentric cylinder geometry (Couette flow)

In this geometry, the sample is used to fill the gap between a cylindrical cup and a cylindrical bob suspended within it. By using a small gap in relation to the cup and bob radii, it is possible to ensure that the shear rate is nearly constant across the gap and only dependent on the rotation rate. However, this makes mechanical design difficult, if high shear rate values are required, and increases the likelihood of viscous heating if prolonged measurement periods are necessary.

The shear stress (τ) at the surface of the inner bob is given by:

$$\tau = \frac{M}{2\pi R_b^2 \, h} \, . \tag{14.7}$$

If the sample is known to be non-Newtonian, then the shear rate at the bob surface can best be determined by the formula of Moore & Davies [11], using the first and second derivative of the experimental stress — rotation speed plot, i.e.

$$\dot{y} = \frac{\Omega}{K} \left[1 + \frac{K}{a} + \frac{K^2}{a^2} \left(1 - \frac{b}{a} \right) + \ldots \right] \tag{14.8}$$

where

$$K = \ln(R_c/R_b)$$
$$a = d \ln \tau / d \ln \Omega$$
$$b = da/d \ln \Omega \; (\equiv d^2 \ln \tau / d \, (\ln \Omega)^2)$$
$$\Omega = \text{rotation speed (radian sec}^{-1})$$
$$M = \text{measured torque}$$
$$R_b = \text{the inner (bob) radius}$$
$$R_c = \text{the outer (cup) radius}$$
and $\quad h = \text{depth of immersion of the bob in the sample.}$

Corrections still have to be made for end effects; the review by Oka [12] should be consulted for details.

Taylor [13,14] showed that for Newtonian liquids, the shear rate at a point, distance r from the centre of the system, is

$$\dot{y}_r = \left[\frac{\Omega R_b^2}{R_c^2 - R_b^2} \right] \left[1 + \left(\frac{R_c}{r} \right)^2 \right] \text{(inner cylinder rotated)} \tag{14.9a}$$

and

$$\dot{y}_r = \frac{2\Omega R_b^2 \, R_c^2}{r^2 (R_c^2 - R_b^2)} \text{(outer cylinder rotated)} \tag{14.9b}$$

Both of these equations reduce for small gaps in relation to cup and bob radii to

$$\dot{y} \simeq \frac{\Omega \, R_b}{R_c - R_b} \simeq \frac{\Omega \, R_c}{R_c - R_b} \, . \tag{14.9c}$$

Whilst the data treatment and theory are more complicated for concentric cylinder geometry, this geometry does have a number of advantages over cone and plate geometry: (1) there is no loss of sample due to centrifugal forces at high rotation speeds, (2) there is less likelihood of dependence on particle size with suspension or dispersion samples, and (3) because the liquid level is below the rim of the cup and the liquid–air boundary area is very small, compared to the measuring surface area on the bob, there is less likelihood of loss of volatile solvent, and any surface skin or thickening will have less effect on the measured torque.

14.4.1.3 Parallel plates

Parallel plate geometry has the advantage over the cone and plate geometry that the gap width can be varied freely. As discussed in section 4.1.1, this is an advantage when measuring suspension or dispersion systems. Against this advantage, the shear rate on the sample varies with the distance from the plate centre, and thus data are more difficult to evaluate.

When oscillatory (dynamic) measurements are being made, parallel plates possess considerable advantages over either cone and plate or concentric cylinder geometries, as will be discussed in section 14.4.2.3.

The maximum shear rate (at the plate rim) is given by

$$\dot{y}_m = \frac{\Omega R}{h} \, , \tag{14.10}$$

and the apparent viscosity (η_a) corresponding to this maximum shear rate ($\dot{y}_m$) may be evaluated, using the following equation [15]:

$$\eta_a = \frac{3M}{2\pi R^3 \dot{y}_m} \left[1 + 3 \frac{d \ln M}{d \ln \dot{y}} \right] \tag{14.11}$$

where M is the measured torque
R is the plate radius
h is the plate separation
and Ω is the speed of rotation.

Like the cone and plate geometry, the parallel plate geometry has the advantage that in theory (if not always in practice) the liquid velocity distribution in the gap is determined by the geometry and not by the liquid properties.

14.4.1.4 Capillary (Poiseuille) flow

If a liquid is caused to flow through a circular section tube or capillary of radius R, then it can be shown [16] that (for a Newtonian liquid and assuming no slip at the capillary wall, i.e. $v_w = 0$),

(1) the velocity distribution across the radius is parabolic (liquid moving fastest at the tube centre), and
(2) the stress varies linearly across the tube radius, being maximum at the wall.

Measurements can be made of the driving pressure (P) and the volume flow rate (Q) of the liquid. For a Newtonian liquid, Q and P are related by the Hagen–Poiseuille equation

$$Q = (\pi R^4 / 8l\eta)P \tag{14.12}$$

where l is the tube length

and η the liquid viscosity.

The maximum stress (at the wall) is given by

$$\tau_w = PR/2l \ . \tag{14.14}$$

Corrections must be made for tube end and kinetic energy effects. For non-Newtonian liquids, the velocity distribution deviates from the parabolic form (see Fig. 14.8), and some assumption must be made about the apparent viscosity shear

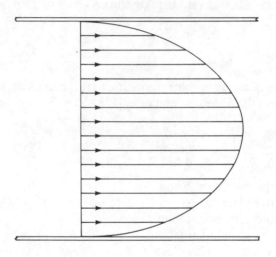

Fig. 14.8 — Liquid velocity distribution for a Newtonian fluid in capillary (Poiseuille) flow.

rate relationship, before the dependence of apparent viscosity on the measured pressures and flow rates can be evaluated [16]. Alternatively, the shear rate at the wall can be evaluated from the volume flow rate — wall shear stress data [17].

$$\dot{y}\,(\tau_w) = \bar{q}\left[\frac{3}{4} + \frac{1}{4}\frac{\mathrm{d}\ln\bar{q}}{\mathrm{d}\ln\tau_w}\right] \tag{14.14}$$

when $\bar{q} = -4Q/\pi R^3$.

14.4.2   Measurement techniques
14.4.2.1   Steady-state
In the definitions given in section 14.3, and the discussion of measurement geometry (section 14.4.1), it has been tacitly assumed that steady-state measure-

ments were being made. This means in practice, for example, that a known shear rate is applied to the sample and the resulting torque is observed. After a few seconds, during which instrument inertia effects die away, the measured torque usually becomes steady in value within the next half to one minute. This steady value is then noted before changing the shear rate value. In the case of thixotropic materials, a steady torque value may not be obtained even after prolonged shearing.

Other measurement techniques are available, and these may give more useful information in some circumstances. They will now be described.

14.4.2.2 Transient techniques

If, in the experiment described in the previous section, the torque is recorded as a function of time, and perhaps also the decay with time of the torque when rotation is stopped, then a transient type of experiment is being performed.

For more solid materials, the more conventional techniques of stress relaxation and creep may be used. In the former, stress (equivalently, torque) is monitored as a function of time when a known deformation is applied to the sample, and vice versa for the latter (creep) experiment. Typical results are shown in Fig. 14.9.

Analysis of the resulting stress or deformation-time plots can give useful information on the elasticity as well as the viscosity of the sample and the dependence of these parameters on the experimental time scale [18]. Such results can supplement the results of dynamic measurements (see next section). However, care must be taken in interpreting the results, owing to artefacts introduced, particularly at short times, by the inertia of the measuring instrument's moving parts.

Such techniques are potentially of great utility in probing rheological structure, such as occurs in thixotropic paints. Because of the greater control exercised on the sample's deformation history, the techniques are much more useful than the thixotropic loop type of experiment described in section 14.3.8.5.

14.4.2.3 Dynamic (oscillatory) techniques

So far, all the experiments described have involved liquid flow in one direction only. If, after torque or rotation has been applied for a short time, the direction of application is reversed for the same time, and this cycle is repeated until a steady response (strain or stress) cycle is observed, then a dynamic experiment is being performed. The applied oscillation is usually sinusoidal with time, although there is no reason (except more complicated theory and mathematics) why sawtooth or square wave may not also be used. Such experiments are usually made with small input amplitudes, and the corresponding (steady) output amplitude and phase shift are measured. The frequency of the cycle is also controlled. From these measurements, values of the dynamic viscosity and dynamic shear modulus of the sample may be calculated, and the variation of these values with frequency studied.

Because of the limited frequency range available with many instruments, it is customary when dealing with polymer solids or melts to vary the temperature, and then use the time–temperature superposition principle to obtain data at one reference temperature over a larger range of frequency [19]. This technique is of limited use in paint measurements, as, apart from practical problems of solvent loss, subtler irreversible changes in the paint may result owing to changes in solubility of

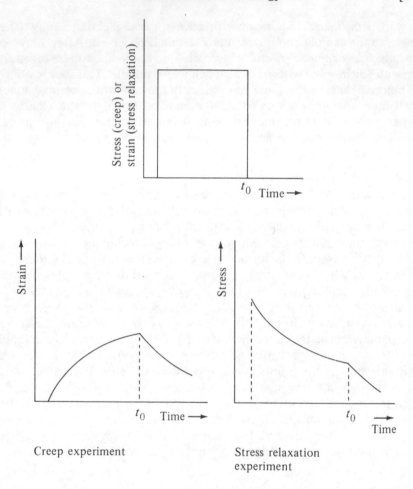

Fig. 14.9 — Transient measurements.

polymers, stability of dispersions etc., brought about by large changes in temperature.

Because the theory of viscoelasticity tacitly assumes linear response, it is essential, particularly with structured or thixotropic paints or suspensions, to check for linearity by varying the input amplitude and monitoring the output waveform. Whilst non-linearity may show as distortion of the output waveform, it may also appear as a change in phase angle and/or change in input/output amplitude ratio as the input amplitude is varied. The theory of non-linear viscoelasticity is less well developed, and experimental results are more difficult to analyse [20].

Whilst capillary flow can be used for dynamic measurements, it is more usual and more convenient experimentally to use torsional flow geometries. Of these, the parallel plate geometry is the most attractive. Not only is it experimentally convenient (for example, the gap may be easily varied without changing any of the components of the geometry), but also the theory is simplest for this geometry, particularly if fluid inertia effects cannot be neglected. (These are severest for low-viscosity liquids and high-frequency measurements) [21].

If the lower plate is driven at a forced (sinusoidal) oscillation of frequency, ω (rad sec^{-1}) and angular amplitude, θ_2 (rad), and, because of the sample between the plates, the upper plate oscillates at the same frequency with angular amplitude, θ_1 (rad) (lagging behind the lower plate by an angle, ϕ (rad)), against the constraint of its suspension (torsion wire or bar of stiffness, E dyne cm rad^{-1}), then the dynamic viscosity (η', poise) and elasticity modulus (G', dyne cm^{-2}) are given by the following equations [22]:

$$\eta' = \frac{-AS \sin \phi}{A^2 - 2A \cos \phi + 1} \tag{14.15}$$

and

$$G' = \frac{\omega AS [\cos \phi - A]}{A^2 - 2A \cos \phi + 1} \tag{14.16}$$

where A is the amplitude ratio (θ_1/θ_2)

$$S = \frac{2h [E - I\omega^2]}{\pi a^4 \omega}$$

with h is the gap between the plates,
a the radius of the plates,
and I the moment of inertia of the upper plate about its axis.

E may be determined by direct measurement and I from the natural frequency of free oscillation of the upper plate in the absence of sample.

In the previous type of dynamic measurement, the strain amplitude is controlled. It is often advantageous, particularly when studying structured paints or dispersions, to control the stress amplitude applied. This may readily be done by fixing the lower plate and applying an oscillatory (sinusoidal) torque to the suspension of the upper plate. Again, amplitude and phase lag of the upper plate are measured in relation to the applied torque, but this time the upper plate motion is constrained by the sample between the plates. In this case, the equations for η' and G' are simpler than equations (14.15, 14.16), [23], but there is no advantage when fluid inertia effects have to be included.

14.5 INTERPRETATION OF RESULTS

Before discussing in detail the rheology of paint (see Chapter 15), a brief review of the factors controlling the rheology of liquid systems will be given to aid the interpretation of results.

14.5.1 Simple liquids

In a low-density gas, the molecules are widely separated in space and the viscosity is determined by momentum transfer between the molecules when they collide. As the density increases, for example, owing to cooling, the momentum of the molecules, due to their thermal energy, is reduced and they come closer together, so that finally they are close enough for intermolecular forces to play a significant role in determining bulk properties like viscosity. When the gas has become a liquid, momentum transfer during molecular collisions contributes an insignificant amount to the liquid viscosity. The liquid viscosity is now determined entirely by the balance between intermolecular forces and the much reduced movements of the liquid molecules, owing to their thermal energy. These movements will be in all directions between existing positions and neighbouring spaces and at random, and against opposition provided by the intermolecular forces. They will be very frequent (in a simple liquid at room temperature, typically, 10^9 to 10^{12} movements or 'jumps' per second).

If a shear stress is applied to the liquid, then the molecular 'jumps' in the direction of the shear stress will be favoured, and those in other directions opposed. The macroscopic consequence of this is that the liquid flows in the direction of shear. Clearly, the stronger the intermolecular forces, the larger the molecules, and the closer the molecules are together (for example, owing to application of pressure), the higher the viscosity of the liquid. Equally, reducing the temperature reduces the thermal energy of the molecules and the frequency of their 'jumps'. So if the shear stress is applied fast enough — in particular, if the time scale of application is equal to, or less, than the time for a molecular jump — then the molecules do not have time for their jumps, and the energy supplied is stored rather than dissipated by molecular movement. In other words, the liquid shows an elastic response. The time scale required for a simple liquid will be very short (10^{-9}–10^{-12} seconds) [1], but will be much longer for a liquid with large molecules such as a polymer melt. Examples of this have already been discussed in the introduction to this chapter.

Finally, when the intermolecular forces predominate over the thermal movements of the molecules, the molecules can only oscillate about a fixed position by means of their thermal energy, and the material has become a solid. Conventionally, the liquid/solid boundary is set at a viscosity of 10^{15} poise.

14.5.2 Solution and dispersion viscosity

If two simple liquids are mixed together and form a homogeneous mixture, then the molecules are of similar size and the viscosity of the mixture may generally be calculated by some sort of mixture law, using mole or volume fractions of the components of the mixture.

However, if the solute molecule or the suspension or dispersion particle size is much greater than the size of the solvent or medium (continuous phase), then the solvent or suspension medium can be treated as a featureless, uniform medium (a continuum) of viscosity, η_s. In this case, it is usual to define a relative viscosity,

$$\eta_r = \eta/\eta_s \tag{14.17}$$

or a specific viscosity,

$$\eta_{SP} = \frac{\eta - \eta_s}{\eta_s} = \eta_r - 1 \ , \tag{14.18}$$

where η is the viscosity of the mixture (solution, dispersion, or suspension). This can be done over a range of sizes from ions of strong electrolytes in aqueous solution through micellar or colloidal solution 'particles' to dispersion particles or very large suspension particles. Microscopically, the viscosity increase over that of the solvent alone is determined by the total 'particle' volume, as well as the interparticle forces and Brownian motion, owing to the impact (thermal energy) of solvent molecules on the 'particle' surfaces.

Einstein [24] formulated a quite general viscosity concentration relationship for such systems. Assuming the particles are (1) spherical, (2) rigid, (3) uncharged, (4) small compared to dimensions of measuring apparatus but large compared to the medium molecular size (continuum hypothesis), (5) with no hydrodynamic interactions between the particles (large interparticle distance or very low concentration), (6) with slow flow (negligible inertia effects) and (7) with no slip at the particle-liquid interface), his equation reads

$$\eta_r = 1 + 2.5\phi \ . \tag{14.19}$$

However, as Goodwin has pointed out, Einstein's working out of this equation also yields [25]

$$\eta_r = \frac{1 + \phi/2}{(1 - \phi)^2} \tag{14.20}$$

which expanded as a polynomial yields

$$\eta_r = 1 + \tfrac{5}{2}\,\phi + 4\phi^2 + \tfrac{11}{2}\,\phi^3 + 7\phi^4 + \tag{14.21}$$

In equations (14.19–14.21) ϕ denotes the dispersed phase volume fraction.

Many workers have attempted since to produce theories which build on Einstein's theory and relax some of the assumptions made (1–7, above). Particular effort has been made to extend the theory to higher concentrations, anistropic particles, and the role colloidal forces play in determining dispersion rheology. These topics have been reviewed many times, but most recently by Goodwin [25], Russel [26,27], and Mewis [28] amongst others.

It is beyond the scope of this chapter to review this work in detail, but brief consideration will be given to concentration effects, as they have important practical consequences for paint rheology, particularly during the drying or flow-out period. The large amount of experimental work on concentration effects has shown that the coefficients of the polynomial terms in the series expansion form of Einstein's equation represent minimum values, obtaining only for very dilute dispersions or suspensions. Departures from spherical shape lead to increases in the coefficient of the linear term. As the concentration increases, so binary and n-tuplet collisions of particles lead to temporary associations between particles, due to hydrodynamic forces (in the absence of any significant colloidal or other interparticle attractive forces), as has been demonstrated by the elegant experiment work of Mason and his co-workers [29] through the last twenty years. These associations lead to some occlusion of medium, and thus the associated group of particles occupies more phase volume than the total volume of the individual particles comprising the group. Vand [10] appears to have been the first to grasp the significance of this effect for dispersion rheology. Equally, the phase volume of individual particles must include a contribution due to the volume occupied by solvating molecules from the medium, adsorbed molecules of surfactants, and polymeric dispersants, as well as the effective volume resulting from strong electrostatic fields generated by ionization at the particle surface (e.g. aqueous colloidal dispersions).

As the concentration increases, it becomes experimentally more difficult to fit polynomials accurately to the data. For this reason, many workers have sought to establish theoretical and empirical relationships based on fractional, exponential, or power law forms. Much of the variability in experimental results used to test these relationships can be attributed to limited understanding and characterization of the systems studied, e.g. imperfect stabilization leading to an unknown and uncontrolled degree of flocculation or aggregation, wide particle size distribution, etc., as well as experimental difficulties due to the onset of non-Newtonian effects at high concentrations. A common factor to these alternative forms is the introduction of a factor which appears in most cases to be related to the critical packing phase volume fraction, ϕ_m. This is because, as the concentration increases, the particles come closer and closer together, until finally they are too close together to be able to move. At this point, the concentration has reached ϕ_m and the dispersion has become solid and will exhibit fracture if stressed too much. The precise value of ϕ_m will depend on the packing geometry, but will be in the range $\phi_m = 0.52$ to 0.74. This implies that for hard particles, the viscosity of the suspension increases very sharply in a limited region of the concentration range around $\phi = 0.60$, as shown in Fig. 14.10, and ultimately reaches very high values. As the particles become closer, so ordering processes become essential for flow, and these require a longer and longer time scale to accomplish. Such processes in the presence of a shear stress are now assisted by Brownian motion (at least for particles below about 2 μm in diameter). The longer time scale implies the appearance of non-Newtonian and viscoelastic effects in exactly analogous fashion to the case of molecular liquids discussed above, except that now the thermal or kinetic energy of the medium molecules impacting the particles provides the energy for the particle movement (Brownian motion) in the absence of a shear field. If a significant number of particle aggregates or floccules are present when the system is at rest, then the sharp increase in viscosity, as particle concentration is increased, will occur at a lower phase volume, owing to the volume contribution of occluded medium. Also non-Newtonian effects may occur at lower concentrations than usual, as the shear field may be strong enough to break down some of the aggregates or floccules, releasing occluded medium and thus reducing the effective volume and the relative viscosity. This breakdown process will be balanced by the rebuilding process when particles collide in the flow, owing to the net attractive interparticle forces resulting from imperfect stabilization, for example.

If the particles are liquid droplets (emulsion systems), then when the droplets are close-packed, they will compress and deform. Even if the stabilizer layer on a hard

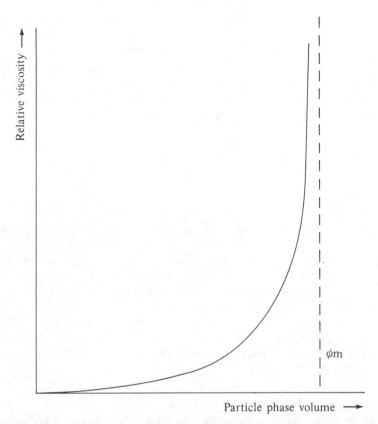

Fig. 14.10 — Relative viscosity — phase volume relationship for hard particle suspension and dispersion systems.

particle forms a significant part of the effective particle volume, then the strong dilatancy and shear fracture effects observed at or close to ϕ_m will be reduced or disappear [30]. However, unlike emulsion droplets, they cannot phase invert, which is what happens finally to emulsions when the concentration is pushed beyond ϕ_m. As a result of this, the disperse phase volume of the emulsion decreases sharply with a corresponding drop in relative viscosity (Fig. 14.11). Just as with hard particles, increases in the time scale lead to the appearance of non-Newtonian and viscoelastic effects in the vicinity of the phase inversion concentration. Equally, these effects can be seen with mixtures of partly miscible simple liquids [32].

Summarizing this discussion, the relative viscosity–disperse phase volume relationship requires the use of the effective phase volume (ϕ_e) instead of the nominal or physical phase volume fraction (ϕ), and the introduction of a critical packing volume fraction (ϕ_m) beyond which the dispersion changes its nature (becoming a solid for hard particles and phase-inverting with liquid droplets). It is the author's judgement that of all the theoretical treatments of this relationship so far published only those which try to take account of hydrodynamic interactions between particles have any predictive utility, for example, those based on cell models, such as the early theory of Simha [33] and the later theories of Yaron & Gal-Or [34] and Sather & Lee [35].

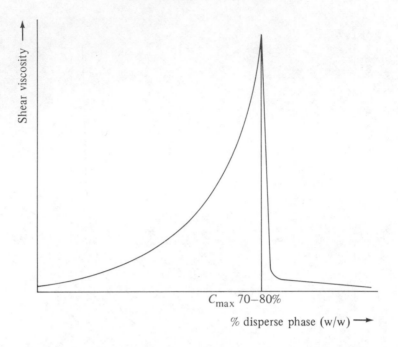

Fig. 14.11 — Relative viscosity — phase volume relationship for an emulsion (liquid dispersion) system.

14.5.2.1 Polymer solution viscosity

The relation between the relative viscosity of a polymer solution and the concentration of polymer in solution shows a more gradual increase with increasing concentration, than is observed for dispersion systems. In general, three concentration regions can be distinguished; namely, dilute, semi-dilute, and concentrated regions (see Fig. 14.12 and cf. Fig. 14.10). The viscosity behaviour derives from considering the polymer molecules as loose, randomly arranged coils, and from considering the balance of three types of intermolecular forces, i.e. those between the polymer molecules, between the solvent molecules, and between the polymer and the solvent molecules.

In the dilute region, the coils are widely separated and not able to interact hydrodynamically. If the solvent is a thermodynamically 'good' solvent for the polymer (polymer–solvent intermolecular forces are strongest), then solvent molecules penetrate the coils, causing them to approach their maximum possible size. Consequently, the effective volume occupied by each coil and the relative viscosity of the solution both approach a maximum. As the thermodynamic quality of the solvent is reduced (for example, by adding non-solvent or changing the temperature), intramolecular forces between neighbouring parts of the polymer coil become stronger in relation to the other intermolecular forces, and the coil contracts with a consequent reduction in effective volume and relative viscosity. Finally, the 'theta state' is reached where the coil volume is reduced to its minimum fully saturated size. Thereafter, further reduction of solvent quality leads to coil–coil association and finally to precipitation (bulk phase separation) of the polymer from solution.

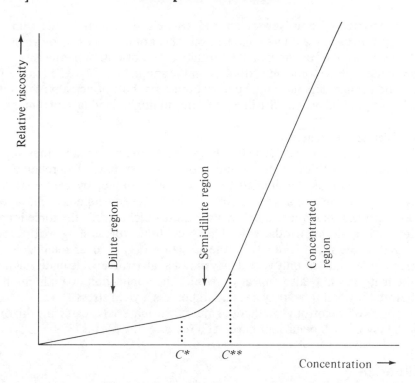

Fig. 14.12 — Relative viscosity — concentration (phase volume) relationship for a polymer in solution.

In the semi-dilute region, the relative viscosity starts to change more rapidly with concentration. In this region, the coils are closer together and interact with each other. Put crudely, they compete for space. In 'good' solvents, this results in the coils reducing in effective volume as the concentration increases until at the upper boundary of this region the coil's effective volume is again at a minimum (unperturbed dimensions). In a 'good' solvent, this region tends to cover a wider range of concentration for a given polymer, than for the same polymer in a 'poor' solvent, because of the greater volume of coil contraction available in the good solvent.

In the concentrated region, further contraction of effective volume is obtained by overlap of the coil volumes. The polymer chains are now forced close together, and intermolecular forces between them are increased, particularly if the solvent is a 'poor' one for the polymer. This results in a much more rapid increase in viscosity with concentration, and, unlike the dilute concentration region, 'poor' solvents give higher viscosity solutions than 'good' solvents.

It should be emphasized that the coils cannot be thought of as discrete entities with definite boundaries, unlike dispersion or emulsion particles. The coil volume has dimensions determined by the statistical average positions of the individual polymer molecular segments in space, averaged over a sufficient length of time. This explains why polymer coil volumes can apparently overlap or interpenetrate in the concentrated region, although they have apparently reached their minimum dimensions at a lower concentration. As before, the interaction of the polymer coils leads to the appearance of non-Newtonian and viscoelastic effects in the semi-dilute region, whilst, even in the dilute region, strong shear or extensional flows can lead to such effects because of polymer coil deformation and orientation effects.

More detail will be found in textbooks on the subject [36,37]. The range of each concentration range depends on a number of factors, but, principally, the polymer molecular weight and weight distribution (determining individual coil dimensions).

14.5.3 Rheological structure

Liquids possessing rheological structure have this feature in common: they appear to be solid-like or to be thick (high-viscosity) liquids at rest, but, when relatively small forces or deformations are applied to them (for example, by gentle stirring or shaking) they become relatively mobile or low-viscosity liquids. This apparent breakdown process may reach a steady state quite quickly while the force is applied (pseudoplastic materials), otherwise it may continue for as long as the force is applied and, if the force is removed, the structure may recover relatively quickly (thixotropic materials) or only very slowly or not at all (irreversible materials). Some solid-like materials may also appear to require the application of a minimum force value before they will flow: they are said to possess a yield stress.

The origins of thixotropy, methods of measuring materials possessing thixotropy, and modelling of such behaviour have been reviewed by Mewis [38].

REFERENCES

[1] Lamb, J. *Rheologica Acta*, **12** 438 (and refs. cited in this paper) (1973).
[2] Bingham, E. C., & Green, H. *Proc. ASTM*, **19** 640, (1919).
[3] Glass, J. E. *J. Coatings Tech.* **50** (641) 56 (1978).
[4] Hutton, J. *Rheologica Acta* **8** (17) 54 (1969).
[5] Strivens, T. A. *J. Colloid & Interface Sci.* **57** (3) 476 (1976).
[6] Bauer, W. H. & Collins, E. A. In: F. R. Eirich (ed.) *Rheology: theory and applications*. Vol. Academic Press (1967).
[7] Eliassaf, J. A. Silberberg, & Katchalsky, A. *Nature*, **176** 1119 (1955).
[8] Walters, K. *Rheometry* Chapman & Hall (1975).
[9] Whorlow, R. W. *Rheological techniques* Ellis Horwood (1980).
[10] Vand, V. *J. Phys. Coll. Chem.* **52** 277, 300, 314 (1948).
[11] Moore, F. & Davies, L. J. *Trans. Brit. Ceramic Soc.* **55** 313 (1956).
[12] Oka, S. *Principles of rheometry*, Ch. 2 pp. 18–82, In F. R. Eirich (ed.) *Rheology, Theory and Applications*. Vol. 3. Academic Press (1960).
[13] Taylor, G. I. *Phil. Trans. Roy. Soc.* (*London*), Ser. A, **223** 289 (1923).
[14] Taylor, G. I. *Proc. Roy. Soc.* (*London*), Ser. A, **157** 546, 565 (1936).
[15] *Loc. cit.* ref. [8], p. 52.
[16] *Loc. cit.* ref. [12], pp. 21–9.
[17] *Loc. cit.* ref. [8], p. 100.
[18] Ferry, J. D. *Viscoelastic properties of polymers*. 2nd edn. John Wiley, pp. 9–12, 38–43 (1970).
[19] *Loc. cit.* ref. [18], pp. 641–3.
[20] *Loc. cit.* ref. [8], pp. 140–151.
[21] *Loc. cit.* ref. [8], pp. 125–140.

[22] *Loc. cit.* ref. [8], pp. 125–7.

[23] Strivens, T. A. (To be published).

[24] Einstein, A. *Ann. Physik* **19** (4) 289; (1906) **34** (4) 591 (1911).

[25] Goodwin, J. W. In: *Specialist periodical reports: colloid science*, Vol. 2. Ch. 7. The Chemical Society (London) (1975).

[26] Russel, W. B. In: Re. E. Meyer (ed.) *The theory of multiphase flow.* Wisconsin U. P. (1982).

[27] Russel, W. B. *J. Rheology*, **24** (3) 287 (1980).

[28] Mewis, J. *Adv. Coll. Interfac. Sci.* **6** 173 (1976).

[29] Goldsmith, H. L., & Mason, S. G. In: *loc. cit.* ref [6].

[30] Strivens, T. A. (Unpublished results).

[31] Barry, B. W. *Advances in Colloid & Interface Science* **5** 37 (1975).

[32] Fixman, M. *J. Chem. Phys.* **36** (2) 310 (1962).

[33] Simha, R. *J. Applied Physics* **23** (9) 1020 (1952).

[34] Yaron, I., & Gal-Or, B. *Rheol. Acta.* **11** 241 (1972).

[35] Sather, N. F., & Lee, K. J. *Progr. Heat & Mass Transfer* **6** 575 (1972).

[36] Yamakawa, H. *Modern theory of polymer solutions.* Harper & Row (1971).

[37] Bohdanecky, M., & Kovar, J. *Viscosity of polymer solutions.* Elsevier (1982).

[38] Mewis, J. *J. Non-Newtonian Fluid Mech.* **6** 1–20 (1979).

15

The rheology of paints

T. A. Strivens

15.1 INTRODUCTION

In this chapter, the rheology (flow properties) of paint will be described both for the bulk paint and the film of paint after application, as well as methods for measuring the paint rheological properties in both states. Brief mention will be made also of other areas in the paint industry, where rheology is important, for example in various important stages of paint manufacture, such as pigment dispersion (wetting), transport through pipelines, and mixing operations.

Inevitably, because of the complexity of most practical paint-flow situations, both the theory and the experimental evidence provided to understand paint flow will appear sketchy or incomplete in many areas. However, the importance of understanding and, by consequence, the ability to control, paint rheology remains paramount to the successful utilization of such products. New legislation to control emission to the atmosphere (air pollution) and safety at work (lower and lower maximum levels of solvent vapours which may be toxic or may present fire hazards) is increasing. Such legislative pressure is increasing in industry throughout the world. The consequence of such pressure is to limit the freedom of choice of both paint formulators and users. Two trends which have emerged from such limitations are the move towards higher and higher-solids paints (less solvent to be eliminated in the formation of the solid paint film) and towards water-based paints (less toxicity hazard in principle). The inevitable result of such trends is paint with more complex rheological properties than before, both in the bulk and in the film.

15.2 GENERAL CONSIDERATIONS ON PAINT RHEOLOGY — PAINT APPLICATION PROCESSES

The assertion has just been made that paint rheology control is essential to the successful utilization of paint. This assertion can be justified by consideration of some examples. An important example is provided by the application of conventional paints to a surface. Whatever application technique is used, be it spray gun, brush, roller, etc., the process has three stages:

(1) transfer of paint from bulk container to applicator;
(2) transfer of paint from applicator to the surface to form a thin, even film; and
(3) flow-out of film surface, coalescence of polymer particles (emulsion paints), and loss of medium by evaporation.

In each of these stages, the paint rheology has a strong controlling influence on the process.

15.2.1   Transfer stage

In the bulk container, the paint will normally be low in viscosity. This is so that it can be readily utilized in the chosen applicator; for example, it can readily penetrate the spaces between the bristles of a brush or the porous surface of a hand roller where it will be held by capillary/surface tension forces during the transfer to the surface to be painted. Of course if the total weight of paint loaded onto the brush by dipping it into the paint in the bulk container is sufficient to overcome the capillary forces, then the paint will drip or run off the brush (definitely not an attractive property for the user!). If the brush-load, without incurring the dripping penalty, is too low, then only a thin film, or a thicker film over a smaller surface area, will be obtained on brushing out. In the former case, solvent loss may be too rapid, and the consequent increase in film viscosity too quick to allow proper flow-out after application (see below). In the latter case, the painting operation becomes too laborious and time-consuming to be satisfactory. What determines the optimum film thickness? Apart from flow-out properties (see below), the required colour density and covering power (the ability of the paint to mask previous coatings — primers, old paint — or surface texture and colour) are important considerations, as well as the protective properties of the final solid film.

Increased brush-loading may be achieved by increasing the bulk paint viscosity or by introducing rheological structure. Both strategies include the useful bonus of slowing down or eliminating pigment settlement by gravity during storage of the bulk containers. However, increasing the bulk viscosity carries the penalty of increasing the mechanical effort required to spread the paint into a film (an important consideration with hand-operated applicators). To be satisfactory, rheological structure must break down quickly under the relatively low shear stress or strain of dipping a brush into the paint or the higher stresses imposed by a hand roller on paint in a tray. This breakdown facilitates the loading/penetration processes, but then the structure must reform quickly to prevent dripping, running, etc.

Turning to industrial application techniques such as spraying or roller-coating, the considerations are similar. Thus, in the case of spraying, the viscosity of the bulk paint must be low enough to allow the paint to be pumped through the fine jet of the spray gun with minimal application of pressure. Usually, such paints are thinned from higher-solids bulk immediately before use, so settlement is less of a problem. The mechanism of droplet formation is reasonably well understood for simple liquids [1], but the influence of such factors as the presence of pigment or polymer particles in suspension or dispersion and the presence of polymer in solution, on the droplet size and size distribution in the spray is largely unknown. However, it can be expected that lowering the bulk viscosity will decrease the droplet size, possibly increasing the danger of wastage by small droplets being carried past the object to be sprayed by the air currents originating from the spray gun. Equally, evaporation of medium from the large surface area of the droplets will be influenced by droplet size and consequent total droplet surface area, and so will influence the initial rheology of the film built up by impaction and coalescence of spray droplets at the surface of the object being sprayed. In the case of industrial roller-coating processes, the paint must be considerably thicker, being able to flow under gravity or low pumping energy to the surface of the application roller, where it may be spread into an even layer by the action of the doctor blade or by another roller, etc. In this situation, the mechanical work required to cause the paint to flow is much less important. However, the paint must be viscous enough to prevent it running off or being thrown off the roller by centrifugal force.

The important feature of both these processes is the very high fluid flow rate and operating speeds, respectively, and the consequent high stresses and strain rates of deformation applied to the paint. However, it should be noted that the paint remains in the spray gun jet for such a short time (or in the 'nip' between the rollers) that a steady state is never attained and, therefore, only transient (for high-frequency oscillatory) measurement methods are likely to produce relevant rheological parameters. Such methods require complex equipment and techniques, particularly at the high stress and deformation rates attained in the application process. Schurz [2] quotes shear rates of $100\ 000\ s^{-1}$ applied for 1 ms in high-speed roller-coaters. Such high values also have a further possible consequence if polymer is present in solution in the paint formulation. At such values, the presence of polymer in solution at a concentration practicable for modern paint fomulations and at a molecular weight above about 10 000 can lead to the development of a high extensional viscosity component, both in the fluid jet from a spray gun nozzle and in the splitting film at the rear of a roller coater nip. Glass [3] has shown that the extensional viscosity of 'thickened' water-based emulsion paints influences such application properties as tracking, spattering, etc., during hand-roller application of such paints. It is reasonable to expect that the appearance of such a high extensional viscosity can interfere with the process of filament or jet rupture to form spray droplets. By Trouton's law, the extensional viscosity of a simple liquid is three times the shear viscosity. However, the presence of a few tenths of a percent of high-molecular-weight polymer in solution in this liquid can raise the extensional viscosity to as much as ten thousand times the shear viscosity [3].

15.2.2 Film formation

The loading and transfer of paint by a hand applicator, such as a paint brush, from the bulk container to the surface to be painted, is followed by regular movement of the hand applicator over the surface to transfer the load of paint from the applicator to the surface and spread it out in an even layer. During this process, hand pressure on the applicator causes shearing and compression of the brush bristles or fibres or of the rubber foam or fibrous mat typically covering the surface of a hand roller. Such shear and compression pushes paint out of the interstices in the applicator. The flow processes involved are very complex and probably impossible to analyse quantitatively. However, attempts have been made to measure the corresponding shear rates appoximately for brush application either directly or by correlation with subjective assessments of brushability of Newtonian paints. Ranges of 15–30 s^{-1} for brush dipping and 2500–10 000 s^{-1} for brush spreading are quoted in the literature [4]. Kuge [5] has attempted to measure the forces exerted during brushing and to relate these to the rheological properties of paint, using simple theoretical equations.

15.2.3 Flow-out of paint film

Over twenty years ago, the suggestion was made that the characteristic irregularities produced on the paint film surface by brushing, the surface striations produced when paint is applied by drawdown bars, and (by implication) also those produced by roller-coating application, have a common origin in hydrodynamic instabilities produced by the unstable flow resulting from the paint film being forced to split between the substrate and the receding applicator edge [6]. Whilst the physical process is understood, it is not always possible to quantify it. Theoretical treatments of the effect have been attempted by Pearson [7] and more recently by Savage [8].

In a previous section, the possible importance of extensional viscosity in the application transfer process from rollers was mentioned. Glass [3] made measurements of extensional viscosity on water-based emulsion paints, thickened with water-'soluble' polymers, such as various cellulose derivatives, acrylamide/acrylic acid copolymers, and polyethylene oxide polymers, and he tried to relate the phenomena of paint spatter and surface tracking to the extensional viscosity measurements. Unfortunately, because of the limitations of the spinning-fibre technique that he used for measuring extensional viscosity, he was forced to use concentrations and molecular weights of the thickening polymers, which gave unacceptable amounts of web formation, filament formation, and surface irregularity when the paint formulations were applied by roller. He did also measure both the elasticity recovery [9] and the normal force under steady shear conditions. The former was done using a technique first described by Dodge [10] in which the sample is first sheared at high shear rate (2600 s^{-1}) and then instantaneously the elasticity is measured as a function of time on cessation of shearing, using a low-frequency oscillatory shear deformation of frequency 0.3 Hz and maximum shear rate of 0.07 s^{-1}. Such measurements may be readily achieved, using an instrument such as the Weissenberg Rheogoniometer. Because of the systems Glass measured for extensional viscosity, whilst acknowledging the importance of elasticity recovery, he tended to ascribe most of the effects he studied to elongational viscosity. However, when this elongational viscosity value was low or moderate, he found the elastic recovery rate correlated well with roll track formation and flow-out.

The critical features of paint application processes are that the paint is first sheared at very high shear rate and then, secondly, forced to split cohesively, either at the spray gun nozzle exit or else on the substrate between the receding edge of the applicator and the layer adhering to the substrate. Both processes take times of only a small fraction of a second. Myers [11] has argued that the high shear rate of the application process will destroy completely any structure present in the paint before application, and that, although the paint film now lacks any rigidity, two factors prevent the normal liquid cohesive splitting by viscous flow:

(1) Rapid separation of the applicator from the substrate leads to tensions which cannot be alleviated by transverse flow of the fluid.

(2) Most paint compositions contain some polymer in solution which can contribute an appreciable elasticity to the solution.

So, whilst the usual split of the liquid into filaments by cavitation takes place, the elastic component of the paint, combined with the speed of the separation, both result in less necking and longer temporal persistence of the filaments than would be expected for a simple viscous liquid. The presence of pigment or polymer particles, present in suspension in the paint, may well assist the nucleation of the cavitation process (see Trevenna [12]). Such an explanation implicitly rules out extensional viscosity (not mentioned by Myers), although it may be equally as important as paint elasticity under some application conditions; and, as Walters [13] has pointed out, extensional viscosity is not *a priori* a function of shear viscosity. Indeed, some recent work with dilute high-molecular-weight polymer solutions [14a] suggests that sharp increases in extensional viscosity at certain extensional deformation rates are due to the polymer molecules becoming fully extended in the deformation field, which is in sharp contrast to the oscillatory average coil configuration probably adopted by such molecules in the oscillatory shear field, customarily used for measuring shear elasticity modulus. Walters [14b] cites an example of a 100 ppm aqueous polyacrylamide solution of shear viscosity 1.4 cP exhibiting an extensional viscosity of 90 P.

Thus, although something is understood about the physics of how surface irregularities arise during the application and formation of wet paint films, there is no agreement about what rheological properties of the paint are relevant to the formation of such irregularities, although there is no dispute about the relevance of such properties.

When one turns to consider the processes of surface levelling, which affect such important practical properties as colour uniformity, hiding power, etc., as well as the more major flow faults such as sagging and slumping, there appears to be an equal lack of understanding of relevant paint rheological parameters. Although the work of Glass [9] and Dodge [10] has clearly indicated a connection between shear elasticity recovery rate and surface irregularity flow-out (levelling), the existing published theories and experimental work on levelling and sagging still regard paint films as Newtonian or pseudoplastic fluids (possibly with a yield point), thus representing no essential advance over the pioneering work of Orchard nearly a quarter of a century ago on the levelling of Newtonian fluids [15]. Indeed, a recent edition [4] of a highly-regarded textbook manages to treat the whole of paint flow and pigment dispersion without once mentioning shear elasticity or extensional viscosity of paint (cf. [16]). There is, however, abundant evidence of viscoelastic behaviour in paints as well as pigment dispersions or millbases [17, 18].

The situation is further complicated by the effects of solvent evaporation. Not only does this affect the rheology of the paint film, but it also affects the surface tension at the wet-film/air interface. Surface tension and gravity forces provide the shear stresses which drive the levelling and sagging processes. The results of evaporation will be to increase solution polymer concentration and to produce cooling at the film surface. Both of these effects will (unless they occur uniformly all over the film surface) lead to a tangential surface shearing force (the Levich–Aris force). Overdiep [19] has recently argued that the hydrostatic pressure gradient in a paint film due to surface tension is insufficient to explain levelling results, as argued by Smith et al. [6] and those that have followed or developed their treatment. He attempts to demonstrate both theoretically and experimentally, using solventborne alkyd paints, that, whilst surface tension tends to produce a flat surface, irrespective of the substrate surface profile underneath, the surface tension gradient developing over the wet paint film surface tends to produce a uniform paint film thickness, i.e. the surface profile of the paint film mirrors exactly the surface profile of the substrate underneath.

In addition, solvent evaporation leads to gradients in solvent content through the film as well as across the surface, and, consequently, to density gradients. Both density and surface tension gradients could contribute to circulatory patterns being set up in the wet paint film. Such circulatory systems may well contribute to the disorientation or randomization of aluminium flake orientation in metallic automotive topcoats, although most publications consider only film viscosity as the controlling factor [20]. In extreme cases, they may lead to the formation of Bénard cell patterns, such as are more commonly seen at the surface of boiling or rapidly evaporating bulk liquid samples. Such effects as a cause of surface irregularities of paint or polymer films have been considered by Anand et al. [21] and Higgins & Scriven [22] amongst others.

Whilst such considerations allow estimates to be made of the forces operative in controlling the rheology of the applied paint film, it is important to realize the potential complexity of the film's rheology because of its physical and chemical composition. From what has been said in the previous chapter, it can be expected that not only will the film material be viscoelastic but also, even at the low stress values involved, highly non-linear and with time-dependent effects deriving from the high shear during application. Something of the complexity of rheological behaviour of bulk systems of analogous composition may be gauged by referring to the published experimental work over many years of Onogi and his co-workers on polymer particle dispersions in polymer solutions and melts [23]. In view of this, attempts to model the levelling behaviour of paint films by considering the material as a Newtonian or pseudoplastic liquid seem to be too simplistic, as do simple rheological measurements based on such concepts. Not only that, the presence of concentration gradients through the film thickness is likely to mean that the rheology will vary through the depth of the film. By comparison, the effect of density gradients in the film is likely to be minor.

A comprehensive review by Kornum & Raaschou Nielsen [24] assesses the balance of factors involved in the levelling and other flow processes of wet paint films, particularly with reference to surface defects. They state the operative forces in levelling to be in the range 3–5 Pa (30–50 dyne cm^{-2}), and in sagging to be about 0.8 Pa (8 dyne cm^{-2}) at the surface of a typical paint film. They also quote estimates of shear rates for levelling processes in paint films of 0.001–0.5 s^{-1}. Because it is shear stress, resulting from gravitational and surface tension forces that controls the flow in levelling, sagging, etc., estimates of the shear rate are irrelevant to the consideration of the flow processes; and, similarly, when trying to measure the relation of paint rheology to flow properties, it is desirable to use controlled stress instruments, rather than the more conventional controlled shear rate instruments.

15.2.4   Desirable rheology for paint application

We can now summarize the desirable rheology for paint application. Initially, the paint must lose its structure at rest and become low in viscosity to facilitate transfer by, or through, the paint applicator. Because of the high shear rates and short timescales involved in the transfer process, both elasticity and extensional flow processes may modify the pattern of surface irregularities on the paint film. These would be expected to arise anyway from unstable hydrodynamic flows produced by cohesive splitting at either the spray gun jet exit or at the interface between the adherent paint film on the substrate and the material on the receding edge of an applicator, such as a roller, moving over the substrate being painted.

The paint must now remain low in viscosity for a sufficient time for the surface irregularities to flow out to an acceptable extent (dependent on whether, for example, gloss or protective properties by evenness of film thickness is the prime property required). However, while the viscosity is low, the paint will flow on vertical surfaces under the influence of gravity. If the thickness (film depth) builds up too much, the effect (sagging) may become noticeable (i.e. offensive!) to the observer, whilst lumps of the thickened paint may slide on the substrate in an irregular fashion to give rise to the effects of curtaining, slumping, etc. In either case, a definite time after cessation of application is required before such effects become noticeable.

Thus the initial low-viscosity pertiod must be followed by a sharp rise in viscosity, resulting from loss of solvent by evaporation, or by a rapid recovery of elasticity (rheological structure) destroyed by the shearing of the transfer process. In either case, the effect is the same, the drying film is virtually immobilized, and the sagging process ceases before it becomes noticeable. The loss of solvent may be affected by differential volatility as well as solvency of components in the solvent mixture forming the paint medium or diluent towards polymeric components in the paint. These solvency effects, in turn, will control the viscosity rise of the drying paint film as it loses solvent. Losses by solvent evaporation will produce cooling of the drying paint film surface (latent heat of evaporation), particularly in the case of fast-evaporating solvents. This cooling may also influence film viscosity.

15.3   EXPERIMENTAL METHODS FOR MEASURING PAINT RHEOLOGY FOR APPLICATION AND FLOW-OUT AFTER APPLICATION

15.3.1   Paint rheology for application

A number of general instruments for rheological measurements, described in the previous chapter, may be utilized for studying paint application and flow-out properties. However, within the limits discussed in the previous sections, many paints may be considered as quasi-Newtonian liquids. Also there is a need for simple instruments for quality control and user viscosity adjustment before application.

This need has been largely satisfied by specialized instruments, developed within the paint industry.

This group of simple instruments consists principally of various flow-cups of different designs, and simplified rotational instruments, like the ICI cone and plate instrument. Flow cups are used in other industries, e.g. the petroleum industry. The principle of the instrument is illustrated in Fig. 15.1. A known volume of paint is held

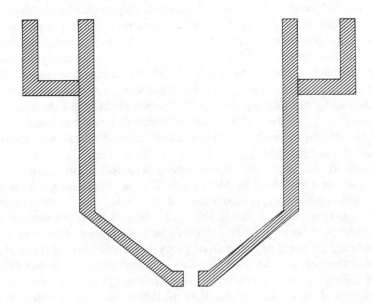

Fig. 15. 1 — Section of a flow-cup (approx. full scale).

within a vertical cylindrical cup, whose bottom has a short capillary of controlled length and diameter. The paint is released to flow through the hole in the bottom of the cup (usually by the operator removing his finger!), and the time for the liquid to flow out of the cup is measured with a stop-watch. The flow end-point is normally taken as the point at which the continuous liquid jet breaks up into drops. A number of points emerge in considering this type of measurement. Firstly, because the liquid height varies during the test, the force (due to gravity) driving the liquid through the capillary also varies. So, if the paint is non-Newtonian, the viscosity result may be very misleading. Secondly, the capillaries are always short, so stable flow conditions within the capillary are not obtained, and this, as well as entry and exit errors, may also affect the result, particularly if the material is slightly structured (elastic). Thirdly, the presence of abrasive particles in the paint may lead to wear of the metal capillary, and so flow-cups should be checked frequently with Newtonian liquids of known viscosity. The cup bottoms may be conical or flat; the former is likely to reduce entry errors. Finally, tests are normally run at ambient temperatures, but it is preferable to control the paint temperature carefully before and during the test to produce accurate and comparable results. This type of test should be used only with near-Newtonian paints. The various types of flow-cup have been reviewed in detail by McElvie [25].

A different type of instrument is exemplified by the ICI cone and plate instrument, first described by Monk [26, 27]. This instrument consists of a fixed lower plate, thermostatted at $25.0 \pm 0.1°C$ by a frigistor system, and an upper cone driven by a motor at 900 rev/min via a torsion spring. The operating shear rate is 10 000 s $^{-1}$ and the measurement range is 0–5 poise with a reading accuracy (by pointer moving over a scale) of better than 0.1 poise. The cone is truncated to reduce wear and to avoid particles jamming in the gap. Only a small sample of paint is required (less than one millilitre), and the instrument is quick and easy to use. Clearly, this instrument is superior to the flow-cup in that the sample temperature is carefully controlled and the sample viscosity is measured under conditions (high shear rate) relevant to application conditions. However, the instrument still provides only a single point measurement, and consequently gives no indication of the paint rheology at rest or during film flow-out. To assess this, a measurement at low shear rates or stresses is necessary, for example, with the ICI (relaxation) low-shear viscometer to be described in the next section.

Other methods are available for measuring the application viscosity of paints, and these have been reviewed by McGuigan [28]. A particular problem group of paints are thixotropic or highly-structured paints. Some idea of the complexity of rheological behaviour of such materials can be gained from the discussion in the previous chapter, but finding simple test methods for quality control is very difficult. Measuring of such paints at rest is relatively easy — some sort of vibrational method, as discussed in the next section under rheology during storage, may be used — but it is extremely difficult to measure breakdown and recovery of structure accurately (the crucial factor in determining application and film flow-out properties of such materials). An excellent review of the methods available has been given by Walton [29]. In the present author's experience, the only satisfactory method is to destroy all the structure by shearing the sample at high shear rate for a sufficient time (this also eliminates errors due to rheological 'history' produced by the sample-handling and loading into the rheometer), followed by monitoring the recovery of structure with time. This can be done by reducing the shear rate to a very low value, and recording apparent viscosity as a function of time; or better still, using a small oscillatory stress or strain, and measuring both dynamic viscosity and elasticity modulus as a function of time (see also [10]). This type of measurement can be done by means of the Low Shear Viscometer (LSV) range of instruments, developed by ICI, and particularly easily on a routine basis by the (relaxation) LSV instrument of this range — to be described in the next section.

Standard methods of viscosity measurement of paints put out by official standards organisations, such as the British Standards Institute (BSI), tend to specify flow-cups [25]. However, it is perhaps significant that there are in increasing number of method specifications using rotational viscometers. For example, the American Society for Testing of Materials method D4287–83 specifies the ICI cone and plate viscometer, whilst method D562–81 specifies a Sturmer viscometer. The Deutshes Institut für Normung DIN 53 214 (1982) method describes the determination of 'rheograms and viscosities' of paints and varnishes, using rotational viscometers, as do French and Czech standards [30, 31].

15.3.2   Paint film flow-out (levelling and sagging)
15.3.2.1   Direct measurements of rheology during flow-out
To summarize the main problems involved in direct measurement of paint film rheology during flow-out (levelling):

(1)   The rheology of the paint film material is extremely complex (not only viscoelastic, but extremely non-linear).
(2)   The rheology can change rapidly with time, owing to compositional changes (solvency balance changes, solids changes, etc., as solvent is lost, as well as rheological structure recovery).
(3)   Small volumes, possibly also inhomogeneous in composition, through the film depth.

The emphasis must therefore be on rapid methods (short individual determination time and, therefore, high repeat rate) and maximizing the information obtained from each determination (for example, by deconvoluting the response curve from an instantaneous stress application to obtain data over a wide frequency range). For this reason we will start by surveying impact and high-frequency oscillatory methods before proceeding to a consideration of other methods.

(a)   Impact method (bouncing ball)
Impact methods have been used widely in many different forms for testing polymers in the form of cylindrical or disk specimens and for testing the mechanical properties of solid (cured) paint films during their service life (see Chapter 16). However, their use for studying paint film flow-out and curing processes appears to have been overlooked, apart from a brief reference by Snow [32] to the bouncing ball method. For bulk polymer samples, a similar method has been briefly analysed by Flom [33] and, in more detail, by Tillett [34], Jenckel & Klein [35], and Pao [36].

The present author [37] has confirmed the original findings reported by Snow. A typical set of results is shown in Fig. 15.2, where rebound height is plotted as a function of paint drying time for a 0.5 cm diameter (0.5 g weight) steel ball dropped on to a $\frac{1}{2}$ in thick glass slab coated with the paint under test. As the paint film dries, the viscosity increases owing to solvent loss; and, consequently, the energy dissipated by the film during the ball impact with the glass increases also, and the ball rebound height decreases initially with drying time. However, when curing starts, either by autoxidation or by 'lacquer-type' drying, as in the example shown, the film develops some elasticity and the rebound height increases again. Whilst it is possible to derive a simple theory relating rebound height to film viscosity, based on ball momentum and energy losses, a number of important factors are neglected in this simple treatment. It is thus better to calibrate with Newtonian standard oils and use an empirical formula to relate the rebound height to viscosity. Amongst these factors is the hydrodynamic force, which will prevent the ball ever actually touching the substrate surface, as well as liquid elasticity effects at the short times of impact (a few tenths of a millisecond).

In spite of these drawbacks, the method is simple and easy to use. Slabs of glass, metal or even wood, together with a range of balls of different size or density and a graduated 'fall' tube for measuring the rebound height, are the only apparatus

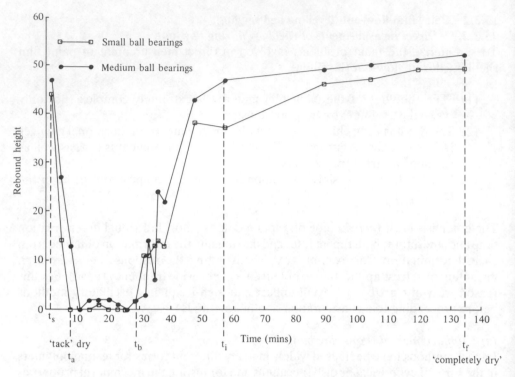

Fig. 15.2 — Time vs rebound height for a 200 μm refinish paint film on glass.

required. By temperature programming a metal slab it ought to be possible to use the technique to study the curing of thermoset systems in a similar fashion to that used by Gordon & Grieveson [38] for polymers. By contrast with the rolling-ball technique (to be described below), this technique does have the capability of measuring the development of film elasticity during curing.

(b) High-frequency (impedance methods)
The mechanical impedance of an elastic shear wave propagating through a medium is changed by the presence of a viscoelastic layer at the surface of the medium. If the elastic wave is completely damped in this layer, the change in characteristic impedance can be related to the rheological parameters of the layer material. This is difficult to achieve with many paint films, but nevertheless, the method can be used to follow changes in the paint-film rheology during drying and curing.

In practice, pulses of high-frequency oscillations are generated by means of a suitably-excited piezoelectric crystal attached to the support medium. After propagation through the support, the attenuated pulses are again transformed into electrical signals by a piezoelectric crystal attached to the support. The phase angle and attenuation of the received pulses are measured, and often changes in the values of either, or both, of these quantities are used to compare rate changes in drying and curing of different paint films. Either the pulses are directed at the surface of the support at a shallow angle, and the reflected pulses detected by a separate receiver crystal, as in the apparatus (Fig. 15.3) of Myers [11, 39–42], or the train of pulses is reflected normally from a face of the support along the same path so that it can be detected by the same crystal as in the torsional quartz rod method of Mewis [43]. In either case, the paint to be studied is coated onto the support.

Although it is not explicitly stated, the present author has the distinct impression that these methods are limited by the sensitivity of the equipment to a limited range of viscoelastic parameter values, and that, therefore, in many paint systems, the technique can be applied to measure film properties during only a limited part of the total drying/curing process of the paint film. Furthermore, the adhesion of the drying paint film to the support material can be expected to have a strong (and possibly, unknown) effect on the results [42]. Myers [11] used frequencies of 2–100 MHz, gated as 4 μs pulses, and measured attenutation from the signals collected as a series of exponentially decaying echoes. Mewis [43] used somewhat lower frequencies around 100 kHz for his torsional-rod method.

(c)   Rolling-ball method
The rolling-ball method is, like the bouncing-ball method, simple in conception and execution. A coated panel is inclined at a convenient angle, and the time taken for a small steel ball to roll over a measured distance on the coated surface is recorded as a function of drying/curing time. Alternatively, the distance rolled by the ball as a function of time is recorded. The method was introduced by Wolff & Zeidler [44], who used it to study the effect of different solvents and plasticisers on the viscosity of nitrocellulose finishes during drying. Quach and Hansen [45] used it to correlate viscosity with surface flow-out in emulsion paints; Taylor and Foster [46] used it to study the reactivity of stoving enamels in the temperature range 100–140°C and Göring *et al.* [47] used it to study viscosity changes in high solids and electrodeposited paints as well as fillers. The latter authors [47] have derived a simple theory, based on force balance considerations, which the present author has found to be inadequate to explain results.

Recently, van der Berg & de Vries [48] from the TNO Institute in the Netherlands have described an ingenious automated version of this technique. The coated panel under study is mounted on a revolving inclined table mounted in a temperature-controlled box. A sphere is placed on the edge of the coated panel, and this is illuminated by a light source which also falls on an array of photocells. The output from these photocells is used to control the table rotation speed accurately, so that the ball remains stationary in relation to the light beam. The authors have improved the theory without giving details, but admit that factors like variations of surface tension and consequent wetting of the ball, paint pick-up during rolling, and the flow pattern around the sphere (bow wave and wake) have not been properly treated. Even so, the results obtained are impressive, although they relate only quantitatively to viscosity changes.

Again, parallel work on rolling friction on solid polymers as well as mechanical test methods for hardness of paint films, such as the Sward Rocker (see Chapter 16), would give the impression that elasticity values could also be derived as with the bouncing-ball technique, but so far this has not been done. Unlike the bouncing-ball, this would require a modification of the experimental technique.

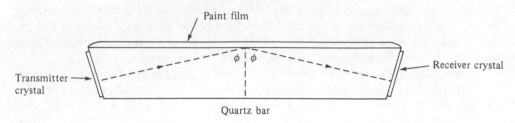

Fig. 15.3 — Myers' impedance technique.

(d)   ICI (relaxation) low-shear viscometer
This instrument was designed to simulate the conditions of high shear rate, followed by flow under low (decreasing) stress forces, which obtain when, for example, paint is applied by a brush. A description of the operating principles of the instrument has been given by Colclough, Smith & Wright [48]. The theory of this instrument and its variants has been given by Strivens [49].

The principle of the instrument is illustrated in Fig. 15.4. It is based on a torsion pendulum. If the pendulum is twisted through a small angle and released, the twist in the coil spring suspension unwinds, driving the bob of the pendulum back towards its initial (equilibrium) position. The inertia of the bob now carries the bob past its initial position, re-twisting the spring in an equal and opposite sense. The bob motion then reverses, and the back-and-forth oscillation continues *ad infinitum* in the absence of energy-dissipative processes (Fig. 15.4b) with an amplitude and frequency of oscillation determined by the mass of the bob (inertia) and the material elasticity modulus of the spring (and its physical dimensions). In a real situation, energy losses always occur, therefore the oscillation amplitude decreases (decays) with time, i.e. the oscillation is damped (Fig. 15.4c). If the bob is placed in contact with a viscous liquid during its oscillations, the oscillation will be still further damped. If the viscosity of the damping liquid is increased, the damping will be increased still further until, even at quite low viscosity, the bob no longer oscillates and the deflection decreases smoothly with time from its first maximum deflection position towards its initial (equilibrium) position. The system is said to be over-damped (Fig. 15.4d), and it is this condition that is used in the instrument.

In the instrument, the liquid (paint sample) is held between parallel plates (S in Fig. 15.5), the lower one being fixed and thermostatted, the upper one with its supporting rod (F) and measurement vane (L) forming the bob of the pendulum system. The supporting rod passes through an air bearing (D), which locates it and prevents any motion except the required torsional movement with minimal frictional loss. The measurement vane forms the moving part of an air condenser, which forms part of a tuned quartz oscillator circuit, so that the slightest movement of the vane produces an off-balance voltage, which may be used to measure deflection. At the top of the rod is the coil spring (K), whose outer end is secured to the frame of the instrument head. The whole assembly may be raised and lowered on a vertical column (C) by means of the spring-loaded arm at the side of the instrument.

In operation, in the initial state the upper plate is deflected electromechanically through an angle of 5°, the sample is inserted, allowed to equilibrate, and then the

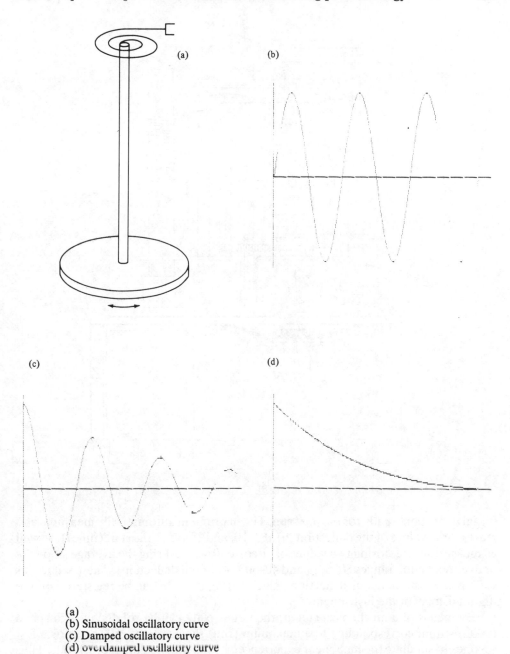

(a)
(b) Sinusoidal oscillatory curve
(c) Damped oscillatory curve
(d) overdamped oscillatory curve
Fig. 15.4 — The principle of the low-shear viscometer (LSV).

start button pressed. This releases the plate, allowing it to move back towards its zero angular deflection position under the influence of the spring and against the resistance offered by the sample. The resulting deflection–time curve can be recorded over a period of about 30 s before the plate is automatically reset to its

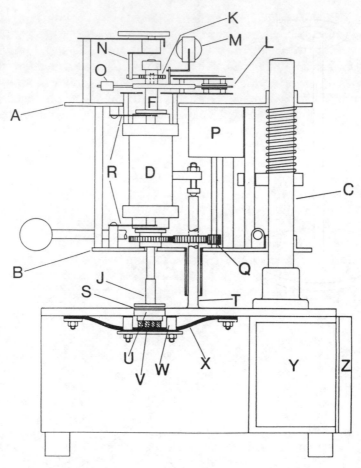

Fig. 15.5

initial deflection, ready for the next run. The instrument automatically measures and stores three values of the deflection (θ) at 3, 10, and 25 s after the start (time t), as well as measuring and storing two values of viscosity (by measuring the average slope $d\theta/dt$ over deflection ranges 80–60% and 40–30% of initial deflection). These values, as well as the instantaneous deflection values during the run, can be registered on the digital display of the instrument.

By means of a mechanical clutch, the upper plate can be rotated at high speed (900 rev/min, corresponding to a maximum (rim) shear rate of 2500 s^{-1}) for a few seconds to simulate the high shear experienced by the sample during brush-out. Thus families of 'relaxation' curves can be generated from the sample at rest, immediately after shear and at timed intervals afterwards to study structure recovery in thixotropic paints. As will be seen from Fig. 15.6, these families of curves correlate very well with the results of subjective assessments of paint flow-out, based on large-scale brush-outs by skilled painters. This is probably due to thixotropic structure recovery having a stronger influence on the initial stages of brushmark flow-out than solvent loss, at least in the paints tested.

By using the theory derived by Orchard for brushmark flow-out and the equation of motion of the torsional oscillator, Colclough *et al.* [48] were able to choose suitable coil springs to give the right level of initial stress (3–5 Pa). Furthermore, as the spring unwinds and the upper plate moves towards zero deflection, the driving force is reduced, i.e. the stress exerted on the sample decreases, in an analogous fashion to what happens in practice, where the driving forces produced by surface tension are proportional to the wavelength and amplitude of the brushmarks, and so, as the brushmarks flow out, the driving force is reduced.

Two variant instruments have also been developed [49]. In one of these, the coil spring is wound up at controlled speed through an angle of 180°C, thus applying to the sample a stress which increases linearly with time. During this wind-up process, the instantaneous upper plate deflection is recorded as a function of time on a chart recorder. A Newtonian liquid shows a deflection which varies as the square of the time from start. However, if there is weak elastic structure in the paint, the initial deflection–time plot will be linear rather than parabolic, with slope proportional to elasticity; then, at some critical stress, the structure starts to break down and the deflection–time plot changes towards that of the Newtonian liquid. Thus with this instrument the kinetics of structure breakdown can be studied.

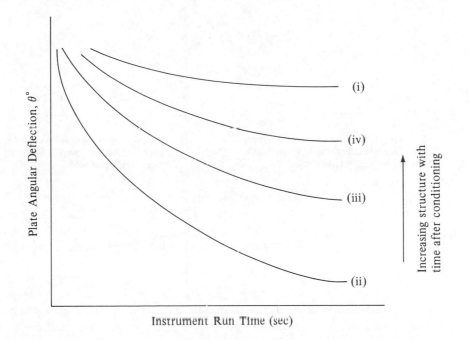

(i)   Initial condition of paint, before subjecting the sample to a predetermined shearing regime.
(ii)   Sample sheared at 2500 s^{-1} for 4 sec; then measurement made immediately upon cessation of shear.
(iii)   Sample sheared at 2500 s^{-1} for 4 sec; 1 min allowed to elapse before making measurement.
(iv)   Sample sheared at 2500 s^{-1} for 4 sec; 5 min allowed to elapse before making measurement.

Fig. 15.6(a) — Typical pattern of results, indicating method of use of the low shear viscometer (LSV).

In the second variant, the coil spring is oscillated by a reciprocating drive of variable amplitude and speed. By this means, an oscillatory (sine-wave) stress of variable frequency and amplitude may be applied to the sample. By measuring the phase angle and the amplitude of the upper plate motion in relation to the drive motion, it is possible to derive the dynamic viscosity and elasticity modulus of the sample, as a function of frequency (0.02–3 Hz range) and stress amplitude (four discrete values 0.6, 1.5, 5, and 10 Pa). Studies of the viscoelasticity of concentrated suspensions, using this instrument, have been published by Strivens [50, 51], and a detailed description of the instrument [52] is in preparation. Because this instrument is fitted with the high-speed shear facility as described above for the relaxation LSV instrument, the kinetics of structure build-up can be followed in more detail than with that instrument, and, as both viscosity and elasticity may be measured at frequent intervals as a function of reformation time, there is potentially more information available than is the case for the relaxation instrument. The relaxation instrument is easy and simple to use and suitable as a development or quality control instrument, whilst the oscillatory instrument is meant to be a research instrument.

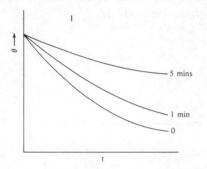

I   Good flow; moderate recovery rate; good flow-out of brush marks; good 'sag' resistance.

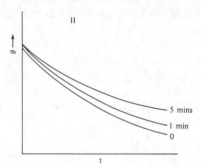

II   Excellent flow; slow recovery rate; excellent-brush mark flow-out; poor 'sag' resistance.

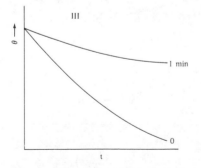

III   Good breakdown with rapid recovery; after ~30 s brush marks cease to flow-out, therefore only moderate performance, but good 'sag' resistance.

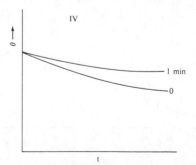

IV   Little breakdown; poor flow; poor brush marking; excellent 'sag' resistance.

Fig. 15.6(b) — Typical curves for paints having different brushing characteristics and flow-out.

(e) Sled method

By sandwiching the paint film between its substrate and a thin plate (say, a small microscope slide), the simplest shear geometry is obtained. Attaching a weight to the slide via a pulley, and recording the motion of the slide as a function of time, allows determination of the paint film viscosity.

Such a simple technique has been described by Kornum [53], amongst others. Practical difficulties with maintenance of shear geometry and edge effects, which can be foreseen, are discussed.

(f) Torsional pendulum (torsional braid)

Whilst these methods can be used for studying early stages of curing of paint, while the paint is still liquid or semi-solid they are more properly described in the next chapter, as being more suitable for studying fully cured paint films.

(g) Other methods

Within the limitations of this chapter, it has been possible to describe only a few of the vast number of techniques available. Inevitably, the author's choice has been dictated by personal experience working within a desire to concentrate on methods that yield absolute values of rheological parameters, and, apart from those, to concentrate on methods commonly used in the industry.

A more comprehensive survey of viscometers of all types used in the paint industry has been given by McGuigan [28].

15.3.2.2 Direct measurement of film flow-out

As an adjunct to the measurement of film rheology during flow-out, the direct measurement of the flow-out or other flow phenomena, such as sagging, has also been attempted. To an extent, this has been in an effort to confirm, or otherwise, predictions based on the observed rheology.

If an even coating of paint can be spread on a flat substrate, and, by some means, a series of regular striations can be produced on the coating surface, resembling brushmarks, but rather more regular, then it should be possible to observe the course of flow-out directly with drying time and to make measurements of amplitude and wavelength directly as a function of time. Such measurements have been made by using light interference methods, by Wapler [54]; whilst Kheshgi & Scriven [55] have used Moiré topography. Wapler has compared his results with the theoretical predictions of Biermann [56]. Kheshgi & Scriven, in addition to studying levelling, has also used their technique to study surface profiles in liquid flowing down a slope, liquid curtains falling vertically from a slot, and disturbances in surface profiles due to encounters with surface-active particles. Whilst less sensitive than interference methods, Moiré fringe techniques are more flexible in that they cover a wide range of displacements (from several micrometres to several centimetres) over which their sensitivity is good.

By contrast, there are a number of technological methods for measuring both levelling and gross flow defects, such as sagging. These are comprehensively surveyed by McGuigan [28]. The usual pattern of sag test applicator consists of a bar-type spreader whose trailing edge is cut in a series of slots of different depth. Moving the spreader across a glass plate leaves a series of parallel stripes of paint of different thickness. The test plate is then placed in a rack usually inclined at an angle of 60°, and the edge profile is observed visually during drying. Sagging resistance is assessed by the minimum film thickness, whose edge profile deforms significantly during the period of observation. Such an instrument has been described by Schaeffer [57] amongst others. Again, Biermann [56] has attempted theoretically to predict the onset of sagging, curtaining, and similar phenomena.

15.4 PAINT RHEOLOGY DURING MANUFACTURE AND STORAGE

Four areas, where rheology control or measurement can be useful in paint manufacture, can be listed:

(a) milling of pigments (preparation of millbases);
(b) transport of paint or intermediates used in paint production through pipelines, etc;
(c) mixing operations; and
(d) storage.

15.4.1 Pigment millbase production

Pigments are delivered to the paint manufacturer in the form of (more or less) dry powders which, for various reasons, contain numbers of pigment particle agglormerates and aggregates. The objectives of the pigment-dispersion process (ball milling, triple rolling, sand milling, etc.) are to:

(a) break down pigment particle agglomerates and aggregates, and
(b) to ensure complete wetting of the pigment surface by the paint medium, in particular by surface-active agents present in the medium, which will prevent subsequent flocculation of the pigment.

The development of the required optical properties of the final paint film will depend strongly on the quality of the pigment dispersion, so it is important to optimize the conditions of pigment dispersion.

Whilst the basic science of the pigment-dispersion process is well understood, the application of such scientific principles to optimize dispersion conditions is often, in practice, rather more difficult to achieve. Parfitt [58] has reviewed this basic science, whilst Kaluza [59] has reviewed the causes and effects of pigment flocculation in paints. Tsutsui & Ikeda [60] have reviewed methods of evaluating particle size and distribution and degree of pigment dispersion. Patton [61] has dealt with the practical aspects of pigment millbase formulation and assessment.

Starting from the basics of dispersion rheology, a number of principles can be derived. Firstly, the effectiveness of the milling will depend on the amount of (mechanical) energy dissipated. To maximize this dissipation, the overall viscosity of the dry pigment/medium mixture should be as high as possible, i.e. the pigment content must be high. The limit to this is an effective volume fraction of around 0.64 where critical packing of pigment particles occurs and the mixture rapidly becomes like a solid. In practice, the onset of dilatancy, with accompanying solid-like behaviour and fracture, may lower this limit still further, as will the presence of anisotropic pigment particles, agglomerates, or aggregates.

The initial effective volume occupied by the pigment particles will be larger than the total effective volume of the individual pigment particles, owing to volume occluded within the pigment particle aggregates and agglomerates. As these are broken down by the dispersion process, the pigment effective volume is reduced, so the viscosity of the pigment/medium mix falls. Also, some of the mechanical energy is dissipated as heat, so unless adequate cooling is used, this reduces the viscosity still further. The net result of this viscosity decrease is to reduce the dissipation of mechanical energy in the dispersion, and thus to reduce the effectiveness of the dispersion process. Thus the attainment of constant viscosity, as a function of dispersion time, could be used as an indicator of the end of the effective part of the dispersion process. However, changes in particle size and size distribution could still take place without significantly altering the effective volume and, hence the dispersion rheology. To measure such changes, methods other than rheology are more appropriate; for example, fineness of grind gauges, as discussed by Tsutsui & Ikeda [60].

A number of authors have used rheological techniques to study aspects of the pigment-dispersion process. Oesterle [62] has used such techniques to study the course of pigment dispersion, as well as the effectiveness of various dispersion machines. McKay [63] has studied the effectiveness of various dispersants in organic pigment dispersion processes; Heertjes & Smits [64] have used rheological measurements to study the dispersion of titanium dioxide in alkyd media, particularly in relation to alkyd molecular weight and the effect of the presence of saturated fatty acid molecules. Strivens [17], Mewis & Strivens [65], and Zosel [66] have studied the viscoelastic properties of pigment dispersions, the former during the course of ballmilling, the latter to assess the nature of the pigment particle interactions and, by implication, the effectiveness of the dispersant and the dispersion process used. Strivens used an oscillatory stress amplitude controlled rheometer (see section 15.3.2.1), and Zosel used his own design of creep apparatus.

15.4.2   Pipe flow
The early literature on the flow of pastes and paints in pipelines has been reviewed by Weltmann [67]. Patton [68] has also devoted a chapter to the subject.

In general, the problem is to calculate the pressure required to move a given paint or intermediate through a pipe at a required flow rate. Although, either in laminar or turbulent flow, this can be done reasonably accurately for Newtonian liquids, non-Newtonian liquids present more of a problem. By measuring apparent viscosity as a function of shear rate over a relevant range of values, using a rotational viscometer and fitting some empirical viscosity-shear rate equation such as the Casson or Bingham equation, approximate predictions of required pressure can be made which may suffice for the purpose of engineering design. However, time-dependency effects (thixotropy) may vitiate some of these calculations, particularly if flow rates are low. Also, strong interactions in the material, giving rise to strong elasticity, as well as dilatancy, due to high-disperse solids, may also lead to unacceptably high initial pressure values being required to start flow of the material. In this case, measurements with a pipeline rheometer (simiar to a capillary viscometer, but with a wider bore) may be more useful.

15.4.3 Mixing
In this area, rheological characterization of the materials to be mixed has an important role to play in designing plant and studying the efficiency of mixing processes. Something of this role may be judged by consulting some of the more recent general textbooks on this subject, for example, that of Oldshue [69].

15.4.4 Storage
The measurement of changes in rheological structure during storage presents a particularly delicate problem, as exemplified by thixotropic paints. If a can of such a paint is opened and a sample withdrawn and placed in a rheometer, some of the structure will be broken down (cf. the dipping of a paint brush into the can). So what one is trying to measure is being altered by the sampling process, and it would thus be desirable to measure the structure of the paint in its container.

An instrument to achieve this, known as OSCAR (acronym for oscillating can rheometer), was developed by ICI (Paints Division) some years ago [70]. Its principle of operation, briefly described by Strivens [49], is similar to that of the oscillating variant of the low-shear viscometer (LSV, discussed in section 15.3.2.1), also developed by ICI. A circular table on top of the instrument is caused to oscillate around its centre, by means of a reciprocating drive operating through a coil spring at a closely-controlled frequency (close to 10 Hz). The amplitude and phase angle of the table motion are measured in relation to that of the drive by a similar condenser system to that of the LSV. If a solid cylindrical can or container is placed on the table, the phase angle difference between the table and the drive is zero, whilst the ratio of the amplitudes has a certain value, dependent on the container mass. This is taken as unity on the instrument scale, and all subsequent measurements are referred to this value for a given container mass. When a can containing a viscoelastic material, such as the thixotropic paint under test, is placed on the table, the amplitude ratio and the phase angle difference alter, because energy is now being dissipated, owing to the viscoelastic property of the can contents. These alterations can now be used to calculate the dynamic viscosity and elasticity of the paint. In practice, because of the complexity of the mathematics of the cylindrical geometry used, this is done numerically, and a graph is produced to allow the results to be read off directly from the instrument's readings. Variations in can weights can be accommodated by means of a potentiometer dial on the front of the instrument. Because the shear wave dies away quickly from the can side, and, as the maximum deflection angle is 5° for a 250 ml can of diameter 9 cm, the maximum shear strain is small and the bulk of the sample is virtually undisturbed.

Measurements can be made very rapidly. Storage samples in cans are weighed and placed on the oscillating table, the weight is dialled on the potentiometer, and normally within half a minute, the digital display of the instrument shows steady readings of phase angle difference and amplitude ratio. Using these readings, values of dynamic viscosity and elasticity modulus may be read off from the standard charts provided. Changes in rheological structure show as changes mainly in elasticity and, to a lesser extent, in viscosity. The instrument may also be used to detect hard settlement, as well as to follow the kinetics of structure formation in bulk gelation processes.

REFERENCES

[1] Richardson, E. G., in: Hermans, J. J. (ed.) *Flow properties of disperse systems.* North-Holland Ch. VI pp. 266–98, (1953), Green, H. L. *Loc. cit.* [1] Ch. VII pp. 299–322.

[2] Schurz, J. *EUCEPA 20th Conf. Budapest,* Paper 11/13 (1982) *World Surface Coatings Abstr.,* 83/8306.

[3] Glass, J. E. *J. Coatings Technol.* **50** (641) 56 (1978).

[4] Patton, T. C. *Paint flow and pigment dispersion:* A rheological approach to coating and ink technology. 2nd edn. Wiley-Interscience pp. 365–7 (1979).

[5] Kuge, Y. *J. Coatings Technol.* **55** (701) 59 (1983).

[6] Smith, N. D. P., Orchard, S. E., & Rhind-Tutt, A. J. *J. Oil & Col. Chem. Assocn.* **44** 618–33, part. pp. 621–3 (1961) Patton, T. C. *Loc. cit.* [4], pp. 553.

[7] Pearson, J. R. A. *J. Fluid Mech.,* **7** 481 (1960).

[8] Savage, M. D. *J. Fluid Mechanics* **80** 743 (1977).

[9] Glass, J. E. *J. Oil Col. Chem. Assocn.* **58** 169 (1975).

[10] Dodge, J. S. *J. Paint Technol.* **44** (564) 72 (1972).

[11] Myers, R. R. *J. Polymer Sci., Pt. C,* (35), 3 (1971).

[12] Trevenna, D. H. *J. Phys., D: Appl. Phys.,* **17** 2139 (1984).

[13] Walters, K. *Rheometry* Chapman & Hall, pp. 219 (1975).

[14] (a) Keller, (b) *Loc. cit.* [13] p. 233.

[15] Orchard, S. E. *J. Appl. Sci. Res.,* A11, 451 (1962).

[16] Wu, S. *ACS, Division of Org. Coatings & Plastics Chem., papers,* **37** (2) 315, 323 (1977).

[17] Strivens, T. A. (Unpublished results).

[18] Amari, T. & Watanabe, K. *Polymer Engineering Rev.* **3** (2–4), 277 (1983).

[19] Overdiep, W. S., in: Spalding, D. B. (ed). *Physicochemical hydrodynamics. V. G. Levich Festschrift.* Vol. II. Advance Publications Ltd, 683 (1985).

[20] See, for example, Wojtkowiak, J. J. *J. Coatings Technol.* **51** (658), 111 (1980).

[21] Anand, J. N. & Balwinski, R. Z. *J. Coll. & Interfac. Sci.,* **31** (2), 196 (1969). Anand, J. N. *Ibid.,* 203, Anand, J. N. & Karam, H. J. *Ibid.,* 208

[22] Higgins, B. G. & Scriven, L. E. *Ind. Eng. Chem., Fundamentals* **18**(3), 208 (1979).

[23] Matsumoto, T., Segawa, Y., Warashina, Y., & Onogi, S. *Trans. Soc. Rheol.* **17**(1), 47 (1970).

[24] Kornum, L. O. & Raaschou Nielsen, H. K. *Progress in Organic Coatings* **8** 275 (1980).

[25] McElvie, A. N. *Progress in Org. Coatings* **6** 49 (1978).

[26] Monk, C. J. H. *J. Oil & Col. Chem. Assocn.* **49** 543 (1965).

[27] Manufactured by Research Equipment (London) Ltd, 64 Wellington Rd. Hampton Hill, Middlesex, UK.

[28] McGuigan, J. P. Viscosity and Consistency Ch. 3.2. pp. 181–212. In: Sward, G.

G. (ed.) *Paint testing manual (ASTM Special Technical Publication* No. 500). 13th ed. American Society for Testing of Materials (1972).
[29] Walton, A. J. *Paint Manufacture,* 47(5) 13–7 (1977) *Ibid.* [6], 16–22, 24.
[30] Association Francaise de Normalisation, NFT 30–029 (1980).
[31] *Urad pro Normalisaci* CSN67 3016 (1981).
[32] Snow, C. I. *Official Digest* (392) 907 (1957).
[33] Flom, D. G. *J. Appld. Physics* 31 306 (1960).
[34] Tillett, J. P. A. *Proc. Phys. Soc.* B67 677 (1954).
[35] Jenckel, E. & Klein, E. Z. *Naturf.* 7a 619 (1952).
[36] Pao, Y. H. *J. Appld. Phys.,* 26 1082 (1955).
[37] Strivens, T. A. (Unpublished results).
[38] Gordon, M. & Grieveson, B. M. *J. Polymer Sci.* 29 9 (1958).
[39] Myers, R. R. *Official Digest* 33 (439) 940 (1961).
[40] Myers, R. R. & Schultz, R. K. *Official Digest* 34 (451), 801 (1962).
[41] Myers, R. R. & Schultz, R. K. *J. Appld. Polymer Sci.* 8 755 (1964).
[42] Myers, R. R., Klimek, J., & Knauss, C. J. *J. paint Technol.,* 38 (500) 479 (1966).
[43] Mewis, J. *FATIPEC IX Congr. Brussels. Proc.* pt. 3 pp. 120–4 (1968).
[44] Wolff, H. & Zeidler, G. *Paint & Varnish Production Manager* 15(8) 7 (1936).
[45] Quach, A. & Hansen, C. M. *J. Paint Technol.,* 46 (592) 40 (1974).
[46] Taylor, R. & Foster, H. J. *J. Oil Col. Chem. Assocn.* 54 1030 (1971).
[47] Göring, W., Dingerdissen, N., & Hartmann, C. *Farbe und Lack* 83(4) 270 (1977).
[48] Colclough, M. L., Smith, N. D. P., & Wright, T. A. *J. Oil Col. Chem. Assocn.* 63 183 (1980).
[49] Strivens, T. A. *Quarterly Report of the Paint Research Assocn.* (79/4), 11 (1979).
[50] Strivens, T. A. *Colloid and Polymer Sci* 261, 74 (1983).
[51] Strivens, T. A. *Colloids and Surfaces* 18 395 (1986).
[52] Strivens, T. A. (In preparation).
[53] Kornum, L. O. *FATIPEC Congr. XIV Budapest Proc.* pp. 329–36 (1978).
[54] Wapler, D. *Farbe und Lack* 81(8) 717; (9) 822; (10) 924 (1975).
[55] Kheshgi, H. S. & Scriven, L. E. *Chem. Eng. Sci.* 38(4) 525 (1983).
[56] Biermann, M. *Rheol. Acta* 7(2) 138 (1968).
[57] Schaeffer, B. *Official Digest* 34 (453) 1110 (1962).
[58] Parfitt, G. D. *FATIPEC Congr. XIV Budapest Proc.* pp. 107–117 (1978).
[59] Kaluza, U. *Progress in Org. Coatings* 10 289 (1982).
[60] Tsutsui, K. & Ikeda, S. *Progress in Org. Coatings* 10 235 (1982).
[61] Patton, T. C. *Loc. cit.* ref [4], ch. 7–12.
[62] Oesterle, K. M. *FATIPEC Congr. XIV Budapest. Proc.* pp. 329–36 (1978).
[63] McKay, R. B. *FATIPEC Congr. XIII Cannes. Proc.* pp. 428–34 (1976).
[64] Heertjes, P. M. & Smits, C. I. *Powder Technology* 17 197 (1977).
[65] Mewis, J. & Strivens, T. A. Paper given at *8th Int. Congr. of Chem. Eng., Chem. Equipment & Apparatus (CHISA),* Prague (1984).
[66] Zosel, A. *Rheol. Acta* 21 72 (1982).
[67] Weltmann, R. N. In: Eirich, F. R. (ed.) *Rheology: theory and applications,* Vol. III Ch. 6. Rheology of pastes and paints. pp. 236–40 (1960).

[68] Patton, T. C. *Loc. cit.* ref. [4]. Ch. 6. pp. 156–183.

[69] Oldshue, J. Y. *Fluid mixing technology. Chem. Eng.* McGraw–Hill Publications Co. (1983).

[70] This instrument is currently being commercially developed by Ravenhead Designs Ltd., Gregge St, Heywood, Lancs., UK.

16

Mechanical properties of paints and coatings

T. A. Strivens

16.1 INTRODUCTION

The mechanical properties of paint and coatings are of paramount importance in maintaining the important protective, as well as to a lesser extent, the decorative functions of such treatments during their service life. Paint films are subject to a great variety of mechanical forces and deformations. Thus they may suffer large forces concentrated over a small surface area for very short times, as in the impact of stones, gravel, etc. on car body paints, or they may suffer a succession of slow cycles of deformation, as happens to decorative gloss paints on wood, for example, on house window frames, as the wood expands and contracts in response to changes in atmospheric moisture and temperature. Such forces and deformations can be large: of the order of giga pascals per unit area in the impact case, or 10–15% strain in the case of wood expansion and contraction (such deformation is also anisotropic, owing to the grain structure of the wood). So it is the ultimate mechanical properties of paint films, which is of most practical importance, i.e. the stress or strain that leads to plastic yield (irreversible deformation) or failure by cracking of the film.

Not only do paint films suffer a wide variety of mechanical stresses and strains during their service life, but also their mechanical properties change during it. This affects the ultimate mechanical properties such as yield and fracture, and in the end determines how long a coating film can preserve its physical integrity and can fulfil its protective role satisfactorily. The continual exposure to air and water condensation (dew) on the paint surface leads to steady leaching of low-molecular-weight species, such as retained solvent, plasticizer, or low-molecular-weight polymeric species as well as degradative reaction products which might otherwise soften the coating and increase its resistance to brittle failure (cracking). Equally, exposure to oxygen in the air and to light (particularly the ultraviolet radiation content) can lead to various photolytic reactions, which may generate free radicals and peroxides which can increase the degree of crosslinking of the film and, consequently, its brittleness. Ultimately, the coating film fails by cracking, either in the film matrix or, less frequently, at an interface, i.e. adhesive failure. The permeation of water into the film may be beneficial because it often 'softens' or plasticizes the film. However, if this is combined with permeation of oxygen or anions, this may lead to mechanical failure by blistering or accumulation of solid corrosion products at a coating/metal interface.

Most paints and coatings are based on organic polymers. They are thus viscoelastic and highly non-linear in their response to mechanical stress or strain. Not only this, but the mechanical properties of the matrix polymer is substantially modified by the presence of dispersed polymer particles, as well as phase-separated (insoluble or incompatible) components of the polymer matrix which separate during application or curing, pigment and extender particles, etc. The role played by pigment or polymer particles in modifying the mechanical properties of the coating is analogous to the role played by fillers and disperse polymer phases in determining the mechanical properties of bulk polymer composite systems. However, there is little evidence that this concept has been grasped, and that any attempt has been made to apply the considerable body of ideas and theory developed for polymer composites to interpreting the mechanical properties of paint films.

In addition, most coating systems are multilayered. Although the polymer matrix may be constant, the pigment nature and content, etc., will usually vary from layer to layer. Not only that, but the processes of solvent loss and spatially non-uniform crosslinking during cure may lead to gradients or heterogeneities in mechanical properties, both through the thickness and across the surface area of the film. Crosslinking and solvent loss processes usually lead to shrinkage of the polymer matrix with a consequent build-up in internal stress within the film. The presence of such internal stresses will modify the mechanical properties of the coating on its substrate. This will in turn affect the decision as to whether to test attached or detached coating samples; and indeed, in extreme cases, could totally invalidate the results of the latter.

Enough will have been said to give an idea of the complexity and importance of coating mechanical properties as well as the importance of monitoring such changes during actual or simulated environmental exposure tests, such as accelerated weathering.

The chapter will be organized as follows. After a brief general survey of the viscoelasticity and ultimate mechanical properties of polymers, consideration will be given to methods of determining such properties, particularly in relation to coating specimens attached to their substrates. After a survey of practical mechanical test methods for coatings and their interpretation in terms of more fundamental properties, the chapter will close with a section on the application of acoustic emission to monitoring changes in the mechanical properties during weathering.

16.2   VISCOELASTIC PROPERTIES OF POLYMERS

In Chapters 14 and 15 of this book, the concepts and experimental techniques for measuring viscoelastic liquids have been outlined, and reference should be made to the first of those chapters for an explanation of the rheological terms used here. Many accounts of polymer viscoelasticity are available; books by Ferry [1], Nielsen [2] and McCrum *et al.* [3], as well as review articles by Graessley [4], Hedvig [5], etc. These should be consulted for more detailed accounts than will be given here.

The characteristic property of most polymers of sufficiently high molecular weight or degree of crosslinking is that they are elastic solids at room temperature. If a constant mechanical stress (creep experiment) or mechanical strain (stress relaxation experiment) is applied to a specimen of a viscoelastic solid polymer, the response is predominantly elastic. In other words, there is insufficient time for the polymer molecules or their component units to move in relation to each other. Consequently they move by change of configuration in response to the applied mechanical disturbance, thus stretching and changing the original bond lengths and angles. When the mechanical disturbance is removed, this stretching and distortion result in the molecules recovering their original configurations. Thus mechanical energy has been stored and released (elastic response). A similar oscillatory energy storage and release process occurs if an oscillatory or sine wave (dynamic experiment) mechanical stress or strain is applied to the specimen, provided that the frequency is high enough (i.e. the timescale of the oscillation is short enough).

Now, if the timescale is lengthened (or equivalently, the frequency of an oscillatory experiment is reduced), there is now sufficient time for a significant number of the polymer molecules to rearrange their positions in relation to each other, and for component units to achieve new equilibrium bond length and angle positions. Thus, when the mechanical disturbance is removed, there is no driving force to return the polymer molecules from their new to their original positions, consequently a significant amount of energy has been dissipated. Although there may not have been a change in sample shape, there still will have been irreversible deformation on a microscopic scale. Continuing this process will lead to irreversible sample deformation. Thus, even a sheet of glass or a glass fibre will bend or stretch irreversibly under a load, given sufficient time.

If the sample temperature is increased, then these irreversible rearrangements are facilitated by the extra translational, rotational, and vibrational energies possessed by the polymer molecules. In gross terms, the sample changes as temperature is raised from a glassy material through a rubbery intermediate state (particularly if the polymer molecules are linked by chemical bonds (crosslinking)) through to a viscous liquid. An equivalent change from predominantly energy storage response to energy dissipative response is observed, if the frequency is lowered or the time scale is increased. This equivalence was recognized and formulated as the Williams–Landel–Ferry (WLF) time–temperature superposition principle [6]. This WLF principle has great experimental importance. Most polymers have wide molecular weight distributions, and, as a result, a wide range of characteristic time scales for rearrangements. The transition in mechanical properties can thus cover many decades of frequency or time. As many instruments for determining mechanical properties cover only comparatively small ranges of frequency or time, the proper characterization of such transitions in mechanical properties (the 'glass transition temperature' or T_g) presents problems, unless a number of instruments with different frequency or time ranges are available. However, if one instrument is used and the sample measured at a number of different temperatures, the data can be reduced, using the WLF principle, to one standard temperature so as to cover a much wider frequency range than the instrument range. Of course, particularly in the case of coatings, due consideration must be given to the effect of temperature alone on the specimens (solvent or plasticizer loss, additional crosslinking, or thermal degradation) before applying such a procedure to determine T_g. This determination, and particularly its change during environmental exposure, is of great practical importance for assessing and predicting coatings performance.

If sufficient mechanical stress or strain is applied over a time scale where the response is predominantly elastic, these irreversible rearrangements may be forced to take place. The behaviour of the sample is now non-linear, the measured response is no longer proportional to the mechanical input, and, taken far enough, mechanical yield and fracture on a macroscopic scale are observed. Thus the ultimate properties of yield and brittle fracture are also time- (frequency) and temperature-dependent. Put another way, as most ultimate property determinations are made by deforming the specimen at a controlled strain rate and measuring the instantaneous force as a function of strain, the results of such measurements are a function of both strain rate and temperature.

The ease or difficulty of such irreversible rearrangements are determined by a number of factors. The chain flexibility of the polymer molecules is of great importance, thus a flexible hydrocarbon polymer such as polyisobutylene will rearrange more easily than a stiff polymer chain with bulky side chains or rigid ring structures in the side chains or main chain, e.g. polystyrene, celluloses, etc.

Other important factors are the presence, number, and flexibility–length relationships of chemical bonded linking molecules between the polymer molecules (thus a vulcanized rubber is 'harder' than the corresponding unvulcanized rubber). Regular arrangements of polymer chains in crystalline materials increases the Tg considerably, as does the presence of other intermolecular forces due to hydrogen bonding, ionic interactions, etc. The presence of low-molecular-weight species, such as plasticizer molecules, or bulky flexible side chains, as in a alkyl methacrylates, increases the volume between polymer molecules, thus making rearrangement easier and lowering the Tg. Also, the polymer molecular weight plays an important role in determining Tg. Put crudely, the longer the polymer chains, the more entangled they become and the higher the Tg. For most polymers the Tg does not increase much over a molecular weight of 20 000. Many coatings polymers have been formulated precisely in the range 5000 and 20 000 in order to achieve low-viscosity solutions suitable for optimum application processes, so the increase of Tg, as the coating dries or crosslinks, is crucial to its performance.

The presentation of viscoelastic data for a solid is analogous to that for a viscoelastic liquid already described in Chapter 14. Thus an apparent modulus (or compliance) is plotted as a function of time with temperature as a parameter for controlled stress or strain experiments. Equally, the real and imaginary components of the modulus (usually referred to as the elasticity and the loss moduli respectively) are plotted as a function of frequency in dynamic experiments, again with temperature as a parameter. However, many commerical instruments operate at a single fixed or approximately constant frequency, and, in this case, it is more usual to plot the real component of the modulus and the ratio of the imaginary to the real components of the modulus (the 'loss tangent', written as $\tan \delta$) as a function of temperature. Most of these instruments operate by some form of temperature scan, and Tg is determined by locating the temperature at which $\tan \delta$ reaches a peak value. However, again, perception of the Tg value must take due account of the effects of temperature scan rate.

In other types of dynamic experiments, in particular a.c. impedance and dielectric measurements, much use is made of Argand diagrams, where the imaginary component is plotted as a function of the real component. Smooth or distorted semicircular or circular arc plots are obtained with points along the plot corresponding to each frequency value used. The distortion of the arc or the limitation of the arc to less than 180° (part of a semicircle) is related to the presence of more than one characteristic time or the existence of a distribution of characteristic times. The perception of these parameters is perhaps easier on the Argand diagram (known as a Cole–Cole plot in dielectric, and a Nyquist plot in impedance, work) than in the more conventional form of presentation. No doubt the generally more limited frequency range available for dynamic mechanical measurements has restricted its use. However, as Havriliak & Negami [7, 8] have demonstrated, it can be very powerful, particularly when comparisons are required with other dynamic data on polymers, in particular, dielectric data. More recently, Laout and co-workers [9, 10, 11] have made extensive use of Argand diagrams in presenting their dynamic mechanical measurements for drying alkyd and emulsion paints. They have fitted a simple equation to these results and have shown by comparison with the results of electron microscopy and other physical techniques, that the equation parameters are related to the degree of crosslinking and integration of the films respectively. Toussaint & Dhont [12] have also recently reviewed the theory of these Argand type plots in relation to dynamic mechanical measurements on paint films.

Examples of these modes of presentation are given as idealized curves in Fig. 16.1(a) and (b). Fig. 16.2 illustrates some experimental results, derived from some of the author's own unpublished data on a polyester film. (δ'' and δ' are related to, but not identical with, G'' and G'.)

16.3   ULTIMATE MECHANICAL PROPERTIES OF POLYMERS

Again, as with the linear viscoelastic behaviour of polymers, only a very brief introduction can be given, and more detailed accounts must be sought in books and review articles. In the context of ultimate properties of polymers, failure of two principal types is considered, namely, ductile failure (yielding) or brittle failure (cracking). Fig. 16.3(a) and (b) show idealized stress–strain curves for the two types of failure.

If a force is applied for a short enough time to a sample, for example, by a blow or

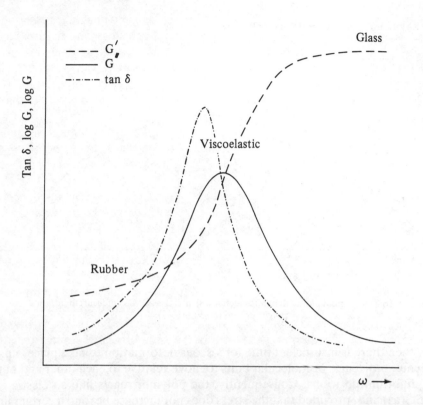

Fig. 16.1(a) — Idealized representation of dynamic mechanical properties of a solid (crosslinked) polymer.

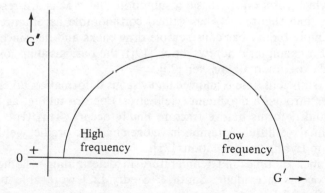

Fig. 16.1(b) — Idealized Argand diagram for mechanical properties of a viscoelastic solid.

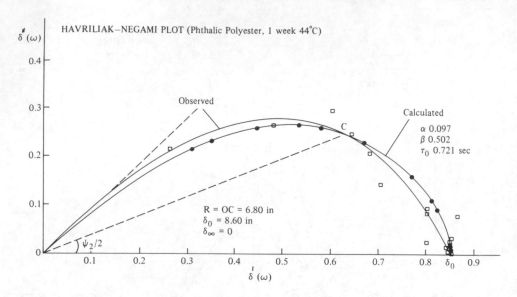

Fig. 16.2 — Argand plot of dynamic mechanical analysis (DMA) measurements on a polyester film (cf. Fig. 16.5).

impact, then there is insufficient time for viscous deformation to take place, i.e. for the polymer molecules or molecular units to move relative to each other and obtain new equilibrium positions. Consequently, the polymer reacts like an elastic glass (Fig. 16.3(a)) and, provided that the stress does not increase beyond a certain limit, brittle fracture does not take place. In principle, it is possible to calculate this limit (brittle strength) from a knowledge of bond strengths and intermolecular forces (see, for example, [17]. In practice, values several orders of magnitude lower than the theoretical are nearly always found. This is generally attributed to structural irregularities in the sample, which serve to concentrate the stress locally. Such structural irregularities include cracks and fissures as well as imbedded particles of foreign bodies or degraded material. The Griffith's theory [18] of the strength of brittle solids, which rests on the basic assumption that cracks are responsible for strengths lower than theoretical, has gained considerable acceptance. However, although this simple theory has considerable drawbacks and has since been much modified (see, for example, Andrews' book [19]), the basic assumption is undoubtedly sound and experimentally proven [20].

If, however, sufficient time is allowed and viscous deformation takes place, then the stress passes through a maximum (yield stress) before falling, as the polymer sample yields and deforms before fracture finally occurs. This type of failure is referred to as 'ductile failure'. Attempts have been made to predict yield stress using theories based on Eyring type equations [21].

The importance of time and temperature in determining whether brittle or ductile failure occurs is evident. Smith & Stedry [22] were able to verify the applicability of the WLF principle to ultimate properties as well. Toussaint [23] has discussed both theory and experimental work relating to the effects of pigment content on the ultimate mechanical properties of paint films.

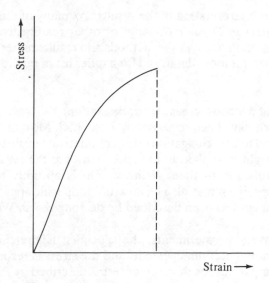

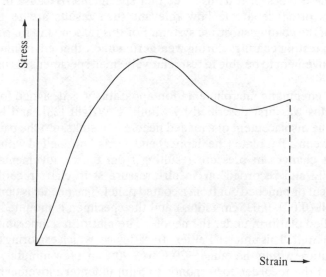

Fig. 16.3 — Idealized stress–strain curves for (a) brittle and (b) plastic yield failure of solids.

16.4 EXPERIMENTAL METHODS FOR DETERMINING MECHANICAL PROPERTIES OF COATINGS

16.4.1 Introduction

In the previous section a general view of polymer viscoelastic properties and their interpretation has been given. A number of reviews are available relevant to, or covering, both experimental methods for determining viscoelastic properties of coatings films and interpretation of the results. As many of these have appeared in the journal *Progress in Organic Coatings* or other readily accessible sources, it is proposed in this section to cover the methods and results rather briefly. The reviews should be consulted for more detail and for a fuller list of published source material.

16.4.2   Transient methods (creep, stress relaxation)

Transient methods have been reviewed by Zosel [24]. Most of the methods use free film specimens and tensile elongation using commercial tensile testing equipment. In creep tests, a weight is applied to the specimen, and the resulting deformation is measured as a function of time. If the weight is abruptly removed, the elastic recovery of the specimen may also be measured. Specific apparatus for making this type of measurement has been described by de Jong & van Westen [25] and Zosel [26].

In stress relaxation experiments, the specimen is stretched rapidly, again in tensile mode to a predetermined extent, and the stress is recorded as a function of time. A typical apparatus for this type of test is described by Pierce [27].

So far these types of tests use free film specimens. Because of internal stress factors, it is often debatable how relevant such results are to the mechanical properties of the coating/substrate system. For this reason, and for convenience and relevance in testing coatings during weathering and other environmental exposure tests, it is convenient to be able to test relatively small specimens of coatings *in situ* on the substrate.

The ICI pneumatic microindentation apparatus was designed for this purpose. The apparatus was first described by Monk & Wright [28], and is commercially available. The displacement of a loaded needle as it sinks into the painted film alters the gap between a flat plate (the flapper) and a small nozzle fed with a restricted air supply. The changes in pressure resulting from these movements are amplified pneumatically and recorded on an air-pressure strip chart recorder. A painted specimen is cut or punched out from a coated panel. The needle is tipped by a steel or sapphire ball (0.001–0.025 cm radius) and the specimen is mounted on a temperature-controlled platform under the needle. The platform temperature is controlled by a Frigistor unit. This unit is a Peltier effect device, which can bring the specimen to any temperature within the range $-20°C$ to $+90°C$ in a few minutes. The maximum deflection on the recorder corresponds to $6\,\mu m$ indenter movement. The recorder chart can be read to about $0.1\,\mu m$, and measurement accuracy is about double this value. Examples of the curves obtained with this instrument for alkyd films as a function of temperature and weathering time are given in Monk & Wright's paper [28]. Interpretation of the shapes of the curves in terms of the coatings' predominant mechanical characteristics, i.e. whether plastic, rubbery, viscoelastic, or glassy, are also given. The theoretical interpretation is a little more difficult. It can be done by applying the basic Hertz theory of spherical indentation [29] as modified by Lee & Radok [30] and Radok [31] for viscoelastic materials, to obtain a compliance as a function of time and temperature in the usual way. However, it should be borne in mind that the stress field beneath the indenter is extremely non-linear and varies with the depth of indentation, as well as the load applied. Also, if the film thickness is too low or the indentation depth too great, this stress field may interact with the substrate and modify the indentation–time curve. Morris [32] has discussed these aspects of the instrument's results critically. In spite of this, the instrument is very useful: disk specimens about one inch in diameter can be cut at intervals during the course of panel weathering, or the indentation temperature can be changed systematically for each of a series of different locations on the same disk and a steady or fixed time delay indentation depth measured as a function of temperature to give a glass transition (T_g) value. Either procedure allows the change in mechanical properties during weathering to be accurately assessed.

16.4.3   Dynamic methods

The subject of dynamic mechanical testing of paint films has been reviewed in considerable detail by Ikeda [33]. This review covers not only experimental methods but also the interpretation of results, the calculation of coating properties from those of the coating plus support (particularly if supported coating specimens are used), the effects of formulation variables (in particular, pigmentation), and the correlation of these results with impact strength.

Four basic methods are available:

(a) free oscillating torsion pendulum methods;
(b) resonance methods in forced vibration;
(c) forced non-resonance vibration methods (longitudinal deformation); and
(d) ultrasonic impedance methods.

Method (a) is normally a low-frequency single-value method, operating around 1 Hz. Methods (b) and (c) normally operate in the range1 Hz to 10 kHz with single frequency values only for the resonance methods. Method (d), as its name suggests, utilizes a high frequency range (a few hundred kHz to ten MHz).

16.4.3.1   *Free oscillating torsion pendulum method*

The basic arrangement resembles the rotary pendulum of some clocks: the coating specimen forms the suspension, and a weight is hung from it. The weight is given a slight horizontal twist and released. The weight then vibrates back and forth, twisting the coating specimen (usually in the form of a narrow strip) back and forth about its vertical axis. Energy dissipation in the specimen and its suspension leads to a decrease in the amplitude of successive 'twists', whilst the mass of the weight (more accurately, the moment of inertia of the pendulum bob) and the elasticity of the specimen (or suspension) determine the frequency of the twisting oscillation. Thus, by measuring the frequency and the amplitude decreases of successive oscillations the shear elasticity and loss moduli (G' and G'') can be found.

The fragility or ductility of many coatings specimens as free films make it difficult to use this single frequency measurement over a sufficiently wide range of temperatures, whilst the tension in the film, due to the weight, may affect the results. Two solutions to this problem exist: (a) either the pendulum is inverted and the bob weight counterbalanced so as to minimize the tensional force applied to the specimen (shown schematically in Fig. 16.4), or (b) the coating is applied to a metal foil or woven glass fibre braid which forms the suspension of the simple pendulum

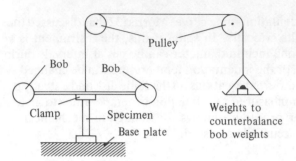

Fig. 16.4 — Schematic diagram of the inverted torsion pendulum.

arrangement and bears the weight of the bob. If a metal foil is used, the moduli of the coating can still be derived if those of the foil are also determined. However, a serious problem in accurate determination of the moduli is the geometry factor for this type of deformation, as the equation for the modulus contains the difference of cubes of the thicknesses for the coating and foil, and the foil alone (Inone & Kobatake [33]). As the thickness is often difficult to determine with sufficient accuracy and may vary with temperature, owing to curing, solvent loss, etc., this can represent a serious drawback. A less obvious disadvantage is that the operating frequency also changes with temperature (because of changes in the coating elasticity, which may be significant even if the coating is supported). These difficulties may be overcome by using forced extensional or flexural vibration, as described in the following sections.

In spite of these drawbacks, the use of glass braids as a support has achieved some popularity, particularly as a method of studying cure reactions of polymer and coating compositions, under the name of 'torsional braid analysis' with commercial equipment available [34, 35]. With the irregular geometry of the impregnated braid, it is not possible to derive the moduli of the coating, but for following curing reactions it appears to be satisfactory. Here again, if temperature scanning is used (as is tempting with modern computer control techniques available), the results may be affected by the temperature scan rate. Far more serious, in the author's view, is the question of wetting and adhesion of the liquid and solid coatings respectively, to the fibres of the braid, which may affect the results. Also, there are scattered reports of spurious loss peaks due to such causes as rubbing of the braid fibres together.

16.4.3.2 *Resonance methods in forced oscillation*

The resonance technique consists of applying a sinusoidal force to the coating specimen and measuring the amplitude of the specimen's response as a function of frequency. The resonance frequency is identified as the frequency corresponding to the maximum or 'peak' amplitude. By measuring this parameter and the 'peak width', the moduli can be calculated. The peak width is defined as the frequency range over which the amplitude has a value equal to the peak amplitude divided by the square root of two. It is sometimes possible to identify higher frequency resonance peaks due to more complex deformation modes, but these are generally smaller in peak amplitude than the normal or fundamental deformation mode.

The method is simple to use. A reed-shaped piece of free coating film or coated substrate is clamped firmly to a vibrator at one of its ends and made to vibrate laterally. The resulting swing amplitude of the free end can be measured by means of a microscope equipped with an eyepiece graticule. Firm clamping is essential, or spurious resonances will occur. Alternatively, if a strip of thin steel shim (for example, razor blade steel) is coated and suspended or supported by fine threads, flexural oscillations can be excited electromagnetically.

In the author's experience, these methods all suffer from the same disadvantage: that, unless the coatings are hard or glassy (i.e. energy dissipation is low), it is difficult to identify the resonance frequency peak precisely enough for a free film, or it will not differ significantly from the value of the substrate alone if a supported film is used.

Again, as with the free oscillation torsional pendulum methods of the previous section, the method is a single-frequency method, and anything which alters the coating elasticity modulus, e.g. temperature, or progress of cure, will shift the resonance frequency. The thickness measurement is slightly less critical for free films — it enters as a square term; however, for supported coating specimens the equations are more complex and involve linear, square, and cubic terms in the coating/substrate thickness ratio [36,37]. To be valid, the theory also assumes perfect adhesion between coating and substrate.

16.4.3.3 *Forced non-resonance vibration methods*

Better than the methods described above, is measurement of the amplitude as a function of frequency, as with the resonance method, and the phase difference between the applied force or strain and the response waveform. This method is the exact analogue for the solid films of the method used for determining the viscoelastic properties of liquids and paints described in the previous two Chapters 14 and 15. It allows a range of frequencies to be covered at constant temperature, supplemented by use of the WLF principle, or a range of temperature at constant frequency. The former is a more accurate way of determining T_g, whilst the latter is a more critical way of studying curing processes. Such methods are collectively referred to as 'dynamic mechanical analysis' (DMA). It is significant that one of the leading commercial models of DMA equipment controls both frequency and temperature. Gearing [38] has published illustrative results obtained with this equipment on paint samples (using flexural deformation).

Potentially, the frequency range of the technique is large (10^{-5} and 10^4 Hz). However, a number of factors limit the frequency range accessible with a particular piece of equipment. Specific limiting factors are the range of frequencies that can be generated, the sensitivity of the response measuring equipment, etc. A general limiting factor for the upper frequency limit is that above the resonance frequency, the response becomes more and more dominated by the inertia of the moving parts of the apparatus, whilst the contribution of the sample inertia increases also with frequency. Thus the frequency range can be extended only by making the moving parts as light as possible. By using longitudinal (tensile) extension of strip samples, the coating thickness measurement becomes less critical, as it now enters as a linear term.

Some years ago the author built a dynamic test apparatus for paint-free and supported film samples, in which he tried to incorporate some of these principles. The coating sample strip was clamped between two small aluminium screw clamps, each carried on two light suspensions, firmly attached to a rigid steel cage. The driver suspension consisted of a loudspeaker coil, attached by a light yoke to the clamp. The yoke was supported by a cruciform arrangement of fine piano wires to allow the yoke to move only vertically. The magnet surrounding the coil was firmly attached to the base of the apparatus. The other clamp was carried on a cruciform phosphor bronze leaf spring system, at the centre of which was mounted the core of a linear displacement transducer (LVDT). The coil of the LVDT and the leaf spring system were attached firmly to a plate that could be raised and lowered by a fine-pitch screw mounted in the top of the apparatus, so that a minimum constant tension could be maintained in the sample. The apparatus was surrounded by a temperature-controlled air or inert gas enclosure. Samples of model alkyd coatings were studied to establish the effects of chemical structure and weathering on the dynamic mechanical properties of these coatings. Some typical results are shown in Fig. 16.5. Even with

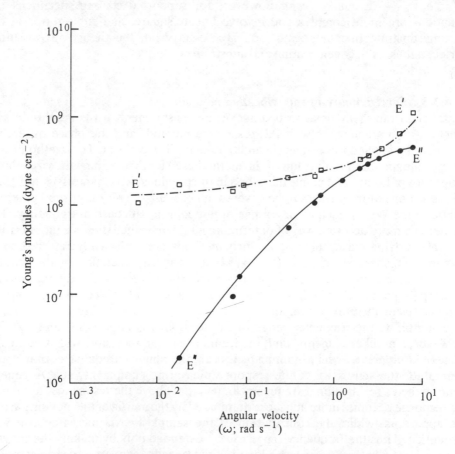

Fig. 16.5 — Phthalic polyester (1 week exposure, 44°C test temperature).

very soft films on aluminium foil substrates, the apparatus sensitivity was good enough to determine the modulus values of the coatings. Similar equipment has been reviewed by Whorlow [40].

Other geometries, such as torsional or flexing, can also be used with forced, non-resonance measurements. In particular, it is possible to produce a dynamic micro-indenter [41]. However, there may be problems if the film surface is not highly elastic or soft enough for adhesion or contact to be maintained over a constant surface area, during the oscillation cycle. However, this does represent an attractive way of measuring film viscoelasticity when the coating is *in situ* on its substrate. The oscillation amplitude should be kept small to minimize these effects and to prevent substrate interference. More recently, an interesting modification of the thermal mechanical analysis technique (TMA) for this purpose has been described by Reichert & Dönnebrink [42].

16.4.3.4 *Ultrasonic impedance methods*
These methods applied to studies of coating cure have been developed by Myers [43]. Myers describes the development and use of the technique.

The principle of the technique is shown schematically in the previous chapter, Fig. 15.3. A shear wave generated in a plane layer passes across the interface with another layer in proportion to how closely the shear mechanical impedances are matched. If a sensor is mounted to detect the reflected energy, the echo suffers an attenuation directly proportional to the impedance of the substrate. A trapezoidal section substrate forms the support for the coating under test. Fused quartz of known impedance was found to be a suitable substrate for propagation of shear waves into an adhering film, and 11° was the optimum angle of incidence [44]. Steel bars can also be used satisfactorily [45].

Four microsecond gated pulses were generated at frequencies ranging from 2.0–100 MHz, applied to a piezoelectric transducer attached to one sloping face of the bar, propagated and reflected off the coated surface, and collected as a series of exponentially decreasing pulses, and the attenuation was measured and recorded. However, from this measurement only the real component of the complex shear modulus (G') can be derived. The imaginary component (G'') requires measurement of the phase relationships at the interface. However, G' is relatively insensitive to any measurement parameter except attenuation [46], provided that the phase angle on reflection is finite.

In practice, a sigmoidal shaped curve is obtained when attenuation is monitored during the liquid to solid(!) conversion during cure and plotted against curing time [47].

Using a similar principle, but utilizing trains of torsional oscillatory pulses down a fused quartz rod, Mewis [48] described an apparatus operating at a frequency of 100–150 kHz. Again, the coating to be tested was applied to the rod surface. In this apparatus, both attenuation and reflected phase angle were measured, so both components of the shear modulus could be calculated. As Mewis [49] points out in a later review, the frequency range really determines the sensitivity of the technique to different portions of the drying process: thus, his own apparatus is really more suited to monitoring the initial stage of the drying or curing process, whereas the higher-frequency equipment is better for the later stages of the process.

However, the electronic equipment required for this technique is considerable, and it is debatable whether rather simpler techniques such as the bouncing-ball (described in Chapter 15), or torsional braid analysis, described in previous sections of the present chapter, may not be just as effective for monitoring cure or drying processes.

16.4.4 Measurement of ultimate (failure) properties

The ultimate properties of a coating may be measured in the same way as for polymers, i.e. by tensile extension at a controlled rate of a rectangular free film specimen of the coating in a commercial instrument, and by the plotting of stress–strain curves at several different rates. Temperature and humidity should be controlled during such tests. The use of these measurements has been reviewed by Evans [50]. Methods for attached films are less frequently described, apart from the technological tests to be described shortly, where test conditions may not be adequately controlled or definable. However, there is recent evidence of more interest in instrumenting and more critically controlling experimental conditions in impact tests. Simpson [51] has designed a cyclical flexure machine, in which coatings applied to metal shims can be tested before and during accelerated weathering tests. Failure is assessed visually by the appearance of surface cracking. Such tests can be used to predict durability or to monitor changes in mechanical properties during weathering. A rather more sensitive technique for detecting failure by cracking (and which will detect subsurface cohesive or adhesive failure, as well) is provided by the acoustic emission technique, which will be described in the last section of this chapter. This technique may be used in a number of deformation geometries, and the coatings are applied to metal foil, shear, or sheet substrates.

The major problem with using free film specimens for failure property tests, apart from the internal stress factors discussed above, is the difficulty of obtaining perfect free film specimens, as the presence of edge defects, for example, will seriously affect the accuracy and repeatability of the measured test parameters. In general, it is better to design and use tests on coatings *in situ* on their substrates.

16.5 DISCUSSION OF EXPERIMENTAL METHODS

The choice of experimental method from the array of test methods available is difficult to specify; apart from obvious factors, such as time and equipment available, and expense, the purpose of the tests is the most important determining factor. However, a few general principles of choice can be listed:

(1) If it is experimentally practicable, a test on a coating *in situ* on its normal substrate is always preferable to a free film test.
(2) Mechanical properties always depend on temperature, and the time scale of the test (frequency, time, strain rate, etc.), and frequently on the humidity. It is therefore absolutely essential to control the first two closely, and desirable to

control the humidity also during the test. Failure to do so represents a major inadequacy and source of unreliability in most of the technological tests used in the industry for assessing mechanical properties (to be described in the next section).

(3) If an assessment or prediction of coating durability is required, then ultimate properties will have to be measured.

However, if the purpose is to study drying/curing processes or the effect of chemical or physical variations in the coating structure (formulation variables, changes during weathering, etc.), then there is little doubt that the transient or dynamic tests using small forces or deformations provide the most information and are the most easy to interpret in terms of the basic physics and chemistry of the coating system. They are also often more reliable because accidental defects in the coatings are less critical under the small force or deformation conditions used.

16.6 TECHNOLOGICAL TESTS FOR MECHANICAL PROPERTIES

There is a wide range of hardness, flexibility, and other types of test used in the coatings industry for measuring the mechanical properties of a coating. As these have been very adequately described in reviews [52, 53, 54], only brief comments will be given here, with a more detailed account of some of the better-defined tests, which give a better chance of relating their results to fundamental viscoelastic properties.

16.6.1 Hardness tests

As Sato [54] has pointed out, hardness is a very imprecise term. It is in commonsense terms the rigidity of the substance, i.e. its resistance to deformation by externally applied force; but, as has already been discussed, it is not a unique property, as it will depend both on the magnitude of the force applied and the rate at which is it applied. There are three basic types of hardness measurement: indentation hardness; scratch hardness; and pendulum hardness.

16.6.1.1 *Indentation hardness*

Indentation hardness testers have been developed from the Brinell or Rockwell hardness testers used in the rubber and plastics industry. To minimize substrate interference, penetration depths must be small. ASTM D1474 specifies the Tukon indentation tester for determining the Knoop indentation hardness of coatings. The diamond indenter has a carefully cut tip of narrow rhombohedral shape, and the coating, on a solid block of glass or metal, is deformed by the indenter under a load of 25 g for 18 ± 2 s. After releasing the load, the long diagonal of the indentation is measured and the Knoop hardness number (*KHN*) calculated. Mercurio [55] plotted *KHN* and Young's modulus with temperature for polymethyl methacrylate coatings, and demonstrated that they both were of similar shape and showed identical T_g values (110°C). Inone [56], using his own equipment, showed that the indentation hardness for a number of thermoplastic coatings was proportional to the Young's modulus of the coating. Probably, the most sensitive of these instruments is the micro-indenter, developed at ICI (Paints Division) and described earlier in this chapter. An objection to using pointed conical or prismatic indenters is that these lead to discontinuous stress fields in the coating, i.e. the stress becomes infinite in the vicinity of points and edges. The spherical indenter of the ICI instrument is not open to this objection, whilst the maximum penetration depth of 6 μm minimizes substrate interference effects.

Most workers with these instruments take the penetration depth of a fixed time after penetration commences. However, it should be borne in mind that with a spherical indenter it is possible to use the penetration–time curve to evaluate the shear compliance–time relationships at each temperature, as in more conventional transient measurements.

16.6.1.2 *Scratch hardness*
Scratch hardness tests vary from the more or less qualitative pencil scratch hardness tests, commonly used in the industry, to the use of pointed loaded indenters drawn across films at a constant rate, and measurement of the groove width (see, for example [57]). Mercurio [55] pointed out that the pencil hardness of the film is related to the elongation at break, i.e. the coating is broken only when the maximum stress, due to the pencil or indenter scratching, exceeds the tensile strength of the coating film.

16.6.1.3 *Pendulum hardness*
Pendulum hardness relates to an indenter performing a reciprocating rolling motion on a horizontal coating. The motion is thus an example of a free oscillating pendulum with the amplitude of oscillation decaying with time. The initial driving force is provided by the initial deflection and the resultant rotational moment, provided by the force of gravity, when the pendulum is released. The energy of motion is dissipated by the coating, thus the frequency of oscillation and the decrement of the oscillation are related to the viscoelastic properties of the coating, when these are compared to the values for the uncoated glass or metal block used as the substrate. Pendulum hardness may be expressed in a number of ways: the time for the amplitude to decrease to half (or some other fixed fraction) of its original value, or as time in seconds, or as a number of swings, or relatively, as a percentage of the corresponding time, measured on a standard glass plate. The assumption that the hardness measured is inversely proportional to the damping capacity of the coating is, however, false, as Inone & Ito [58] have demonstrated. It is thus misleading to compare coatings with different viscoelastic properties by hardness alone, although the comparison is valid if the viscoelastic properties are similar.

Two basic kinds of pendulum are available: those popular in Europe (the Koenig and Persoz pendulums), which use an indenter as the pivot for the pendulum, and those like the Sward rocker, popular in Japan and the USA, where a circular cage, containing an internal pendulum, rolls back and forth across the coated substrate.

The advantages and disadvantages of the two types have been listed by Sato. The pivot type is superior in accuracy and reproducibility, and has a smaller area of contact with the film.

In terms of the basic physics, the pendulum hardness test has been analysed by Persoz [59] and Inone & Ikeda [60] for the pivot type, and Baker *et al.* [61], Roberts & Steels [62], and Pierce *et al.* [63] for the Sward rocker.

Comparisons of the temperature-dependence of the oscillatory decrement of free torsional oscillatory measurements and pivot pendulum decrement were made by Sato & Inone [64,65] for melamine/alkyd coating systems. Good agreement was obtained between the two methods. However, it must be remembered that the damping will depend on the pivot surface area in contact with the film, and that this will increase as the film softens with temperature and the indenter pivot sinks further into the film. In spite of this, good results can be obtained for the viscoelastic properties, and the simplicity of technique makes it attractive. Sato [54] has surveyed all this work in some detail.

16.6.2   Flexibility tests

Flexibility tests are of two basic types: the bend test, in which painted panels are bent around mandrels, and the Erichsen test in which the panel metal is deformed by a large hemispherical-ended indenter.

In the bend test it is customary for thin coated metal panels to be bent around mandrels of varying diameter with coating on the outside of the end in the panel. The object of the test is to discover the smallest diameter mandrel at which cracking occurs in the coating. Clearly, it is important to control the temperature and the bending rate to obtain comparable results.

In the other types of test, the Erichsen film tester, a hemispherical indenter is forced into the panel (again from the metal side), and the indentation depth at which the film starts to crack is measured. Again, it is important to control both temperature and deformation rate.

Both tests could be made more sensitive by using a conventional tensile tester to control (and, if required, vary) the deformation rate, and by using an acoustic emission detector as a more sensitive detector of cracking than visual observation. That this is possible for the bend test has been shown by Strivens & Bahra [66], using a three-point bend test adapter with fixed mandrel size, and an industrial tensile tester. Although not so sensitive as tests done in the conventional tensile test mode with acoustic emission, it is still possible to differentiate between coatings of different performance.

16.6.3   Impact tests

It appears to be difficult at present to predict impact resistance from the viscoelastic properties of the coating, owing to the complicated stress and strain profiles produced and the lack of reliable data concerning the mechanical properties at very short times. Timoshenko & Goodier [67] have derived a theory of impact which allows the magnitude of the most significant impact parameters to be estimated. Using this theory, Zorll [68] estimated the impact time to be some tens of microseconds. This is in agreement with experimental figures. For example, Kirby [69] found a figure of $18 \pm 3\,\mu s$ at room temperature for an 8 mm steel ball striking a thick glass block at a terminal velocity of $1.70\,\text{m s}^{-1}$, whilst Calvit [70] found a figure of $100\,\mu s$ for a block of pMMa below its T_g (5 mm ball, $0.70\,\text{m s}^{-1}$ impact velocity).

A particular requirement for automotive coatings is resistance to gravel impact. Zosel [71] built an apparatus to stimulate gravel impact, where a spherical steel indenter of low mass is forced against the panel at velocities of between zero and $25\,\text{m s}^{-1}$. The minimum velocity at which damage becomes visible at the impact point and the area A of the impact mark, are measured. From these measurements there appears to be a temperature range around the coating T_g for which impact resistance appears to be optimum. Similar conclusions follow from the results of Bender [72, 73] comparing Gravelometer results with the viscoelastic properties of coatings.

Intuitively, it would be expected that the impact behaviour will also depend critically on the shape of the impacting object and its angle of approach to the panel. This has been verified by, among others, Breinsberger & Koppelmann [74] who compared conical and spherical impactors. They again verified the conclusions of Zosel and Bender regarding the relation of coating T_g to optimum impact resistance.

Also important in determining impact resistance are the mechanical properties of the individual layers of the total coating system and their relationship to each other, as the work of Bender demonstrates. Technological tests rely on dropping graded objects down pipes onto panels: for example, dropping 1/4 in hexagonal nuts down a 4.5 m-long pipe (Brit. Std. BS AU 148); or carefully graded gravel or steel shot is blown against the panel by a strong air blast, as in the Gravelometer apparatus.

16.7 ACOUSTIC EMISSION

The use of acoustic emission to detect cracks in engineering structures under stress, such as North Sea oil platforms, high-pressure vessels, aircraft wings, etc., has been commonplace for a number of years. However, it is a novel application to study the acoustic emission of coatings under stress. This technique, pioneered by ICI (Paints Division), has proved to be of considerable use in monitoring and even, in some cases, predicting the durability of coatings during environmental testing, such as natural weathering, accelerated weathering, salt spray corrosion tests, etc. In addition, it has proved very useful in evaluating the effects of formulation variables on the ultimate mechanical properties of coatings, and in evaluating these properties for the individual layers of a coating system, as well as elucidating the ways in which these properties interact in producing the total properties of the full system. The technique is, in principle, extremely simple. Any sudden microscopic movement in a body, e.g. crack formation and propagation, may give rise to acoustic emission. For example, strain is concentrated at the growing tip of a crack. As this crack propagates, strain energy is released in two main forms: as thermal energy and as acoustic energy. The acoustic energy radiates as a deformation wave from the source, and is refracted and reflected by solid inclusions and interfaces until it reaches the surface of the body. Here the surface waves may be detected by sensitive detectors: usually piezoelectric or capacitive transducers. The amplified signal from the transducer is then analysed. Familiar examples of acoustic emission, at frequencies and intensities audible to the human ear, are the cracking of ice on a pond or the creaking of the treads of wooden stairs under the weight of a human body.

The paint is coated onto one side of a metal foil strip, and this is then inserted in the jaws of a tensile tester; the transducer is attached and the sample stretched. The noise emitted is analysed, and some noise characteristic is plotted as a function of total strain. Although tensile testing is more usual, there is no reason why bending or any other form of deformation may not be used. The only essential is that there should be no spurious noise generated by slippage between the specimen and the instrument's clamps. This is why, in practice, slow deformation rates are used. Apart from this source of noise, there is no need to shield the apparatus acoustically. Strivens *et al.* [75, 76, 77] have found narrow-band resonant piezoelectric devices (resonant frequency around 150 kHz) to be satisfactory for this purpose. Good acoustic coupling between the transducer surface and the specimen is essential to maximize detector sensitivity: this is readily achieved by means of a thin connecting layer of silicone grease.

The methods of analysis available for characterizing the acoustic emission are numerous. Because of the simultaneous occurrence of many noise sources, often of different kinds, as well as the modification of the waveforms both by propagation through the body of the specimen and by the response characteristics of the detector itself, it is very difficult with acoustic emission from paint specimens to analyse the complex signal forms to obtain information about the original signal source. There is also too little theory or experimental work with 'model' systems relating waveform characteristics to source mechanism. Thus complicated frequency analysis or amplitude analysis techniques are not generally useful, although amplitude analysis can be revealing if the failure mechanism changes drastically, for example, if there is a change from micro- to large-scale cracking at a particular strain value. Strivens *et al.* have concentrated on using simple analysis techniques, such as plots of 'ring-down' or event counts against total strain to characterize the coating, and have then used these on a comparative basis, for example, to monitor changes in the failure properties of coatings during weathering, to study the effects of changes in formulation or chemical structure on the ultimate mechanical properties of the coatings, etc. This simple utilization of the technique has proved extremely useful. Not only can the strong effect of moisture in alkyd films be clearly seen (Fig. 16.6), but also, in most cases, adhesion failure can be seen as abrupt, stepwise discontinuities in the plots, whilst cracking tends to produce uniform smooth changes.

The principle of 'ring-down', and event counting, can be seen in Fig. 16.7. If the amplified voltage output of the transducer, corresponding to a single event, is idealized as a sinusoidal decay curve, then in ring-down counting, one count is registered every time the voltage rises above a threshold voltage (this is imposed to stop random electrical noise affecting the analysis). In event counting, a preset delay is imposed after the first count before another can be registered. By proper choice of the delay time, comparison of ring-down and event counts will give a crude estimate of the average amplitude of the signals. A cumulative plot of total counts against strain is then produced for each.

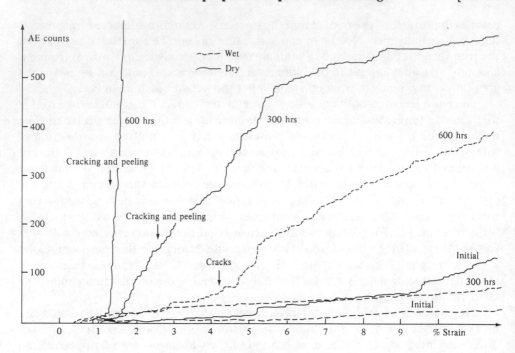

Fig. 16.6 — Effect of humidity and accelerated weathering on noise emission with
deformation (exterior alkyd gain system).

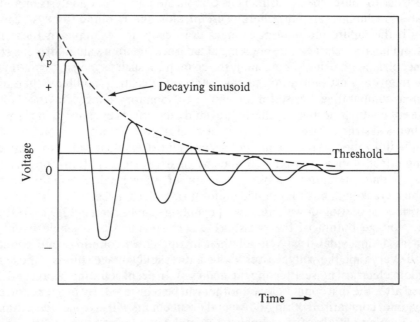

Fig. 16.7 — Method of counting in acoustic emission.

Further details and examples of results will be found in the publications already cited.

REFERENCES

[1] Ferry, J. D. *Viscoelastic properties of polymers*. J Wiley, New York, 2nd ed. (1970).
[2] Nielsen, L. E. *Mechanical properties of polymers*. Reinhold Publishing Corpn., New York (1962).
[3] McCrum, N. G., Read, B. E. & Williams, G. *Anelastic and dielectric effects in polymeric solids*, J Wiley, New York. (1967).
[4] Graessley, W. W. *Adv. Polymer Sci.* (1974) (16), 1.
[5] Hedvig, P. *J Polymer Sci., Macromol. Rev.* **15** 375 (1980).
[6] Williams, M. L., Landel, R. F. & Ferry, J. D. *J. Amer. Chem. Soc.* **77** 3701 (1955).
[7] Havriliak, S. & Negami, S. *Brit. J. Appld. Physics* Ser. 2 **2** 1301–15 (1969).
[8] Havriliak, S. & Negami, S. *Polymer* **8** 161–210 (1967).
[9] Laout, J. C. & Duperray, B. *XIVe Congrès AFTPV, Aix-les-Bains* 143 (1981).
[10] Laout, J. C. *XVIth. FATIPEC Congress, Liège* I, 165 (1982).
[11] Duperray, B., Laout, J. C. & Mansot, J. L. *Double liaison* (330), 61 (1983).
[12] Toussaint, A. & Dhont, L. *Prog. in Org. Coatings* **11**(2) 139 (1983).
[13] Ward, I. M. *Mechanical properties of solid polymers*, Wiley-Interscience (1981).
[14] Andrews, E. H. *Fracture in polymers* Oliver & Boyd (1968).
[15] Kinloch, A. J. & Young, R. J. *Fracture behaviour of polymers* Applied Science Publishers (1983).
[16] Kinloch, A. J. *Adv. in Polymer Sci.* (1985) 72, 46.
[17] Cuthrell, R. E. In: Chompff A. J. & Newman, S. *Polymer networks: structural and mechanical properties* Plenum Press, p. 121.
[18] Griffith, A. A. *Phil. Trans. Roy. Soc. (London)* **A221** 163 (1920/1).
[19] Loc. *cit ref.* [14], pp. 123–31.
[20] Berry, J. P. *J. Polymer Sci.* **50** 313 (1961).
[21] Loc. *cit ref.* [13].
[22] Smith T. L. & Stedry, P. J. *J. Appld. Physics* **31** 1982 (1960).
[23] Toussaint, A. *Proc. 5th Int. Conf. on Organic Coatings Sci. and Tech.*, Athens (1979) 489.
[24] Zosel, A. *Prog. Org. Coatings* **8**(1) 47 (1980).
[25] de Jong, J. & van Westen, G. C. *J. Oil Colour Chem. Assoc.* **55** 989 (1972).
[26] Zosel, A. *Farbe Lack* **82** 115 (1976).
[27] Pierce, P. E. In: Myers, R. R. & Long, J. S. (eds.), *Treatise on coatings*, Vol. 2, Pt. I, Ch. 4, Marcel Dekker, New York, p. 112 (1969).
[28] Monk, C. J. H. & Wright, T. A. *J. Oil Col. Chemists Assocn.* **48** 520 (1965).
[29] Hertz, H. *J Reine Angew. Math.* **92** 156 (1881).
[30] Lee, E. H. & Radok, J. R. M. *Proc. 9th Internat. Congr. Appld. Maths.* **5** 321 (1957).
[31] Radok, J. R. M. *Quart. Appld. Maths* **15** 198 (1957).

[32] Morris, R. L. J. *J. Oil Col. Chem. Assocn.* **56** 555 (1973).
[33] Inone, Y. & Kobatake, Y. *Kolloid-Z* **160**, 44 (1958).
[34] Roller, M. B. & Gillam, J. K. *J. Coatings Tech.* **50** (636), 57 (1978).
[35] Roller, M. B. *Metal Finishing* **78** (3) 53; (4), 28 (1980).
[36] van Hoorn, H. & Bruin, P. *Paint Varnish Prodn.* **49** (9) 47 (1959).
[37] Yamasaki, R. S. *Official Digest* **35** 992 (1963).
[38] Gearing, J. W. E. *Polymer Paint Col. J.* **172** (4081) 687 (1982).
[39] Strivens T. A. (Unpublished results).
[40] Whorlow, R. W. In: *Rheological techniques*, Chap. 5., Ellis Horwood (1980).
[41] Smith, N. D. P. & Wright, T. A. (Unpublished results).
[42] Reichert, K-H. W. & Dönnebrink, G. *Farbe u Lack* **86** (7), 591 (1980).
[43] Myers, R. R. *J. Polymer Sci.*, C, (35), 3. (1971).
[44] Mason, W. P., Baker, W. O., McSkimmin, H. J. & Heiss, J. H. *Phys. Rev.* **75** 936 (1949).
[45] Peck, G. (Private communication).
[46] Myers, R. R. & Knauss, C. J. *Polymer colloids*, Plenum Press, New York (1971).
[47] Myers, R. R., Klimek, J. & Knauss, C. J. *J. Paint Technol.* **38** 479 (1966).
[48] Mewis, J. *FATIPEC IX, Brussels* (1968) 3–120.
[49] Mewis, J. In: Walters, K. (ed.) *Rheometry: industrial applications*, John Wiley Ch. 6, p. 323 (1980).
[50] Evans, R. M. In: Myers, A. R. & Long J. S. (eds.) *Treatise on coatings* Vol. 2, Pt. 1, Ch. 5. Marcel Dekker, New York (1969).
[51] Simpson, L. A. *FATIPEC XVI, Liège* **1** 33 (1982).
[52] Corcoran, E. M. In: Sward. G. G. (ed.) *Paint testing manual* (13th edn.) ASTM Special Publication No. 500, Pt. 5, Ch. 1, Hardness p. 281 (1972).
[53] Schurr, G. G. In: *loc. cit.* ref. [52], Pt. 5, Ch. 4, Flexibility, p. 333.
[54] Sato, K. *Prog. Org. Coatings* **8** (1) 1, (1980).
[55] Mercurio, A. *Official Digest* **33** 987 (1961).
[56] Inone, Y. *Seni Gakkaishi* **6** 147 (1950).
[57] Inone, Y. *Loc. cit.* ref. [56] p. 150.
[58] Inone, Y. & Ito, Y. *Shikizai Kyokaishi* **27** 37 (1954).
[59] Persoz, B. *Peint., Pigm. Vernis* **21** 194 (1945).
[60] Inone, Y. & Ikeda, K. *Kobunshi Kagaku* **11** 409 (1954).
[61] Baker, D. J., Elleman, A. J. and McKelvie, A. N. Official Digest **22** 1048 (1950).
[62] Roberts, J. & Steels, M. A. *J. Appld. Polymer Sci.*, **10** 1343 (1966).
[63] Pierce, P. E., Holsworth, R. M. & Boerio, F. J. *J. Paint Technol* **39** 593 (1967).
[64] Sato, K. & Inone, Y. *Kogyo Kagaku Zasshi* **60** 1179 (1957).
[65] Sato, K. & Inone, Y. *Kobunshi Kagaku* **15** 421 (1958).
[66] Strivens, T. A. & Bahra, M. S. (Unpublished results).
[67] Timoshenko, S. & Goodier, J. N. *Theory of elasticity*, McGraw-Hill, New York, p. 383 (1951).
[68] Zorll, U. *FATIPEC IX Congr. Brussels* (1968) p. 3.
[69] Kirby, P. L. *Brit. J. Appld. Physics* **7** 227 (1956).
[70] Calvit, H. H. *J. Mech. Phys. Solids* **15** 141–150 (1967).
[71] Zosel, A. *Farbe Lack* **83** 9 (1977).

[72] Bender, H. S. *J. Appl. Polymer Sci.* **13** 1253 (1969).
[73] Bender, H. S. *J. Paint Technol.* **43** (552) 51 (1971).
[74] Breinsberger, J. & Koppelmann, J. *Farbe u Lack* **88** (11) 916 (1982).
[75] Strivens, T. A. & Rawlings, R. D. *J. Oil Col. Chem. Assocn.* **63** 412 (1980).
[76] Strivens, T. A. & Bahra, M. S. *J. Oil Col. Chem. Assocn.* **66** (11) 341 (1983).
[77] Strivens, T. A., Bahra, M. S. & Williams-Wynn, D. E. A. *J. Oil Col. Chem. Assocn.* **67** (5) 113; (6) 143 (1984).

17

Appearance qualities of paint — Basic concepts

T. R. Bullett

17.1 INTRODUCTION

One of the most useful properties of paint is an almost infinite capacity to modify the appearance of a substrate. Camouflage and ornamentation have always been two of the main themes in the use of paint. Craftsmen employed great pains and much ingenuity to reproduce, with paint, the appearance of naturally occurring materials; graining and marbling to simulate wood and polished marble were major subjects in the days when painters served long apprenticeships. The term 'enamel' came into the paint industry when paintmakers sought to reproduce the appearance, and hardness, of vitreous enamelled jewellery; they were so successful that vitreous enamellers now insist that the paint products must be qualified as 'enamel paint'. But paint is also a medium in its own right which can be prepared and manipulated to yield smooth surfaces, ranging from highly glossy to full matt, textures of many kinds, and a vast range of colour effects.

This chapter will be concerned, largely, with the basic physics underlying the more important appearance qualities, namely gloss, opacity, and colour. Whilst reflection, scattering, and absorption of light are subject to the laws of physics, appearance is also a function of the observers, including their physiology and, in many instances, their psychology.

17.2 PHYSICS OF REFLECTION BY PAINT/AIR INTERFACES

17.2.1 Plane surfaces

When a beam of light reaches an interface between two materials of different optical density a proportion of the light is reflected, the remainder travelling on, with change of direction (refraction), into the second material (Fig. 17.1). The proportion

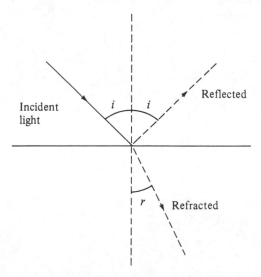

Fig. 17.1 — Reflection and refraction at air/paint interface.

reflected depends on the refractive indices of the two media and on the angle of incidence. Quantitative description of reflection is complicated by the fact that light polarized in the plane of the surface is reflected more easily than light polarised perpendicularly; this is somewhat analogous to the way that a flat stone will bounce off water if thrown horizontally but will penetrate and sink if its long axis is vertical. Mathematically [1], reflectivity for light polarized in the plane of the surface is:

$$R_s = \left(\frac{\sin(r-i)}{\sin(r+i)}\right)^2 \qquad (17.1)$$

and for light polarized at right angles to this plane is:

$$R_p = \left(\frac{\tan(r-i)}{\tan(r+i)}\right)^2 \qquad (17.2)$$

Fig. 17.2 shows the variations of R_s and R_p with angle of incidence for a refractive index of 1.5 for the second material, a typical if slightly low figure for a paint medium.

For unpolarized light, reflectivity is the average of R_s and R_p and increases steadily, for $n_2 = 1.5$, from 0.04 (4%) at normal incidence to 1 (100%) at grazing incidence ($i = 90°$).

Two further consequences of equations (17.1) and (17.2) are of interest. Firstly, when $r + i = 90°$, $\tan(r+i)$ becomes infinite, and thus, $R_p = 0$, which means that the reflected light is completely polarized in the plane of the surface. For $r + i = 90°$, $\sin r = \cos i$, so that from Snell's expression, $n = \sin i/\sin r = \tan i$ for this condition. This is the basis for the Brewster angle method for measuring refractive index in

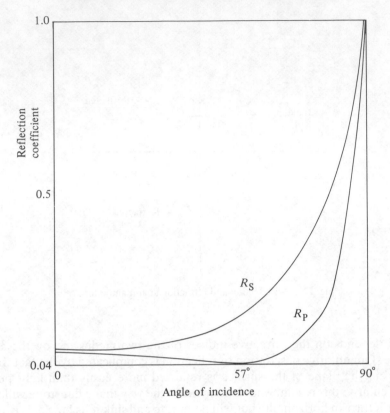

Fig. 17.2 — Variation of R_S and R_P with angle of incidence ($n_2 = 1.50$).

which the angle of incidence is found for which the reflected beam can be completely cut off by a polarizing filter rotated to the correct orientation. The technique requires high-quality apparatus for precision, but is useful for determining the refractive index of black glass (used for gloss standards) or of resin or varnish films on black glass. It can also be used with some loss of sensitivity for glossy paint films.

Secondly, for normal incidence ($i = 0$) reflectivity reduces to a simple expression:

$$R = \left(\frac{n_1 - n_2}{n_1 + n_2}\right)^2 \tag{17.3}$$

where n_1 and n_2 are the refractive indices of the first and second materials. Fig. 17.3 shows how R increases with refractive index ratio n_2/n_1 over the range 1.2 to 2.0. It will be seen that the intensity of specular reflection increases sharply with refractive index, so that much brighter reflection is possible from a paint or varnish based on a high refractive index resin (e.g. a phenolic resin) than from one based on a low

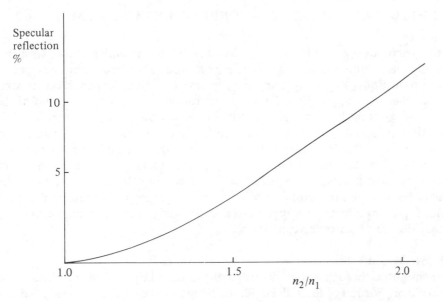

Fig. 17.3 — Increase in specular reflection with refractive index ratio.

refractive index resin (e.g. polyvinyl acetate). These observations are very significant to both the practical operation of glossmeters and the formulation of high-gloss paints.

17.2.2 Effects of surface texture

The discussion in 17.2.1 assumes an optically plane interface. When the surface is distorted, for example by uneven shrinkage over pigment particles or by residual texture from irregular application, individual facets of the surface present different angles to the incident beam. The reflected light thus becomes spread over a wider range of angles, and clear mirror-like reflection is destroyed. The scale of texture necessary to break up specular reflection is related to the wavelength of light and to the angle of incidence. For normal incidence and for angles up to about 45°, surface roughness on a scale and depth equal to the wavelength of light (0.4–0.7 μm) is sufficient to give at least a veiling effect on specular reflection; for grazing incidence much larger texture is necessary to destroy low-angle sheen. Thus, when the surface of a gloss paint begins to erode on weathering the first effect is loss of gloss when viewed at high angles to the surface, whilst it is not until the film has begun to craze or micro crack that all grazing incidence sheen is lost. Also a fully matt paint film can be produced only by incorporating particles that are coarse, relative to the wavelength of light. Typically, particles of 10–15 μm diameter are necessary in thick films.

Appearance variation is not a simple scale from fully glossy to matt. Large-scale ripples of low amplitude, such as residues of brushmarks, give visible disturbance of the specular image if the peak-to-trough amplitude exceeds 0.5 μm. Disturbance of the surface by large flat particles under the surface can give a diffused specular reflection, resulting in a pearly appearance.

17.3   LIGHT-SCATTERING AND ABSORPTION BY PAINT FILMS

17.3.1   General

Light refracted into a paint film is partly absorbed by the medium (resin or varnish), but mainly encounters particles where it is scattered, absorbed, or transmitted in various proportions. Light-scattering by a pigment particle depends on its size, relative to the wavelength of the light, and its refractive index ratio to that of the medium. Absorption by a pigment particle depends on the path length of light through the particle and on the extinction coefficient (absorbance) of the pigmentary material for the particular wavelength. Most of the theoretical treatments of light reflection by paint film, such as that of Kubelka & Munk [17], assume that repeated scattering by large numbers of particles sets up a completely diffused flux of light, such as is encountered in a cloud or thick mist. It is important to realize that such treatments are imperfect when applied to thin or lightly pigmented films where the necessary degree of scattering cannot occur.

17.3.2   Reflection at interfaces

Light transmitted through a paint film is partly absorbed by the substrate and partly reflected; a proportion of the reflected light eventually re-emerges, so that the film appears lighter over a more reflective substrate.

A full mathematical treatment of these phenomena is complicated even for the simplified case of a glossy paint film of uniform thickness over a smooth substrate. This is because a considerable proportion (usually over one half) of the light scattered back, diffusely, from a pigment layer is internally reflected at the paint/air interface (Fig. 17.4), and thus is attenuated again by absorption by pigment or

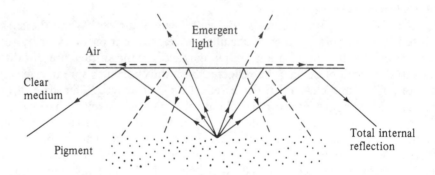

Fig. 17.4 — Internal reflection at paint/air interface.

substrate before reaching the interface for a second time. A similar situation occurs at the paint/substrate interface where light is inter-reflected between the substrate and the pigmented layer [3]. Fig. 17.5 indicates some of the infinite series of inter-reflections that occur until all the light has been absorbed or has emerged finally from the film.

The effect of the air/paint interface can be calculated fairly simply by summing the geometric progressions of inter-reflections. The result obtained is:

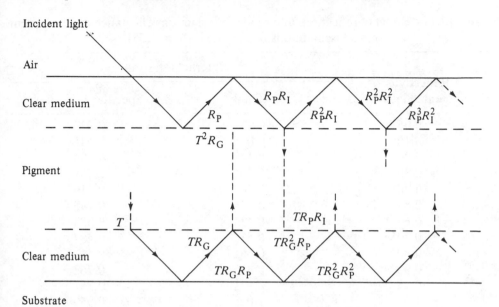

Fig. 17.5 — Inter-reflections at surface and substrate.

$$R_T = R_E + \frac{(1 - R_E)(1 - R_I)R_P}{(1 - R_I R_P)} \qquad (17.4)$$

where R_T is the reflectivity of the glossy paint film,
         R_E the external reflection coefficient of the interface,
         R_I the internal reflection coefficient of the interface,
         R_P the reflectivity of the pigmented layer over its substrate.

Table 17.1 indicates how R_T varies with R_P for $R_E = 0.05$ and $R_I = 0.55$, typical values for 45° illumination on to a film with a medium of refractive index 1.5. It will be seen that the effect of the interface is to reduce reflection markedly, espcially for the medium range of reflectivity. A consequence of this relationship is that most non-glossy materials (including matt paint films) are darkened appreciably if varnished over. For surfaces that are very dark the effect of varnishing depends on whether the observer picks up the externally reflected (specular) light; if he does, as, for example, when the surface is indirectly illuminated by reflection from a cloudy sky or from walls, the original colour will appear lighter. This apparent lightening of dark colours is often greater then would be suggested by Table 17.1 because for completely uniform illumination from all directions R_E rises to a value of about 0.1.

One further consequence of reflection at interfaces is important in considering the optical properties of paint films. A detached film or a film applied to transparent foil has an air/paint interface on its underside as well as on the side which is illuminated. This second interface also has an internal reflection coefficient of about 0.55 for diffuse light. Thus a detached film of non-hiding thickness will show the same

Table 17.1 — Effect of reflections at air/paint interface on reflectance of glossy paint films (assumptions: $R_E = 0.05$, $R_I = 0.55$).

Reflectivity of pigmented layer (R_P)	External reflectance	
	Total	Specular excluded
0.05	0.072	0.022
0.10	0.095	0.045
0.20	0.148	0.098
0.30	0.204	0.154
0.40	0.269	0.219
0.50	0.345	0.295
0.60	0.433	0.383
0.70	0.537	0.487
0.80	0.661	0.611
0.90	0.812	0.762
0.95	0.901	0.851
1.0	1.0	0.950

reflectivity as it would have if painted over a light grey substrate; if laid loosely over a black-and-white checkerboard the contrast will be much reduced from that of the same film painted over the board.

17.3.3 Scattering by white pigments

White pigments are made from transparent, almost colourless materials used in paint as fine particles. The relationship between particle size and light-scattering was investigated by Mie in 1908 [4] who showed that maximum scattering per unit amount of material occurred for a particle diameter rather less than the wavelength of light. Fig. 17.6 indicates the variation of scattering with particle size for uniform spheres of various diameters. Strictly, this curve refers to single scattering, that is light scattered only once by each particle, but in practice the optimum size is not greatly modified in paint films, except at very high levels of pigmentation where scattering by more closely spaced particles is considerably reduced. Commercial white pigments are developed to the particle diameter that gives maximum scattering of green light (for maximum film opacity); this is about 0.25 μm for rutile titanium dioxide. Particles of this size are less efficient in scattering yellow or red light, so that thin white paint films show distinctly orange transmission.

Work by the rutile pigment industry has assessed the extent to which scattering coefficients are reduced at high pigment concentrations. For particles of the optimum size of scattering at low concentrations the scattering per particle is approximately halved at a *PVC* of 30%. For this particle size, increasing concentration of pigment beyond 30% gives no further gain in opacity; indeed, opacity may actually fall over a range of concentrations where the gain in scattering from an increased number of particles is less than the loss from closer packing. At very high

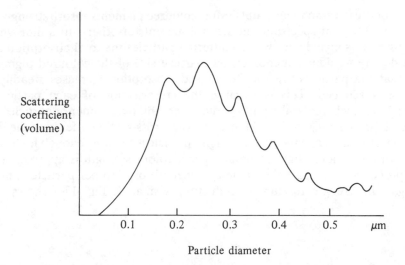

Fig. 17.6 — Scattering vs particle size for single spheres (calculated for rutile TiO_2 in linseed oil).

PVC there is insufficient medium to fill the spaces between pigment particles, so that air/pigment interfaces give dry hiding and increased opacity. There is some evidence that at and above 30% *PVC* rather larger pigment particles, say 0.4 µm instead of o.25 µm, give better opacity. It has also been shown, by Stieg [5] and others, that replacement of a proportion of the rutile pigment particles by small extender particles of low refractive index (such as fine silica or calcium carbonate) greatly increases opacity at high *PVC*.

These findings are consistent with the idea that at high pigment concentrations scattering can be considered as being from the edges of holes between particles rather than from isolated particles. In theory a scattering system based on cavities of low refractive index in a continuum of higher refractive index could be as effective as the reverse system; indeed the opacity of much biological material, for example dry bones, derives from cavities. Many attempts have been made to utilize cavities for developing opacity in paints. A simple method is to emulsify an immiscible liquid, such as white spirit into an aqueous gelatin solution; on drying and eventual evaporation of the white spirit an opaque white film, filled with minute spherical cavities, is left. Other more sophisticated techniques have given similar results with better film properties [6, 7, 8]. Possibly the most successful industrial development has been the production of conglomerate pigment particles consisting of small lumps of resin incorporating both small cavities and small particles of high refractive index pigments [9]. Theory also suggests that if particles of high refractive index material of optimum size were coated with a shell of low refractive index, an opacifying pigment less sensitive to concentration or to quality of dispersion would be obtained [10].

17.3.4   Absorption by pigments
All pigments absorb radiation of some wavelengths, but for white pigments the absorption becomes strong only in the ultra-violet region. Black pigments absorb at all visible light wavelengths but may be transparent in the infrared, a property of

importance for use in camouflage paints. Most coloured pigments absorb strongly for parts of the visible light spectrum but are transparent for others. In a film where coloured pigment is mixed with white scattering particles the total absorption and hence the depth of colour depends on the particle size of the coloured pigment. Provided that the particles are fully dispersed, absorption increases steadily as particle size is reduced. This is because the cross-section of each particle is proportional to d^2, where d is the particle diameter, and the number of particles per unit volume is proportional to $1/d^3$. Hence the total cross-section offered to the light is proportional to $1/d$. While the path through a particle is still long enough for a large proportion of the incident light to be absorbed, colour strength is approximately inversely proportional to particle diameter; when absorption per particle is much less, the gain in strength for further reduction is smaller. Fig. 17.7 shows data

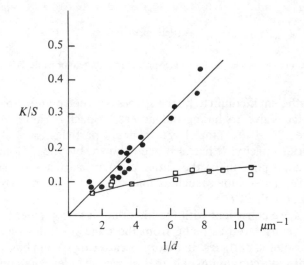

Fig. 17.7 — Absorption coefficient vs reciprocal particle size for organic pigments (data of Carr [11]). ● Phthalocyanine Blue; □ Pigment Green B (both measured as 1:12.5 reductions with titanium dioxide).

reported by Carr [11] for paints containing organic pigments after varying periods of grinding replotted against $1/d$. For the strongly absorbing phthalocyanine blue the proportionality between K and $1/d$ holds down to a particle size of $0.15\,\mu m$, but for the weaker Pigment Green B it does not.

17.4 COLOUR OF PIGMENT MIXTURES AND PIGMENTED FILMS

17.4.1 Perception of colour

A full account of colour vision would be inappropriate in this manual. Readers are referred to other textbooks [12–16]. For present purposes it is sufficient to recognize that in daylight the human eye analyses the light of wavelengths 0.4–$0.75\,\mu m$ in terms of three primary sensations (approximately to blue, green, and red), and that the colour perceived results from the balance of these sensations. Analysis of colour matching results with lights of known spectral energy distribution showed that each of the primary sensations was caused, with varying effectiveness, by light over a wide band of wavelengths. Fig. 17.8 illustrates the distribution of this sensitivity in terms

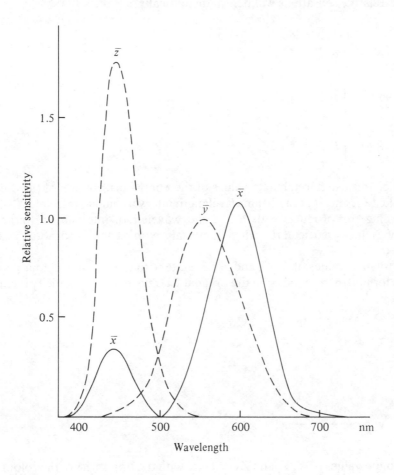

Fig. 17.8 — CIE spectral sensitivity distributions: equal energy stimulus.

of the three primaries X, Y, and Z of the internationally recognised CIE system. It will be noted that the red sensation, X, is stimulated by light of 0.43–0.45 μm, which is the violet part of the spectrum, as well as by light of 0.55–0.65 μm, which is the yellow, orange, and red part of the spectrum. These 'mixture curves' were obtained from observations by a group of observers, all judged to have normal colour vision, working under standardized conditions. The perception of colour by any single individual is likely to differ somewhat from this average performance, and is certainly influenced by surrounding colours, stimuli to which the eyes have just been exposed, and other factors. However, most practical colour measurement is based on the mixture curves as shown in Fig. 17.8, which were obtained with colorimeters where the coloured patch subtended a 2° angle at the eye or, on slightly different data, obtained with a 10° field.

From a set of response curves such as those in Fig. 17.8 it is possible to calculate for any energy distribution of incident light (E_λ) and distribution of reflection (R_λ) what the relative sensations will be. Mathematically,

$$X = \int R_\lambda E_\lambda \bar{x}_\lambda d\lambda$$

$$Y = \int R_\lambda E_\lambda \bar{y}_\lambda d\lambda \qquad\qquad (17.5)$$

$$Z = \int R_\lambda E_\lambda \bar{z}_\lambda d\lambda \ .$$

In practice it is usual to tabulate values of the energy distribution of the illuminant (E_λ) and $\bar{x}_\lambda, \bar{y}_\lambda$, and $\bar{z}_\lambda$ at say 10 nm wavelength intervals, and to calculate X, Y, and Z by summing the products with R_λ. If the wavelength intervals are large, some accuracy is lost, particularly when materials with sharp absorption peaks are measured.

The relative values of X, Y, and Z correlate with the depth of colour perceived, which is termed 'chromaticity' in the original CIE system. Chromaticity is expressed as x and y where

$$x = \frac{X}{X+Y+Z} \qquad\qquad (17.6)$$

and

$$y = \frac{Y}{X+Y+Z} \ .$$

The absolute values of X, Y, and Z correlate with the brightness of the colour, that is the proportion of the incident light reflected. The CIE primaries were chosen such that the Y value corresponds to the amount of light reflected, X and Y simply indicating the extent of variation from neutral white or grey. Thus Y, x, and y constitute a complete specification of the colour of a surface under a particular illumination. If the illumination is changed, say from north sky daylight to tungsten filament lamplight, Y, x, and y will all change, the extent of change depending on the particular reflection curve of the surface. Because the response curves to the three primary stimuli (Fig. 17.8) are broad it is possible for two widely different spectral reflection curves to give the same integrated values of X, Y, and Z for one particular illuminant. Such a colour match is destroyed by a change in illuminant because the new tristimulus values under the second illuminant will no longer be identical. Two colours of this type are termed 'metameric', and the phenomenon of matching under one illuminant but not under another is termed 'metamerism'. A colour which exhibits a marked change in hue with change in illuminant is termed 'dichroic'. Fig. 17.9 shows reflection curves for three colour chips prepared by the Munsell Color Co.; A and B match under Macbeth 7500°K daylight lamp; B and C match under a cool white fluorescent lamp; none match under tungsten lamps.

17.4.2   Colour of pigmented films

The objective quantity determining the colour of a paint film is the amount of the incident light reflected at each wavelength in the visible spectrum, or to be precise, the amount of that light received by the observer. The first successful attempt to relate this reflection to light scattering and absorption in the film was that of Kubelka & Munk [17, 18]. They made the assumption that a thin horizontal slice of a film (Fig. 17.10) would

(a) scatter light equally backwards and forwards in proportion to a scattering coefficient S and the slice volume, and
(b) absorb light in proportion to an absorption coefficient K and the slice volume.

From these simple assumptions the flux of light transmitted or reflected by the film can be integrated. For a film of finite thickness, h, the resultant expression for R is complicated:

$$R_P = \frac{S[1 - e^{-2h\sqrt{K(K+2S)}}]}{K+S+\sqrt{K(K+2S)} - K+S-\sqrt{K(K+2S)}\ e^{-2h\sqrt{K(K+2S)}}} . \quad (17.7)$$

For films of infinite thickness this simplifies to,

$$R_P = \frac{S}{K+S+\sqrt{K(K+2S)}} , \quad (17.8)$$

which may be transformed to,

$$\frac{K}{S} = \frac{(1-R_P)^2}{2R_P} . \quad (17.9)$$

The ratio K/S thus determines the reflection from a pigmented layer of sufficient thickness to give complete hiding. Because of reflections at the air/paint interface, not considered in the original Kubelka–Munk treatment, R_P is not the same as the actual reflection from the paint film (see 17.3.2 above). However, in dealing with the colour of pigmented films it is convenient to calculate R_P at each wavelength and then to make appropriate corrections to give R_λ.

The second development in the theory of colour of pigmented films was the assumption by Duncan [19] that absorption and scattering coefficients of different pigments could be added in proportion to the concentrations of the pigments.

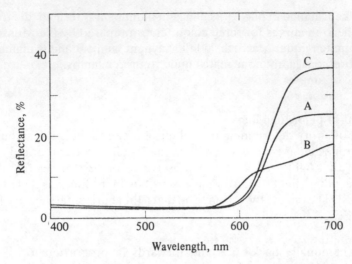

Fig. 17.9 — Reflection curves of metameric matching colours.

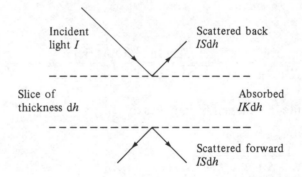

Fig. 17.10 — Kubelka–Munk absorption and scattering in slice of paint film.

Mathematically,

$$\frac{K}{S} = \frac{C_1 K_1 + C_2 K_2 + \ldots}{C_1 S_1 + C_2 S_2 + \ldots}. \tag{17.10}$$

This assumption, which amounts to saying that the presence of other pigments in the film does not affect the scattering or absorbing power of a pigment particle, proved, like several similar rules in physics, e.g. Dalton's law of partial pressures, to be a useful working rule. It enables R_P, and thus R, to be calculated for any mixture of pigments once the appropriate absorption and scattering coefficients are known for the individual pigments.

17.4.3   Colour prediction

The prediction of the mixture of pigments that will produce a desired colour in a paint film, obtained by computation, has largely replaced trial and error formulation by mixing pigments on a slab. The first colour prediction programmes were for the simplified case of films of hiding thickness, but as more powerful computers became available calculation has been extended to films of non-hiding thickness over backgrounds of specified reflectivity. These more complex calculations also deal with absolute concentrations of pigments rather than the relative proportions predicted by the earlier computations. The accuracy of the predictions must depend on the degree to which the underlying Kubelka–Munk theory applies to scattering and absorption in the film, although shortcomings of the theory can be partly confounded by using K and S values determined for each pigment, in the media used in the paint, and at concentrations close to those to be used in the final mixed system.

Many different colour prediction programmes have been produced. They can be based on calculating the amounts of selected pigments that will give a close match to the colour values X, Y, Z of the desired colour, in which case metameric solutions are possible, or on a close match to a series of R_λ values. For simple tristimulus matching a unique solution can be found only for three tinting pigments (more can give an infinite number of solutions); spectrum matching at more than three wavelengths accommodates more variables, but as more pigments are introduced the calculations become much more lengthy and more difficult to resolve to the optimum solution. One compromise is to calculate formulations for several selected sets of tinting pigments, and to choose between the results on the basis of cost, light-fastness, or other known properties and optical factors such as degree of metamerism, which can be computed easily.

A useful spin-off of the methods for predicting formulations is that the effect on the colour of the film of small changes in the concentrations of each pigment can be calculated without difficulty; indeed, this is often a necessary part of the method of successive approximations used to arrive at a solution of the main problem. From these tinting differentials or tinting factors it is a matter of simple algebra to calculate the additions necessary to correct a batch initially off colour by small amounts ΔX, ΔY, and ΔZ. Again, these corrections will be accurate only if the paint system behaves in accordance with the Kubelka–Munk and Duncan assumptions. Major disturbances such as flotation of one pigment of flocculation in a mixed system cannot be accommodated.

17.5   CHANGES IN PAINT FILMS

17.5.1   General

Discussion of factors controlling appearance is not complete without considering the changes that occur during film formation and subsequently. For solution type paints the main change is a rapid shrinkage during the drying stage and, often, a continuing slow shrinkage as residual voltatile material is lost over a period of days or months. Oxidizing media also shrink slowly (by as much as 10% over 3 months) owing to scission and loss of volatile breakdown products; these processes also result in a rise in refractive index of the residual binder, thus reducing scattering by white pigment particles.

Some basic pigments such as zinc oxide and white lead react with acidic products of oxidation, so that metal soaps build up around the pigment particles. This process, combined with general shrinkage, results in roughening of the surface with severe loss of gloss if the particles or clusters of particles are large; the 'hazing' of zinc oxide paints is an example.

Latex paints form films by aggregation and partial coalescence of the polymer globules as water is lost from the film. This process involves the loss of light-scattering from the latex particles and concentration with some clustering of the pigment particles. The result is again reduction in light-scattering, leading to reduced opacity of the dried film and increase in depth of colour, because tinting pigments are less affected by the changes. It is difficult to achieve high gloss from latex paints because pigment tends to cluster between groups of coalesced polymer globules, giving some surface roughness (see Fig. 17.11). One method used to preserve more

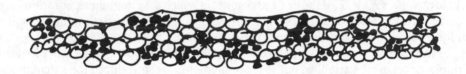

Fig. 17.11 — Diagrammatic structure of poor quality latex paint film. ○○ Partially coalesced polymer particles; ●● Clusters of pigment particles.

uniform pigment dispersion is to incorporate a proportion of water-soluble, film-forming material that can hold the particles apart and plasticize the film shrinkage process.

17.5.2   Opacity changes in films
For the reasons dicussed in 17.5.1, and others, white or lightly coloured paint films can show very considerable changes in opacity as the films dry. There are two main factors:

(a) *Changes in refractive index*
Scattering is proportional to

$$\frac{(n_1 - n_2)^2}{(n_1 + n_2)^2}$$

where n_1 is the refractive index of the pigment and n_2 that of the medium around it. When a long oil alkyd resin thinned with white spirit dries and oxidizes, n_2 may rise from perhaps 1.46 to 1.53. For a pigment of refractive index 2.7 (rutile titanium dioxide) the reduction in scattering is 14%. For a latex paint where the medium is thinned to at least 60% water ($n = 1.34$) the reduction in scattering may be even greater. It is important to realize that a reverse effect occurs with a lightly bound latex paint or a distemper when after film formation some of the pigment is effectively in air. A rutile particle in air will, theoretically, scatter 158% more than one in a medium of refractive index 1.5; in practice not all this gain is achieved because in the dry film the particles are not completely separated, but there is a major gain in opacity. This phenomenon is sometimes termed 'dry hiding'; it explains why a cheap, underbound emulsion paint film can be more opaque than an expensive, fully-bound paint containing more prime pigment.

(b) *Changes in concentration*
Pigments scatter most effectively when each particle is at a considerable distance from its neighbour. Experiments have shown that scattering coefficients, calculated on the amount of pigment, remain reasonably constant up to a volume concentration of 5%, and then fall rapidly (Fig. 17.12) [20, 21, 22]. It follows that the increase of

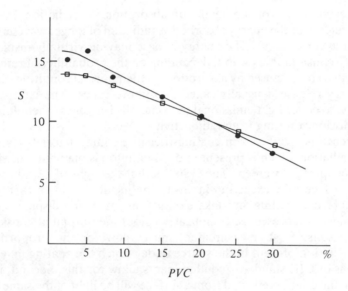

Fig. 17.12 — Scattering coefficient (calculated on volume of pigment)/*PVC*. Rutile TiO_2 in alkyd resin. ⊙ 0.34 μm particle size. □ 0.44 μm particle size.

concentration that occurs as solvent evaporates from a film, where the pigment concentration in the dried film exceeds 5%, must result in some decrease in opacity. This effect is often intensified by some flocculation of the pigment or by crowding into regions of the film, resulting from Bénard cell movements. The overall result is to produce a greater fall in opacity than that due to refractive index changes [23].

17.5.3 Colour changes
Reduction in scattering for the reasons discussed in 17.5.2 usually results in significant increase in the depth of colour as a paint film dries; the exception is where pigmentation is high enough to give dry hiding, including also application to a porous substrate where binder can be lost by penentration. For most paints the colour

change on drying is reproducible, so that it is possible to control the colour of the dry film by adjusting the wet paint colour. This makes possible fully automated colour control during manufacture of repeat batches of paint, by blending liquid tinting paints to give the required wet bulk colour.

17.6 FLUORESCENCE AND PHOSPHORESCENCE

Discussion so far in this chapter has been concerned solely with reflection of incident light or its absorption by films and substrates. Consideration of optical properties of coatings is incomplete without reference to changes in colour of reflected light that can occur with some materials, and stimulation of emission of light by other forms of energy.

Fluorescence is the phenomenon of absorption of radiation by a material followed by release of the energy absorbed as radiation of longer wavelength (that is, as quanta of lower energy). The energy release may be virtually instantaneous or phased over a considerable period, depending on the probability of return of atoms from the excited state, caused by absorption, to their normal condition. Substantially delayed energy release is usually called phosphorescence. The latter term is also used, loosely, to cover light emission due to chemical changes (chemiluminescence) and light emission resulting from radioactivity.

Fluorescence is significant in coatings because some materials transform near ultraviolet radiation which is present in daylight into visible light emission, or blue light (for example) into green. The result is that the light of certain wavelengths leaving the surface may exceed in intensity the incident light of that wavelength. Daylight fluorescent paints and inks exploit this effect for display purposes and hazard warnings. Fluorescence complicates colour measurement considerably, for two main reasons. First, measurement must allow for the proportion of near ultraviolet radiation present in the source under which the coating may be viewed; later revisions of CIE standardized illuminants allow for this. Second, it cannot be assumed that the energy reflected from a surface will be light of the same wavelength as the incident light, so that it is necessary to illuminate with the complete spectrum and then to analyse the light reflected rather than to illuminate with narrow wavelength bands in succession. Phosphorescence presents even greater problems of measurement, but fortunately is confined to a small proportion of materials used only for very specialized purposes.

17.7 COLOUR APPRECIATION

17.7.1 General

It is a truism that colour exists only in the mind of the observer; nobody can ever know how another person perceives colour, only what he accepts as the same colour or judges to be different. Experimental evidence does, however, indicate the extent to which people agree and the factors that govern the average person's judgement. It is also possible to infer some relationships between perception of colour and the physical factors of wavelength distribution and intensity that influence perception (see 17.4.1). Some of these relationships and some of the variations between individuals will be discussed; for more detailed consideration further reading is recommended.

17.7.2 Anomalous colour vision

At least 10% of human males, but a considerably smaller proportion of females, show colour perception markedly different from the normal described in 17.4.1. The higher male figure is because most anomalous colour vision results from a sex-linked recessive genetic deficiency that is transmitted through females but rarely shown by them. The anomalies may be of several types, the most common being protanomaly (ascribed to reduced sensitivity of the red receptor) and deuteranomaly (ascribed to reduced ability to separate the responses of the red and green receptors). The differences are easily shown by an anomaloscope in which a mixture of red and green lights is used to match a spectral yellow; the protanomalous observer needs more red than a normal observer, and a deuteranomalous observer more green. A quick check can also be made with Ishihara test plates [24] which consist of patterns of variously coloured spots that either a normal or an anomalous observer, respectively, may recognize as clear numbers or patterns or 'Colorules' — slide rules with which highly metameric coloured patterns are matched [25, 26]. Much more rarely, observers are found with no colour discrimination (presumably seeing only shades of grey) or with tritanopia, that is a reduced blue response so that some blues and greens and some pinks and yellows are confused.

Severe colour vision anomaly is a serious deficiency for anyone engaged in a colour-using industry so that it is desirable to screen workers for any sensitive department. But the widespread incidence of anomalous vision also means that metameric matches are likely to be unacceptable to a significant proportion of observers, which is a strong argument in favour of complete spectral matching.

Even the normal observer analyses differently light received on different parts of the retina, the light-sensitive coating on the back of the eye. This is because of anatomical variations; near the central, most sensitive area the cone receptors are closely packed to give the highest spatial discrimination, but further away the cones are interspersed with rod receptors that operate largely in dim light and are responsible for twilight vision characteristics (when yellows become dark and blues relatively light). When under normal daylight an observer focuses on an object the image of which covers only the foveal area the response approximates to that of the tritanope. In extreme cases metameric matches that are acceptable when viewed closely, i.e. by a large area of the retina, will be utterly rejected when seen from a few yards away; often with such matches, when the two patterns are placed in contact, a red spot can be seen on one pattern as the eye scans the boundary, balanced by a blue green or greyish spot on the other.

17.7.3 Fatigue and contrast

The photochemical changes in the retina that result from light-absorption are communicated to the brain by electrical nerve pulses. This complex mechanism succeeds surprisingly well in conveying information on changes in intensity and wavelength distribution, but fatigues, at some stage, under extreme conditions or where extreme contrasts occur. This results in reduced discrimination when very bright and saturated colours are present in the visual field, and anomalies in colour perception, especially when contrasting colours are presented. These factors have some bearing on visual colour matching. Thus, samples are best judged against standards with a background similar in lightness to that of the standard, and whilst a considered judgement may be made on light or unsaturated colours, for bright oranges and reds a snap judgement on a match is likely to be more discriminating.

17.7.4   Appreciation of colour differences

The CIE system of colour measurement in terms of X, Y, and Z can be looked upon as establishing a position in a colour solid in which X, Y, and Z are measured along rectilinear axes. The distance between two colours represented by X_1, Y_1, Z_1 and X_2, Y_2, Z_2 is thus:

$$\sqrt{[(X_1 - X_2)^2 + (Y_1 - Y_2)^2 + (Z_1 - Z_2)^2]} \; . \tag{17.11}$$

Unfortunately, equal distances in this colour solid do not represent equally perceptible colour differences. This was shown by MacAdam [27] who studied the variation of colour matches, and thus of just perceptible colour differences, for different colours. Fig. 17.13 shows MacAdam ellipses for colours of constant lightness but

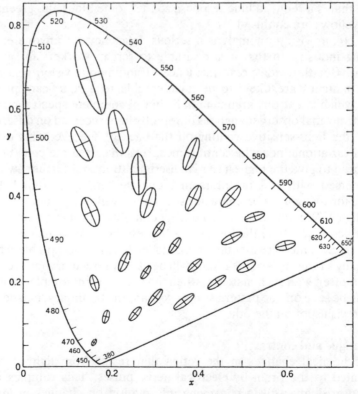

Fig. 17.13 — MacAdam ellipses indicating preceptibility of chromaticity differences on the x/y diagram. The ellipses correspond in size to 10 times the standard deviation of matches to a colour represented by the central point.

varying x and y. Clearly, distances in this x/y plot exaggerate colour differences in the green region of the field, and minimize those in the blue and orange/brown. Many attempts have been made to transform the XYZ colour space into one giving more uniform correspondence with visual appreciation. A perfect answer is not possible because visual appreciation varies with field size and colour of surrounding field and to some extent illumination levels, but much better approximations have been found. Thus the response of the eye to lightness change, represented by the luminance factor Y in the CIE system, is not a linear relationship; that is, if Y is 100 for white, the grey visually equidistant from black and white will be considerably darker than that for which Y is 50. Again, x and y can be transformed by linear equations to new coordinates α and β which convert the MacAdam ellipses into something nearer circles of more uniform size. Of the many transformations proposed the L^* a^* b^* system published by the CIE in 1976 has proved to be one of the most useful in practice [16, 28], and its use is being standardized in ISO 7724 Paints and Varnishes — Colorimetry. This system is complicated, but basically employs cube root functions of X, Y, and Z for all light colours, and linear functions for dark colours. The colour difference is calculated as $\sqrt{(\Delta L^{*2} + \Delta a^{*2} + \Delta b^{*2})}$. It is recommended that for near-white specimens the colour difference should be described in terms of ΔL^*, Δa^*, and Δb^*, but for colours the difference can be analysed into ΔL, a chroma difference ΔC^* (representing a difference in depth of colour), and a hue difference ΔH^*.

17.7.5 Effects of Illumination

The effect of variation in the energy distribution of light sources on perceived colour has already been indicated. These effects are of great importance in all specification and measurement; even the opacity of a white paint film will be greater when measured with a light containing a high proportion of blue radiation than with one where red predominates. International specifications of illuminants have generally been related to sunlight or to a tungsten-filament lamp. The CIE initially standardized Illuminant A, corresponding to a tungsten lamp at a colour temperature of 2854°K, Illuminant B, corresponding to sunlight, and Illuminant C, corresponding to daylight. Either B or C was used generally for colorimetry until it was found that measurements on rutile titanium dioxide paints were not in accord with visual experience. This led to more careful examination of response curves and to the adoption of new standardized illuminants with greater short wavelength contents. The CIE standardized D_{65} based on natural daylight at a colour temperature of 6500°K is now the most widely used; its energy distribution is tabulated from 320–780 nm (see Table 17.2) and compared with those of Illuminants A and C.

17.8 FURTHER READING

Judd & Wyszecki [13] is probably the best general review for the paint technologist; the earlier 1962 edition by Judd alone is shorter and more easily read. Billmeyer & Saltzman [14] is a clearly presented state of the art summary with a useful appendix on more specialized publications. McLaren [16] is particularly valuable on colouring properties of dyes and pigments. Wright [12] written by a pioneer of colour physics is

Table 17.2 — Energy distribution of standard illuminants (normalized at $E_{560} = 100$).

Wavelength (nm)	Relative spectral energy		
	D65	Illuminant C (daylight)	Illuminant A (tungsten)
320	20.2	0.01	
340	39.9	2.6	
360	46.6	12.3	
380	50.0	31.3	9.8
400	82.8	60.1	14.7
420	93.4	93.2	21.0
440	104.9	115.4	28.7
460	117.8	116.9	37.8
480	115.9	117.7	48.2
500	109.4	106.5	59.9
520	104.8	92.0	72.5
540	104.4	97.0	86.0
560	100	100	100
580	95.8	92.9	114.4
600	90.0	85.2	129.0
620	87.7	83.7	143.6
640	83.7	83.4	158.0
660	80.2	83.5	172.0
680	78.3	79.8	185.4
700	71.6	72.5	198.3
720	61.6	64.9	210.4
740	75.1	58.4	221.7
760	46.4	55.2	232.1
780	63.4	56.1	241.7

possibly the best source book on colour vision. Wyszecki & Stiles [15] has the most comprehensive treatment of colour perception and the mathematics of presentation of colour space and calculation of colour differences, but is perhaps a book for the specialist rather than the general technologist. Hunter [2] is recommended for general background on perception and measurement of appearance properties.

REFERENCES

[1] Heavens, O. S. *Optical properties of thin solid films*, Butterworth (1955).
[2] Hunter, R. S. *The measurement of appearance*, John Wiley & Sons (1975).
[3] Ross, W. D. Theoretical computation of light scattering power, *J. Paint Tech.* **43** 50 (1971).
[4] Mie, G. *Ann. Physik* **25** 377 (1908).
[5] Steig, F. B. Effect of extenders on hiding power of titanium pigments, *Off. Dig.* **31** 52 (1959).
[6] Siener, J. A. & Gerhart, H. L. *XI Fatipec Congress 1972, Proceedings* 127.
[7] PPG Industries US patent 3 669 729.
[8] Chalmers, J. R. & Woodbridge, R. J. Air and polymer extended paints, *Paint R.A. 5th International Conference, P.R.A. Progress Report* **3** 16 (1983).

[9] Kershaw, R. W. A new class of pigments, *Australian OCCA Proceedings and News* **8** No 84 (1971).

[10] Loney, S. T. Scattering of light by white pigment particles, *Paint R.A. Technical Paper* No 213 (1960).

[11] Carr, W. Dispersion — the neglected parameter, *J. Oil Col. Chem. Assoc.* **65** 373 (1982).

[12] Wright, W. D. *The measurement of colour* 4th ed. Adam Hilger (1969).

[13] Judd, D. B. & Wyszecki, G. *Color in business, science and industry* 3rd ed. John Wiley (1975).

[14] Billmeyer, F. W. & Saltzman, M. *Principles of color technology 2nd ed.* John Wiley (1981).

[15] Wyszecki, G. & Stiles, W. S. *Color sciences — concepts and methods* 2nd ed. John Wiley (1982).

[16] McLaren, K. *The colour science of dyes and pigments*, Adam Hilger (1983).

[17] Kubelka, P. & Munk, F. *Z. Tech. Physik* **12** 593 (1931).

[18] Kubelka, P. *J. Opt. Soc. Amer.* **38** 448 (1948).

[19] Duncan D. R. The colour of pigment mixtures, *J. Oil Col. Chem. Assoc.* **32** 296 (1949).

[20] Vial, F. On the dependency of scattering coefficients on PVC, *IX Fatipec Congress* 1968 1–40.

[21] Tioxide Plc *Opacity with tioxide pigments*, BTP Leaflet 107.

[22] Steig, F. B. Hiding power of titanium pigments, *Off. Dig.* **29** 439 (1957).

[23] Bullett, T. R. & Holbrow, G. L. Wet and dry opacity of paints, *J. Oil Col. Chem. Assoc.* **40**, 991 (1957).

[24] Ishihara, S. *Test for colour blindness*, Tokyo, Kanehara Shuppan (1973).

[25] Glenn Colorule — obtainable from AATCC, Research Triangle Park, North Carolina, USA.

[26] Davidson & Hemmendinger Color Rule — obtainable from Munsell Color Co., Newburgh, N.Y., USA and Kohlmorgen (UK) Ltd., Bridgewater House, Sale M33 1EQ.

[27] MacAdam, D. Visual sensitivity to color differences in daylight, *J. Opt. Soc. Amer.* **32** 247 (1942).

[28] Robertson, A. R. The CIE color difference formulae, *Color Res. Appl.* **2** 7 (1977).

18

Specification and control of appearance

T. R. Bullett

18.1 GLOSS

18.1.1 Classification

The gloss of paint films is classified according to the degree to which they exhibit specular reflection. The broad descriptions are:

full gloss — showing clear specular reflection at all angles of view
semi-gloss — showing specular reflection when viewed at low angles to the surface but only a hazy reflection at higher angles
eggshell — showing hazy reflection for all angles of view with, possibly, clear specular reflection near grazing incidence
flat (matt) — showing no specular reflection even at grazing incidence, and, for full matt, no preferential reflection around the specular angle

The term 'oil gloss' is sometimes used for a level between full gloss and semi-gloss, and represents the typical appearance of old-fashioned oil paints when first applied. In BS 2015 : 1965 *Glossary of Paint Terms* 'eggshell' is further divided into 'eggshell gloss' and 'eggshell matt'. 'Silk' or 'satin' are also used, particularly for emulsion paints, for the eggshell range.

Gloss is not a simple property that can be assigned a value on a linear scale. There are several different characteristics that must be considered:

(1) Intensity of specular reflection or brightness or reflection at or close to the specular angle.
(2) Distinctness of images, that is the detail that can be resolved in a pattern reflected in the surface.

(3) Grazing incidence sheen, that is preferential reflection near the specular angle for light at near-grazing incidence.

Even these three characteristics are not sufficient to differentiate between all surfaces that are visually different, particularly in the semi-gloss range where it is necessary to analyse the distribution of reflected light over a wide range of angles, and in the matt range where a rough surface causes preferential back reflection, so that some films may appear much brighter when viewed from the direction of incidence than from the usual direction of specular reflection.

To understand the reasons for these differences in appearance it is useful to consider the distribution of light reflected from some typical films. In Figs. 18.1 and 18.2 illumination by a very narrow beam at 60° is assumed. For a perfect mirror all the

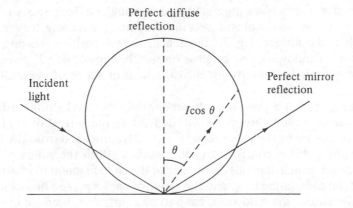

Fig. 18.1 — Distribution of reflected light from perfect mirror and perfect diffuser.

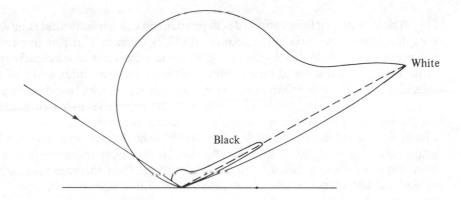

Fig. 18.2 — Distribution of reflected light from semi-gloss black and white films.

light is reflected at the specular angle. For a perfect diffuser the intensity of reflected light varies as the cosine of the angle with the normal, giving the circle shown in Fig. 18.1; for an extended surface this is equivalent to looking equally bright from any direction of observation. A semi-gloss white paint shows a circular plot of reflection, from the underlying pigment, with a spike of reflection around the specular angle; the narrowness of the spike and its height at the specular angle are a measure of the gloss of the paint. A semi-gloss paint shows an almost insignificant circular plot with a spike again varying in shape with the gloss (Fig. 18.2).

18.1.2 Specular reflection gloss measurement

A great deal of attention has been given to the measurement of gloss by specular reflection. For many years the standard method in the UK was that detailed in BS 3900 : Part D2 : 1967 in which specular reflection at 45° is compared with that from a glossy black glass standard. This method has now given place, largely, to BS 3900 : Part D5 : 1980 [3], which is essentially the international standard ISO 2813—1978, in turn based on a much older ASTM standard [2]. Part D5 also involves comparison with reflection from a black glass standard, but angles of incidence of 20°, 60°, or 85° are used according to the gloss levels involved and the reasons for measurement. A high angle to the surface, e.g. 20° incidence, gives a sensitive measure for high-gloss paints, but for following loss of gloss on weathering 45° or 60° gives a more open scale; 85° gloss is useful only for measurement of sheen of eggshell or near-matt paints.

From Fig. 18.2 it is clear that the amount of reflected light recorded as specular reflection from a semi-gloss paint will depend on the spread of angles around the specular picked up by the glossmeter. Fig. 18.3, reproduced from BS 3900 : Part D5, illustrates how this is controlled in the method. Both the range of angles in the incident beam, which depends on the size of the lamp filament in relation to the focal length of the collimating lens, and the range of angles picked up, which depends on the angular size of the field stop, have to be controlled. Increasing either of these angles reduces discrimination for high-gloss and opens the scale at the low-gloss end [4].

There are several other points to remember:

(1)   The intensity of light reflected is dependent on the refractive index as well as on the planarity of the surface (see equation (17.3) Chapter 17). For this reason the refractive index of the black glass used for comparison has to be clearly specified. Additionally, a paint based on a medium of high refractive index will give a higher specular reflection value than a paint of equal surface planarity based on a medium of lower refractive index. Table 18.1 indicates the refractive index of some typical media and the specular reflection from perfectly plane surfaces, compared with the black glass standard of BS 3900 : D5 as 100. The human eye also tends to be impressed by the higher lustre of reflection from a high refractive index surface (diamonds are distinguished from glass!), but imperfect surfaces may be wrongly rated by glossmeters because of refractive index differences.

(2)   Directional defects such as residual brush marks strongly affect gloss measured across the direction of the defects, but have much less effect on measurements along the direction.

(3)   Substrates must be very flat because of the small tolerances in angles of

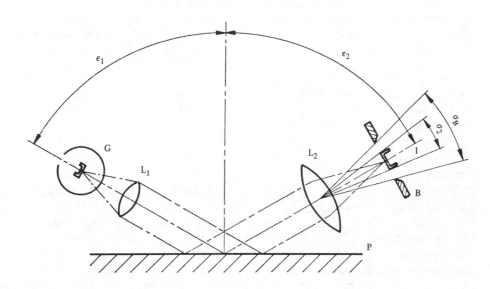

Fig. 18.3 — Optical specification of 60° glossmeter BS 3900 : Part D5. G = lamp; L_1 and L_2 = lens; B = receptor field stop; P = paint film; $\varepsilon_1 = \varepsilon_2 = 60 \pm 0.2°$; σ_B = receptor aperture angle = $4.4 \pm 0.1°$; σ_2 = source image angle = $0.75 \pm 0.25°$; I = image of filament.

incidence and reflection. It is usual, therefore to spread films for measurement on glass plates. When transparent films are measured the glass must be either opaque or the underside must be coated to avoid a second specular reflection from the underlying glass/air interface; merely placing the glass plate over a black surface is not sufficient.

18.1.3 Distinctness of image gloss
There are several shortcomings to the specular reflection method for assessing gloss of very glossy surfaces. Anomalies due to refractive index differences have already been mentioned; these can result in a rise in specular reflection with film age for alkyd gloss paints, which is not correlated with any improvement of glossiness. A second problem is that accurate specular reflection measurements can be made only on a plane substrate; the method is inapplicable on the curved surfaces of a car body. Perhaps even more important is that specular reflection glossmeters to the ASTM and BS specifications do not enable significant differences in gloss resulting from improved dispersion of very fine pigments to be measured; these differences involve the clarity of mirror reflection without substantial change in the amount of light reflected in a cone of 1° or 2° about the specular angle. These various shortcomings are avoided when gloss is assessed in terms of distinctness of image.

Table 18.1 — Refractive indices of paint media and comparative materials with relative specular reflection intensity.

Material	n_{D25}	Specular reflection (cf. BS 3900 : D5 Standard = 100)
Glass standard (BS 3900 : D2)	1.523	88.1
Glass standard (BS 3900 : D5)	1.567	100
Polytetrafluorethylene (PTFE)	1.35	45.5
Polyvinylidene fluoride (PVF$_2$)	1.42	61.7
Polybutyl acrylate	1.466	73.2
Polyvinyl acetate	1.466	73.2
Polyester resins	1.523–1.54	88.1–92.6
Long oil alkyd resin films	1.53–1.55	89.9–94.7
Epoxy resins	1.55–1.60	94.7–109.1
Chlorinated rubber	1.55	94.7
Polyvinylidene chloride	1.60–1.63	109.1–117.6
Phenolic resins (unmodified)	1.66–1.70	126.1–137.8

Direct visual methods for assessing distinctness of image have been based on assessing the finest patterns of lines or figures that can be clearly resolved under controlled conditions [5]. The technique can be sensitive but is subject to variations of judgement between individuals common to all such sensory methods. Nevertheless, such methods are most useful for assessment of gloss achieved in production, for example on car bodies. Many attempts have also been made to measure the physical equivalent of distinctness of image. This is essentially the sharpness of the specular spike as indicated in Fig. 18.2. In Fig. 18.4 the specular reflection from two equally bright points in the object at slightly different angles to the surface is indicated. The two points will be distinct in the image if the minimum in the $R_A + R_B$ curve is significantly below the heights of the peaks corresponding to reflection from A and B. Direct measurement of this discrimination using a twin source requires very precise equipment. A rather more straightforward method is to scan one reflected beam with a photometer that is swung across the image plane of the lens and to record, for example, the angular width between peak and half-peak recordings. A cruder technique that was tentatively introduced by ASTM was to compare specular gloss readings obtained using a very small and a larger field lens aperture; this technique eliminates refractive index anomalies in the medium range of gloss but is insufficiently sensitive, at least with relatively inexpensive equipment, for the high-gloss range.

18.1.4 Assessment of sheen

Specular reflection at low angles to the surface is the source of glare on table tops under bright light, and of stray light in optical instruments; and it is often undesirable in interior decoration, where it shows up unevenness of walls and ceilings. One

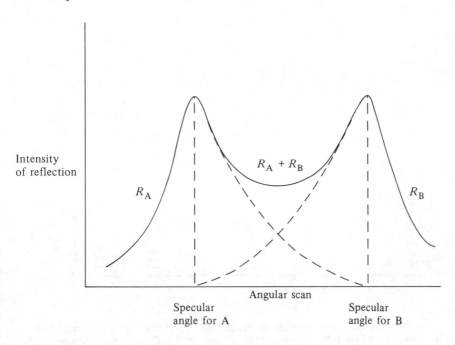

Fig. 18.4 — Distinctness of images from point sources.

method for assessing this so-called sheen effect is to measure specular reflection at a near-grazing angle, e.g. 85° to the normal [3]. Results obtained by this method are sensitive to planarity of test specimens and to correct alignment of the glossmeter, but the technique can be used as a go/no-go method to exclude finishes showing significant sheen even if its value for exact measurement is doubtful. Visual methods might seem simpler and, therefore, more attractive. From about 1940 to 1980 the standard UK method, described in its final form in BS 3900 : Part D3 : 1975 [7] (withdrawn June 1983), was to assess the highest angle from the surface at which an image of a pattern of black and white bars, reflected in the surface, could be resolved. The technique gave reproducible results for the same observer; but, inevitably, there were frequent differences of opinion as to whether a pattern was just visible or not. Eventually, despite many attempts to improve the test apparatus and to standardize viewing conditions more closely, the method was dropped by the committee responsible for BS 3900 methods, being replaced, effectively, by the 85° specular reflection method of BS 3900 : Part D5 : 1980 (Fig. 18.5).

18.2 OPACITY OF PAINT FILMS

18.2.1 General
One of the practically important characteristics of an applied paint film is the uniformity of colour over the surface. Unless films are applied at a thickness sufficient to give complete opacity, which is rarely the case than for very dark or metallic paints, colour variation will result either for thickness variation or from

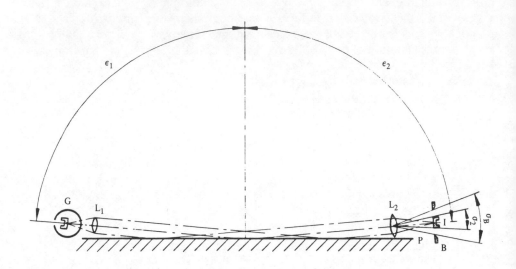

Fig. 18.5 — Optical specification of 85° glossmeter BS 3900 : Part D5. G = lamp; L_1 and L_2 = lens; B = receptor field stop; P = paint film; ε_1 = ε_2 = 85 ± 0.1″; σ_B = receptor aperture angle = 4.0 ± 0.3″; σ_2 = source image angle = 0.75 ± 0.25″; I = image of filament.

variation in colour of the substrate. Most measurements of paint film opacity or hiding power are evaluated in terms of ability to obscure the colour variation of the substrate, and are expressed as contrast ratio, e.g. reflection over black divided by reflection over white; contrast ratio may be expressed either as a fraction or a percentage. For strongly coloured paints the variation in colour due to film thickness variation over a uniform substrate may, in practice, be more significant. Examples are the effects of heavy brush marks and the effects of retraction from sharp edges, which are often evident on painted mouldings. These effects are linked with the flow properties of the paint and are probably best assessed by practical trials on standardized substrates. Their significance can be reduced in practical painting by appropriate choice of undercoats. One practical method for assessment of film thickness dependence for emulsion paints is to paint a large blackboard with one coat overall, a second coat over the lower half, and a third over the right-hand half; the solidity of areas carrying one, two, and three coats can then be compared. Tests of this kind, carried out by experienced painters, enable the relative hiding characteristics of different paints to be compared under a simulation of practical application.

18.2.2 Contrast ratio determination

The most widely used method for assessing the opacity of paints is to apply a film of controlled thickness over a background consisting partly of black areas and partly of white. Initially, checkerboard patterns such as Morest charts were used, and the contrast over black and white areas was judged visually. This subjective judgement is not reproducible, and has largely given place to photometric measurements. There are several aspects of contrast ratio assessment that merit discussion; some are practical points involved in the measurement, some are theoretical, and some involve interpretation of the results obtained in relation to practical painting.

18.2.2.1 Measurement of contrast ratio

Meaningful results are only obtained on uniform films of known thickness. Drawdowns using doctor blades on flat black-and-white glass panels can be effective. An alternative is to spread the paint on transparent plastic foil, e.g. polyester of uniform thickness, and to measure with the painted foil in optical contact with glass plates (a liquid of similar refractive index to that of the foil gives optical contact) [8]. With both methods the thickness of the paint film can be found from its weight per unit area. White paints are much more transparent to red light than to blue, so that to achieve correspondence with visual assessment the combination of light source and recording system must approximate in spectral response to that of the eye in daylight. Failure to achieve correct spectral response is one of the chief causes of error in contrast ratio measurements.

18.2.2.2 Theoretical basis of contrast ratio

If the paint film is considered to have reflectivity R_0 over a non-reflecting background and R over a background with reflectivity R_G, and the transmission of the film is T, then it can be shown that:

$$R = R_0 + \frac{T^2 R_G}{1 - R_0 R_G} \tag{18.1}$$

This expression is not exact because it does not allow for the dependence of T on the distribution in angle of the incident light, or the fact that R_G may be modified by painting over, which eliminates a glass/air interface on a glass standard. However, the expression is useful when considering the effect of variation in R_G on contrast. Table 18.2 shows the variation of R with R_G for two typical cases: a white paint film ($R_0 = 0.7$, $T = 0.25$), and a tinted paint of good opacity ($R_0 = 0.5$, $T = 0.1$). In both cases R is seen to be much more sensitive to change of R_G for light substrates than for dark substrates. The conclusion from these calculations is that to make reproducible contrast ratio measurements, close specification of the black substrate is not necessary, but the reflectivity of the white substrate must be very closely controlled.

The calculations in Table 18.2 also show how much absorption in the film contributes to hiding; in case (a) only one twentieth of the incident light was assumed to be absorbed in one passage through the film, but for case (b) 40% was absorbed.

18.2.2.3 Relation to practical painting

The argument in 18.2.2.2 might seem to suggest that it is much easier to hide variations in a dark substrate than those in a lighter one. However, when the variations in R that can arise from film thickness variation is also considered, the use

Table 18.2 — Effect of luminance factor of substrate on reflectance of white and light grey paint films.

Substrate reflectance	(a) White paint $(R_0 = 0.7, T = 0.25)$		(b) Grey paint $(R_0 = 0.5, T = 0.1)$	
	R	Contrast ratio (R_0/R)	R	Contrast ratio (R_0/R)
0	0.7	1	0.5	1
0.1	0.707	0.99	0.501	0.998
0.2	0.715	0.98	0.502	0.996
0.4	0.735	0.95	0.505	0.990
0.6	0.772	0.91	0.509	0.982
0.8	0.814	0.86	0.513	0.975
0.9	0.852	0.82	0.516	0.969
1.0	0.908	0.77	0.520	0.961

of dark substrates becomes less attractive; the optimum substrate (or undercoat) colour is generally close to that of a thick film of the coat to be applied.

With dark-coloured paints, which are usually much more opaque than lighter paints, contrast ratio measurements of little value; with these paints failure to hide the substrate is usually due to thin streaks in the film that relate more to flow properties and application defects than to intrinsic poor opacity.

For white paints, possibly the most useful parameter is the degree of hiding that will be obtained for application at an average spreading rate. ISO 3906–1980 published by BSI as BS 3900 : Part D6 : 1982 [9] describes a method for determining contrast ratio corresponding to a spreading rate of $20\,m^2/l$ (i.e. wet film thickness $50\,\mu m$, reasonably typical of brush application). The method involves measurement on six films spanning the desired spreading rate and interpolation.

An alternative approach is to determine the film thickness or weight that will give a specified degree of hiding, by plotting contrast ratio against film weight and interpolating or extrapolating. When this method was first developed the required level specified was a contrast ratio of 0.98. This figure was chosen because a 2% difference in luminance factor was taken to be the smallest contrast that the eye could recognize clearly. Such a high contrast ratio is probably too extreme a requirement for normal painting because, in practice, paint is rarely applied over a black and white substrate. There are also serious experimental difficulties in the accurate determination of high contrast ratios because of inevitable errors in the measurement of luminance factors. For both these reasons it is preferable to work with a contrast ratio of 0.95 over black and white as equivalent in practice to satisfactory hiding for a white paint. To be acceptable a white paint should achieve this figure for an application rate not below $10\,m^2/l$, i.e. in two coats each of $20\,m^2/l$. Calculation shows that for a white paint with almost negligible absorption a contrast ratio of 0.95

in two coats corresponds to approximately 0.85 for a single coat. Thus a figure of 0.85 attained in the BS 3900 : D6 method implies satisfactory hiding in two coats, uniformly applied.

18.2.3   Paint film opacity and scattering coefficients
Opacity can be related to absorption and scattering using Kubelka–Munk theory. This is the basis of the method developed in the USA for assessing the hiding power of paints, which was standardized as ASTM D2805 and in Germany as DIN 53162. The technique was adopted as an international test method by ISO and published as ISO 6504/1–1983, and by BSI as BS 3900 : Part D7 : 1983 [10]. The working form of the Kubelka–Munk relationships used is:

$$R = \frac{1 - R_G(a -- b\coth bSt)}{a + b\coth bSt - R_G} \tag{18.2}$$

where

$$a = \frac{1}{2}\left(\frac{R_\infty + 1}{R_\infty}\right) , \qquad b = a - R_\infty ,$$

R is the reflectance of a paint film of thickness t applied over a substrate of reflectance
    R_G,
S is the scattering coefficient per µm,
t is the film thickness in µm,
R_∞ is the reflectance of a film so thick that further increase in thickness does not
    increase the reflectance, and
a and b are parameters, mathematically related to R_∞, introduced for simplification
    of the working equation.

It will be seen that the reflectance over a black substrate ($R_G = 0$) reduces to

$$R_B = \frac{1}{a + b\coth bSt} . \tag{18.3}$$

Equation (18.3) can be rearranged to give

$$St = \frac{1}{b}\text{arcoth}\left(\frac{1 - aR_B}{bR_B}\right) . \tag{18.4}$$

To simplify calculation the product St is usually determined from published graphs relating St to R_B and R_∞. Once St has been determined it is then possible to calculate the film thickness or spreading rate corresponding to a contrast ratio of 0.98.
    The method requires only the determination of R_∞, which is most easily done for light-coloured paints by applying a thick film over an opal glass or white ceramic tile, and the measurement of R_B for a film of known thickness. For the R_B measurement,

films can be supplied directly, by doctor blade, to black glass. Alternatively, they can be spread on polyester foil and laid over black glass, a few drops of white spirit being first applied to the glass to give optical contact. Despite the complexity of the theory and the apparently unwieldly form of equations (18.2), (18.3), and (18.4), the method when used with all the aids to computation given in BS 3900 : Part D7 is not unduly difficult to use.

For the simple case where absorption in the film can be neglected in relation to scattering, a much simpler expression for R_B can be derived:

$$R_\mathrm{B} = \frac{St}{St+1},$$

which rearranged gives

$$St = \frac{R}{1-R}. \tag{18.5}$$

Hence a plot of $R/(1-R)$ against t should give a straight line plot of slope S. Fig. 18.6

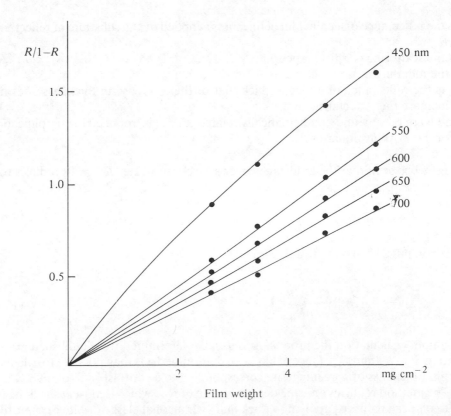

Fig. 18.6 — Scattering coefficients from plot of $R/(1-R)$ vs film weight (ZnO in linseed oil 20% *PVC*).

shows results of measurements at various wavelengths for films of a white paint applied to a black glass substrate. It will be seen that at the longer wavelengths the linear relation holds, and that S increases with decreasing wavelength. At 450 nm, where absorption by the pigment cannot be neglected, linearity is lost.

18.3 SPECIFICATION AND CONTROL OF COLOUR

18.3.1 General

Colour can be controlled purely by reference to standard colour cards by visual judgements of matching. It can be argued that ultimately the eye is the only true arbiter, but this begs many questions, of which the most difficult is, possibly, 'Whose eye?' Apart from the genetic abnormalities discussed earlier it is known that colour vision changes with age due to build up of yellow macular pigmentation in the eye, and that the standard observer is a statistical abstraction. There is thus a strong argument for basing all colour control on physical measurements. Equally, these measurements and their interpretation must be related closely to the responses of visual observers. Colour control in practice, accordingly, has two branches, visual and instrumental.

18.3.2 Visual colour control
18.3.2.1 Colour systems and colour standards

Many attempts have been made to set up comprehensive visual colour systems. The most universal is the *Munsell book of color* first published in 1929 [11]. The complete system has 40 pages, each of a different hue running around the spectrum to red and on through purple back to violet (PB in the Munsell notation). The colours on each page are arranged in rows of equal *Value* (corresponding to Y value) and in columns of equal *Chroma* (corresponding to saturation or depth of colour) (see Fig. 18.7). Each colour has three references corresponding to hue, value, and chroma, e.g. 5YR/5/10 is a saturated orange. A wide range of Munsell colours are available as small chips (either glossy or matt), but the numbering system allows for interpolation or extrapolation. The Munsell system has also been standardized by reflection measurements and some smoothing of spacing in the original system, so that a Munsell book can be used for visualization of tristimulus values. Bearing in mind that the human eye can distinguish at least half a million colours, under optimum viewing conditions, it is not surprising that whilst the Munsell system enables a colour to be specified approximately, it does not replace the use of individual colour cards for precise specification. A further reason for the use of colour cards for industrial purposes is that visual colour matching is difficult to standardize unless the gloss and texture of the surfaces to be compared is also similar. Thus one approach to control of colour of successive batches of paint is to match a first master batch very carefully to a standard, and then to use colour cards prepared from this master batch as working standards.

18.3.2.2 Visual colour matching

Apart from selection of observers with normal or average colour vision the most important factor in visual judgements of matches is the illumination. It is only necessary to look at the colour change of blue or purple flowers in a shaft of sunlight

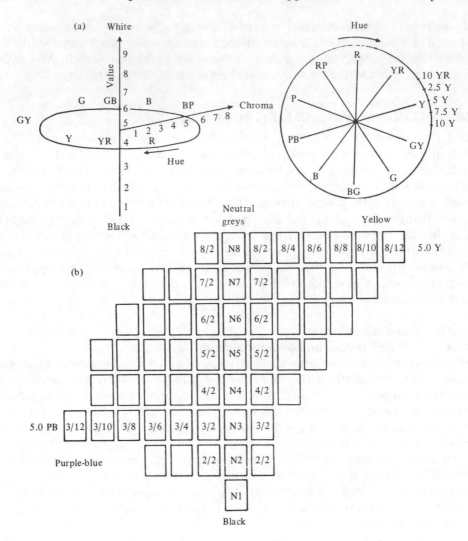

Fig. 18.7 — Arrangement in Munsell book of color.

to realize how greatly colour can change with illumination. The old method in paint factories was to arrange the colour-matching bench under north sky daylight, possibly the most constant natural source. Most matching is now done in booths with carefully selected fluorescent lamps and controlled conditions of viewing. Standard conditions for visual comparison of the colour of paints are laid down in ISO 3668–1976, reproduced as BS 3900 : Part D1 : 1978 [12]. This standard covers both daylight matching under north sky daylight of at least 2000 lux intensity, and artificial light (D 65) matching with illumination between 1000 and 4000 lux. A background of a neutral grey of about 15% luminance factor (Munsell N4 to N5) is recommended for general use, but for whites and near-whites a higher level of 30% (Munsell N6) is preferred. Specimens to be compared are best positioned with a long touching or overlapping edge, and should be viewed from a distance of about 500 mm. Metamerism should be checked by, for example switching to a tungsten lamp or another source of radically different spectral distribution. Often, difficulty in deciding on the quality of a match is an indication of metamerism because, where there are marked differences in spectral reflection curves, the visual response from the foveal region of the retina indicates a mismatch, while that from the surrounding areas may correspond to matching. In extreme cases an observer may actually see a reddish area on one side of the touching line and a blue green area on the other when strongly metameric colours are compared.

One problem with visual colour standards is that they may fade or otherwise change in colour. Regular checking by instrumental means is imperative to guard against drift in standards. Working standards that show significant change must be replaced.

18.3.3 Instrumental colour control

18.3.3.1 General

There are two basic methods for measuring the colours of surfaces. The first is to simulate the supposed analysis made by the eye in terms of responses to three stimuli (as discussed in 17.4.1). This technique is known as 'tristimulus colorimetry', and it sets out to measure X, Y, and Z directly. The second method is to determine reflectance (R) for each wavelength band in turn across the range of the spectrum to which the eye is sensitive, and then to calculate the visual responses by summing products of R and the standard values for distribution of the sensitivity of the three colour responses ($\bar{x}$, $\bar{y}$, and $\bar{y}$). The tristimulus method has theoretical advantages where the materials to be measured are fluorescent, but there are serious practical problems in assuming that a tristimulus colorimeter exactly matches human vision, that is, in eliminating colour blindness from the instrument.

18.3.3.2 Tristimulus colorimetry

A tristimulus colorimeter has three main elements:

(1) a source of illumination, usually a lamp operated at a constant voltage;
(2) a set of three combinations of filters used to modify the energy distribution of the incident or, better, the reflected light;
(3) a photoelectric detector that converts the reflected light intensity into an electrical output.

The general requirement is that the product of energy distribution from the source (E_λ), filter transmission ($\mathbf{F}_\lambda$), and detector sensitivity (S_λ) shall match the product of the spectral distribution of the sensitivity of the eye and the energy distribution of the illuminant (e.g. D 65) to which the tristimulus values are to be referred. A perfect match is impossible, but the best tristimulus colorimeters achieve a reasonable compromise. Usually no attempt is made to match the two peaks of the $\bar{x}$ distribution; instead, the shape of the short wavelength peak is assumed to be close to that of the $\bar{z}$ distribution, and the measured X value (X_M) is increased by an equivalent proportion of the measured Z value, e.g. the recorded tristimulus values may be:

$$X = X_M + 0.18Z_M$$
$$Y = Y_M$$
$$Z = Z_M .$$

Measurements made on a tristimulus colorimeter are normally comparative, the instrument being standardized on glass or ceramic standards. Because the correct responses are not always attained, or maintained during use of the instrument, for best accuracy standardization should be carried out using calibrated standards of similar colours to the materials to be measured. This 'hitching post' technique enables reasonably accurate tristimulus values to be obtained even when the colorimeter is demonstrably colour blind. However, tristimulus colorimeters are most useful for quick comparison of near-matching colours. When, as on most modern instruments, the electrical output is digitalized, colour differences can be automatically computed in L, a, b or L, C, and H units, for use in quality control systems. As with all digitalized recording, it is necessary to remember that the accuracy of results depends on the input rather than the computer; if the spectral response of the lamp/filter/photodetector system is wrong, the recorded colour differences are most probably wrong, despite the precision of the computation and the number of figures in the printout.

18.3.3.3 *Spectrophotometry*

For precise measurement of colour in absolute terms it is advisable to use a spectrophotometer, that is, to measure the reflectance for each wavelength in turn and then to calculate tristimulus values. The advantage over tristimulus colorimetry is that sufficient information is obtained to calculate colour values for any illuminant and that metamerism is automatically detected. The disadvantages are that high-quality spectrophotometers are very expensive and that measurements take longer (although this disadvantage has been greatly reduced by instrument development). As with colorimeters, a built-in or add-on computer can be used to process readings to give tristimulus values under a range of illuminants, colour differences from standards, and variance of colour between repeat specimens or over parts of a surface.

In a spectrophotometer the light is usually split into a spectrum by a prism or a diffraction grating before each wavelength band is selected in turn for measurement. Instruments have also been developed in which narrow bands are selected by interference filters. If fluorescent materials are to be measured, the specimen must be illuminated with the complete spectrum and the reflected light split up for analysis [13]. The spectral resolution of the instrument depends on the narrowness of the bands utilized for each successive measurement. For most paint work a 10 nm bandwidth gives sufficient resolution, but where there are sharp-edged absorption bands, as with some dyestuffs, sharper resolution is desirable. In theory, a spectrophotometer could be set up to compare reflected light directly with incident light, but it is more usual to calibrate against an opal glass standard that has been calibrated by an internationally recognized laboratory. Checks must also be made on the optical zero, e.g. by measurements with a black light trap, because dust or other problems can give rise to stray light in an instrument, which would give false readings.

18.3.3.4 *Illumination and viewing conditions*
With both types of colour-measuring equipment, results obtained and their correlation with visual observation depend on illumination and viewing angles; the effects are greatest for dark, glossy specimens. Colorimetry of paints and varnishes is covered by an international standard ISO 7724† which is to be reproduced in BS 3900. This standard describes six different illuminations and viewing conditions (see Table 18.3). Of these the first, 45/0, and fourth, d/8, specular reflection

Table 18.3 — ISO standard illumination and viewing conditions. Paints and varnishes — colorimetry.

Measurement conditions		Designation (abbreviation)
Illumination	Viewing	
directional 45° ± 5°	directional 0 ± 10°	45°/normal (45/0)
directional 0 ± 10°	directional 45° ± 5°	normal/45° (0/45)
hemispherical integrating sphere	directional 8° ± 2°	diffuse/8° (d/8)
hemispherical integrating sphere with gloss trap	directional 8° ± 2°	diffuse/8° (d/8) specular reflection excluded
directional 8° ± 2°	hemispherical integrating sphere	8°/diffuse (8/d)
directional 8° ± 2°	hemispherical integrating sphere with gloss trap	8° diffuse (8/d) specular reflection excluded

NOTE   In the last four conditions 8° is preferred to 0° because with normal illumination or viewing there may be difficulties with inter-reflections between glossy specimens and the illuminating or viewing optics; specular reflection from glossy specimens can largely be trapped with an 8° angle.

excluded, probably correspond most closely to visual examination in a colour matching cabinet. However, the third and fifth conditions where the specular reflection is included have the advantage of minimizing the effect of gloss differences on colour measurements, and, of course, give results that correspond to colour seen under a cloudy sky or by indirect illumination in a room.

Many colour-measuring instruments now incorporate integrating spheres for

---

† In course of publication.

illumination or viewing. The sphere is coated internally with matt white paint, which must have a uniform reflectivity throughout the visible spectrum (pure barium sulphate is the usual pigment for the topcoat). There are ports for the specimen to be measured, to admit the incident beam, and for viewing; the specimen port should represent only a small proportion of the total surface area, since otherwise repeated reflections from the specimen may increase the saturation of the colour in the sphere. By careful design of the sphere geometry, removable caps or light-trapping cones can be inserted to eliminate specular components and thus enable measurements to be made of diffuse reflection only or of total reflection.

18.3.3.5 Colour tolerances

Under optimum conditions the eye can detect extremely small colour differences, about 1% in Y value or luminance factor and about 0.1 in ΔE on the CIE $L\,a\,b$ system [14]. Industrially, it is unpractical to control colour to within such limits even when sufficiently precise methods for measurement are available. For routine control, therefore, acceptable limits of colour tolerance should be set. The best correlation with visual judgements is possibly given by setting limits on the differences in psychometric lightness L^*, chroma, C^*_{ab} and hue H^*_{ab} as defined in the CIELAB 1976 formulae. These limits can be calculated back to allowable variations in X, Y, and Z for any particular colour, but with dedicated computer facilities available on most colour-measuring equipment it is simplest to work directly in L, C, and H. Some investigations have shown, however, that acceptability is not always simply related to perceptibility of colour difference [14]. Thus a customer may tolerate departure of a neutral grey paint towards blue but not towards yellow.

18.4 COLOUR CONTROL IN PAINT MANUFACTURE

Reference has already been made to calculation of colour of pigment mixtures and computerized formulation and colour correction. Such systems work to maximum advantage if close quality control is kept on all pigments and media used [15]. Metameric batches are avoided if the number of tinting pigments used is kept to a minimum, usually no more than three, and if these pigments have reproducible colouring properties. Given adequate control, most paints can be prepared from single-colour bases and tinting paints by volumetric blending. This is also the logical approach to retail paint supply where a large number of colours are demanded, for example for car refinishing. Such mixing systems are widely used in retail outlets in Scandinavia and North America, although they have not been so generally accepted in the United Kingdom.

ACKNOWLEDGEMENTS

The author wishes to thank the staff of the Paint Research Association Library for general assistance, and Mr David Cain of the Paint Research Association for help with the section of colour measurement.

Extracts from BS 3900 are reproduced by permission of the British Standards Institution. Complete copies of the document can be obtained from BSI at Linford Wood, Milton Keynes, MK14 6LE, England.

REFERENCES

[1] BS 2015 : 1965 : *Glossary of Paint Terms.*

[2] ASTM D523 – 1939 revised as D523 – 67.

[3] BS 3900 : Part D5 : 1980 *Measurement of specular gloss of non-metallic paint films at 20°, 60° and 85°.*

[4] Bullett, T. R. & Tilleard, D. L. Optical properties of paints — the basis of instrumental measurements, *J. Oil Col. Chem. Assoc.* **36** 545 (1953).

[5] Hunter, R. S. Gloss evaluation of materials ASTM Bulletin No. 186 48 (1952).

[6] ASTM D1471 *Test for 2-parameter 60° specular gloss.*

[7] BS 3900 : Part D3 : 1975 *Assessment of sheen* (withdrawn June 1983).

[8] BS 3900 : Part D4 : 1974 *Comparison of contrast ratio (hiding power) of paints of the same type and colour.*

[9] BS 3900 : Part D6 : 1982 *Determination of contrast ratio (opacity) of light-coloured paints at a fixed spreading rate, using polyester foil.*

[10] BS 3900 : Part D7 : 1983 *Determination of hiding power of white and light-coloured paints by the Kubelka-Munk method.*

[11] Munsell, A. H. *Munsell book of color*, Munsell Color Co., Baltimore, Maryland, USA.

[12] BS 3900 : Part D1 : 1978 *Visual comparison of the colour of paints.*

[13] Billmeyer, F. W. *Colorimetry of fluorescent specimens. A state of the art review*, Technical Report NBS – GCR 79–185. National Bureau of Standards Washington DC (1979).

[14] Tilleard, D. L. Tolerances in colour matching *J. Oil Col. Chem. Assoc.* **40** 952 (1957).

[15] Johnston, R. M. Color control in the small paint plant, *J. Paint Tech.* **41** 415 (1969).

For general reading on colour measurement and control see textbooks, references [12–16] in Chapter 17 and on gloss Hunter ref. [2] Chapter 17.

19

Durability testing

R. Lambourne

19.1 INTRODUCTION

Durability may be defined as the capacity of a paint to endure; that is, to remain unchanged by environment and events. The 'events' we are concerned with are those that impose stresses and strains on the paint system that may be of short duration (e.g. impact) or of long duration (e.g. slow expansion or contraction of the film and substrate). Effects of environmental conditions have an enormous effect on durability, and test methods for developing and monitoring the performance of paint systems are always designed to simulate conditions of usage. They are usually designed to accelerate the degradative processes to which paints are subjected. The need for this acceleration of the degradation processes is to provide early warning of paint failure. Another aspect of durability is the capacity of the paint to withstand abuse under conditions of use, and in this case the requirements for different markets dictates different methods of test. Broadly, these tests are of two types: chemical resistance tests and physical or mechanical tests for all types of paint. The durability of a paint system is often dependent on the nature of the substrate, and this must be taken into account when designing appropriately realistic test methods.

The extent and range of tests that may be applied vary according to a number of circumstances. In the industrial paint markets the paint manufacturer may have to meet specifications laid down by the user. Test method specifications have been drawn up by a number of national and international organizations such as the American Society for Testing Materials (ASTM), the British Standards Institution (BSI), Deutsche Industrie Normal (DIN), and the International Standards Organization (ISO). In the United Kingdom other organizations have developed and are responsible for maintaining certain standard test specifications. Examples of these organizations are the Society of Motor Manufacturers and Traders (SMMT) and the Ministry of Defence (MOD DEF Specifications). Establishments such as the

Building Research Station and the Paint Research Association have contributed to the development of test methods and influence the standardization of test methods through representation on technical panels of the national and international standards organizations.

19.1.1   Why do paints fail?

The reasons for paint failure arc legion. Nevertheless some reasons for failure are readily identifiable, and attempts can be made to combat them. Architectural paints based upon autoxidizable binders have the seeds of degradation within them. The oxidation process does not stop when the film has dried. Oxidation proceeds, giving an increasingly crosslinked film. Oxidation products are lost from the film, the net effect being embrittlement ultimately to the stage that changes in the substrate cannot be accommodated. The final effect is the cracking and flaking of the film. The adequacy of durability of modern exterior gloss paint is due to a careful choice of binder which aims to keep the oxidizability of the film to the minimum, maximizes the extensibility of the film whilst maintaining adequate hardness, among other considerations.

Failure of an industrial finish, e.g. on a washing machine, may not be a problem during the lifetime of the appliance, unless the paint is unsatisfactory with respect to detergent or alkali resistance. In this case very different criteria are applicable, and the methods of test reflect these differences. With motor vehicles not only does the paint system have to retain an attractive appearance on exposure to ultraviolet radiation and weathering generally, but it must be resistant to stone chipping. If stone chipping does occur the system should be designed to resist the spread of rust from the damaged site. These examples serve to indicate the extent of the problem of maintaining adequate standards of durability. Formulation of paints for different markets to meet more or less exacting requirements has been discussed in Chapters 9–13. It will be clear that the requirements to be met include both chemical resistance and optimum mechanical properties. Failure of paints will be due to either of these factors or to a combination of them. We shall now discuss an extensive range of tests that have been devised to give an indication of the probable performance of a paint appropriate to its everyday use.

19.1.2   Methodology

Durability testing is carried out for a number of reasons. It may be used as a quality control on standard products, or it may be applied to new products under development. It is done to establish that certain criteria with respect to performance are met. This means that the paint manufacturer may have to test very large numbers of samples, usually applied to test panels which also are required to meet some standard which will be specified. Because of the large number of individual samples it is often not possible to carry out the tests with statistically significant sample numbers in any one comparable series. At best, sample panels may be duplicated, and it is essential to include a standard or control sample of established performance within each series. This standard is used as a basis for comparison. If the standard composition performs less well than expected the whole series must be suspect and should be repeated. This can happen when some unusual circumstance has arisen such as overheating or contamination of a test solution has occurred, which might otherwise have gone unheeded.

19.1.3 Test sample preparation

The type of panel is frequently specified. It is common to use mild steel panels (6 in × 4 in, 150 mm × 100 mm). For UK Goverment specifications these panels must conform to BS specification 1449, which designates EN2A deep drawing quality. For Government specifications the panels are prepared and the paint applied by an appropriate method as given in DEF Standard 1053. Essentially this means that one should ensure that the panel is free from surface imperfections, such as rolling marks, scores, and corrosion. It should be degreased thoroughly with trichloroethylene and dried. The panel is then abraded on the test side(s) with 180 grade silicon carbide paper and wiped with petroleum spirit (SBP No. 3) to remove any contaminants. The material to be tested is then applied in accordance with the appropriate product or test specification, including any necessary pretreatment of the panel and the application of primer and/or undercoat. Care should be taken that at no time between degreasing and painting, the prepared surfaces are touched by hand or otherwise contaminated.

The coated panels are air-dried or stoved, as required. In some cases the paint system extends to both sides of the panel. In others only one side is coated, and the back of the panel may be protected (e.g. against corrosion) by a different material. This 'backing' may be a quick air-drying paint or in some cases a wax coating. Usually the backing will be specified in the test method.

In testing products under development it is the practice to have only one experimental paint as part of the system. For example, if a finishing coat is being tested it should be applied over a standard pretreatment, primer, and/or undercoat as the case may be. Similarly if an experimental primer is under test it should be tested in combination with standard pretreatment, undercoat, and finishing coats.

Mild steel panels are not used exclusively, and in some tests aluminium or tinplate panels may be preferred; wood, hardboard, chipboard (particle board), asbestos cement, and glass may be used for some tests, particularly in connection with mechanical testing (see section 19.3) and weathering.

19.2 CHEMICAL RESISTANCE TESTING

It is not proposed to reproduce in this book detailed specific test methods. Those that are commonly adopted and are derived from (or identical to) specifications drawn up by one of the standards institutions will be identified by their reference numbers. They will be described in sufficient detail for the reader to appreciate the applications which they are designed to meet, but the reader should consult the original specification if intending to make use of them. Unless one adheres strictly to specifications it is unlikely that reproducible results will be obtained, and it is likely that a user may reject experimental data unless it has been obtained by laid-down methods. In many cases tests are developed by an individual paint manufacturer to suit his needs, or in conjunction with a user. In any of these cases, the methods of test may form part of a contractual agreement between the paint manufacturer and the customer.

19.2.1 Specific tests

19.2.1.1 Water resistance
The purpose of this test is to assess the resistance of a surface coating to immersion in distilled water. It is sometimes referred to as 'blister resistance'. The method is set out in SMMT 57 (Standards for the British Automobile Industry). It is applied to a wide range of industrial products.

The test is carried out in a thermostatically controlled water bath equipped for mechanical stirring. A rectangular laboratory water bath 15 l capacity is suitable; the water is heated electrically to 38°C ± 0.25°C. The panels, prepared as described in section 19.1.3, are supported in panel racks made of material inert to water (e.g. Perspex). The panels are packed in pairs, back-to-back vertically in the panel racks. The racks are placed across the tank so that the water, which is circulated by a propeller situated at one end, can pass across the face of the panels. After 24 hours' immersion the panels are removed from the tank and gently wiped dry with a dry soft cloth. They are examined immediately for blistering and for loss of gloss. Blistering within 12 mm of the edge of the panel is usually disregarded. After examination the panels are replaced in the bath and the immersion continued until the specification limit is reached, usually 7 days at least. The panels are examined every 24 hours up to this point. At the end of the test the panels are removed from the bath, and after examination are allowed to dry at ambient temperature so that they may be subjected to other tests as desired. Most commonly they are examined for adhesive failure, and it can usually be established whether failure has taken place between the coating and the substrate or is an intercoat failure.

Blistering is commonly assessed by using photographic standards published in ASTM Standard method D714-56. The photographs enable the classification of blisters by size and number or density. Thus each blister size is categorized by four levels of density, designated 'few', 'medium', 'medium dense', and 'dense'. Loss of gloss is also estimated. Where significant blistering has taken place it is not practicable to do this instrumentally so that the gloss is estimated on a numerical scale in comparison with a control that has not been subject to test. A 1–10 rating is used where 10 = excellent, i.e. no loss of gloss, to 1 = complete loss of gloss.

19.2.1.2 Moisture resistance: BS 3900 Part F2
This is similar to DEF–1053 Method 25. This test differs from the water resistance test in that it is concerned with behaviour of paints under conditions of temperature and humidity cycling such as may be encountered on motor vehicles or other coated metal objects in many climates. It is a test frequently specified in government contracts. In this test the humidity cabinet itself is specified. It consists of a closed cabinet in which the relative humidity is maintained at approximately 100% and in which the temperature cycles between 42° and 48°C to ensure condensation on the panels. Heating is by immersion heater in a water reservoir at the bottom of the cabinet. The air temperature cycles continuously between the two extremes in

60 ± 5 minutes. The air circulation (by fan) is such that the air temperatures at any two points in the air space do not differ by more than 1.0°C at any given moment.

Test panels, prepared as for the water immersion test, are examined visually every 24 hours for signs of deterioration. When they are replaced in the humidity cabinet they are put in different positions to minimize any effects that might be due to their position in the cabinet. At the end of the test they are removed and allowed to stand at room temperature for 24 hours and then examined for loss of adhesion, change of colour, or embrittlement. In some cases, a strip of paint may be removed from the panel with a non-corrosive paint stripper and the metal substrate examined for corrosion.

19.2.1.3 Resistance to salt spray

Salt spray tests are probably the most common tests applicable to corrosion resistance, and the most controversial. It is well established that salts such as sodium chloride can cause rapid corrosion of ferrous substrates, and it is useful to have information on the behaviour of a particular system in protecting such substrate from corrosion both with intact and damaged paint films. Controversy arises largely from the interpretation of the data because of the poor reproducibility of the tests. However, they are well established, and, despite the problem of reproducibility, are quite a useful guide to performance in the absence of longer term corrosion data. They are thus unlikely to be discarded. They are considered to be unrealistic by some workers because of the degree of acceleration of the corrosion process that they achieve and the variability of the extent of 'damage' that is inflicted in some of the tests.

Two tests are in common use: the continuous salt spray test and the intermittent.

The continuous test:
This test was originally developed in the UK as a government specification (DEF-1053 Method No. 24). It was developed subsequently into the British Standards Institution Method BS3900: Part F4: — 1968 (Salt-Spray). The test is of the 'pass' or 'fail' type, whereby the coating is subjected to treatment for a specified time and then examined for failure. Care must be exercised in interpreting the results of the test; it is not intended to be used as an accelerated test for normal weathering.

The panels are prepared by the method described in BS 3900: Part A3 and Part A4. This is essentially the same as described earlier in section 19.1.3. The back and the edges of the panel are coated with a good protective air-drying material. The panel(s) is aged for 24 hours before starting the test. If the panel is aged for a longer period the time of aging should be recorded since this might influence the results. Using a scalpel, the coating is cut through to the metal, starting 1 inch (25 mm) from the top of the panel and finishing 1 inch (25 mm) from the bottom. The cut should be parallel to the longer side of the panel, and it is important that the surface of the metal should be scored. The test is carried out in a chemically inert container (e.g. glass or plastic) with a close-fitting lid. A salt mist is produced by spraying a synthetic sea-water solution through an atomizer. The panels are supported on nonmetallic racks so that their faces are approximately 15° to the vertical. The spray is so arranged that it does not impinge directly onto the panel surfaces. The solution which drains from the test panels is not recirculated. Panels are examined after 48 hours, 1,

2, and 3 weeks. They are rinsed in running tap water and dried with absorbent paper and examined immediately for blistering, adhesion, and corrosion from the cut. Blistering is assessed as previously described, with reference to photographic standards. In this case, however, particular attention is paid to the extent that blistering occurs in the vicinity of the cut. It is reported as extending 'under $\frac{1}{8}$ inch', 'under $\frac{1}{4}$ inch', or 'under $\frac{1}{2}$ inch'. Loss of adhesion is estimated by testing the ease of removal of a coating with a finger nail, and how far in from the cut the poor adhesion extends. The observations are reports as 'no change', 'slight loss of adhesion', or 'bad loss of adhesion' as appropriate. For reasons of reproducibility a standard salt solution prepared from 'analytical quality' reagents is used. The test solution composition is specified as follows:

Sodium chloride (as NaCl)	26.5 g
Magnesium chloride (as $MgCl_2$)	2.4 g
Magnesium sulphate (as $MgSO_4$)	3.3 g
Potassium chloride (as KCl)	0.73 g
Sodium hydrogen carbonate (as $NaHCO_3$)	0.20 g
Sodium bromide (as NaBr)	0.28 g
Calcium chloride (as $CaCl_2$)	1.1 g
Distilled water	to 1000 ml

It is specified that the calcium chloride should be added last and that the temperature of the solution during the test should be $20 \pm 2°C$.

The intermittent salt spray
This test is similar to the continuous test except that the mist is produced each day for 8 periods of 10 minutes at intervals of 50 minutes. It is carried out for 5 consecutive days and then 'rested' for 2 days. It is normally confined to government contract specifications and is clearly not as severe as the continuous salt spray method.

A number of variants of these tests are in use. They have often arisen through the development of test methods by major paint users such as the motor manufacturers. In some of these tests the panels are scored in a different way to that specific in the BS standard. They are scored diagonally, as in the diagram:

It is claimed that the loss of adhesion is easier to detect (and is probably manifested at an earlier stage) at the centre of the X. It is also common practice to specify the type of surgical scalpel and the force that should be exerted in scoring the panel. The latter is done by placing the panel on a laboratory balance and pressing down to achieve 1 kg force whilst scoring. Panels prepared in this way and showing different degrees of corrosion are shown in Plates 19.1a and 19.1b.

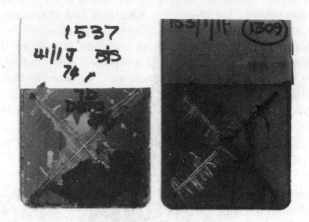

Plate 19.1a — Salt spray tests using mild steel panels. The left-hand panel illustrates a film that has poor adhesion, resulting in ready detachment from the substrate and severe rusting under the film. The right-hand panel shows a film that has good adhesion. Rusting has occurred at the score mark, and whilst staining is apparent, the rusting does not extend under the film which remains firmly attached despite attempts to scrape it away from the score mark.

Plate 19.1b — Salt spray tests using aluminium panels. The left-hand panel illustrates a coil coating composition that gives excellent protection to the aluminium substrate. The right-hand panel shows a typical industrial paint, not specifically designed for good corrosion resistance, which illustrates gross failure resulting from blistering and adhesive failure from the region of damage.

19.2.1.4 Alkali and detergent resistance

At least three alkali resistance tests are in common use. These are based upon the use of trisodium orthophosphate, soda ash (anhydrous sodium carbonate), and sodium hydroxide respectively. They are commonly applied to industrial finishes such as may be used on domestic appliances such as washing machines. In addition a detergent test may also be used for these types of product. These alkalis represent different degrees of aggressiveness. The panels are prepared as described previously for all three tests. It is usual for the panels to be coated on both sides (and edges), but if only one side is coated the back and edges can be sealed with molten paraffin wax (mp 49°C–60°C). When it is intended to coat both sides of the panel with the paint system it is convenient to have a small hole drilled near one end of the panel so that it can be suspended in the oven during stoving, or air-drying. In tests which have been carried out in ICI Paints Division laboratories for many years the concentration of solute and temperature of test are as follows:

> trisodium phosphate   1% w/w in distilled water;   75°C
> soda ash   10% w/w in distilled water;   65°C
> sodium hydroxide   5% w/w in distilled water;   25°C

In these tests the panels are partly immersed in the solutions, to a depth of 3–4 inches. As in previously described tests, they are supported in the tank in nonmetallic racks such that they are close to the vertical. The effect of partial immersion is to provide an opportunity to make direct comparisons of the effect of the solution on the paint system. It is essential if comparisons with an accepted standard are to be made to include a control in the series under test. The panels are removed for examination after the first 4 hours' immersion and then subsequently at 24-hour intervals as appropriate. After removal from the tank the panels are rinsed with tap-water, dried with a chamois leather, and examined for blistering, softening, loss of gloss, and erosion. They are then allowed to dry at room temperature for 1 hour and examined for cracking and peeling. They are then re-immersed in the solution and re-examined after 24 hours (and at subsequent intervals), using the same procedure. Blistering is assessed as described previously for water resistance, using the ASTM method. Hardness or degree of softening may be assessed by comparing the pencil hardness (section 19.3.2) of the immersed portion with that of the non-immersed portion of the paint system. Adhesion loss is normally evident by 'lifting' and 'shrivelling' of the coating. The film may be easily lifted off with the finger nail. If a more accurate method of assessing loss of adhesion is required the 'cross hatch' method (section 19.3.5) may be applied. Loss of gloss, discolouration, or staining may all be reported on a crude scale as 'none', 'slight', or 'bad'.

The detergent test is carried out in a similar manner to the alkali tests. In the author's laboratories a 5% solution of 'Deepio' (ex Procter and Gamble) is used for the test. The test is carried out at 74°C ± 0.5°C. The period of test is shorter than the alkali tests, the panels being examined after 1, 2, 4, 6, 24, and 48 hours from the start of the test. The same criteria for the assessment of deterioration of the paint are adopted as for the alkali tests. These tests are abitrary, but nevertheless they can give information that is relevant to the performance of the paint in practice. Once accepted as standard methods they are seldom changed, although the proliferation of tests is indicative of other criteria being sought by paint manufacturers and industrial users alike. In most cases the tests which are used are derived by collaboration between paint manufacturer and user.

19.2.1.5 Solvent resistance

Solvent resistance may be tested for very different reasons. Tests of resistance to petrol and diesel fuel are carried out on compositions that may be expected to encounter contact or intermittent splashing with these liquids, e.g. motor vehicle finishes, storage tanks, etc. The use of polar solvents such as ketones is often used to assess the degree of cure of a cross-linkable composition.

In the case of the petrol resistance test it is preferable to use a synthetic petrol of known composition, because of the wide variation in the composition of the commercial petrols and their incorporation of materials such as antiknock additives. A typical synthetic petrol has the following composition;

> SBP petroleum spirit No. 3 40%
> SBP petroleum spirit No. 4 40%
> Industrial methylated spirits 20%
>   (74° over proof)

For the diesel fuel test, DERV is suitable. It should have an aniline point between 60°C and 70°C as measured by the Institute of Petroleum Method No. 2.

For solvent resistance, methyl isobutyl ketone is recommended. The panels are prepared as previously described and are immersed to a depth of 4″ in a tank of the appropriate liquid, whilst being maintained in a near-vertical position. All of these tests are carried out at ambient temperature. The panels are removed for examination 1, 2, 4, 6, 24, and 48 hours from the start of the test. In the case of the volatile solvents they are examined immediately after removal for blistering, hardness, adhesion, and discoloration, and again after allowing 5 minutes for the evaporation of the solvent. The reason for this procedure is to detect the initial degree of softening and to observe the recovery of hardness on evaporation, if it occurs. In the case of the diesel fuel, evaporation does not take place significantly in the time allowed, and the panels are allowed to drain for one minute before testing.

This type of test may, of course, use any solvent, depending on the type of information sought. In general the test will be useful to assess the performance of a particular system in use, the solvent chosen representing a real contaminant for the cured system which may be the solvent itself or solutions of other materials, e.g. adhesives or mastics. Alternatively the test may serve to give an indication of the degree of cure, in which case it is common to use the solvent or solvent mixutre that carried the original paint.

In addition to immersion testing, solvent resistance may be assessed by a solvent–rub test. In this case the surface of the panel is rubbed with a piece of soft cloth moistened with the solvent. Solvent resistance may be measured by the number of rubs necessary to disintegrate the film or rub it through to the next coat. Alternatively a specified number of rubs may be carried out and the appearance of the film assessed against an accepted scale, say 0–5, where 0 indicates that the solvent has had no effect and 5 indicates complete failure of the film either by disruption or dissolution; in between these extremes, varying degrees of loss of gloss and film attack will be observed.

19.2.1.6 Resistance to staining
It is particularly important that kitchen equipment such as refrigerators, washing machines, etc., should be coated with materials that are resistant to staining by a wide range of household products. It is thus necessary to test new finishes and compare them with existing standards before introducing them. Panels are prepared as described in section 19.1.3. In the case of these tests it is however much more common to use 12 in × 4 in panels because of the number of contaminants used.

A typical range of contaminants might include lipstick, red wax crayon, black and brown shoe polishes, mustard pickle, tomato ketchup, and a slice of lemon. The contaminants are applied to the surface of the flat panel and are individually covered with watch glasses. Contact time is normally 24 hours, after which the contaminants are wiped off with a clean soft cloth. The extent of staining is assessed visually and reported accordingly. It is essential that a standard paint is included for comparison to establish a point of reference.

19.3 TESTING MECHANICAL PROPERTIES OF PAINTS
The significance of the mechanical properties of paint films and their measurement has been discussed in detail in Chapter 16. The testing, on a routine basis, of specific mechanical properties, in many cases related to the end use of the paint, has resulted in the establishment of standard tests. A selection of some commonly used tests will be discussed here.

Although the paint film performance is the primary consideration, several of the standard tests involve damage to the substrate as well as the paint. Ideally it should be possible to cause considerable substrate deformation before paint failure by cracking, peeling, and loss of gloss occur. However, in practice it is necessary to compromise between hardness, flexibility, and extensibility in order to achieve adequate performance. In these cases some arbitrary limits of acceptance are laid down. In this section we shall deal with resistance to impact, scratching, and bending; adhesion measurement; and indentation. Although these tests are primarily of the 'pass' or 'fail' type according to predetermined accepted standards and are applied to new (or at least non-weathered) films, they may also be applied to weathered films. In this case they may be used as an adjunct to accelerated or natural weathering.

19.3.1 Resistance to impact
Three types of impact test are in use, a pendulum test, a falling weight impact test and the Erichsen Indentation test.

19.3.1.1 Pendulum test
This method was developed as DEF-1053 Method 17(b). It calls for the use of a special piece of test equipment which is also specified in the method. The apparatus is available from Sheen Instruments (Sales) Ltd, 9 Sheendale Road, Kingston, Surrey, UK.

   The apparatus consists of two swinging arms, each supporting a piece of steel tubing coated with the paint system under test. One of the test pieces is pivoted on a horizontal axis, while the other piece is held horizontally. The mechanism is such that when the arms are swung towards each other, the tubes strike and the pivoted test piece slides across the other. The damage to the coating is then assessed visually. This is one of the few tests that does not use a flat steel panel. The test pieces are tubes, $1\frac{1}{2}$ in external and $1\frac{1}{8}$ in internal diameter, 5 in long. The normal methods of sample preparation, using the sequence of operations as described previously for flat panels (section 19.1.3) are used. Whilst the test is simple to carry out, some experience is needed in interpreting the results. The extent of damage at the point of impact is assessed in terms of loss of adhesion.

19.3.1.2   Falling weight test
This test was devised to determine the degree to which a paint on a metal surface could accommodate rapid deformation. The panel on which the surface coat has been applied is deformed rapidly by allowing a weighted indenter to fall on it from a fixed height. The impact may be onto the coated surface or onto the back of the panel (reverse impact) according to the specification requirements for the material. The test is published as BS 3900: Part E3 (Impact). The apparatus is available from Sheen Instruments (Sales) Ltd. It consists of a steel block which slides vertically between two guides. Mounted under the block is a tool holder in which is fixed an indenter. The block and tool fall under gravity onto two die blocks with a hole in the centre. The test panel clamped between the die blocks is impacted by the tool. The depth of the indentation is varied by inserting washers of known thickness between the indenter and the tool holder. The test panels are prepared as described previously. However, in this test it is important to use films of standard thickness, 0.001 in, unless otherwise specified. The thickness of the steel panel must also be constant. In this case the panel is somewhat smaller than those used for chemical resistance testing. Panels 4 in × 2 in are specified having a thickness of 0.0495 in or 1.257 mm. Failure of the paint is shown by cracking and by loss of adhesion at the deformed portion of the panel.

19.3.1.3   The Erichsen indentation test
This test is concerned with the degree to which a paint is able to accommodate slow deformation of the substrate. A panel coated with the paint system under test is deformed slowly by forcing a ball-shaped indenter into the uncoated side. The distance the indenter has moved when the coating first shows signs of cracking is taken as a measure of its deformation characteristics. Two machines are available for carrying out the test, these are: The Erichsen Lacquer Testing Machines, Model 229/E (Manual) and 225/E (Mechanical). They are available from Elcometer Instruments Ltd, Fairfield, Manchester, England.

   Both instruments require a panel 70 mm wide to fit the instrument. The panels, prepared from standard panels as previously described, are cut to size after coating and just before testing. The panel is clamped painted side upwards by a circular boss against a fixed face. A hole in the face allows a 20 mm diameter hemispherical indenter to be moved into the back of the panel. In Model 229/E a manually operated wheel is used to exert hydraulic pressure on a piston connected to the indenter. A self-illuminating microscope placed at the centre of the boss enables the operator to observe the deformation taking place and to stop the indenter when the first break occurs in the coating. A dial gauge micrometer shows continuously the depth of the indentation. In Model 225/E the piston is similarly hydraulically operated but actuated by an electric motor instead of by hand. The motor drives the indenter at 0.2 mm per second into the back of the panel.

Adhesional failure is shown by a clear-cut circular break around the apex of the deformation or radial breaks often combined with circular breaks. In such cases the coating is unable to accommodate the dimensional changes in the substrate. This type of break generally occurs quite suddenly. Cohesional failure is indicated by finely divided cracks. In this case the coating adheres well to the substrate, but the cohesion of the film breaks down as the deformation is increased. This does not occur in as clear-cut a way as with adhesional failure, and careful observation is necessary. With the type of steel used for test purposes, fracture of the panel occurs between 9 and 10 mm. Failures of the test film may be due to such fracture, and the damaged area must therefore be examined very carefully (see Plate 19.2).

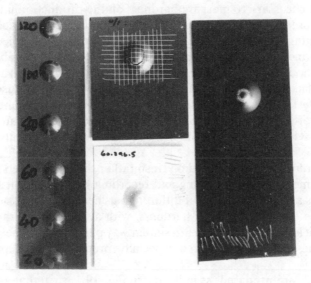

Plate 19.2 — Typical Erichsen Indenter test results, including a pre-crosshatched example, which also shows failure of the metal substrate.

19.3.2 Hardness testing

It may seem inappropriate to consider hardness testing in the context of durability, since the general mechanical properties of paint films are covered in Chapter 16. However, here we shall cover only some of the standard methods of test that have been developed. Firstly, they can be considered in relation to adequacy of cure, which will have a profound effect on ultimate durability. Also, in some cases changes in hardness as a result of weathering give an important insight into the rate and nature of the weathering process.

A number of tests have been developed over the years to give information on the hardness of coatings. Hardness is very difficult to define in absolute terms, except to say that it is a composite function of the mechanical properties of a material that is related to resistance to deformation. This definition is, however, too simple, since materials can be brittle, plastic, elastic, etc., and it will be perceived that two materials although deformed to the same extent under load may differ in their behaviour on removal of the load. For example, one material may be permanently deformed, the other not. These two materials will have undergone plastic deformation in the first case and elastic deformation in the second. The paint technologist has to take a more pragmatic view of hardness, and for this reason, simple, standard methods of measurement are adopted. Since one is concerned with measuring the properties of a thin film on a variety of substrates it is important to recognize that the substrate can influence the apparent hardness of the film. It is common therefore to make hardness measurements on rigid substrates like glass or steel and to accept that one may be compressing the test material between the substrate and the probe, or whatever device may be used.

It may be necessary to measure hardness on the production line, so that simple tests and the use of portable equipment will be required in some cases. Carrying out hardness testing in the laboratory, under standard conditions, with carefully prepared samples is always to be preferred whenever this is possible.

Four methods of determining the hardness of paint films, (a) Pencil, (b) Sward Rocker, (c) Tukon tester, and (d) ICI Indenter, are in common use. Pendulum and indentation tests have been described in Chapter 16. In the context of durability studies it is worth noting that the Sward and Tukon test methods are of little value on eroded films. The ICI Indenter can, however, give useful information on almost any type of film.

Pencil hardness is mainly used on fresh rather than weathered films. It is one of the simplest methods of measuring and recording hardness (albeit on an arbitrary scale). It uses a range of pencils of different hardness as a basis for comparison. As geologists use the Mohr scale of hardness, with a well defined range of 'standard' minerals, so it is possible to relate in a similar way paint film hardness to conventional pencil hardness, using a series of specially prepared pencil points of varying hardness.

The pencils are prepared, as indicated in Fig. 19.1, such that approximately $\frac{1}{4}$ in of lead is exposed from the wood. The lead is cylindrical in shape with the end squared off with a fine grade abrasive paper (e.g. 400 grade). In use the pencil lead is pushed firmly along the surface of the paint film at 45° to the surface, using the maximum force, i.e. just less than sufficient to break off the lead. The hardness of the film is equated with that of the pencil that just *fails* to damage the coating. With practice a high degree of reproducibility can be achieved. A good-quality set of pencils ranging from 6B to 9H should be used,. (Eagle Turquoise 'chemi-sealed' pencils are recommended.) It is not necessary to work through the whole range of pencils once one is familiar with the typical characteristics of the paint under test. Thus for checking low-bake refinishes only the four pencils ranging from 2B to F are

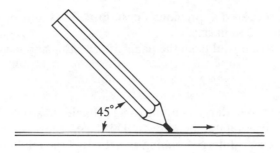

Fig. 19.1 — Pencil hardness testing.

likely to be required. Atmospheric conditions (temperature and moisture) can affect surface hardness, and ideally all films should have been prepared, aged, and tested under exactly the same conditions.

The original Wolff–Wiborn method upon which the above description is based differs only in specifying the hardness of the *softest* pencil that just marks the surface.

Slightly better reproducibility can be achieved by using a pencil-holding device marketed under the name 'Erichsen Scratch Pencil Hardness Tester Model 291' (available from Elcometer Instruments Ltd, Fairfield, Manchester, England).

19.3.3 Scratch resistance

The purpose of this test is to assess the ability of a surface coating to withstand scratching. A needle with a spherical steel point (of specified diameter), carrying a predetermined weight, is lowered onto the paint film and drawn across the surface at a set speed.

The method has been used in the paint industry (and in government specifications) for many years. The test may be operated in either of two ways, as a 'pass or fail' test using a specific weight for the material under test, or (in assessing new materials) with an increasing load until failure occurs. The method was standardized in BS 3900: Part E32 (Scratch), itself based upon DEF-1053 Method No. 14.

The apparatus consists essentially of a horizontal sliding panel to which the test panel is clamped, coated face upwards. The sliding panel is moved beneath the needle point at a speed of 3–4 cm per second. The 'needle' itself consists of a shank to which has been soldered a 1 mm diameter grade A1 steel ball. It is fixed into a holder at the end of a counterpoised arm, which is kept horizontal by adjusting the length of the needle in the holder. Weights are placed over the needle in the holder and the panels set in motion. The forward travel lowers the needle gently onto the surface. At least 6 cm of scratch is required for the test.

Although manual and motordriven apparatus are available for the test, the latter is preferred since more reproducible results can be obtained. (The apparatus is available from Sheen Instruments (Sales) Ltd, 9 Sheendale Road, Richmond, Surrey, England, or REL Ltd, 64 Wellington Road, Hampton Hill, Middlesex, England.)

The panels are prepared as previously described. They may be of steel, tinplate, or aluminium, 5 in × 2 in in size.

The test panel is removed from the panel after testing and examined visually.

19.3.4 Bend tests

The ability of a paint to undergo bending (on a suitable substrate) has called for the development of several related bend tests. This type of test is applied to materials that may be expected to undergo bending as a result of the method of fabrication of the articles to which they are applied. Thus the tests find greatest use in connection with industrial finishes.

Most commonly, the paint system to be tested is applied to a suitable panel which can be bent through 180° around a cylindrical mandrel of known diameter. The coating is then examined for cracking or loss of adhesion. Although usually regarded as a test of flexibility, the test is a composite one which encompasses adhesion, flexibility, and extensibility. One form of the test [developed originally as DEF-1053 Method No. 13 and subsequently as BS 3900* Part E1: 1966 (Bend)] uses a 'hinge' into which the mandrel is incorporated. The hinge consists of two rectangular metal flaps hinged at the extremities of one of their shorter sides with a long pin. The mandrel is fitted around this pin, such that it rotates freely around it. It is so positioned that when the hinge is open there is sufficient clearance to insert the test panel between the mandrel and the flaps. Each mandrel is fitted in its own hinge so that it is necessary to have a complete range of hinges to apply the test in increasing severity as the diameter of the mandrel is reduced. The normal range of mandrels is $\frac{1}{8}$ in, $\frac{3}{16}$ in, $\frac{1}{4}$ in, $\frac{3}{8}$ in, $\frac{1}{2}$ in, $\frac{3}{4}$ in, and 1 in diameter. It is important that the test panel meets certain dimensional requirements. For government specifications 4 in × 2 in aluminium panels conforming to BS 3900: Part A3 are used. In non-specification testing tinplate or aluminium panels may be used, but they should not be thicker than 0.012 in.

The hinge is used in the following manner: the panel is inserted into the hinge with the coated side outward from the direction of bending. The hinge is closed evenly, without jerking, in not less than one second and not greater than $1\frac{1}{2}$ seconds, bending the panel through 180°. The coating is examined immediately after bending and before removing it from the hinge for evidence of cracking or loss of adhesion. Cracking closer to the edge of the panel than $\frac{1}{4}$ in is ignored. The test should be carried out in a constant temperature room (e.g. at 25°C), allowing at least two hours for the panels and hinges to come to temperature.

This extremely simple test can be very informative since it may be related to conditions of use of many of the compositions which it may be used to test. It can be developed in different ways to give information of the acceptability or otherwise of materials that may be subject to environmental changes after a coated substrate has been formed. Thus compositions that pass the test at a given mandrel diameter at 25°C may crack and lose adhesion when they are heated (subject to a second stoving sequence) or subject to excessive cooling, when the coating becomes glassy.

A variation of the previously described test uses rod mandrels which are supported in a steel cradle. The method is used for test panels which are too thick to fit into the hinges described above. However, the thicker the substrate the smaller the range of mandrels that can be successfully employed. Thus 26 gauge (0.018 in) mild steel can be successfully bent over all mandrels, whereas 20 gauge (0.039 in) mild steel can be bent only over mandrels greater than $\frac{5}{16}$ in diameter.

19.3.5   Adhesion: The cross hatch test

Adhesion, as will already have been established in Chapter 16, is a very difficult property to measure. Nevertheless it is a very important property of a paint system, and some useful empirical information can be obtained by the cross hatch test. In this test a die with a number of close-set parallel blades is pressed into the test successively in two directions at right angles to each other. The second pressing is superimposed on the first, giving a pattern of squares. A strip of self-adhesive tape stuck over the pattern is removed sharply, and the adhesion of the film is assessed from the amount of the coating removed.

In the test applied in the laboratories of ICI Paints Division the die, made of hardened carbon steel, consists of nine parallel blades $\frac{1}{16}$ in apart and 1 in long. The tip of each blade has a radius of 0.003 in–0.004 in, and the shoulders of each make an angle of 60° with one another. The die is applied to the coating with an hydraulic press at 2000 p.s.i. The tape is applied over the pattern left by the die and pressed firmly down, using a soft rubber eraser. It is left in contact for 10 seconds and then stripped rapidly by pulling the tape back on itself at an angle of approximately 120°. The adhesion is reported as good if there is little or no removal of the coating; moderate, where there is some removal, small particles of coating still left adhering in the middle of all or most of the squares; poor, if almost complete removal of the coating occurs. In this test it is important to use a 'non-release' cellulose type of tape. A suitable type is Sellotape NR Cellulose 1101.

If a suitable die and hydraulic press is not available, a variant of the test which uses a scalpel to cut an equivalent pattern into the coating can give useful if not reproducible data. In this case one should attempt to score through the film to the substrate, using a surgical scalpel. In some cases of poor adhesion some detachment of the film may occur during the scoring and before the attachment of the adhesive tape. In these circumstances it is preferable to score the film more carefully in another place so as to make comparable assessments between different coatings.

19.4   ACCELERATED WEATHERING

The durability of modern paint systems is such that on exposure to natural weathering they may show little signs of deterioration for periods well in excess of a year. Indeed users of coil coatings may expect guarantees that the paint system will not break down in less than ten years. In formulating these types of coating, the more durable they are the more difficult the testing becomes. Means of accelerating the processes have been sought for many years, and some standard methods have been developed. However, a major problem arises in assessing how reliable a particular test method may be. Incorporation of known standards against which experimental paints may be compared helps to overcome this problem to a certain extent. It does not follow that a good correlation will be obtained between accelerated and natural weathering in all cases, and results from accelerated tests must always be treated with extreme caution.

The deterioration of organic coatings on exposure to the elements is due to the effects of radiation (particularly ultraviolet radiation), moisture, and temperature. The degradation processes are the result of chemical change (oxidation) and the effects of mechanical stresses. The accelerated weathering methods seek to intensify these effects so that film breakdown occurs in a fraction of the time that it would do naturally. To achieve this a number of weathering machines have been designed in which radiation/moisture cycles are maintained to achieve perceptible change within periods of up to 2000 hours exposure.

The type of weathering machine and the test cycle may be specified by the customer, and this is commonly the case with large customers such as the motor car manufacturers. Specifications have been drawn up by bodies such as the British Standards Institution and the American Society for Testing Materials. These specifications may be the basis of tests agreed between paint manufacturer and customer, but in many cases the customer may have designed his own tests to which the suppliers' product must conform.

Several types of weathering machine are in common use. Examples are the Marr Carbon Arc, the extensive range of Atlas Weatherometers, (registered trademark of the Atlas Electric Devices Company, Chicago, Illinois), and more recently, the Atlas UVCON and the QUV Weathering Tester. Each accelerates the breakdown of paint films, but results are not directly comparable. The Marr Carbon Arc conforms to BS 3900, Part F3, *Resistance to artificial weathering (enclosed carbon arc)*, when the following cycle is used:

> 4 hours wet — atomizers on and fan off
> 2 hours dry — atomizers off and fan on
> 10 hours wet — atomizers on and fan off
> 2 hours dry — atomizers off and fan on
> 5 hours wet — atomizers on and fan off
> 1 hour stopped for servicing and assessments.

The Marr (illustrated in Plates 19.3 and 19.4) consists of a circular tank with the radiation source, a 1600 watt enclosed carbon arc, suspended at the centre. The painted panels (standard 6 in × 4 in prepared as described in section 19.1.3) are secured on the periphery of an inner concentric circular supporting ring which rotates slowly during the humidity/radiation cycle. The cycle may be varied to meet different needs, but usually includes intermittent periods of wet and dry.

The Atlas XW Weatherometer conforms to ASTM E42-57. It uses slightly smaller panels (6 in × 3 in) which are prepared as described in section 19.1.3. The machine is constructed in the form of a rectangular cabinet, and the panels are mounted on an interior circular supporting track accessed through a door with an observation window. The radiation source is an enclosed carbon arc UV lamp. In this machine it is common for the panels to be exposed continuously to radiation with an intermittent water spray for 3 minutes every 20 minutes.

The Atlas XWR Dew Cycle Weatherometer provides a more aggressive environment. It uses an unfiltered carbon arc, with a high UV output. The cycle is UV exposure for 1 hour followed by dew deposition/darkness for 1 hour, the duration of the test being decided by the originator. It is often unnecessary to continue the cycle

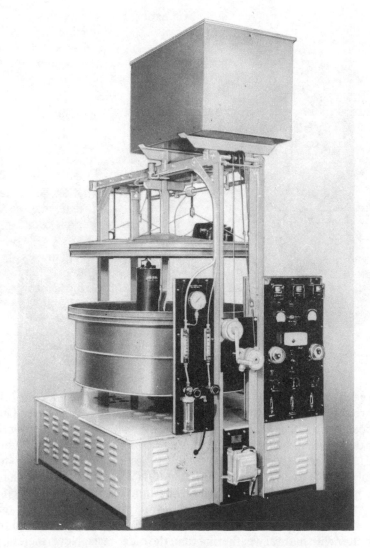

Plate 19.3 — The Marr Carbon Arc Weathering Machine.

for more than 1000 hours. This machine and cycle have been approved by the National Coil Coaters Association.

The Atlas Company provides the most comprehensive range of weathering machines. One of the most recent designs, the Ci65, Controlled Irradiance Exposure System, is illustrated in Plate 19.5. The radiation sources can be varied to suit many applications. The dual enclosed carbon arc lamps employed in the cDMC model are shown in Plate 19.6, and the 6500 watt Xenon arc used in the Ci65 model is shown in Plate 19.7.

The QUV Weathering Tester differs from the two types previously described in three respects. It uses fluorescent tubes as the radiation source, the panels are static

Plate 19.4 — Interior of the Marr, showing test panels in place.

within the cabinet, and it relies on condensation of water on the panels resulting from temperature changes, rather than a water spray. Care has to be exercised in ensuring that the positions of the fluorescent tubes are rotated to eliminate variations in behaviour due to small but significant differences in the output of the tubes. Also a prescribed sequence of replacement of the tubes has to be adopted to achieve reproducible results over a long period. The QUV and the UVCON have gained popularity in the testing of automotive finishes and are sometimes specified rather than the Atlas or Marr Carbon and Xenon arc machines in recent times. It is debatable whether they are more reproducible or that the data derived from them correlate better with natural weathering than the older tests. Nevertheless, the QUV type of test is rapidly becoming established as one of the criteria that automotive systems must satisfy before gaining customer acceptance, particularly in North America.

19.4.1 Assessment methods: definition of criteria
Paints subjected to accelerated (or natural) weathering are assessed for changes in certain properties such as gloss and colour, and for specific types of breakdown under the following headings:

(a) chalking (b) bronzing (c) cracking and checking
(d) blistering (e) flaking (f) rusting and corrosion of substrate
(g) corrosion from a cut (h) erosion (i) water spotting
(j) water marking (k) dirt retention and ingrained dirt

Plate 19.5 — The Atlas Ci65 Weatherometer.

(l) mould growth (m) chalking into film (n) efflorescence

These terms are defined in BS 2015 (1965), *Glossary of Paint Terms*.

For the most part the assessment of change and breakdown is done visually, assisted in some cases by the use of a low-power magnifying glass or microscope. Occasionally, gloss or colour measurements are made instrumentally, but in many cases this is not essential. Comparisons are made with unexposed panels or in some cases (particularly with accelerated tests) with parts of the panel that have been covered as a result of the method of securing the panels in the machine. An arbitrary 0–10 scale is adopted in most cases. Thus in assessing *gloss* a rating of 0 signifies no loss of gloss, and rating 10, corresponding to a completely matt surface, signifies a total loss of gloss. In making judgements of gloss it is common to assess the sharpness of an image reflected by the paint film. A 'gloss box' has been developed to improve the method [1].

Plate 19.6 — Dual enclosed carbon arc lamps employed in the cDMC Weatherometer.

Colour change can take many forms, and often the type of change is as important as the degree. A colour may become darker or faded, duller or cleaner. Fading may be real or apparent. For example, chalking (q.v.) gives an appearance of fading, but in this case the original colour can be restored by cleaning to remove chalking in the surface layer. 'Duller' would indicate an increase in greyness, i.e. a loss of purity of colour, not necessarily associated with fading. 'Cleaner' indicates the development of a purer colour and is the opposite of 'duller'. Specific changes in colour may be recorded as such, e.g. redder or bluer than standard.

Gloss and colour assessments may be made before and after washing, i.e. wiping with a water-wet cloth, leathering off, and drying.

Chalking is defined as 'the formation of a friable powdery coating on the surface of the paint film caused by the disintegration of the binding medium due to disruptive factors during weathering'. Chalking can occur in any colour, but is more usually

Plate 19.7 — The 6500 watt xenon arc system employed in the Ci65.

associated with pastel shades where the white component gives it prominence. In deeper shades, particularly blues and maroons, the chalking exhibits a lustrous effect and is described as *bronzing*. Chalking can be rated easily and with a reasonable degree of accuracy by the 'finger stroke' method. The end of the index finger is drawn with light pressure along the film and the amount of 'chalk' removed related to photographic standards. Bronzing is assessed in the same way. *Cracking* is specifically a case of film breakdown in which the cracks penetrate at least one coat and may propagate through the whole paint system. *Checking* is a lesser form of cracking where the cracks do not penetrate the topcoat. Three types of cracking and checking are defined, (i) micro (mc): confirmed by examination under a low-power binocular microscope at ×16 or ×32. (ii) minute: fine cracks, but clearly visible to the naked eye; (iii) cracking or checking, where the effect is larger and immediately obvious. Cracking and checking form patterns which can be classified according to their type. These are as follows:

Fine, or linear with approximately parallel cracks.
Pattern, where the angle between cracks is random and the cracks can join up to
form a continuous network.
Crowsfoot, where cracks radiate from a point after the manner of a bird's foot.
Hairline and grain cracking specifically in wood finishes.

All of these defects are rated according to photographic standards, as are the
majority of those other defects listed. The use of photographic standards for
blistering has already been noted, in connection with chemical resistance testing. An
example of blistering of a gloss paint on soft wood is shown in Plate 19.8.

Plate 19.8 — Blistering of a decorative gloss paint on softwood.

The defects of *flaking*, *rusting*, and *corrosion* from a cut are likewise judged
against photographic standards for the type of finishes involved. Erosion is akin to
chalking, but is a more severe defect which might lead ultimately to exposure of the
underlying surface. It may be due to the action of rain drops, the abrasive action of
wind borne particles of grit, or a combination of these effects. *Water spotting* and
water marking are different phenomena. The former is the spotty appearance of a
paint film caused by drops of water on the surface which remain after the water has
evaporated. The spots usually appear to be lighter in colour than the surrounding
paint. They may or may not be permanent. Water marking, on the other hand, is due
to a distortion of the paint film by water droplets. No colour change or whitening
occurs. Both of these defects can be related to the photographic standards for blister
as an indication of abundance and size. Dirt retention, ingrained dirt, and mould
growth are more likely to occur on natural weathering than under accelerated
weathering conditions for obvious reasons. Ingrained dirt and mould growth can
sometimes be confused, and it is often necessary to resort to microscopic examination to distinguish them. Panels are assessed after wiping one half of the panel

lengthwise, with a damp soft cloth, taking care not to remove or disturb the dirt on the remaining half. The panel may then be replaced for a further period of weathering. The panel is assessed for colour changes against an unexposed standard. Mould growth can readily be distinguished from dirt by using a ×16 binocular microscope. It is characterized by spidery tentacles radiating from a dark nucleus. Even after washing the remains of the tentacles can still be recognized, and it is common to observe a regrowth of mould on the washed area when the panel is re-exposed.

'Chalking into film' is one of the less common defects. It is an effect that looks like chalking, but which cannot be removed by washing. Efflorescence is the development of a powdery deposit on the surface of brick and cement due to the migration of water-soluble salts through to the surface and their subsequent deposition by the evaporation of water. In some cases the deposit may be formed on the top of any paint film present, but normally the paint film is detached and broken by the efflorescence under the coat.

19.5 NATURAL WEATHERING

By exposing paint systems to the elements one might assume that one would collect the most reliable data on paint performance. This is true only up to a point, since there are so many variables involved, and it is extremely important to take into account all significant variables. Because natural weathering is a slower degradative process than can be achieved artificially it is particularly important to design the experimental series with care as any fault in the experimental design may not become apparent for one or several years.

Good documentation of the exposure series is called for because it may extend over many years and there may be changes in staff who are likely to be assessing the panels, and the staff who may be seeking information from the series. The geographical location of the exposure site may be important. If the paint is to be used in an industrial environment (e.g. structural steelwork) it is probably best to weather the test panels in an industrial rather than rural environment. On the other hand it might be useful to use some sites that are known to give rise to more rapid breakdown than the location of anticipated use. Thus many technologists regard the North American sites in Florida and Texas favourably in this way, and regard them as offering 'accelerated' weathering compared to sites in the UK. This is because of the higher UV radiation levels impinging on the panels in these areas coupled, with, in some cases, large variations in temperature and humidity. Typical 45° south facing panel racks are shown in Plate 19.9.

19.5.1 Choice of test pieces
Not only does the nature of the substrate affect the performance of panels in a test series, but the physical shape and method of construction can also do so. The substrates most commonly used for tests are wood, mild steel, aluminium, and some forms of masonry material, e.g. asbestos cement panels. The choice will depend upon the market for which the paint system has been developed. Thus architectural paints may be tested on wood, metal, or masonry, whereas automotive finishes and refinishes will be tested only on metal panels. Exceptions do occur in some cases, for

Plate 19.9 — Panel racks for exterior exposure of panels.

example with compositions specifically designed for application to plastic body shells, where it is important to use the appropriate substrate. Wood is probably the most variable of all substrates. It is anisotropic and subject to change as a result of variations in temperature and humidity, and is particularly sensitive to the ingress of liquid water.

With the exception of wood, most products are tested on flat panels of a convenient size. They are assessed, usually at quarterly or half-yearly intervals, for periods of up to five years. Some compositions, e.g. coil coatings, may be exposed for longer periods because of their expected long life as protective coatings. The methods of assessment are akin to those described under accelerated weathering, section 19.4.

For many years it was the practice to collect weathering data for decorative paints on flat panels, 12 in × 6 in, of best quality British Columbian Pine. This type of panel was chosen because of the good reproducibility of data that could be obtained, and it is less demanding than other more variable substrates. In recent years, however, a number of research establishments and paint manufacturers have attempted to devise test methods that are more demanding of the coating and yet more realistic of 'in use' performance. For example, Whiteley (Building Research Station) has developed test methods based upon a test house, in which glazed window frames are used as test pieces. The orientation of the test house is such that maximum exposure to direct sunlight is obtained. The interior temperature and humidity are cycled such that condensation of water occurs on the interior of the glass surfaces and this runs down onto the (interior) uncoated wooden frame test pieces. This test method simulates the worst conditions of use for decorative paint systems in housing, i.e. in kitchens and bathrooms. This is a rather expensive way of collecting data, and other workers have attempted to achnieve the same objective in alternative ways. Thus the Paint Research Association devised weathering cabinets that accommodated test pieces that were similar to those used by Whiteley, but in which the glass was replaced by a rectangular piece of marine plywood. Softwood frames were used as before.

Both of these methods simulate the problem of maintaining film integrity over joints between different pieces of wood with grain running in different directions at the junction between individual parts of the frame. The differences in dimensional change on the absorption or loss of water are considerable. Across the grain of a softwood, water absorption may cause a dimensional increase of about 8%. Along the grain, under the same conditions, only about 1% dimensional change may be observed. These dimensional changes can cause the opening or closing of the joint at the junction between the two pieces. Thus the paint film is under considerable stress at the joint, and this is one of the main causes of breakdown, at or around joints in standard softwood joinery. In window frames the lower two joints and the lower rail are the first sites for paint failure because of these considerations. The formulation of paint systems to meet these demanding situations has been discussed in Chapter 9.

A simpler method for simulating these effects has been devised by ICI Paints Division. It too recognized the need to induce effects due to water absorption through the end grain of joinery sections. Because of its simplicity and hence low cost we shall consider this method in more detail.

19.5.2 ICI joinery section test method
This method seeks to obtain weathering data on paints, using a realistic approach designed to incorporate some of the most demanding conditions that such systems are likely to encounter. It was arrived at after careful consideration of the factors most likely to contribute to the rapid breakdown of paint systems. The most important factor was found to be the ingress of water through the end grain, usually through joints, e.g. at the junction of pieces of wood forming a window frame. Also, the prolonged contact with water that arises when liquid water collects at joints or between frames and sills was found to have an important bearing on paint performance. The test piece (see Plate 19.10) is thus designed with a 'water trap' which the

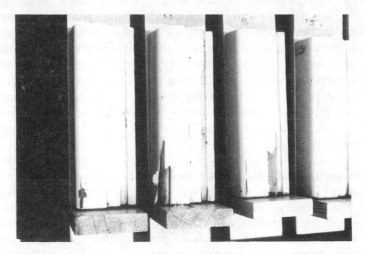

Plate 19.10 — ICI joinery section test, showing part of a series of experimental paints after 2 years' exposure.

joinery section abuts. The end of the joinery section abutting the water trap is unpainted. The test pieces, mounted on a backing panel, are exposed (in the northern hemisphere) at 45° facing south, so that they have the maximum incident UV and actinic light energy impinging upon them. Rainfall on the test piece drains into the water trap which is therefore the last part of the test piece to dry out. The test has been found to be reliable and cost effective. It should, however, be realized that there are many variables that can influence reproducibility of results in weathering, particularly on wooden substrates.

19.5.2.1 Selection of panels for a given series

It should be emphasized that the nature of the wood has a major influence on the performance of the coating, which in many cases will surpass effects arising from the coatings themselves. This must be allowed for in the design of the experiment. For reference purposes and later correlation, the ratio of heartwood to sapwood should be recorded. This may be estimated using a stain test [2]. The grain orientation may also be recorded by pressing a representative specimen of the end grain of the wood onto an ink pad and then transferring an imprint to a piece of paper. Because weathering is a process which leads to an inherent distribution of results, some replication will be necessary if valid comparisons are to be made between different systems. The degree of replication is dependent on the magnitude of the difference being sought and the standard deviation within the series. This will vary between series, but for independent systems it is unlikely that anything less than fourfold replication will be sufficient. By 'independent system' is meant any two systems which have no intrinsic relationship to each other, i.e. there are no known common factors. If the systems have a relationship, e.g. a *PVC* ladder, then statistical methods can be used to analyse the results, and the degree of replication may be reduced.

In a small series it will be possible to cut all the test pieces from a single piece of wood. This will minimize the variability, though as a precaution the pieces should be taken in random order. It must be remembered, however, that the results of the exposure will be relative, and care must be taken in drawing any absolute comparisons with other series. The total credibility of the series can be increased by replication on other test pieces cut from another piece of joinery section. For a large series the number of joinery sections (test pieces) is taken from two or more pieces of wood. The test pieces are pooled and used in random order. Alternatively, a note may be kept of the source of each test piece and the original lengths of wood treated as an experimental variable.

19.5.2.2 Preparation of joinery section test pieces

The test pieces ($7\frac{1}{2}$ in long) are cut from lengths of deal joinery section and are numbered consecutively. The joinery section is cut to the profile indicated in Plate 19.10 from average quality (HMS grade, i.e. Swedish Brack Group 1, Fifth Quality), shake-free lengths of red deal, excluding the first eight growth rings. Sections that are damaged, have poor sharp edges, dead knots, bad knotting at sharp edges, or any other major defect are not used. The top end of the section is sealed with a single coat of aluminium primer sealer and then finished with the system under test. The system to be tested is applied to all surfaces of the test piece with the exception of the bottom end which is to be in contact with the water trap. At least the minimum recommended drying time should be allowed between coats. The backing panel (12 in × 3 in) and water trap (3 in × 3 in) are constructed out of flat panels withdrawn from previous series. The backing panel and water trap are screwed to the test piece. Both are painted with standard exterior paint but are not subject to assessment. A coat of gloss paint is given to the upper face of the water trap which will be in contact with the unpainted end given of the joinery section.

19.5.2.3 Examination and assessment procedure
The joinery sections are evaluated by considering two distinct regions of the test piece, the bottom 2 in adjacent to the water trap and the top $5\frac{1}{2}$ in. The latter is assessed in the same way as plane (flat) panels for all of the appropriate defects listed in section 19.4. The region close to the water trap is examined specifically for signs of cracking and flaking since it is to be expected that these defects will appear more rapidly in this region. The joinery sections are part washed at examination times, on the raised 1 in wide $7\frac{1}{2}$ in long strip and on the adjacent side. The criteria of assessment are as previously discussed (section 19.4.1).

19.5.3 Recording of weathering data
The recording of weathering data takes place over such long periods, and may be so voluminous and diverse, that it must be done systematically. A systematic approach to origination and data-recording calls for the adoption of standard methods both of testing and reporting, so that data retrieval is facilitated . This type of data storage and retrieval is readily computerized, but it is important that any system adopted is sufficiently flexible to take into account variables that have not yet been defined, yet may assume considerable importance sometime in the distant future. The system should also include a detailed record of the compositon under test. Typically the weathering data should, in addition to the results obtained during the period of test, contain at least the following information:

Series identification; originator; purpose of the series;
date of initial exposure; site, substrate; product type;
system, indicating number of coats of each paint;
pigmentation; binder; minor compositional variants, if important to the series;
application conditions (e.g. temperature, humidity, film thickness, and recoat time).

It is also useful to record weather conditions so that if any abnormal conditions are encountered that might affect particular series they will not be overlooked.

19.5.3 'Natural' versus 'artificial' durability testing
The purpose of testing durability is to establish the probable performance of a product that is designed to meet a particular need. In most cases paints are sufficiently durable to survive many months or even years of exposure to the elements before showing the first signs of breakdown. This period may in the absence of other data preclude the marketing of a product until confidence in its performance has been established. If this period can be shortened, the response of the develop-

ment group can be more rapid and potential problems identified and overcome. The benefits of an accelerated test procedure can very readily be seen in terms of a reduction in development costs and possibly a greater lead time over competition for the manufacturer developing a new product.

Acceleration of durability testing can be carried out in several ways. If the main causes of degradation and failure are due to ultraviolet irradiation, heat, and moisture, it may be convenient to submit a product in a part of the world that will provide more intense irradiation and higher temperatures for longer periods than can be expected in the UK. If mould growth is a problem, regions more conducive to mould growth, i.e. where the temperature and humidity are high, such as in Malaysia, may be ideal for testing a given product. More often than not, however, the researcher may wish to accelerate degradation by a factor far greater than can be achieved by using a natural tropical exposure site, and he will resort to the type of equipment described in section 19.4. In doing so he may risk the possibility that the more aggressive test will bear little relationship to the performance of the test paint in reality. In these cases it is often assumed, with some justification, that if an experimental paint fails before a comparable standard paint of known performance it is unlikely to perform better than the standard paint in practice. If the experimental paint performs better than the standard on the accelerated test there is no guarantee that it will do so in practice. In general the test conditions will be programmed to simulate as closely as possible the type of exposure to which the paint may be exposed. The comparative distribution of irradiance for sunlight and for various artificial sources has been quoted by Scott [3] and are shown in Table 19.1. The 6500

Table 19.1 — Comparative distribution of irradiance (from Scott [3])

Wavelength (nm)	Sunlight %	6500 W Xenon %	Carbon arc (open flame) %	Fluorescent source %
300	0.01	0.01	0.5	14.0
300–340	1.6	1.5	2.5	70.0
340–400	4.5	5.0	11.0	13.0
Total to 400	6.1	6.5	14.0	97.0
400–750	48.0	51.5	34.0	3.0
750	46.0	42.0	52.0	0.0
Total to 750	94.0	93.5	86.0	3.0

watt xenon source, with a borosilicate glass inner and outer filter, provides irradiation with a distribution closest to that of sunlight. The open carbon arc has slightly more than double the radiation in the 340–400 nm band than sunlight. Fluorescent tubes have a very much more intense output in the region below 400 nm, 97%, compared to 6.1% in sunlight. It is to be expected that this intense UV output will be much more aggressive, and this is borne out in practice. This makes data from weathering testers like the QUV and the UVCON much more difficult to interpret than those obtained from machines utilizing less aggressive radiation sources. Nevertheless, some major paint users such as the motor manufacturers may demand a certain standard of performance under these particularly aggressive conditions, although they may not correlate well with conditions of use.

Scott [3] has pointed out that laboratory testing may be designed to accelerate the natural processes of degradation or to simulate a wide range of environmental conditions. The former represents the traditional use of weathering machines that has led to standard sets of conditions, i.e. specifications of performance according to predetermined cycles. The latter more recent approach is more concerned with simulating, under controlled laboratory conditions, the natural environment in any part of the world.

Sophisticated weathering machines now available are capable of being programmed to provide environmental cycles of radiation, temperature, wet/dry cycling, and atmospheric pollutants such as NO_2, SO_2, and ozone. This type of equipment enables a more rigorous approach to the study of the effect of the environment on paint degradation.

SUPPLIERS OF TEST EQUIPMENT

Accelerated weathering machines	Standard
Atlas Electrical Devices Co., 4114 N. Ravenswood Avenue, CHICAGO, Illinois, 60613, USA (UK Agent: Westlairds Ltd., North Green, Datchet, SLOUGH, Berks., SL3 9JH)	ASTM G-53, ASTM E42-57
J. B. Marr & Co., 53 Portland Road, KINGSTON-UPON-THAMES, Surrey, KT1 2SH, England	BS 3900, part F3
QUV Accelerated Weathering Tester, Q Panel Company, 102 Taylorson Street South, Ordsall, SALFORD, M5 3CL, England	ASTM G-53

Salt spray test equipment	
Industrial Filter and Pump Mfg. Co., 5900 Ogden Avenue, CHICAGO, USA Illinois, 60650, USA. (UK Agents: Westlairds Ltd.)	BS 287, BS 117

The Harshaw Chemical Co.,
1945 97th Street,
CLEVELAND,   ASTM B117
Ohio, 44106, USA

Research Equipment (London) Ltd,
64 Wellington Road,
HAMPTON HILL,   BS3900, part F4
Middlesex, TW12 1JX, England

REFERENCES

[1] A visual method of gloss estimation *JOCCA* **47** (11) 867 (1964).
[2] *J. Wood Sci.* **5** (6) 21–24 (1971).
[3] Scott, J. L. *JOCCA* **65** 182 (1982).

ACKNOWLEDGEMENTS

The author is indebted to T. K. Batty of Westlaird Ltd for the provision of plates
19.5, 19.6, and 19.7, and other helpful information; to J. B. Marr and Co for
permission to publish plate 19.3. Also to his former colleagues at ICI Paints
Division, notably M. Bahra, C. W. A. Bromley, A. Doroszkowski, and D. Farrow
for supplying the test panels that formed the basis of the remaining plates illustrating
various aspects of paint failure under test.

Index